互聯互通與
香港新經濟融資創新

Mutual market connectivity and
Hong Kong market innovations
for new-economy financing

香港交易所首席中國經濟學家
巴曙松教授　主編

Edited by Prof. BA Shusong,
HKEX Chief China Economist

商務印書館

互聯互通與香港新經濟融資創新

主　　編：巴曙松

副 主 編：蔡秀清　巴　晴

編　　委：朱　曉　羅得恩

責任編輯：張宇程

封面設計：涂　慧

出　　版：商務印書館(香港)有限公司

　　　　　香港筲箕灣耀興道 3 號東滙廣場 8 樓

　　　　　http://www.commercialpress.com.hk

發　　行：香港聯合書刊物流有限公司

　　　　　香港新界大埔汀麗路 36 號中華商務印刷大廈 3 字樓

印　　刷：美雅印刷製本有限公司

　　　　　九龍觀塘榮業街 6 號海濱工業大廈 4 樓 A 室

版　　次：2019 年 10 月第 1 版第 1 次印刷

　　　　　© 2019 香港交易及結算所有限公司

　　　　　ISBN 978 962 07 6634 3

　　　　　Printed in Hong Kong

Mutual market connectivity and Hong Kong market innovations for new-economy financing

Chief editor: BA Shusong

Deputy chief editors: Essie TSOI   Qing BA

Editors: Jennifer ZHU   Daniel LAW

Executive editor: Chris CHEUNG

Cover design: Tina TU

Publisher: The Commercial Press (H.K) Ltd.,
8/F, Eastern Central Plaza, 3 Yiu Hing Road,
Shau Kei Wan, Hong Kong

Distributor: The SUP Publishing Logistics (H.K.) Ltd.,
3/F, C & C Building, 36 Ting Lai Road,
Tai Po, New Territories, Hong Kong

Printer: Elegance Printing and Book Binding Co. Ltd.
Block A, 4th Floor, Hoi Bun Building
6 Wing Yip Street, Kwun Tong, Kowloon, Hong Kong

© 2019 Hong Kong Exchanges and Clearing Limited
First edition, First printing, October 2019

ISBN: 978 962 07 6634 3
Printed in Hong Kong

目錄

Contents

Part 3   The Connectivity platform for onshore and offshore cross-border asset allocation

風險與免責聲明

買賣證券的風險

證券買賣涉及風險。證券價格有時可能會非常波動。證券價格可升可跌，甚至變成毫無價值。買賣證券未必一定能夠賺取利潤，反而可能會招致損失。

買賣期貨及期權的風險

期貨及期權涉及高風險，買賣期貨及期權所招致的損失有可能超過開倉時繳付的按金，令投資者或須在短時間內繳付額外按金。若未能繳付，投資者的持倉或須平倉，任何虧損概要自行承擔。因此，投資者務須清楚明白買賣期貨及期權的風險，並衡量是否適合自己。投資者進行交易前，宜根據本身財務狀況及投資目標，向經紀或財務顧問查詢是否適合買賣期貨及期權合約。

免責聲明

本書所載資料及分析只屬資訊性質，概不構成要約、招攬、邀請或推薦買賣任何證券、期貨合約或其他產品，亦不構成提供任何形式的建議或服務。書中表達的意見不一定代表香港交易及結算所有限公司（「香港交易所」）或本書其他作者所屬的機構（「有關機構」）的立場。書中內容概不構成亦不得被視為投資或專業建議。儘管本書所載資料均取自認為是可靠的來源或按當中內容編備而成，但本書各作者、香港交易所和有關機構及其各自的附屬公司、董事及僱員概不就有關資料（就任何特定目的而言）的準確性、適時性或完整性作任何保證。本書各作者、香港交易所和有關機構及其各自的附屬公司、董事及僱員對使用或依賴本書所載的任何資料而引致任何損失或損害概不負責。

Risk statements and disclaimer

Risks of securities trading

Trading in securities carries risks. The prices of securities fluctuate, sometimes dramatically. The price of a security may move up or down, and may become valueless. It is as likely that losses will be incurred rather than profit made as a result of buying and selling securities.

Risks of trading futures and options

Futures and options involve a high degree of risk. Losses from futures and options trading can exceed initial margin funds and investors may be required to pay additional margin funds on short notice. Failure to do so may result in the position being liquidated and the investor being liable for any resulting deficit. Investors must therefore understand the risks of trading in futures and options and should assess whether they are suitable for them. Investors are encouraged to consult a broker or financial adviser on their suitability for futures and options trading in light of their financial position and investment objectives before trading.

Disclaimer

All information and views contained in this book are for informational purposes only and do not constitute an offer, solicitation, invitation or recommendation to buy or sell any securities, futures contracts or other products or to provide any advice or service of any kind. The views expressed in this book do not necessarily represent the position of Hong Kong Exchanges and Clearing Limited ("HKEX") or the other institutions to which the authors of this book belong ("Relevant Institutions"). Nothing in this book constitutes or should be regarded as investment or professional advice. While information contained in this book is obtained or compiled from sources believed to be reliable, the authors of this book, HKEX, the Relevant Institutions or any of HKEX's or the Relevant Institutions' subsidiaries, directors or employees will neither guarantee its accuracy, timeliness or completeness for any particular purpose, nor be responsible for any loss or damage arising from the use of, or reliance upon, any information contained in this book.

序 言

在國際大變局中探索香港的
獨特戰略定位

　　當前的全球經濟金融體系，可以説正處於百年未有之大變局中，而推動這一
變局的重要力量，就是東西方的互動。香港作為與世界金融體系融為一體的國際
金融中心，在思考自身的戰略定位時，必然需要將其放在這樣一個宏大的變革背
景下來考察。

　　實際上，香港經濟金融體系的成長史，也可以説是在全球化大浪潮的風雲激
蕩中不斷創新探索的歷史。過去 40 年，香港伴隨着中國改革開放一路走來，為中
國內地的經濟起飛卓有成效地做了三件大事：第一，70 年代以後香港作為連接中
國與世界的重要轉口貿易港，為打開國門的中國帶來第一桶金；第二，80 年代開
始，香港為內地產業發展提供了大量海外直接投資（FDI），自己也成為內地最大
的海外直接投資來源地；第三，香港資本市場大發展，源源不斷地為內地輸送了
發展經濟的寶貴資本。從 1993 年青島啤酒第一家內地國有企業在香港上市以來，
香港已然成為內地企業境外集資的最主要國際市場。自那時起，內地企業在港上
市首次公開發行的集資總額已超過 29,900 億港元，上市後增發集資額達 34,200
億港元，總集資額超過 64,000 億港元；目前內地企業佔香港市場總市值已超過
68%。香港上市融資不僅為內地企業提供了自身發展的重要資金，更在制度建設、
管理效率等方面幫助內地企業向國際準則靠攏、提升其公司治理水平和透明度，
為內地企業成功轉型升級，也為全球領先的科技、電訊、能源、銀行企業打下扎
實基礎，進而也成為國際資本進入中國市場的重要門戶市場。

　　人民幣國際化必然是香港作為國際金融中心發揮戰略性推動作用的領域，也必然是會給香港帶來巨大發展新動力的領域。在人民幣國際化的進程中，也許香港早已不再是中國連接世界的唯一窗口，但仍能創造不可替代的戰略價值。如果人民幣想真正成為一個在國際市場完全自由流通的交易媒介、一個被國際金融體系廣泛接受的價值儲藏載體和投資計價工具，就需要有一個高度國際化的金融市場環境來支持。這個市場不但能提供專業化、市場化的服務，還要具備中外各方均能接受和認可的規則體系與制度安排，這恰恰是香港重要的競爭優勢所在。

　　香港的最大價值就在於擁有「一國兩制」的獨特優勢，以及它與國際高度接軌的金融市場體系與制度環境。在內地與國際金融規則與運作體系存在明顯差異的情況下，香港可以最小的制度成本連接中國與世界，擔當中國資產和資本與國際資產和資本之間的轉換器和連接器，加速推動人民幣融入世界經濟金融體系。

　　這也正是近十年來香港交易所推動的「互聯互通」體系以及推出多項制度改革、不斷進行人民幣計價的產品創新的主要戰略着眼點：讓兩地市場在充分保留各自監管規則、市場結構和交易習慣的前提下，實現兩地市場完全市場化的交易互聯互通；在不增加資本流出風險的情況下，實現內地市場最大效果的對外開放，讓國際的產品、價格或投資流量更快地「請進來」，也讓中國自己的產品、價格或投資流量更便利地「走出去」。

　　在香港交易所過去的三個「三年戰略規劃」期內，我們不斷升級和擴容風險可控的「互聯互通」機制 —— 從「滬港通」到「深港通」再到「債券通」，推動內地資本市場的有序開放，也逐步把香港從傳統的離岸集資中心轉型成一個為股票、債券、外匯、商品多產品類別提供全方位服務的國際金融中心，連接中國與世界。

　　隨着人民幣國際化的持續推進，我們需要站得更高、看得更遠，更加全面和系統地籌劃人民幣融入國際金融體系的路徑和方向。如果說以往香港交易所的戰略規劃還主要側重於單個「互聯互通」項目或產品的開發，為人民幣國際化的全面展開而試水，那麼，未來規劃將會進一步整合這些突破點，把它們連成線、形成面·聚成體，依託「互聯互通」框架，從錢（資金）、貨（產品）、場（功能）三個方面立體推進人民幣國際化：讓越來越多中國的「錢」和國際的「錢」積聚於香港市場，讓越來越豐富的中國「貨」、亞太「貨」和國際「貨」擺在香港的貨架上，在香港構建離岸人民幣金融產品生態圈，發展風險管理、交易後託管結算和質押品管理服務能力，讓中外投資者在香港「互聯互通」這個既符合中國金融開放的要求、又符合國際認可規則的交易體系中，自由地實現國際與國內、人民幣與外幣、資金與資產等多個層次的資源互換，更快推動人民幣的國際使用。

我們也相信，未來全球經濟轉型的關鍵載於創新型經濟的發展，而未來十年內地和香港將會有大量的創新科技企業出現，這些新經濟企業的公司組織、管治架構和發展軌跡與傳統的企業有很大差異，需要突破傳統的上市框架，引入金融資源。香港交易所 2018 年 4 月宣佈實施的上市制度改革，為不同股權架構的新經濟企業和未有收入的生物科技公司上市融資開闢了路徑，也將為中國企業提升國際地位和中國經濟結構轉型提供新的支點。

經濟崛起中的中國已經有了上海、深圳和廣州，放眼未來，她同樣還需要一個與眾不同、擁有「一國兩制」優勢的香港。香港的繁榮不應離開中國的發展；中國的發展更應利用好香港的獨特優勢；過去如此，今後亦然。

是為序。

李小加

香港交易及結算所有限公司　集團行政總裁

2019 年 10 月

Preface

Exploring Hong Kong's unique strategic positioning in the face of international changes

The global economic and financial systems are facing their biggest changes in the past century, driven by the strong interaction between East and West. Hong Kong, as an international financial centre and integral component of the world's financial system, needs to examine its strategic positioning amid this macro-level transformation.

The development of Hong Kong's economy and financial system is one of continuous and unrelenting exploration. Over the past 40 years, Hong Kong has made three key contributions to the rise of the Mainland economy through its path of reform and opening-up. First, Hong Kong has been a significant entrepôt for Chinese trade since the 1970s, bringing China its first bucket of gold. Second, it has channelled capital flows into various industrial sectors in the Mainland through foreign direct investment (FDI) from 1980s onwards, and it has itself become the single largest source of FDI to the Mainland. Third, the prosperous development of the Hong Kong capital market has helped fund the Mainland's economic development. From the first H-share listing by a state-owned enterprise, Tsingtao Brewery, back in 1993, Hong Kong has become the primary international market for Mainland enterprises raising funds offshore. Since then the city has helped Mainland enterprises raise more than HK$2.99 trillion through initial public offerings (IPOs), and HK$3.42 trillion through follow-on equity financing, adding up to a total of HK$6.4 trillion. Today, Mainland enterprises account for over 68% of Hong Kong's total market capitalisation. Listing in Hong Kong not only provides Mainland enterprises with the necessary funding for growth, but also helps them align with international practices in terms of institutional improvement and managerial efficiency, and enhances their corporate governance and transparency. All in all it supports the transformation and upgrading of Mainland enterprises, helping them become world leaders in the industries of technology, telecommunications, energy, and banking. In the process, Hong Kong has emerged as an important gateway through which international capital can access China.

The internationalisation of the RMB is naturally an area in which Hong Kong, as an international financial centre, will play a strategic role and be a key driver for Hong Kong's future growth. In the process of RMB internationalisation, Hong Kong may no longer be the only window through which China connects with the world, but the strategic values it presents are still irreplaceable. Before the RMB becomes fully convertible and widely accepted internationally as a store of value and a unit of investment valuation in the international financial system, the backing and support by a highly internationalised financial market is indispensible. Such a market shall serve the RMB well, by providing market-based professional services and with rules, systems and institutional setups that are widely accepted and recognised by market participants. This is exactly where Hong Kong's strengths and competitiveness lie.

The biggest value of Hong Kong is its unique "One Country, Two Systems" governance framework and its financial market structures and systems that are highly in line with international practices. Given the vast differences between the Mainland and the rest of the world in terms of financial rules and operating systems, Hong Kong is able to bridge the gaps with the minimum institutional costs, linking up Chinese assets and capital with international counterparts, like a converter, and to help accelerate the RMB's integration into the world economy and finance.

This precisely points to the strategic focus underlying HKEX's efforts over the years in implementing the Mutual Market Access (MMA) programme and in introducing institutional reforms and launching innovative RMB products. We are looking to enhance market-based trading connectivity between the two markets across the border without major alterations to any of the rules and regulations, market structures or trading practices of either side; to maximise the benefits of the opening-up of the Mainland market without adding risk of capital outflows, so as to allow the speedy "coming-in" of international products, pricing and liquidity while facilitating the convenient "going-out " of Mainland products, pricing and liquidity.

HKEX has been enhancing and expanding the MMA programme through the implementation of the last three "3-Year Strategic Plans", from Shanghai Connect to Shenzhen Connect, and then to Bond Connect. This has helped facilitate an orderly opening-up of the Mainland capital market, and in the process, has transformed Hong Kong from a traditional offshore fundraising centre to an international financial centre offering comprehensive services across multiple asset classes ranging from equities, debts and foreign exchange to commodities, linking up China and the world.

As the internationalisation of the RMB continues to progress, we need to stand higher and look farther to plan and map the path and direction of the RMB's integration into the global financial market. If HKEX's strategic plans in the past were considered as focused on developing individual MMA schemes or products, like testing the water before the RMB goes fully internationalised, our future plans could be seen as integrating all these breakthroughs, joining all the dots into lines, lines into faces, and faces into solid figures. In other words, we are trying to facilitate the internationalisation of the RMB with the support of the MMA framework in three dimensions, namely liquidity (money), products, and market (venue). To be specific, we are aiming to (i) tap into both Chinese and international liquidity, (ii) enrich the types of Chinese and Asian and international products available in the Hong Kong market, and (iii) create an ecosystem of offshore RMB financial products, with proper risk management, post-trade custody, clearing and settlement as well as collateral management capabilities. We hope that, with both domestic and overseas investors trading freely in Hong Kong under the MMA framework (which meets China's need to open up while being also in compliance with international rules), exchanges of resources that follow at multiple levels (e.g. between international and domestic, the RMB and foreign currencies, and liquidity and assets) will help boost the global use of the RMB.

Going forward, we believe that the development of innovative economies will dictate how the world economy will transform. Countless innovative new technology companies will emerge both on the Mainland and in Hong Kong in the coming decade. They differ a lot from traditional companies in the aspects of corporate structures, governance framework and courses of development. New alternative listing frameworks are needed to accommodate the funding needs of these newcomers. In light of this, HKEX announced and implemented reforms to its listing regime in April 2018, which allow the listing of new-economy companies with weighted voting rights structures and pre-revenue biotech companies. This shall support Mainland enterprises in improving their international status as well as the structural transformation of the Chinese economy.

China already has Shanghai, Shenzhen and Guangzhou playing important roles in the country's economic rise. It also needs Hong Kong as another pillar, with its distinctive "One Country, Two Systems" framework and the competitive advantages it can leverage. The prosperity of Hong Kong will not be deep rooted without China's support, and China shall make better use of Hong Kong's unique strengths in its course of development. This has always been, and will always be, the case.

Charles LI

Chief Executive,
Hong Kong Exchanges and Clearing Limited
October 2019

第一篇

互聯互通與新經濟融資的新趨勢

第1章

互聯互通：
香港成為連接中國與世界的
金融樞紐

李小加

香港交易所集團行政總裁

引 言

　　香港過去數十載的經濟金融體系發展歷程，充滿着東西方交匯與互動、不斷探索自身定位的傳奇故事。香港把握世界經濟格局變化的重大機遇，以其特殊的地理區位優勢和自由化、國際化的金融體系，為中國經濟騰飛做出巨大貢獻，也順勢而為，實現了跳躍式自我轉型，成為國際公認的國際金融中心和「連接中國與世界」的區域樞紐。香港利用「一國兩制」的獨特優勢，推動互聯互通模式下共同市場的形成，以最小的制度成本幫助內地資本市場取得最大效果的開放。本章將說明香港交易所近十年來圍繞互聯互通和新經濟融資推出的主要改革理念、措施和效果，為推動中國資本市場開放和中國經濟結構轉型提供新模式和新動力。

1 「互聯互通」的時代大背景

中國內地改革開放初期，香港源源不斷地為其輸送用以發展經濟的寶貴資本，也利用內地轉口貿易和外商直接投資的歷史機遇，自身發展為海外及內地籌集資金及投資的重要樞紐。隨後 30 年發展，全球金融及經濟體系與中國市場的互動日益頻繁，中國經濟成為推動世界經濟增長的重要支柱。2016 年人民幣成功加入國際貨幣基金組織（IMF）的特別提款權（SDR）貨幣籃子，成為中國深入融入全球金融體系的標誌性事件，同時中國的資本流動也逐步呈現出多方面的重要轉變，成為全球不容忽視的資本浪潮：

- 第一，中國內地已經從資金短缺向資金富足轉變，甚至有望成為亞洲、「一帶一路」沿線國家新的國際貨幣流動性提供者；
- 第二，內地的對外貿易形式發生變化，由以往的商品輸出為主，走向商品輸出與資本輸出相結合，中國資產的全球配置將掀起全球資本流動的新一輪浪潮；
- 第三，中國將由以往遊戲規則的遵循者，轉為既是遵循者也是遊戲規則的參與者和制訂者，其在全球金融體系中的影響力日益崛起，需要有效的平台和工具與其國際話語權相匹配。

人民幣國際化和中國資本全球化佈局的大格局必然推動跨境資金雙向流動，資源及資產雙向配置的重大趨勢，將國際市場的產品、價格或投資流量「請進來」，也讓中國內地的產品、價格或投資流量「走出去」，使內地市場與國民的財富投資配置更多元化與國際化，以及讓中國更有效地參與和影響國際定價與標準的制訂。

在此時代背景下，香港的定位必將發生巨大變化。香港可以充分利用自身「一國兩制」和作為國際資金自由港的獨特優勢，成為實現國際人民幣跨境投融資、風險管理、金融創新的重要門戶；在內地與國際金融市場的規則體系與運作還存在明顯差異的情況下，香港更可以利用「滬港通」、「深港通」和「債券通」等有助於內地與香港資本市場雙向開放的特殊渠道，加速推動人民幣「走出去」；香港將在中國進一步國際化和經濟轉型中擔當關鍵角色，進一步強化自身作為連接內地和國際市場的「轉換器」的功能。

2 「互聯互通」造就了「共同市場」的創新性雙向開放模式

加速推動人民幣與中國經濟融入世界體系，需要逐步開放境內市場和資本項目。然而今天的中國內地市場與國際金融市場在規則體系、投資習慣、基礎設施運作方面存在相當大程度的差異，這一現狀既有市場慣例的延續，也有創新摸索的產物。中國與國際的差異不分執優執劣，兩者個別發展並於其各自的空間探索。可以預見的是，中國內地金融市場的開放和創新不可能完全改變自己的制度、複製國際規則，而外資「請進來」也不可能放棄現有的國際通行慣例、採用內地模式。因此，有必要以創新思路建立一種新的連接：在一個有序可控、既符合中國金融開放要求、又符合國際認可規則的跨境交易體系中，推動中國內地的資本、資產及資源與國際金融體系之間的良性平穩互動，進而逐步形成中國資本融入世界的大潮流。

香港「一國兩制」的獨特制度設計，為連接中國與全球提供了不可比擬、無法複製的銜接功能，可以讓雙方在不用改變自己的前提下進行大規模的深度互動。在「一國」的框架下，香港與中國內地的監管協作深度可以超越其他國家和地區，為建設全面開放體系提供了有效的可控性。「兩制」又使得香港擁有符合國際金融慣例、又備受信任的市場運行機制和制度框架，具備高度的靈活性來滿足內地與國際市場的連通需求，可以幫助內地實現在現有制度背景下還不能完全實施、卻對金融發展具有重要意義的「走出去」和「引進來」。

這正是我們香港交易所近年來着力發展的「互聯互通」金融市場開放體系的基本立足點。「互聯互通」最重要的意義在於它探索出一種創新性雙向開放模式，創設了一個同時滿足兩地監管機構要求的「共同市場」。通過兩地交易所和結算公司的金融基建聯接，形成一個既開放又風險可控的平台，將國際投資者和企業與中國內地企業和投資者集合於一個閉環式體系內。這個閉環裏既有國際與內地機構投資者羣體，也有中國內地規模龐大的散戶投資者羣體；既有傳統資產管理機構，也有量化交易商。系統的設計可以涵蓋股票、交易所買賣基金（ETF）、固定收益產品、貨幣產品、衍生產品及大宗商品多個資產板塊，無論是國際投資者感興趣的內地投資標的，還是內地投資者需要的國際產品，都可納入在這個系統之內。通過這條透明、安全、可控的渠道，內地企業與居民可以用最小的適應成本，參與到全球權益投資、金融風險管理和世界商品定價等多個市場的投資和交易當中；在不改變雙方的市場制度規則與交易習慣的原則下，實現國際與內地的資金與資產等多個層次的資源流動與配置。

3　實踐證明，「互聯互通」以風險可控的方式連接中國與世界

　　「互聯互通」機制自「滬港通」開通後平穩運行近五年，是經過實踐證明的「放得開」、「管得住」的可靠途徑，可以在風險可控的情況下推動中國內地資本市場雙向開放。它的獨特性體現在以下幾方面：

　　（1）按「本地市場原則」實行，以最小的制度成本，換取最大的市場成效。 在「互聯互通」機制下，兩地投資者通過委託本地證券商，經當地交易所與結算所買賣及交收於對方市場上市的股票，不會改變本地已有的制度規則與市場交易習慣，也就可將雙方的交易制度轉換費用降到最低。內地投資者通過「港股通」能「坐在家裏投世界」，足不出戶就可以通過本地券商、按自己的交易習慣進行多元化的海外配置。國際投資者（包括大量追蹤 MSCI 指數的國際被動型基金）通過「滬股通」和「深股通」（合稱「滬深股通」）「坐在香港投中國」，使他們可以按照國際慣例、透過國際券商投資內地市場，為內地市場引入更多國際流量。

　　（2）全程封閉的結算交收方式實現了既開放又風險可控的資金回流。 「互聯互

通」系統內資金只能用來買賣規定範圍內的股票，賣出套現資金必須通過本地結算系統沿原路返回，而不會以其他資產形式留存在對方市場，造成資本外流。所有的跨境資金流動都是以人民幣進行，在離岸市場完成雙向人民幣換匯，最大程度降低對在岸人民幣市場及在岸外匯儲備造成匯率影響，在加快人民幣國際化步伐的同時，消除對在岸市場可能帶來的風險。

（3）淨額跨境結算實現最小跨境流動。 在「互聯互通」系統內，股票結算的資金先在本地進行對減，然後再以淨額方式與對方市場作最終結算。根據此機制安排，過去四年多，「互聯互通」完成近 15 萬億元人民幣的成交額，而真正跨境結算的人民幣資金只有 1,000 多億元，真正實現了減低資金大進大出，控制資金跨境流動的風險，避免對貨幣市場造成波動。

（4）「互聯互通」投資標的覆蓋廣泛且具有獨特性，特別是「港股通」為內地投資者提供了豐富的資產配置選擇。 截至 2019 年 4 月底，「港股通」標的股票總計480 隻，標的股票總市值約佔香港主板市場的 88%。當中既有以四大內地銀行、中國石油、中國移動等為代表的、在中國經濟騰飛過程中扮演中流砥柱角色的大型中資金融、能源、通信類公司，也有以騰訊為代表的、在全球資訊科技高速發展中崛起的科技類龍頭企業。香港於 2018 年 4 月進一步推出有利於新經濟公司的新上市規則，吸引更多創新型公司來港上市，並有機會被納入互聯互通體系也讓內地投資者投資，有利於各方投資者實現多元化的資產配置，分享經濟增長紅利。

表 1：2018 年「港股通」前十大成交額的股票（買賣盤雙邊成交額）			
滬港通		深港通	
股票名稱	佔滬港通總成交的比重	股票名稱	佔深港通總成交的比重
騰訊控股	10.3%	騰訊控股	8.0%
工商銀行	3.6%	融創中國	3.0%
建設銀行	3.4%	金斯瑞生物科技	2.8%
融創中國	3.1%	中興通訊	2.6%
中國平安	2.7%	金蝶國際	1.8%
吉利汽車	2.4%	建設銀行	1.5%
滙豐控股	2.0%	吉利汽車	1.4%
農業銀行	1.9%	中芯國際	1.4%
招商銀行	1.8%	舜宇光學科技	1.4%
中國太保	1.7%	中國平安	1.4%
前十大股票	32.8%	前十大股票	25.3%
成交總額：1.82 萬億港元		成交總額：1.01 萬億港元	

註：由於四捨五入之誤差，各股票的成交比重相加未必等於十大股票的總比重。
資料來源：香港交易所。

　　可以說，「互聯互通」模式實現了風險可控、可追溯的雙向資本流動和資金配置，在中國資本項目開放進程中具有里程碑式的意義。該模式運作四年多以來，不僅沒有導致資本外流，而且還在幫助中國國民財富實現全球配置中起到了積極作用。自開通至 2019 年 6 月，「互聯互通」以 362 億元人民幣跨境資金流動（南向淨流入）撬動超過 600 倍成交金額，南北向成交金額（雙邊）累計約 22 萬億元人民幣[1]。2019 年 6 月的「港股通」日均成交額（單邊）在港股市場佔比 5.0%，內地投資者利用「南向通」投資港股的持股金額達 9,140 億港元。同期，「滬股通」和「深股通」（「北向通」）累計成交 14.3 萬億元人民幣，境外投資者經「北向通」持有內地股票（滬深股市合計）1.04 萬億元人民幣。得益於「互聯互通」模式，香港逐步成為內地資金進行國際化配置，以及國際資本進入內地資本市場投資的重要平台。（見圖 1 至圖 4。）

1　資料來源：香港交易所。

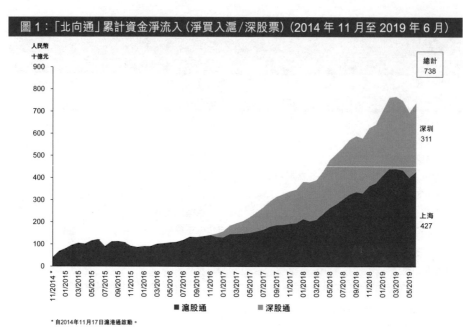

圖 1：「北向通」累計資金淨流入（淨買入滬／深股票）（2014 年 11 月至 2019 年 6 月）

* 自2014年11月17日滬港通啟動。

資料來源：香港交易所。

圖 2：「南向通」累計資金淨流入（淨買入港股）（2014 年 11 月至 2019 年 6 月）

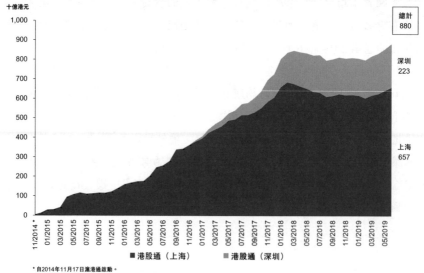

* 自2014年11月17日滬港通啟動。

資料來源：香港交易所。

圖 3：「港股通」每月淨流入/出（滬市及深市）（2017 年至 2019 年 6 月）

資料來源：香港交易所。

圖 4：「港股通」日均交易量及佔香港市場整體交易量的比重（2014 年至 2019 年 6 月）

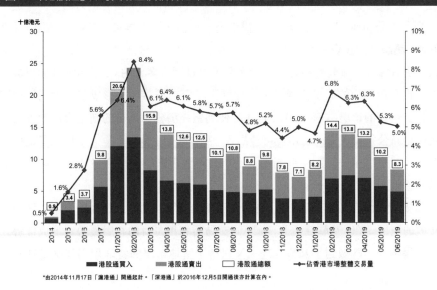

*由2014年11月17日「滬港通」開通起計。「深港通」於2016年12月5日開通後亦計算在內。

註：由 2014 年 11 月 17 日「滬港通」開通起計。「深港通」於 2016 年 12 月 5 日開通後亦計算在內。

資料來源：香港交易所。

實踐證明，「互聯互通」是投資 A 股市場的可靠渠道。指數供應商 MSCI 在考慮將中國 A 股納入指數時，認為「互聯互通」消除了股票買賣和資金進出的羈絆，有利於推動 A 股被納入指數[2]。近年來，內地基金紛紛利用「互聯互通」涉足境外市場，內地的社保基金和保險資金也可利用「港股通」風險可控的渠道進行跨市場投資。以上種種充分體現出「互聯互通」的投資便利性贏得了中外共同信任和國際認可。

對內地投資者而言，即使內地資本項目完全開放，「互聯互通」仍然是內地具有不同風險偏好和投資需求的投資者的重要投資渠道。市場總是由具有不同風險偏好和投資需求的投資者構成。我把他們比喻成研究生、大學生和中學生三個類別：有能力、有資源、有渠道、有意願並早已自行投資海外的好比「研究生」；無論是否開放都對境外市場不感興趣的投資者好比「中學生」；而有較多資金、投資經驗、有需要將一部分資產配置在海外、卻不敢直接投資海外市場的好比「大學生」。儘管中國資本市場會越來越開放，但不同市場的交易制度和文化的巨大差異也將長期存在。只要「互聯互通」能夠不斷豐富產品種類，既有複雜到衍生品交易，也有簡單到被動性指數 ETF 投資，能滿足不同類別投資者的風險偏好和投資需求，「互聯互通」就能不斷吐故納新，成為連接中國和世界的富有活力的雙向渠道。

對上市公司而言，「互聯互通」體系進一步強化了資本市場的融資功能。「互聯互通」讓國內的投資者和資本，與國際投資者和資本對接博弈，促進了一個規模更大、更有深度、流動性更好、投資者群體更為龐大多元的「共同市場」的形成，將大大提升市場內的股份流通效率，促進再融資市場活躍，形成多層次、風格迥異又相互包容的市場結構。2018 年 4 月便利新經濟公司上市的新上市規則推出，讓更多創新公司實現海外上市並可獲納入「互聯互通」體系，極大地擴大了這些上市公司的投資者基礎，對這些新經濟上市公司實現國際擴展和融資具有極為重要的戰略意義。

對監管者而言，「互聯互通」也加快了兩地監管互動，加速內地資本市場與國際接軌的速度。在「互聯互通」下，不同交易所利用自身的相對優勢，形成優勢互補的市場格局，既做到了互聯互通，又保持了當地市場的特殊規則和交易特點，

2　具體參見 "Consultation on further weight increase of China A shares in the MSCI indexes"，MSCI，2018 年 12 月。

使得業界和監管機構對對方市場的運作和文化產生更加深入的接觸和認識，相互學習和借鑒，進一步推動兩地市場的共同變革和進步。

4 「互聯互通」為連通多個資本市場板塊提供了藍圖指引和可行方案

作為有制約、可監控的資本項目開放渠道，「互聯互通」可以進一步延伸和擴容至更多資產類別。中外投資者可以在「互聯互通」這家更全面的「百貨公司」裏購買到從中國到國際的各式金融產品：不僅包括股票，也可有定息及貨幣產品和大宗商品產品；不僅有為國際投資者提供的中國標的，也有為中國投資者提供的國際標的；投資標的不僅包括在香港上市的國際公司，也包括在其他國際市場上發行的產品（借助 ETF 或「債券通」的「南向通」），讓香港的貨架擺上越來越豐富的中國「貨」、亞太「貨」和國際「貨」，讓越來越多中國的「錢」和國際的「錢」來參與香港市場，使香港成為多資產類別、全方位發展的金融資產中心。

於 2017 年開通的「債券通」的「北向通」，是「互聯互通」模式在債券及定息類產品市場上的具體體現。「債券通」通過香港與內地債券市場的基礎設施機構連接，將在岸與離岸債券市場連成一體。境外機構可直接利用境外的基礎設施，以及自己已經熟悉的國際法則和交易慣例，「一點接入」內地債券市場。「債券通」的「互聯互通」體現在它創造性地將國際債券市場通行的名義持有人模式，迭加中國的託管制度下所要求的穿透性模式，實現了「一級託管」制度與「多級託管體系」的無縫連接，在操作層面有效降低了不同市場體系對接的轉換成本。

「債券通」的總體框架設計仿照「互聯互通」實現了相對封閉的設計，在風險可控的情況下加速推動中國債市的對外開放程度。自 2017 年 7 月「債券通」開通以來，截至 2019 年 4 月末，845 家境外機構或基金接入「債券通」，較 2018 年年底的 503 家大幅增加，國際投資者來自全球 27 個國家和地區，日均交易量達 53.1

億元人民幣[3]。圖 5 顯示內地的銀行間債券市場中的外資持有量自「債券通」開通後的快速增長。

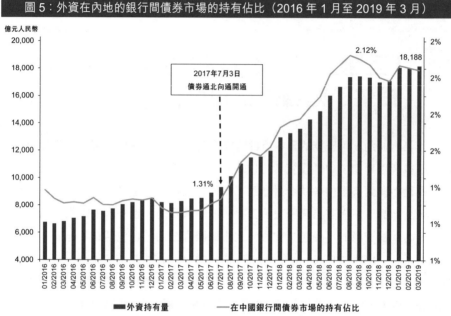

圖 5：外資在內地的銀行間債券市場的持有佔比（2016 年 1 月至 2019 年 3 月）

資料來源：Wind、中國人民銀行網站。

　　目前，「債券通」的業務範圍已經同時覆蓋一級市場和二級市場，中國農業發展銀行等中國內地的政策性銀行已經建立了「債券通」渠道下的海外債券發行機制，並將債券發行資訊在境外「債券通」公司網站掛牌，更加便利境外機構參與境內的人民幣債券發行和投資。我們期待在未來條件成熟時進一步推出「債券通」的「南向通」，在滿足信息透明、安全可控的監管要求的同時，推動人民幣資本走出去，為提高居民資產收益、實現中國資產的全球配置提供重要支援。

　　兩地監管機構還表明日後會將「互聯互通」的投資標的拓展至 ETF，這將為內地帶來更多被動型機構投資者和資金流量，為內地投資者進行全球資產配置提供更多的選擇。「互聯互通」下的 ETF 將並不僅限於追蹤香港上市證券的 ETF，還

3　資料來源：債券通公司。

可以包括那些跟蹤國際（尤其是亞太區）資產標的的 ETF。只要 ETF 發行商受香港的監管機構（香港證券及期貨事務監察委員會）監管，確保相關產品的適當性即可，不需要對 ETF 投資的股票有地域限制，這樣內地投資者可以通過「互聯互通」購買到全球各地有特色、符合內地投資者需求的 ETF 產品（包括全球能源、新興市場等多種類型），更好地滿足內地投資者的全球資產配置策略。

利用「互聯互通」框架，香港可以不斷豐富貨架上具吸引力的國際產品和機制內涵，包括納入及擴闊以海外資產為標的的 ETF、引入大宗交易機制、允許和方便內地機構投資者參與申購國際新股等，為中國企業和投資者參與全球定價和交易提供便利，為內地投資者足不出戶投資海外提供更多選擇，為國際投資者進入中國內地市場提供更多交易便利。

5　構建以「互聯互通」為基礎的離岸金融產品生態圈，打造香港成為離岸人民幣產品交易及風險管理中心

「錢」與「貨」全面擴容也意味着需要全面提升「場」的功能。

「互聯互通」不僅是一個實現跨境投資、雙向循環的載體，更可能成為一個多層次風險管理產品和金融工具的孵化器。這也是我們對「互聯互通」的更深層次設想：帶動離岸人民幣金融產品生態圈的構建，大力發展風險管理和交易後託管清算服務，全面提升「場」的功能。

在人民幣融入世界的大潮流中，我們必然會經歷中國資產價格巨大調整的動盪時期。通過與市場接觸，我們也發現相當一部分有意持有中國資產（包括股票、債券與貨幣）的國際投資者因缺乏對沖工具而對中國市場裹足不前。另一方面，逐步持有海外資產的中國投資者也需要對應的工具管理匯率風險。有市場調研顯示，國際參與者認為，相對於其他因素，匯率波動可兌換性是影響人民幣國際化的主要政策因素。由此可見，圍繞「互聯互通」提升「場」的功能，構建交易、結算、

風險管理的金融生態圈是「互聯互通」向縱深發展的必由之路。

　　一個有深度的、多元而且發達的金融產品市場，必定會在人民幣國際化的進程中成為重要的載體和支持力量。近年來香港交易所基於市場反饋，積極推動產品創新，從 2012 年開始，已推出多項以人民幣計價或人民幣匯率的衍生產品，回應境內外機構對人民幣資產的利率和匯率風險管理的需求（見表 2）。2018 年人民幣貨幣產品合約成交創下歷史新高，達到 180.9 萬張，較 2017 年激增 137%。人民幣貨幣期貨於 2019 年首 6 個月的平均每日成交量為 10,833 張合約，成為國際投資者重要的風險管理選擇（見圖 6）。

表 2：香港的場內主要衍生產品類別		
推出時間	產品	特徵
2012 年 9 月	美元兌離岸人民幣期貨（即美元兌人民幣（香港）期貨）	於人民幣漸趨國際化下，為市場提供貨幣風險管理工具及投資工具
2014 年 12 月起	人民幣計價的大宗商品期貨合約（包括鋁、銅、鋅、鉛、鎳及錫等六種）	是中國境外首批針對人民幣匯率風險的金屬合約產品，對人民幣作為亞洲時區內相關金屬的定價標準起支持作用
2016 年 5 月	離岸人民幣分別兌歐元、日圓及澳元的人民幣計價貨幣期貨：歐元兌人民幣（香港）期貨、日圓兌人民幣（香港）期貨及澳元兌人民幣（香港）期貨，以及美元計價的人民幣（香港）兌美元期貨	進一步豐富人民幣匯率產品線
2017 年 3 月	美元兌人民幣（香港）期權	協助市場參與者更有效地對沖貨幣風險
2017 年 4 月	中國財政部 5 年期國債期貨合約（2017 年 12 月合約到期以後，暫停國債期貨試點計劃，適時再推出新的人民幣利率產品）	首個以試點形式，在離岸市場以中國政府債券為標的的期貨產品

資料來源：香港交易所。

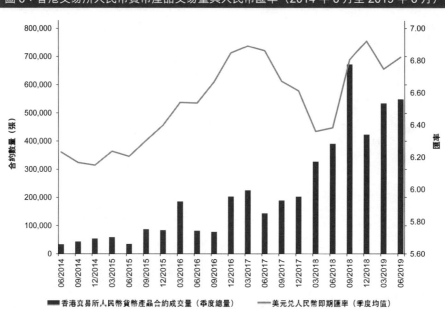

圖 6：香港交易所人民幣貨幣產品交易量與人民幣匯率（2014 年 6 月至 2019 年 6 月）

■ 香港交易所人民幣貨幣產品合約成交量（季度總量）　—— 美元兌人民幣即期匯率（季度均值）

資料來源：Wind。

　　隨着中國 A 股被正式納入 MSCI 新興市場指數，中國債券獲納入彭博巴克萊
等被全球基金廣泛追蹤的債券指數也在 2019 年落地。越來越多追蹤這些全球指數
的中長期機構投資者（包括各國央行、主權基金、國際養老基金等）將根據指數變
化，被動加大配置內地股、債資產，勢必引起全球近 4 萬億美元資金規模的資產
管理行業進行配置轉換，需要有足夠的、流動性好的工具讓他們實現風險對沖。
資料顯示，2017 年 4 月，香港推出人民幣國債期貨產品之後，香港交易所上市的
人民幣債券每日成交量明顯增加，提高了市場整體流動性和定價效率（見圖 7）。
另外，隨着人民幣匯率雙向波動幅度的擴大，香港人民幣貨幣期貨合約的交易量
也在不斷增長（見上文），顯示投資者在人民幣匯率波幅擴大的情勢下，更有需要
對沖人民幣匯率風險。

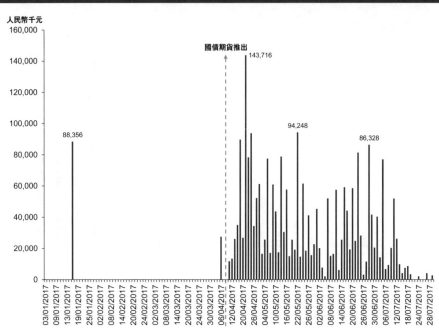

圖 7：香港交易所上市的人民幣債券每日成交額（2017 年 1 月至 7 月）

資料來源：香港交易所。

　　香港作為人民幣離岸市場的體制優勢在於其服務專業、金融基礎設施完善，以及可以為市場提供廣泛而豐富的風險對沖工具。目前香港交易所開發的離岸人民幣衍生品成交活躍，為投資者提供良好的流動性和市場深度。香港場內離岸人民幣產品的機制和結算方式也能有效保證離岸價格最終收斂於在岸市場，使得離岸的人民幣產品交易可以進一步幫助擴大境內價格的國際影響力，將定價權把握在境內。「互聯互通」帶動更多風險管理需求，促進發展和豐富各類離岸人民幣衍生品來配合實體經濟的需要，有利於香港逐步發展為離岸人民幣產品交易及風險管理中心，對促進人民幣廣泛的國際使用和國際化進程起到重要作用。

6 讓「互聯互通」成為支持中國經濟轉型和企業創新的助推器

當前內地多層次資本市場建設已取得相當大的成就，就上市公司市值而言，已成為全球第二大資本市場[4]。但另一方面，以散戶為主的市場結構也令內地資本市場制度改革面臨制肘。現行的發審制度、交易制度、投資者結構等方面與國際標準存有一定差距，還不能充分滿足眾多企業尤其是新經濟、創新企業的融資需求。內地已經推出了上海證券交易所科創板，如果能繼續改革和進一步開放，將為新經濟公司提供充足的融資來源和更大的市場靈活性。

2018 年，香港交易所對上市規則作出修訂，使香港市場緊隨主要金融市場的步伐[5]，該等修訂包括新增為生物科技類公司和同股不同權架構公司而設的上市規則，同時考慮將這些公司也納入「互聯互通」體系，為這些創新型、生物科技企業直接引入更加多元化的投資者。這能讓來自中國內地的香港上市公司、特別是新經濟企業可以接觸到無論來自國際還是內地的更廣泛投資者羣體，擴大其國際影響力。

自上市新規生效後至 2019 年 6 月底，共有 50 家新經濟公司在香港發行新股，融資規模達 1,675 億港元，已佔同期新股融資總額的 52%。當中包括 14 家生物科技公司，總計集資 475 億港元（包括沒有營業收入的 8 家通過新上市規則章節上市，融資 235 億港元）。相信在新科技已經成為驅動世界經濟發展新引擎的現今時代，香港交易所憑着國際化的上市標準和「互聯互通」的制度創新，可以大大推動中國一批創新型企業壯大發展，提升中國企業在全球市場中的話語權和影響力，為中國經濟增長注入新動力。

4　根據國際交易所聯會發佈的 2019 年 5 月底數據，滬、深交易所市場合計的上市公司市值近 7.6 萬億美元，僅次於在美國的紐約證券交易所和納斯達克交易所的上市公司合計市值。

5　目前全球各大資本市場都以多種方式吸引創新型公司的上市。數據顯示，全球最大的 46 個國家股票市場中有 30 個國家有或曾經使用過雙重股權架構的公司上市。（具體參見香港交易所研究報告《雙重股權架構與生物科技行業的上市制度改革》，載於香港交易所網站，2018 年 11 月 15 日。）

7 結語

利用「互聯互通」打造的共同市場，香港能夠充分利用其獨特的市場體系和金融環境，從國際市場為內地資本引入高質量資產作更廣泛的資產配置，同時又深刻理解內地需求，為中國優質公司擴展更廣泛的國際投資者基礎，使他們能在「互聯互通」體系下進行國際投資和融資活動，實現中國資源的全球配置，為具有核心競爭力的優質企業吸納資本，並共享其所創造的財富。在當前開放度和一體化程度如此高的全球金融環境中，香港通過「互聯互通」機制打造的風險可控、雙向開放、真正有廣度和深度的共同市場，為香港引入新的資金流量和持續發展的新動力，更為提升中國企業的國際競爭力和中國經濟體系轉型提供了新的支點與助力。

第2章

新經濟融資創新與香港的探索

陸挺

野村國際 (香港) 有限公司

環球市場部中國首席經濟學家

王立升

野村國際 (香港) 有限公司

環球市場部中國經濟學家

王競

野村國際 (香港) 有限公司

環球市場部中國經濟學家

摘 要

　　科技正在快速地重塑世界經濟。科技發展通過提高生產效率、改善產業結構和提高人民生活水平，帶來了新經濟部門的崛起，讓國際資本市場對新經濟投融資更加青睞。作為世界前三的金融中心，香港在中國內地以及其他亞太地區企業的海外融資中，一直扮演着重要的角色，並且在新經濟融資創新領域不斷探索和嘗試。我們認為在未來十年，香港將會為中國的新經濟部門融資作出更突出的貢獻。本章重點關注新經濟部門近年來的發展狀況、新經濟企業的融資問題癥結，以及香港在為新經濟企業融資方面作出的努力。我們特別探討香港在幫助中國內地新經濟企業國內項目融資、海外併購和對沖匯率風險方面的貢獻。

1　全球範圍內崛起的新經濟部門

根據經濟合作與發展組織的定義[1]，新經濟部門是指在一個經濟體中生產或積極應用創新或新型技術的部門。作為一個較新的概念，新經濟部門主要包括大眾越發依賴電腦、電信和網結絡以完成對於產品和服務的生產、銷售與分配的行業。然而，新經濟部門覆蓋的範圍不只局限於特定的一二領域：一家採用新型數位計算技術的資訊與通信技術產業互聯網公司，或是一家生產新型藥物的生物技術公司，抑或發展提供新型金融產品與服務的金融公司，都屬於新經濟部門。

如今，新經濟部門已經成為全球經濟發展中不可或缺的因素。作為新經濟部門的一個重要表現形式，數字經濟改變了原有的經濟格局，並越發顯著地影響着全球經濟的方方面面。華為與牛津經濟研究院於 2017 年合作發表的一份研究[2] 表明，在過去的 15 年內，數字經濟的發展速度超過全球經濟總量（國內生產總值，簡稱 GDP）增長的 2.5 倍。截至 2016 年，數字經濟規模已達到 11.5 萬億美元，佔全球經濟總量的 15.5%。該研究預測，截至 2025 年，數字經濟規模有望達到 23 萬億美元，屆時將佔全球經濟總量的 24.3%。美國與歐盟在 2016 年全球數字經濟中的佔比分別為 35% 和 25%，而中國的份額也從 2000 年的 4% 迅速提升至 2016 年的 13%。以三年的投資期限為例，該研究預測，考慮到顯著的溢出效應，數字概念相關投資的長期回報率將超過其他非數字概念相關投資 6.7 倍。

2　中國新經濟部門的飛速發展

在加入世界貿易組織的大背景下，伴隨標誌性的全球戰略和國內市場的逐步

1　參見經濟合作與發展組織的統計術語表 (https://stats.oecd.org/glossary/detail.asp?ID=6267)。
2　參見華為與牛津經濟研究院（2017）《數字溢出報告：衡量數字經濟的真實影響力》(https://www.huawei.com/minisite/gci/en/digital-spillover/files/gci_digital_spillover.pdf)。

開放，中國經濟更加積極地參與到全球產業鏈中，並和世界其他地區緊密相連。據世界貿易組織統計，2018 年中國已經成為全球最大的商品貿易出口國，而在 1995 年中國僅排第十一位。世界銀行稱，按 GDP 計算，自 2010 年開始中國已經成為僅次於美國的世界第二大經濟體，排名相對於 1995 年的第八位有了明顯的提升。隨着新科技、新能源和新商業模式的高速發展，中國當下產業結構的重心已經開始向新經濟部門轉移。

新經濟部門對全球經濟的影響是毋庸置疑的，此外，新經濟部門也已逐步成為中國經濟發展中至關重要的因素。中國國家統計局表明，2018 年全國「三新」（新產業、新業態、新商業模式）經濟增速為 12.2%，比同期 GDP 現價增速高 2.5 個百分點，相當於 GDP 的比重為 16.1%（2017 年全國「三新」經濟相當於 GDP 的比重為 15.7%）。中國經濟發展新動能指數自 2015 年開始由國家統計局每年公佈一次。該指數基於固定互聯網寬頻接入用戶、研究與試驗發展（R&D）經費投入以及快遞量等一系列產業層面的指數對新經濟部門發展的趨勢進行了追蹤。2018 年，中國經濟發展新動能指數由 2017 年的 210.1 增長至 270.3，幾乎達到 2014 年（100.0）的三倍，生動詮釋了中國新經濟的蓬勃發展。從主要構成指標來看，網絡經濟指數由 2017 年的 362.1 激增至 2018 年的 605.4，對經濟發展新動能指數增長的貢獻率為 80.8%，貢獻最大。此外，創新驅動指數、轉型升級指數、經濟活力指數和知識能力指數對總指數增長的貢獻率分別為 10.3%、3.9%、2.6% 和 2.4%。

除龐大的國內市場和先進的基礎設施網絡之外，包括財政補貼在內的、中國政府出台的政策也為新經濟部門的穩健發展提供了有力的支援。鑒於中國的金融體系被大型商業銀行主導，當下金融環境對於國有企業比私有企業更為有利。因此，在中國內地的新經濟企業的發展很大程度上需要依賴自有資金與海外融資。然而，這些資金不足以滿足該類企業特殊的融資需求，一個更加市場導向的融資渠道是支撐新經濟部門高速發展的必要條件。

3　融資問題仍是中國新經濟部門的當務之急

不少新經濟企業雖然擁有極具吸引的資本回報率與光明的發展前景，但卻不受傳統的銀行體系與資本市場所青睞。究其原因，這些企業多存在以下特點：缺少抵押品、信用記錄有限、未來的收益情況具有較大不確定性，比起大型企業尤其是國有企業等交易對手，它們幾乎無法獲得政府的隱形擔保。

自 2009 年中國政府允許影子銀行快速擴張以穩定經濟發展後，一些新經濟企業出現了過度依賴通過影子銀行進行融資的情況，其中包括信託貸款、委託貸款、點對點借貸等。然而，自 2017 年中，隨着金融監管和去槓桿力度的加強，新經濟企業在過去兩年間通過這些渠道進行的融資逐漸減少。此外，隨着國內經濟發展放緩以及中美貿易戰所帶來的不確定性加劇，市場避險情緒升溫。這使得中國新經濟企業在外部融資上遇到了更大的挑戰，對於跨境業務為主的企業來說更是難上加難。

以中國 A 股市場融資為例，新經濟企業的創始人需要交出部分控制與投票權，以從外部投資者處獲得資金。通常來講，隨着外部投資的資金增加，創始人失去的控制與投票權也會按比例增加。中國內地股票市場如此的機制在很多情況下削減了企業創始人進行股票融資的動力，特別是當創始人對於企業的前景有很積極的展望時此現象更加明顯。這導致了市場上所提供的高品質的新經濟企業股票十分有限，進一步推高這些股票的價格，令外部投資者望而卻步。

當國內金融環境收緊，特別是在 2017 至 2018 年國家大力推廣金融去槓桿期間，中國新經濟企業在融資方面遇到了更多難題。即便如此，對於國際投資者而言，部分新經濟企業仍保持較高的吸引力。這與這些企業的高新技術、新商業模式與高速發展的前景是分不開的。

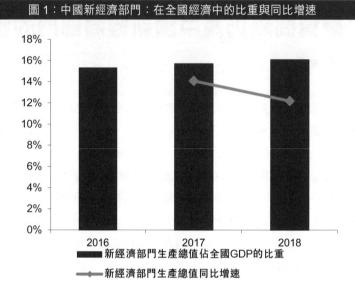

圖 1：中國新經濟部門：在全國經濟中的比重與同比增速

■ 新經濟部門生產總值佔全國GDP的比重

◆ 新經濟部門生產總值同比增速

資料來源：Wind 與野村全球宏觀。

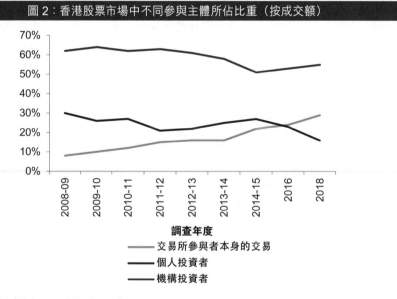

圖 2：香港股票市場中不同參與主體所佔比重（按成交額）

調查年度

—— 交易所參與者本身的交易

—— 個人投資者

—— 機構投資者

註：交易所參與者本身的交易是指以其公司本身賬戶所進行的交易。

資料來源：香港交易所各年度的《現貨市場交易研究調查》。

4 香港在新經濟融資創新領域的積極探索

　　立足於發展成熟的金融體系，大量的金融和法律人才資源，以及與全球市場的高度聯動性，香港已經在新經濟融資創新領域上走在世界的前列。過去十年，最為顯著的成就包括香港交易及結算所有限公司（簡稱「香港交易所」）在 2018 年推出的港股上市制度改革、以「滬港通」、「深港通」和「債券通」為代表的「互聯互通」機制，以及為全球企業，特別是中國內地企業，提供的海外外幣債融資市場。

4.1　港股上市新規助推新經濟企業股權融資

　　為吸引和便捷新經濟企業來港上市，香港交易所於 2018 年 4 月 30 日對其上市制度進行了重大改革。作為改革的主要部分，香港交易所全資附屬公司香港聯合交易所（簡稱「聯交所」）在其主板《上市規則》新增三個章節，並對現行《上市規則》條文作出相應修訂，以：（1）允許未能通過主板財務資格測試、或未有盈利的生物科技公司上市；（2）允許擁有不同投票權架構，即「同股不同權」的公司上市；及（3）為尋求在香港作第二上市的中資及國際公司設立新的第二上市渠道。

　　香港交易所集團行政總裁李小加表示[3]，此次改革意義重大，是自 1993 年引入 H 股機制後香港市場最大的一次上市制度改革。李小加總裁在此後的網誌中提到，得益於此次改革，2018 年新經濟企業在香港通過首次公開招股（IPO）募資 1,360 億港元，佔據了香港市場 IPO 總募資近一半的份額[4]。

　　以生物科技企業為例，在真正獲取營業收入之前，生物科技企業往往需要投入大量的研發經費來支援長時間的臨床試驗和通過相關監管部門的認證。上市新規允許生物科技企業可以在不受主板財務資格測試限制的條件下進行股權融資。但與此同時，聯交所已召集多位業界專家成立生物科技諮詢小組，以協助審閱根

3　見〈明日起航的香港上市新規，祝你一帆風順！〉，載於香港交易所網站的《李小加網誌》，2018 年 4 月 29 日（https://www.hkexgroup.com/Media-Centre/Charles-Li-Direct/2018/Dawn-of-a-New-Era?sc_lang=zh-HK）。

4　見〈明星公司的 IPO 為甚麼也會破發？——回顧 2018 香港上市新規〉，載於香港交易所網站的《李小加網誌》，2018 年 12 月 20 日（https://www.hkexgroup.com/media-centre/charles-li-direct/2018/reflections-on-hong-kongs-ipo-market-to-close-out-2018?sc_lang=zh-HK）。

據新規則提交的生物科技公司的上市申請。

香港擁有一個公認的公平對待投資者的資本市場，並力爭在最大程度上保護投資者的利益。與目前的投資者保護制度一致，上市新規徹底改變了「同股不同權」的創新型企業股東獲得管理權的方式。與過去我們只認可金融資本不同，上市新規承認並接受股東通過其他方式來獲得企業的管理權，例如貢獻人力資本（如知識產權和新商業模式）。但這與保護小股東的既有方式並不衝突。

為幫助在紐約證券交易所、納斯達克市場和倫敦證券交易所已上市的創新型公司（包括無營業收入的生物科技企業和「同股不同權」的創新型企業）獲益於更加靈活的上市要求，上市新規為尋求在香港作第二上市的中資及國際公司設立了新的第二上市渠道。這為已上市的新經濟企業，特別是來自中國內地的新經濟企業，提供了新的融資渠道。與此同時，內地投資者也可通過香港與內地市場之間的「互聯互通」機制投資此類企業。

根據香港交易所的統計，新規改革生效的一年內（2018 年 4 月 30 日至 2019 年 4 月 30 日期間），共有 40 家新經濟企業在香港發行新股上市，大部分為內地企業，融資 1,504 億港元，佔同期 IPO 融資總額的 53%。例如，智慧手機生產商小米在 2018 年 7 月成為在香港上市的第一家擁有雙重股權結構的企業，另一家「同股不同權」的企業，食品配送服務商美團點評，也於 2018 年 9 月在香港上市。與此同時，9 家生物科技公司在香港發行新股融資 320 億港元，包括上海醫藥科技公司藥明康得和北京癌症製藥公司百濟神州。其中，有 7 家無營業收入的生物科技企業通過新上市章節融資 223 億港元。

4.2 「互聯互通」機制將強化上市新規優勢

長期以來，中國經濟的增長為香港發展和完善其資本市場提供了各種機遇。以「滬港通」和「深港通」為代表的股票市場「互聯互通」機制對香港和中國內地有着里程碑式的意義，為兩地投資者提供一個獨特的平台，只需通過各自的股票經紀（券商）來投資彼此股票市場的上市企業。上市新規結合兩地股票市場「互聯互通」機制，將允許內地新經濟企業在兩地股票市場同時進行融資，吸引更多的內地新經濟企業來港上市。

香港既是國際金融中心，又是中國內地連接世界的窗口。與內地交易市場相比，香港的資本市場更國際化和市場化；與其他主要國際交易市場相比，特別是美國市場，香港更接近內地市場，在文化、語言和交易習慣上更接中國地氣。由此，香港成為內地企業境外上市的首選之地。截至 2018 年 12 月，包括 H 股、紅籌股和內地民營企業在內的 1,146 家企業在香港上市，市值達 2.6 萬億美元，佔市場總額的 68%[5]。

　　得益於 2018 年推出的上市新規，香港已成為對中國內地新經濟企業更具吸引力的上市地點。同時，中國內地與香港之間的股市「互聯互通」機制可強化上市新規的優勢。「滬港通」和「深港通」使得新經濟企業可以接觸到多元的全球投資者，其中包括對於內地新經濟公司的發展趨勢及潛力更加了解的在岸投資者，和以機構投資者為主的離岸投資者。根據香港交易所《現貨市場交易研究調查 2018》的數據，機構投資者（本地及海外）佔香港交易所證券市場 2018 年總成交額的 55%。相比之下，根據《上海證券交易所統計年鑒（2018 卷）》，個人投資者佔 2017 年整個上海證券交易市場總成交額的 82%。由於機構投資者與個人投資者相比，傾向於作出更為長期的投資決策，香港證券市場能更好地滿足新經濟企業在其成長軌跡中的融資需求。

4.3　「債券通」機制為境外投資者投資中國新經濟企業提供便利

　　「債券通」是中國內地與香港債券市場互聯互通合作的全新計劃，讓中國內地與境外投資者通過在內地與香港之間聯通的金融基礎設施機構可以在對方市場買賣債券。「北向通」已於 2017 年 7 月 3 日開通，允許香港及其他國家與地區的境外投資者，經由香港與中國內地基礎設施機構之間在交易、託管、結算等方面互聯互通的機制安排，投資於中國內地的銀行間債券市場。中國人民銀行與香港金融管理局（簡稱「香港金管局」）表示未來將適時研究擴展至「南向通」，以便讓中國內地的投資者投資香港及其他國家與地區的債券市場。

　　在「債券通」落地之前，境外投資者需要通過非常繁瑣和複雜的流程來開通交

5　參照香港貿易發展局網頁 http://hong-kong-economy-research.hktdc.com/business-news/article/Market-Environment/Economic-and-Trade-Information-on-Hong-Kong/etihk/en/1/1X000000/1X09OVUL.htm。

易賬戶、申請人民幣購匯額度，以及尋找清算機構來處理相關的國際結算。在該機制落地之後，境外投資者可以直接買賣中國銀行間市場的債券，也因此有了更便捷的投資中國新經濟部門債券資產的途徑。「債券通」下的合格投資者主要為金融機構，銀行、保險公司、證券公司和資產管理公司。截至 2019 年 7 月，中國內地債券市場的託管債券總市值為 62.1 萬億人民幣，其中境外投資者共持有 1.7 萬億人民幣左右，佔總規模的 2.7%[6]。外資所持的主要為政府債與政策性銀行債。這一比例遠低於美國（25%）和日本（12%）等發達國家經濟體，也低於韓國（6%）和巴西（5%）等新興市場經濟體。這也意味着中國在債券市場開放和吸引境外投資者方面仍具有較大潛力[7]。

4.4　快速崛起的中資企業海外外幣債市場

近年來越來越多的中資企業開始湧向海外外幣債市場（特別是香港的海外外幣債市場）發債融資。這些市場為中國內地以及其他國家與地區的企業提供了額外的債券融資選擇，為這些企業在國內項目融資、海外併購和對沖跨境業務匯率風險方面作出貢獻。因為中國的新經濟企業在獲得銀行貸款與直接市場融資方面通常面臨較多困難，海外外幣債融資渠道對他們而言尤為重要。從宏觀角度來看，海外外幣債市場在一定程度上也有助於中國的產業結構向高科技製造業與服務業轉型。

我們將中資企業海外外幣債券定義為非政府及政策性銀行的中資企業在海外債券市場發行的非人民幣計價的債券[8]。中資企業自 20 世紀 90 年代開始發行境外外幣債，但在 2008 年之前的發行規模非常小（以 2008 年為例，全年發行額僅有 84 億美元）。中資企業海外外幣債券的發行自 2009 年以來經歷了三次爆發式增長。我們於以下分節簡單回顧過去三輪債券發行的爆發期。

6　資料來源：Wind 資料庫，由中債估值中心定期發佈。

7　資料來源：中國國家外匯管理局（2019）〈2018 年中國國際收支報告〉(http://m.safe.gov.cn/safe/file/file/20190329/bdcf1afe0d5d41838f9f9554721494c6.pdf)。

8　有關本節所述的海外外幣債市場的數據，參照 Ting Lu、Wendy Chen 和 Lisheng Wang (2018) "China Monthly: Exploring more obscure financing channels"，*Nomura Asia Insights*；Ting Lu、Lisheng Wang、Wendy Chen 和 Jing Wang (2018) "A bumpier ride ahead for China's offshore corporate dollar bonds"，*Nomura Asia Insights*；以及 Ting Lu、Lisheng Wang 和 Jing Wang (2019) "China: An update on corporate dollar bond financing"，*Nomura Asia Insights*。

4.4.1　第一輪：2009 年三季度到 2011 年二季度，由房地產開發商發債驅動

中國政府於 2008 年全球金融危機之後出台了四萬億元刺激計劃來穩定經濟，但在 2009 年年底至 2010 年間逐步收緊。在境內融資環境收緊的大背景下，考慮到海外外幣債的超低融資成本與平穩的人民幣兌美元匯率走勢，中資企業開始轉向海外外幣債市場。中資企業海外外幣債的發行規模自 2009 年年底開始快速攀升，在 2009 年三季度至 2011 年二季度期間季度平均發行量達 86 億美元，遠高於之前四個季度的平均值 7 億美元。主要原因在於樓市繁榮但境內融資收緊，促使亟需資金的房地產開發商在海外大規模舉債。中國政府在 2010-11 年間出台了越發嚴格的政策來給過熱的房地產市場降溫，使得這一輪債券發行高峰期止於 2011 年年底。

4.4.2　第二輪：2012 年一季度到 2014 年二季度，由人民幣升值和中美利差套利驅動

中國於 2008 年末全球金融危機之時將人民幣與美元匯率掛鈎，並在 2010 年中脫鈎。在後危機時期，發達經濟體的借貸成本跌至歷史新低，但反觀中國，對融資成本並不敏感的地方政府融資平台債務激增和居民購房需求強勁，推動內地利率水平持續走高。在相當長的一段時間內，借入離岸美元再換成在岸人民幣的套利交易為中國內地帶來了持續的熱錢流入，也推動了人民幣兌美元強勁升值至今 20 年來的最高點（2014 年 1 月中旬，6.04 左右）。按中國人民銀行的數據，中國的官方外匯儲備規模一路上升至 2014 年中的 4 萬億美元左右。

在這樣的背景下，中資企業境外外幣債發行自 2012 年一季度起迎來了第二次井噴，將大量熱錢引入中國內地。根據我們的測算，中資企業海外外幣債在 2012 年一季度至 2014 年二季度期間的季度平均發行量達 210 億美元，遠高於之前四個季度的平均值 85 億美元。這一輪的發債高峰主要由一般行業的企業驅動。我們定義一般行業為除了政府、政策性銀行、房地產開發商、金融機構和地方政府融資平台之外的其他部門，廣泛地包含了消費、科技、工業和能源等行業，部分新經濟企業也包含其中。與此同時，房地產開發商和金融機構的發債規模也明顯上升。

自 2014 年中起，人民幣對美元開始貶值，在岸借貸成本也因經濟放緩和政策寬鬆而快速下降，中資企業（特別是房地產開發商）開始明顯放緩海外外幣債的發

行節奏，甚至償付他們部分的高收益債券來降低成本。本輪債券發行高峰期也很快隨之結束。

4.4.3 第三輪：自 2016 年二季度至今，部分由政府對企業海外發債限制放鬆導致

中資企業海外外幣債的發行規模於 2016 年二季度再次激增，由 2016 年的 1,298 億美元上升至 2017 年的 2,494 億美元，並且在 2018 年和 2019 年上半年依然維持在相對高位（分別為 2,096 億美元和 1,308 億美元）。在 2016 年二季度至 2019 年二季度期間的季度平均發行量達到 543 億美元，明顯高於之前四個季度的平均值 300 億美元。

我們認為 2016 年二季度開始的中資企業海外外幣債發債高峰，主要由以下因素導致：2015 年 8 月起人民幣開始快速貶值，並帶來了明顯的資本外逃，中國政府顯著地放鬆了中資企業海外發債的限制，以緩解資本外流壓力。2016 年中開始，海外融資成本開始下降但國內融資成本上升，這構成了另一驅動因素。此外，2016 年 9 月之後，由於更加擔心國內的房地產泡沫以及地方政府債務問題，中國政府開始限制對房地產開發商和地方政府融資平台的信貸供給。由於國內信貸環境收緊和融資成本明顯上行，對於中資企業而言海外外幣債市場變得更有吸引力。

與之前兩輪發債高峰期相比，在本輪發債高峰期中一般行業的企業表現依舊搶眼，但主要驅動力來自於金融機構。中資金融機構的海外外幣債存量由 2014 年年底的 548 億美元上升至 2019 年二季度末的 2,805 億美元。中資銀行、券商和金融集團公司因不同的原因轉向海外外幣債市場融資：銀行需要補充資本金，部分為了自己或者客戶的海外併購業務；券商需要為自己在岸和離岸的股權質押貸款業務募集更多的流動性資金；金融集團公司需要拓展自己的國際業務。

根據我們的測算，截至 2019 年二季度末中資企業海外外幣債存量規模已達 8,416 億美元，是其 2014 年年底規模（2,687 億美元）的 3 倍以上，其中一半左右的債券通過香港市場發行。按發行主體分析，2019 年二季度末一般行業外幣債存量為 3,225 億美元，佔全部中資企業外幣債總存量的 40%，金融機構外幣債存量為 2,805 億美元（33%），房地產開發商外幣債存量為 1,668 億美元（20%），地方政府融資平台外幣債存量為 619 億美元（7%）。

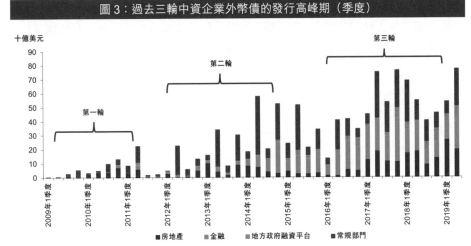

圖 3：過去三輪中資企業外幣債的發行高峰期（季度）

資料來源：Bloomberg 與野村全球宏觀。

4.4.4　新經濟部門或從中資企業海外外幣債市場的崛起中獲益

　　部分來自新經濟部門的發行人已經參與到香港外幣債市場中，如高新技術製造業、電子商務、清潔與可再生能源等行業的企業。值得注意的是，尤其是對中小型新經濟企業來說，發行海外外幣債有潛在積極的品牌效應，這或許可以幫助它們吸引到更多國際投資者並因此而降低融資成本。

　　以綠色債券為例，綠色債券是指為關於氣候和環境問題的項目募集資金的債券。近些年，全球範圍內綠色債券融資呈爆發性增長，主要原因在於中國內地對綠色項目融資有強烈的需求，而國際投資者對關於綠色項目投資有着強烈的興趣。

　　香港作為連接中國內地與世界其他地區的橋樑，連接這兩個市場並把握相關綠色投資機會的優勢與生俱來。得益於此，香港已經發展成為世界最大的綠色債券市場之一，市場中的主要發行人也已經從相關本土機構逐漸轉變成中國內地和海外機構。香港金管局於 2019 年 2 月發佈的《香港綠色債券市場報告》顯示，2018 年在香港安排和發行的綠色債券總額達 110 億美元，相比 2017 年的 30 億美元增長了 237%。其中，中國內地及海外機構於 2018 年在香港安排並發行了總計 90 億美元的綠色債券，佔全年香港綠色債券發行總量的 83%。

　　我們相信隨着全球金融市場的進一步融合、全球經濟發展重心不斷向新經濟部門轉移、跨區域合作日趨密切（特別是通過大灣區相連的中國內地與香港，以及通過中國金融市場開放與「一帶一路」加緊合作與共贏的中國與世界其他地區），新經濟企業將從香港的外幣債市場中獲得更多的收益。

第3章

互聯互通開啟中國資本市場新紀元

洪灝

交銀國際控股有限公司
研究部主管兼董事總經理

摘　要

中國內地與香港股票互聯互通（「滬深港通」）額度的有序開放、投資目標的擴大以及北向看穿式監管機制的開啟，為投資者提供了更多元化的選擇並保障了投資者的利益。滬深港通的發展也體現了互聯互通機制的進一步完善，對中國內地資本市場開放起到舉足輕重的作用。

互聯互通對 A 股市場的影響

- 滬深港通的北向交易（「陸股通」）抑制了 A 股市場的投機性。通過陸股通進入內地市場的境外機構投資者體現出成熟的價值投資理念。陸股通股票標的體現一定的「港股化」傾向（價值投資為主，換手率與波動性較 A 股市場少）。然而，港股通標的「A 股化」（散戶化與追漲殺跌的投機性特徵）的傾向較弱。通過港股通進入香港市場的境內投資者，在港股市場卻表現出一定的類境外機構投資者的特徵。對於 A 股市場，隨着境外投資者參與度的提升，其對資本市場成熟的資訊處理與分析能力將提高股票的定價效率。這將助力 A 股標的的價值回歸，避免股價偏離基本面引發的暴漲暴跌。

- 互聯互通對 AH 溢價收斂作用有限。套利渠道、資金限流以及投資者結構差異是導致 AH 溢價現象的主要原因。再者，陸股通與港股通的投資額度使用率始終低於預期。跨境資金流動的不活躍，對 AH 股估值趨同的推動作用有限。不過，AH 股溢價的收斂是結構性存在的。高分紅、高機構持股比例、低波動與低換手率的 A 股與其對應的 H 股價差較小。隨着互聯互通進程的推進，A 股市場將引入更多注重中長期資產配置的機構投資者，理論上來說可以引導 A 股風格向價值投資轉移。如果中國內地資本市場進一步開放，AH 溢價現象也將隨之消除。

- 互聯互通加強了中國內地與香港市場的聯動性，而聯動性的增強更多體現於滬港通，對深港通的作用較弱。滬港通開通後，滬港兩市主要指數（上證綜合指數與恒生指數）的相關性有所上升，但深港兩市（深證綜合指數與恒生指數）的相關性卻下降。這反映了在宏觀因素相同的背景下，陸股通的啟動是影響兩地市場聯動性變化的因素之一。

互聯互通推進中國內地資本市場改革開放

- 互聯互通是中國內地股票市場國際化的重要探索。近年來，境外投資者進入內地市場投資的渠道日益完善。互聯互通的擴容與機制的優化為 A 股國際化先行鋪路。進入 2018 年以來，伴隨 A 股被正式納入 MSCI 新興市場指數，借道滬深港通「北上」的資金顯著增加。長期來看，海外基金有望為 A 股帶來數萬億元人民幣的增量資金，這些基金將逐漸成為內地資本市場的重要組成部分。A 股市場國際化水平將得到進一步提升。

- 滬深港通是人民幣國際化的重要組成部分。其引入資金以人民幣進行流動、淨額結算的交易方式，在加快人民幣國際化進程的同時，亦有效地控制了跨境資金波動的風險。滬深港通作為實現資金在香港和內地雙向流動的橋樑，有利於海外投資者更便利及全面地投資內地資本市場，在拓展境外人民幣的投資渠道的同時也提升了其流動性。

1 滬深港通歷史進程

1.1 滬深港通機制不斷完善、有序開放

從額度的控制上來看，滬港通最初開通時設立了總額度 2,500 億元（南向）和 3,000 億元（北向）的限制，而每日最高限額為 105 億元（南向）和 130 億元（北向）。隨着 2016 年年底深港通的開啟，總額度限制被取消，但每日最高限額維持不變。2018 年 5 月，為助力 A 股納入 MSCI 指數，滬深港通每日限額擴大四倍，增至 420 億元（南向）和 520 億元（北向）。額度有序的開放體現了互聯互通機制的進一步深化，並在中國內地資本市場開放的進程上起到舉足輕重的作用。

從投資標的及投資者門檻來看，最初滬港通開通時北上資金僅可投資包括中證 180 及上證 380 指數成份股和在上交所上市的 A+H 股在內的 500 餘隻大中盤股，投資者門檻則設為不低於 50 萬元的證券資產餘額。而 2016 年深港通的股票標的較滬港通有所擴大，投資者亦可投資部分中小盤股，投資標的涵蓋深圳成份指數和深圳中小創新指數成份股約 900 檔股票，但同時設有股票市值限定 60 億元以上。投資者門檻也有所提高，創業板僅限專業機構投資者。此舉在為投資者提供更多元化選擇的同時也保障了投資者的利益。

2018 年 9 月，滬深港通北向看穿式機制正式啟動，隨後 10 月北向投資者識別功能也一併開啟，對於不符合要求的交易進行拒單。在滬深港通額度日益擴大、為 A 股市場吸引外資的同時，監管也成為大家關注的焦點。北向看穿式機制的建立提升了跨境市場的監管效率，有利於維護市場穩定運行，為投資者創造一個良好的交易環境。

2　互聯互通對 A 股市場的影響

2.1　陸股通抑制了 A 股市場的投機性

互聯互通旨在促進境內與境外市場雙向開放，不僅吸引更多境外投資者投資 A 股，也為境內投資者進入海外市場提供了便利。市場曾預計，互聯互通將為兩地市場增加新的投資羣體，因此新的投資理念與偏好標的將使本地市場出現一定程度的異質化。不過，現實不盡為然。通過北向交易（「陸股通」）進入內地市場的境外機構投資者，在滬深兩市確實具有穩定市場的作用，體現出成熟的價值投資理念。陸股通股票標的的體現一定的「港股化」傾向。然而，通過港股通進入香港市場的境內投資者，在港股市場卻表現出一定的類境外機構投資者的特徵。港股通標的的「A 股化」（散戶化與追漲殺跌的投機性特徵）的傾向較弱。

對於 A 股市場，境外投資者對資本市場成熟的資訊處理與分析能力提升了股票的定價效率。這將助力 A 股標的的價值回歸，避免股價偏離基本面引發的暴漲暴跌。相對於本地投資者，境外投資者多以中長期資產配置為主要目的，頻繁換手的現象較少。其市場參與度的上升可降低散戶的市場影響，抑制股票波動。與此同時，境外投資者偏好價值投資，傾向選擇資訊公開透明、公司財務狀況良好、股價波動性較低的股票。因此，境外投資者與 A 股波動性之間互為因果、相互促進。

我們通過研究互聯互通開通前後，陸股通與非陸股通股票池中相關 A 股標的的平均年化波動率與換手率的變化，來衡量投資者的交易偏好對異地市場的影響。為分析滬港通開通對上海股票市場投資者行為的影響，我們將考察時段 2005 年 1 月至 2019 年 7 月分為兩個時段。第一個時段為滬港通開通之前，即 2005 年至 2014 年。第二個時段為滬港通開通後，即 2015 年至 2019 年 7 月。我們比較了兩個時段中，滬港通與非滬港通股票池的平均年化波動率與換手率，來驗證滬港通的開通是否能減緩上海股票市場的波動性與過度交易。與此同時，我們也運用同樣的方法來考察深港通對深圳股票市場的影響。

境外資金的流入使得陸股通目標追漲殺跌與頻繁買賣的投機性交易特徵有所

改善。從波動率來看，2014 年 11 月滬港通開通後，滬股通股票池中整體的波動性
小幅上升，但漲幅明顯小於非滬股通標的。在 2016 年 12 月深港通開通後，深股
通股票池中整體的波動性下降，且降幅大於非深股通標的。這顯示互聯互通開通
後，境外資金的進入抑制了滬深兩市股票的波動性，A 股散戶化的追漲殺跌現象
有所改善。在換手率方面，2014 年 11 月滬港通開通後，滬股通的換手率[1] 微幅上
升，但漲幅低於非滬股通標的。與此同時，2016 年 12 月深港通開通後，深股通的
換手率下降，但降幅大於非深股通標的。這顯示陸股通標的的投資者持倉時間較
長，較低的換手率對滬深兩市頻繁買賣的高換手特徵有所壓制。

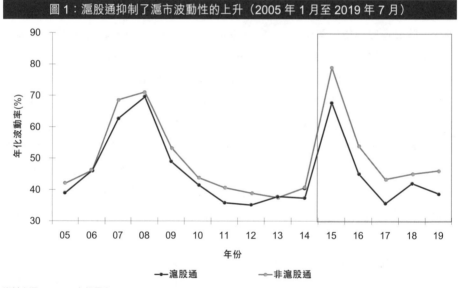

圖 1：滬股通抑制了滬市波動性的上升（2005 年 1 月至 2019 年 7 月）

資料來源：Wind、交銀國際。

1　對於深股與滬股，換手率是運用每年平均日度換手率。

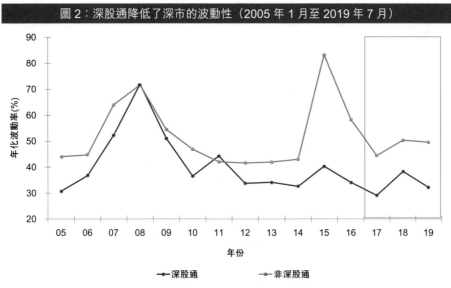

圖 2：深股通降低了深市的波動性（2005 年 1 月至 2019 年 7 月）

資料來源：Wind、交銀國際。

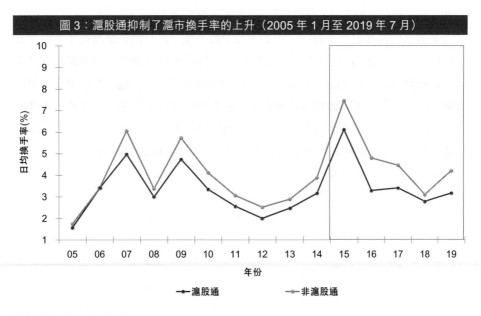

圖 3：滬股通抑制了滬市換手率的上升（2005 年 1 月至 2019 年 7 月）

資料來源：Wind、交銀國際。

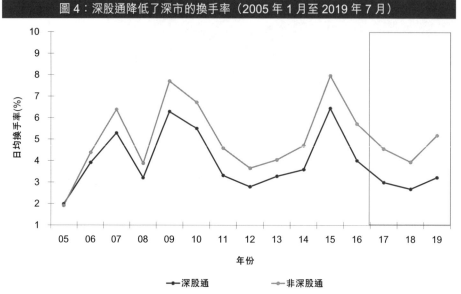

圖 4：深股通降低了深市的換手率（2005 年 1 月至 2019 年 7 月）

資料來源：Wind、交銀國際。

港股通的開放進一步減少了香港市場的波動性，但境內資金高換手率的交易特性仍然存在。2014 年 11 月滬港通開通後，滬市港股通股票池的整體波動性下降，且降幅大於其餘港股。2016 年 8 月深港通開通後，港股通股票池擴容，整體波動性下降，降幅與非港股通標的相當。這說明中國內地資金流入香港市場後，追漲殺跌的投機風格有所收斂。不過，內地資金高換手率的特徵仍然存在。滬、深港通開通後，港股通的換手率[2] 雖有所下降，但降幅不及非港股通標的。

2　對於港股，換手率是運用平均年度換手率。

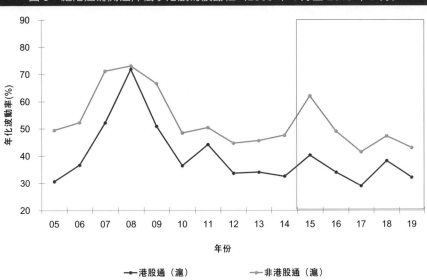

圖 5：滬港通的開通降低了港股的波動性（2005 年 1 月至 2019 年 7 月）

資料來源：Wind、交銀國際。

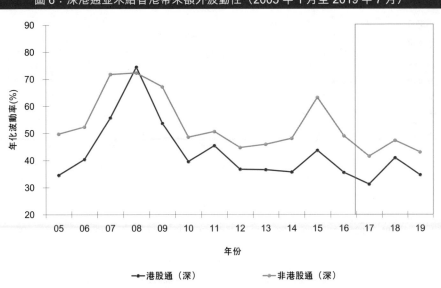

圖 6：深港通並未給香港帶來額外波動性（2005 年 1 月至 2019 年 7 月）

資料來源：Wind、交銀國際。

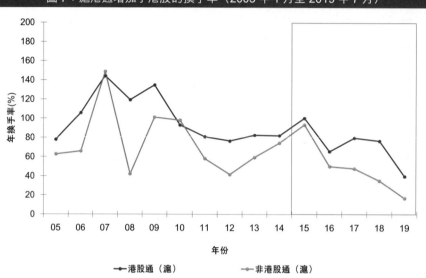

圖 7：滬港通增加了港股的換手率（2005 年 1 月至 2019 年 7 月）

資料來源：Wind、交銀國際。

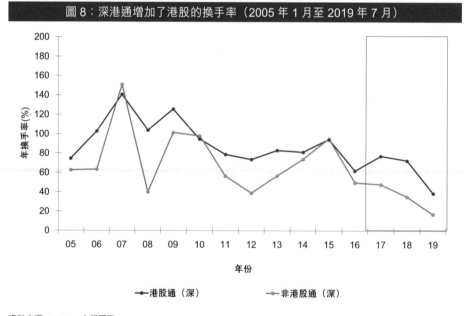

圖 8：深港通增加了港股的換手率（2005 年 1 月至 2019 年 7 月）

資料來源：Wind、交銀國際。

2.2　互聯互通對 AH 溢價收斂作用有限

儘管同股同權，但同一家公司的 A、H 類股票並未回歸「同價」。在滬港通開通前，市場普遍認為 A、H 股票價差會有所收斂。有分析認為，AH 股價差收窄的預期將吸引投資者主動買入 AH 股中價值被低估的一方。在互聯互通開通後，資金在 A、H 股之間的流動也將平抑估值的差異，從而縮小 AH 股價差。然而，隨着互聯互通的開通與擴展，A 股對 H 股始終存在溢價，價差並未如預期般收窄。

套利渠道、資金限流以及投資者結構差異是導致 AH 溢價現象的主要原因。滬、深港通開通後，恒生 AH 溢價指數始終在 110-150 點的區間內運行，A、H 股價差並未如預期般收斂。由於互聯互通僅為中國內地與香港的投資者投資對方市場的渠道，但在中國內地與香港兩地上市的同一家公司的 A、H 股之間並不能跨市場轉換，無法進行價差套利。再者，中國對跨境資本流動的管制較為嚴格，互聯互通項目開通後，陸股通與港股通的投資額度使用始終低於預期。資金流動並不活躍，對 A、H 股估值趨同的推動作用有限。我們也注意到，在 A 股市場包括創業板、中小板在內的中小盤、成長股受到追捧的同時，港股的同類股票卻不受投資者青睞。這是兩地投資者結構大相徑庭、其投資理念存在重大差異所導致。中國內地市場的成交額中有 80% 來自個人投資者，而香港市場 80% 的成交額來自傾向價值投資與基本面分析的機構投資者（包括聯交所參與者的自營業務）[3]。

3　資料來源：香港交易所的《現貨市場交易研究調查 2018》，載於香港交易所網站，2019 年 7 月。

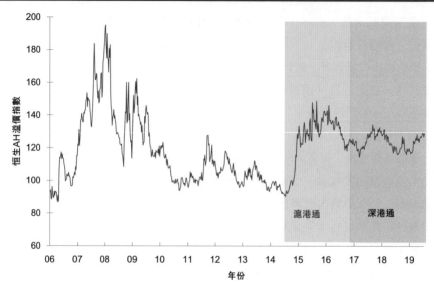

圖 9：滬深港通的開通並未消除 AH 溢價現象（2006 年 1 月至 2019 年 7 月）

資料來源：Wind、交銀國際。

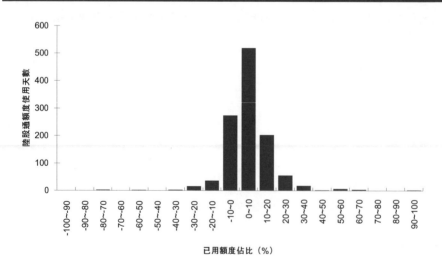

圖 10：滬深港通每日額度使用不及預期（2014 年 11 月 17 日至 2019 年 7 月 31 日）

註：已用額度佔比為負數表示賣盤成交金額大於買盤訂單。

資料來源：Wind、交銀國際。

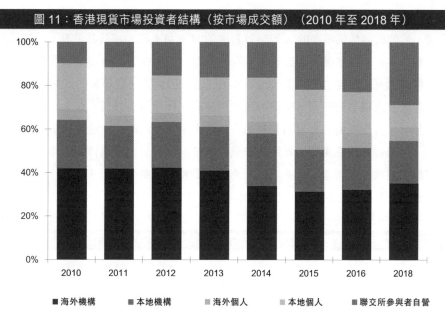

圖 11：香港現貨市場投資者結構（按市場成交額）（2010 年至 2018 年）

■ 海外機構　　■ 本地機構　　■ 海外個人　　■ 本地個人　　■ 聯交所參與者自營

資料來源：香港交易所《現貨市場交易研究調查》。

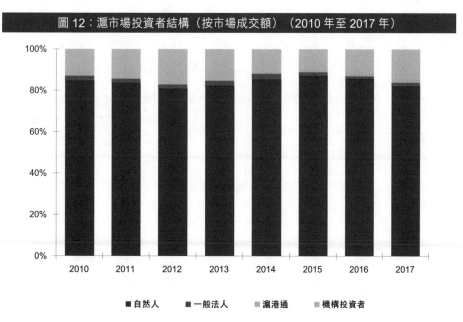

圖 12：滬市場投資者結構（按市場成交額）（2010 年至 2017 年）

■ 自然人　　■ 一般法人　　■ 滬港通　　■ 機構投資者

資料來源：《上海交易所統計年鑒》、Wind。

　　不過，AH 股溢價的收斂是結構性存在的。研究發現，高分紅、高機構持股比例、低波動與低換手率的 A 股與其對應的 H 股價差較小。具體來看，機構持股比例較高，A、H 股價差明顯收斂。有分析認為，由於 A、H 股同股同權，在分紅權與針對公司治理的表決權的行權效用相當，因此買、賣等量的 A、H 股能實現這類大股東權利的套利。機構持股比例越高，即意味着投資者在上市公司的話語權越大，實現機構股東權利套利的空間更大。與此同時，由於機構更偏好價值投資，股票買賣頻率與追漲殺跌的投機性較低，所投資的股票標的通常具有低換手率與低波動性的特徵。因此，換手率與波動性較低的兩地上市股票，其 AH 溢價也應較低。我們的數據分析也同樣印證了這一觀點。

圖 13：機構持股比例與 AH 溢價負相關（2018 年 1 月 1 日至 2018 年 12 月 31 日）

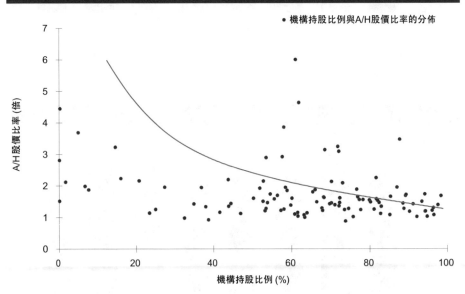

資料來源：Wind、交銀國際。

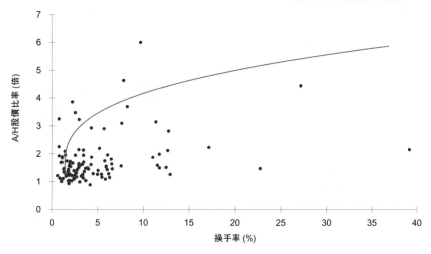

圖 14：換手率與 AH 溢價正相關（2018 年 1 月 1 日至 2018 年 12 月 31 日）

● 換手率與A/H股價比率的分佈

資料來源：Wind、交銀國際。

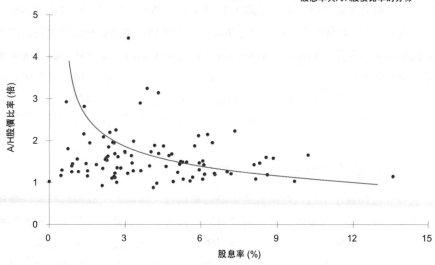

圖 15：股息率與 AH 溢價負相關（2018 年 1 月 1 日至 2018 年 12 月 31 日）

● 股息率與A/H股價比率的分佈

資料來源：Wind、交銀國際。

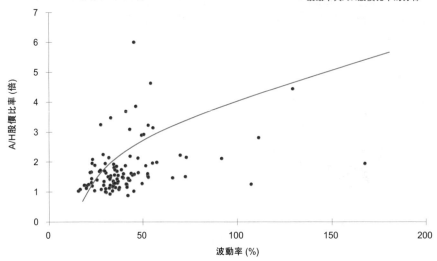

圖 16：波動性與 AH 溢價正相關（2018 年 1 月 1 日至 2018 年 12 月 31 日）

波動率與A/H股價比率的分佈

資料來源：Wind、交銀國際。

　　隨着互聯互通進程的推進，A 股市場將引入更多注重中長期資產配置的機構投資者，理論上來說可以引導 A 股風格向價值投資轉移。如果中國內地資本市場進一步開放，AH 溢價的現象也能隨之消除。滬深港通的啟動促使內地與香港的資本流動性加強，資訊不對稱程度有所緩解。雖然在短期內，受制於投資額度與市場開放程度的影響，A、H 股的估值完全趨同難以實現。然而，隨着未來內地資本市場開放程度提升，滬深港通將有助於內地和香港兩地市場回歸均衡估值。AH 估價差異的收斂，將取決於市場的開放程度以及投資者結構的改善。

2.3　互聯互通強化兩地市場的聯動性

　　互聯互通加強了中國內地與香港市場的聯動性。理論上，股市的聯動性多由共同宏觀經濟變量、市場傳染假說與投資者行為決定。金融市場存在明顯的順週期性和傳染性。在經濟全球化中，宏觀經濟的趨同變化會直接影響股市的波動，而局部極端事件的發生將通過流動性、匯率、情緒等渠道蔓延至其他市場，進而引發股市共振。另一方面，由於資訊不對稱引發的羊羣效應也會同化投資者交易行為，增強股市的聯動性。

互聯互通對兩地股市聯動性的增強更多體現於滬港通，對深港通的作用較弱。整體來看，滬港通開通後，滬港兩市主要指數（上證綜合指數與恒生指數）的相關性有所上升，但深港兩市的相關性卻下降。倘若將樣本縮小至滬、深股通的股票池，兩市相關性的增強與削弱現象體現得更加明顯。恒生指數與深股通股票標的的相關性多次從正相關變為負相關。這反映了在宏觀因素相同的背景下，互聯互通是影響兩地市場聯動性變化的主要因素之一。

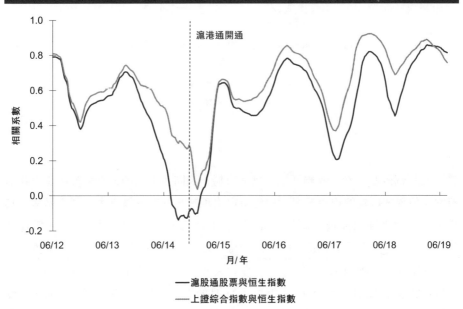

圖 17：滬股通開通後，滬股通目標對滬港兩市的相關性貢獻更為顯著
（2012 年 6 月至 2019 年 7 月）

資料來源：Wind、交銀國際。

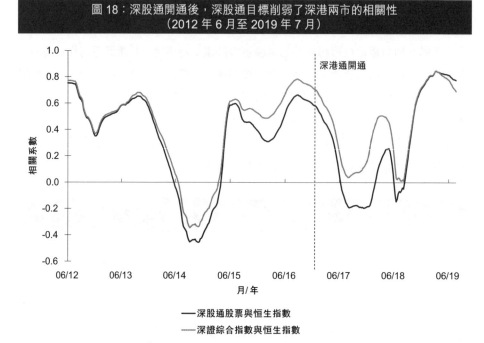

圖 18：深股通開通後，深股通目標削弱了深港兩市的相關性
（2012 年 6 月至 2019 年 7 月）

資料來源：Wind、交銀國際。

　　從尾部效應來看，互聯互通中兩地市場出現同步上漲的概率大於同步下跌的概率。但對於極端的市場情況，兩地市場出現同步暴跌的概率大於同步暴漲。具體來看，滬港通開通後（基於 2014 年 11 月 17 日至 2019 年 7 月 31 日的數據計算），滬市與香港同步上漲的概率為 37%，大於同步下跌的概率 30%。滬港兩市同日漲幅均超過 2% 的概率為 1.6%，跌幅大於 2% 的概率為 2.0%。深港通開通後（基於 2016 年 12 月 5 日至 2019 年 7 月 31 日的數據計算），深市與香港同步上漲的概率為 37%，同步下跌的概率為 31%。深港兩市同日漲幅均超過 2% 的概率為 0.8%，小於同日跌幅超過 2% 的概率 1.4%。

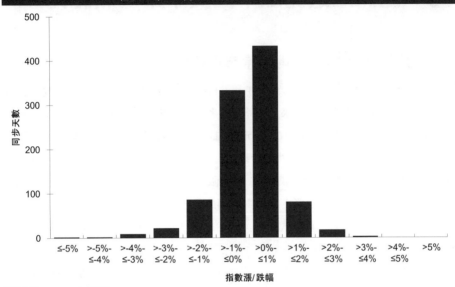

圖 19：滬港通開通後，滬港兩市同步上漲概率大於同步下跌，但暴跌概率大於暴漲
（2014 年 11 月 17 日至 2019 年 7 月 31 日）

資料來源：Wind、交銀國際。

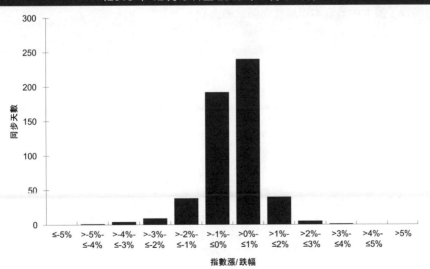

圖 20：深港通開通後，深港兩市同步上漲概率大於同步下跌，但暴跌概率大於暴漲
（2016 年 12 月 5 日至 2019 年 7 月 31 日）

資料來源：Wind、交銀國際。

3 滬深港通開啟中國資本市場新紀元

3.1　滬深港通推進內地股票市場國際化

　　滬深港通、B 股轉 H 股、RQFII 和 QFII 的開啟都是內地資本市場雙向開放過程中穩步行進的足跡。近年來，境外投資者進入內地市場投資的渠道日益完善。RQFII 的規模在不到 3 年間由 100 億元增加至近 3,000 億元，在 2014 年的不到 12 個月內更實現翻番。近期中國外匯管理局亦傳遞出「改革 QFII 和 RQFII、擴大投資範圍、考慮適度放寬甚至取消 QFII 額度限制」的訊息。而滬深港通也在制度不斷完善的情況下，成為內地資本市場對外開放的重要里程碑。

　　滬港通於 2014 年 11 月正式啟動，初始總額度為 5,500 億元人民幣，每日最高總限額（北向與南向合計）為 235 億元人民幣。2016 年 12 月伴隨着深港通開通，滬深港通的交易總額度上限被取消。為助力 A 股順利「入摩」（被納入 MSCI 新興市場指數），2018 年 4 月每日交易額度進一步擴大至四倍。額度的逐步擴大很大程度上提高了市場流動性的空間，也側面反映出投資者對滬深港通的認可和信賴。此外，推出北向交易投資者識別、將不同投票權架構公司納入港股通等制度改革也體現了擴大內地資本市場雙向開放的決心。

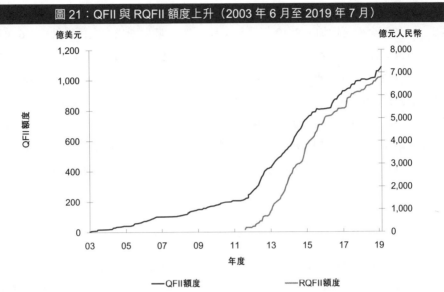

圖 21：QFII 與 RQFII 額度上升（2003 年 6 月至 2019 年 7 月）

資料來源：Wind、交銀國際。

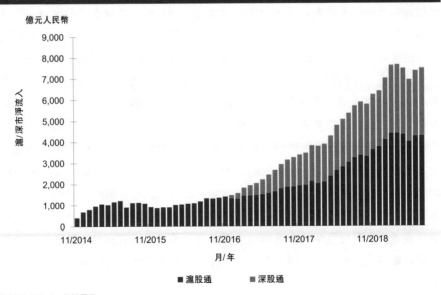

圖 22：境外資金通過陸股通持續流入 A 股市場（2014 年 11 月至 2019 年 7 月）

資料來源：Wind、交銀國際。

進入 2018 年以來，伴隨 A 股被正式納入 MSCI 新興市場指數，借道滬深港通「北上」的資金顯著增加。2019 年截至 7 月 31 日，陸股通成交金額累計達 5.6 萬億元人民幣，日均交易金額 410 億元人民幣，較 2018 年日均成交額翻番。滬深港通的成功啟動是 A 股獲納入全球指數體系錦上添花的一筆，為內地資本市場的開放帶來了歷史性的積極變化。繼 A 股被納入 MSCI 及逐步擴容後，2019 年 6 月 A 股順利「入富」(被納入富時羅素指數)。與此同時，預計同年 9 月 A 股將正式被納入標普道瓊斯指數。長期來看，海外基金有望為 A 股帶來數萬億元人民幣的增量資金，這些基金將逐漸成為內地資本市場的重要組成部分。A 股市場國際化水平將得到進一步提升。

3.2　滬深港通是人民幣國際化的重要組成部分

二次世界大戰後，美元通過布雷頓森林體系成為國際貨幣體系的核心。然後美國通過馬歇爾計劃使境外美元的供應量大幅增加，形成美元在岸 / 離岸流動的循環。儘管布雷頓森林體系於 20 世紀 70 年代的崩潰影響了市場對美元的信心，但石油美元體系最終奠定了美元作為國際貿易定價和清算貨幣的地位。而日圓國際化的最終夭折更多是因為日本未能持續地在亞洲地區行業發展的產業鏈裏維持其主導地位。當然，日本無法僅憑日圓貸款擴充海外日圓的供應，也是主要因素之一。這些其他國家主要貨幣的國際化之路，對中國來說，可以作為他山之石。

開通人民幣跨境貿易結算機制，實現結算便利，使得更多國家願意持有人民幣，是人民幣國際化進程中必不可少的一步。滬港通引入北向資金以人民幣進行流動、淨額結算的交易方式，最大程度上減弱了對在岸人民幣市場的匯率影響。在加快人民幣國際化進程的同時亦有效地控制了跨境資金波動的風險。香港作為最大的離岸人民幣中心，滬港通借其力量得以實現人民幣投資的循環。隨後推出的深港通也同樣運用了這種跨境結算機制。提高人民幣在國際計價中的結算、儲藏價值之外，推動其向金融領域投資貨幣轉化，也是人民幣國際化中重要的一環。滬深港通作為實現資金在香港和內地雙向流動的橋樑，有利於海外投資者更便利及全面地投資內地資本市場，在拓展境外人民幣的投資渠道的同時也提升了其流動性。

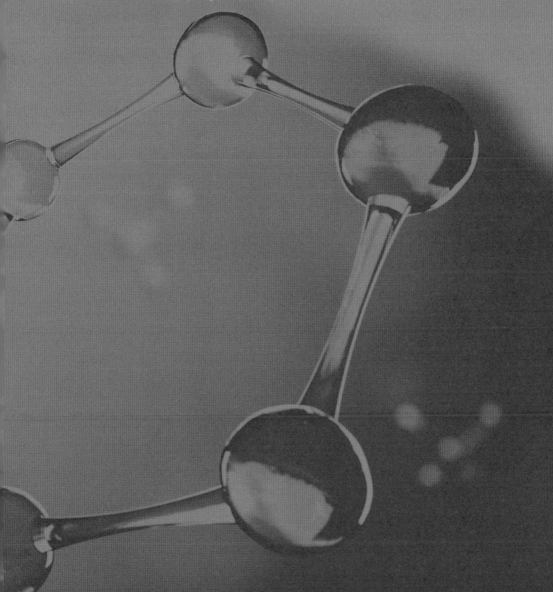

第二篇

新經濟公司的上市制度
改革和融資創新

第4章

香港聯交所的新股上市流程和重點

香港交易及結算所有限公司

首席中國經濟學家辦公室

及

環球上市服務部

摘 要

　　本章簡潔地介紹在香港聯合交易所（即在香港交易及結算所有限公司所營運的證券市場）作首次公開招股及上市的程序。有意尋求於香港上市的公司（特別是中國內地公司）可以從多種股權架構中作出自己的選擇。

　　除了主板與 GEM 的上市條件外，本章亦介紹在首次公開招股及上市的過程中，企業所需的各種專業服務，以及上市後作股權再融資的各種途徑。

1 公司赴港上市的一般要求

　　無論公司經營何種業務，如希望申請在香港聯合交易所（簡稱「香港聯交所」）的主板掛牌上市[1]，一般而言應符合《香港聯合交易所有限公司證券上市規則》（以下簡稱《上市規則》）第八章所規定的一系列條件，以滿足基本的上市資格要求。需要說明，香港聯交所對接納或拒絕具體某家公司的上市申請保留絕對酌情權，即使申請人符合有關條件，也不一定保證其適合上市。

1.1　主板上市條件

　　香港主板市場是為具備一定市值/收入/利潤規模、業務根基穩健的公司而設的市場，上市公司行業眾多，包括綜合企業、銀行、房地產開發公司、互聯網公司及醫療健康公司等。因此，公司作為上市申請人申請在香港主板市場上市，必須符合以下基本條件：

(1)　財務要求（符合以下其中一項測試即可）[2]

盈利測試

- 上市前最近一個會計年度股東應佔盈利≥ 2,000 萬港元；
- 前兩年累計的股東應佔盈利≥ 3,000 萬港元；及
- 上市時市值≥ 5 億港元[3]。

市值/收益測試

- 上市前最近一個會計年度收入≥ 5 億港元；及
- 上市時市值≥ 40 億港元。

市值/收益/現金流量測試

- 上市前最近一個會計年度收入≥ 5 億港元；
- 擬上市業務現金流入前 3 個會計年度合計≥ 1 億港元；及

1　就本章所述「上市」內容，除非特別說明，均指在香港聯交所主板首次公開招股（IPO）並掛牌上市。

2　詳情參閱《上市規則》第 8.05 條。

3　詳情參閱《上市規則》第 8.09(2) 條。

    • 上市時市值 ≥ 20 億港元。

(2) 其他要求

    除以上財務要求以外，擬赴港上市的企業還需在經營歷史、公眾持股、管理層及控制權等方面滿足一系列基本要求。此類要求主要包括：

- 具備不少於三個會計年度的營業記錄（特殊情況下可考慮較短的營業記錄[4]）；
- 最低公眾持股量（一般為 25%，如上市時市值 >100 億港元可酌情減至 15%）；
- 上市時至少擁有 300 名股東[5]；
- 至少前三個會計年度的管理層維持不變；
- 至少最近一個會計年度擁有權和控制權保持不變；
- 至少有三名獨立董事，並必須佔董事會成員人數至少三分之一[6]；
- 如為 H 股上市申請人，至少須有一名獨立非執行董事通常居於香港；
- 至少有兩名執行董事通常居於香港[7]；
- 必須委任一名相關專業人履行公司秘書職責[8]；
- 申報會計報告的最後一個會計期間的結算日期距上市文件刊發日期不超過 6 個月。

1.2　香港上市股權架構

1.2.1　獲接納的註冊地司法轄區

    擬赴港上市的申請人公司的註冊地首先需要屬於香港聯交所「獲接納司法權區」[9] 範圍之內。根據《上市規則》，於香港、中國內地、開曼羣島及百慕達四個司法權區註冊成立的公司符合提出上市申請的資格要求。截至 2019 年 6 月 30 日，

4　詳情參閱《上市規則》第 8.05A 條。
5　詳情參閱《上市規則》第 8.08 條。
6　詳情參閱《上市規則》第 3.10 條及第 3.10A 條。
7　詳情參閱《上市規則》第 8.12 條。
8　詳情參閱《上市規則》第 3.28 條。
9　詳情參閱香港交易所官方網站「獲接納的海外司法地區清單」。

其他 27 個海外公司適用的「獲接納司法轄區」包括英格蘭及威爾斯、英屬維京羣島、法國、意大利、俄羅斯、日本、新加坡、澳洲等。海外公司須證明其註冊成立地的股東保障水平至少相當於香港的相關水平。若未能證明，海外公司可通過修改其組織章程文件以提供相等水平的保障。

1.2.2　赴港上市的內地公司架構

對於主要業務和背景在中國內地的公司，香港聯交所接受靈活多樣的上市架構選擇，主要包括：

- **紅籌架構** —— 上市主體是境外控股公司，以境內股權或資產注入境外控股公司間接實現境外上市。紅籌架構下又包含兩種模式：

  (1) **大紅籌** —— 中國內地的企業、資產或業務的實際控制人通常是某個政府機構，又稱國企紅籌；

  (2) **小紅籌** —— 中國內地的企業、資產或業務的實際控制人通常是某個個人，又稱民企紅籌。

- **H 股架構** —— 上市主體是在中國內地註冊成立的股份有限公司，境內股權直接在香港聯交所上市。H 股上市架構下又包含四種可行的方式：

  (1) **先 A 後 H** —— 已在中國內地上市的公司可憑藉 A 股公司身份赴香港作雙重主要上市；

  (2) **先 H 後 A** —— 如公司先在香港聯交所上市，之後還可以再回到中國內地 A 股上市；

  (3) **A+H 同時進行** —— 公司同時申請在內地 A 股和香港聯交所 H 股上市並同時掛牌交易；

  (4) **新三板 + H** —— 已經在內地新三板掛牌的公司毋須事先除牌即可申請到香港聯交所上市。

- **分拆上市** —— 已經在香港、內地或海外上市的公司也可以分拆子公司或一部分業務在香港上市。

1.2.3 紅籌架構模式

紅籌模式通常指通過在海外設立控股公司，即「特殊目的公司」，將中國內地企業的資產和權益注入境外的控股公司，並以境外控股公司的名義在海外上市募集資金的方式。業界通常將紅籌分為「大紅籌」和「小紅籌」；統計顯示境外上市內地企業中大部分採用的都是紅籌模式，尤其以小紅籌模式實踐較多。

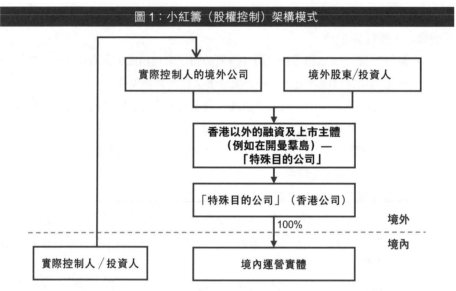

圖 1：小紅籌（股權控制）架構模式

註：上圖僅為示意性表述，實踐操作中可能有所不同。

1.2.4 可變利益實體（VIE）架構模式

對 VIE（又稱「協議控制」）架構下的上市申請審核，香港聯交所以披露為本的監管方針為指引，主要考慮因素包括以下內容：

- VIE 架構採用原因應只限於解決外資擁有權的限制，如這些限制不再存在，公司必須解除結構性合約安排，即「只限切合所需」(Narrowly Tailored) 原則 [10]；
- 應將與相關中國法規出現衝突的可能性減至最低；

---

10 詳情見香港交易所上市決策 HKEX-LD43-3。

- 實際可行的情況下，上市申請人須向監管機構索取適當的監管確認；
- 當法律允許上市申請人毋須採用合約安排方式經營業務時，須終止合約安排；
- 確保合約安排：(1) 包含境內運營主體公司股東授予上市申請人董事及其繼任人行使境內運營主體公司股東所有權力，並確保不存在任何潛在利益衝突；(2) 載有解決爭議的條款；(3) 涵蓋處理境內運營主體公司資產的權力；
- 考慮內地《外商投資法》的最新進展。

審核企業 IPO 申報材料時，香港聯交所需要：

- 由**保薦人**確認公司符合上市條件並按規定在招股書內披露合約安排詳情及相關風險；
- 由**公司法律顧問**出具正面意見；
- 由**申報會計師**確認合併計算境內運營主體公司的財務業績符合現行會計準則。

　　如公司計劃以 VIE 架構申報上市且 VIE 架構涉及不常見事項，香港聯交所建議公司盡早開展上市前諮詢及預溝通（Pre-A1 Submission），並根據溝通情況對上市架構進行調整。

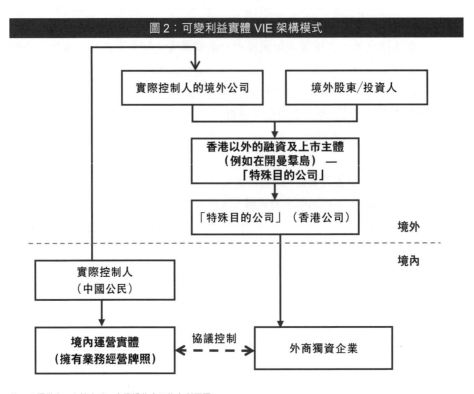

圖2：可變利益實體 VIE 架構模式

註：上圖僅為示意性表述，實踐操作中可能有所不同。

1.2.5　H 股架構模式

　　H 股是指在中國內地合法成立的股份有限公司在香港發行、向境外投資者募集並在香港聯交所上市的股份。中國內地也在《證券法》及《國務院關於股份有限公司境外募集股份及上市的特別規定》等法律法規中，對 H 股的發行作出了相關規定。

　　目前，以 H 股架構申請赴港上市的公司需要向中國證券監督管理委員會（簡稱「中國證監會」）及香港聯交所同時履行境內、境外雙重審批程序。

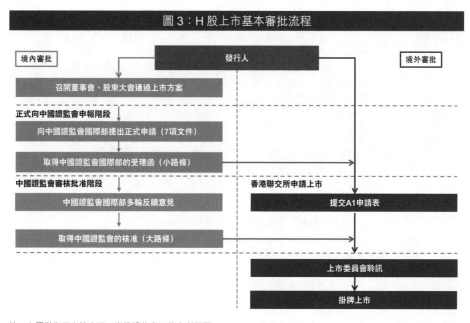

圖 3：H 股上市基本審批流程

註：上圖僅為示意性表述，實踐操作中可能有所不同。

　　中國證監會國際部審核人員在初審過程中發現需要特別關注的問題，將提交回饋會議討論。申請人收到回饋意見後，組織相關中介機構按照要求準備回覆意見，並在規定時間內向受理部門提交回饋材料。審核人員應撰寫審核報告，履行核准或者不予核准境外首次公開發行並上市或境外增發行政許可的簽批程序後，審結發文，經受理部門向申請人發出行政許可核准文件。

　　根據中國證監會於 2019 年 7 月 23 日最新發佈的《股份有限公司境外公開募集股份及上市（包括增發）審核關注要點》，目前關於 H 股上市的具體關注要點如下表 1 所述：

表 1：中國證監會關於 H 股上市的審核要點	
外資准入與宏觀調控及產業政策	1. 發行人及各下屬公司業務範圍是否涉及國家禁止或限制外商投資的領域，境外發行上市前後是否持續符合有關外資准入政策。
	2. 發行人業務範圍是否符合以下情形之一：（1）主營業務為房地產業務，或（2）房地產業務（併表內）佔營業收入比重大於或等於 50%，或（3）房地產業務的收入和利潤均在所有業務中最高，且均佔到公司總收入和總利潤的 30% 以上（包含本數）；在符合上述情形之一的情況下，是否存在違反《國務院辦公廳關於繼續做好房地產市場調控工作的通知》（國辦發〔2013〕17 號）與《國務院關於堅決遏制部分城市房價過快上漲的通知》（國發〔2010〕10 號）的情形，是否存在被住房和城鄉建設部門公示為「違法違規房地產開發企業和中介機構」的情形。
	3. 發行人及各下屬公司是否存在違反《市場准入負面清單》的情形；是否屬於產能過剩行業，是否存在違規和未取得合法手續、不符合重點產業調整和振興規劃及相關產業政策要求、未經批准或違規審批的項目等違反國發〔2013〕41 號、《產業結構調整指導目錄》、國辦發〔2013〕67 號、銀發〔2009〕386 號、國辦發〔2016〕34 號等國務院有關文件及有關部門規章的情形。
合規經營	4. 發行人及各下屬公司近一年是否存在違反《國務院關於進一步加強企業安全生產工作的通知》（國發〔2010〕23 號）的情形；是否存在違反國家和地方環境保護相關要求的情形。
	5. 發行人及各下屬公司近兩年是否存在涉嫌違反《證券法》、《證券投資基金法》、《期貨交易管理條例》、《國務院關於股份有限公司境外募集股份及上市的特別規定》（國務院令第 160 號）及《國務院關於進一步加強在境外發行股票和上市管理的通知》（國發〔1997〕21 號）等證券、期貨法律法規行為的情形。
	6. 發行人及各下屬公司、發行人聘請的相關證券服務機構是否存在因涉嫌違法違規被行政機關立案調查，或者被司法機關偵查，尚未結案的情形；是否存在被中國證監會依法採取限制業務活動、責令停業整頓、指定其他機構託管、接管等監管措施，尚未解除的情形。
	7. 發行人的發起人認購的股份是否繳足，發起人用作出資的財產權轉移手續是否已辦理完畢。發行人是否屬於國發〔2016〕33 號文規定的嚴重失信主體。
股權結構與公司治理	8. 發行人是否在公司章程中載明了《到境外上市公司章程必備條款》所要求的內容。發行人現有股東及本次發行對象（如適用）之間是否存在關聯關係或一致行動關係。請披露單獨或合計持有發行人 5% 及以上的主要股東、本次發行對象（如適用）的股權控制關係（包括但不限於控股股東及實際控制人）。
	9. 發行人及各下屬公司是否建立健全了完備、規範的保密和檔案規章制度並落實到位，是否符合《關於加強在境外發行證券與上市相關保密和檔案管理工作的規定》。
本次發行	10. 本次發行並上市是否履行了完備的內部決策程序，是否取得了必要的內部批准和授權；是否取得了行業監管部門出具的監管意見書（如適用）等必要的外部批准程序；發行對象（如適用）及發行幣種是否符合《國務院關於股份有限公司境外募集股份及上市的特別規定》（國務院令第 160 號）有關要求。
	11. 本次發行募投項目是否取得了必要的審批、核准或備案文件（如適用），是否符合固定資產投資管理有關規定；是否符合國家和地方環境保護相關要求。本次境外發行募集資金是否投向《國務院關於促進節約集約用地的通知》（國發〔2008〕3 號）規定的違法用地項目；是否投向《產業結構調整指導目錄》規定的淘汰類及限制類的產業、《市場准入負面清單》規定的禁止准入事項；如涉及境外投資，是否符合國辦發〔2017〕74 號文規定的境外投資方向、是否依法履行了境外投資核准或備案程序。

(續)

	表 1：中國證監會關於 H 股上市的審核要點
特定對象適用事項	1. 本次發行並上市是否符合《境內企業申請到香港創業板上市審批與監管指引》（證監發行字〔1999〕126 號）。（適用於擬在香港創業板上市的境內公司）
	2. 本次發行並上市是否存在違反《中國證券監督管理委員會關於規範境內上市公司所屬企業到境外上市有關問題的通知》（證監發〔2004〕67 號）的情形。（適用於擬在境外上市的境內上市公司所屬企業）
	3. 本次發行並上市是否存在違反《優先股試點管理辦法》（證監會令第 97 號）、《關於商業銀行發行優先股補充一級資本的指導意見》（銀監發〔2014〕12 號）的情形。（適用於擬在境外發行優先股的境內公司）
	4. 請本次首發前已持有發行人股份的股東及持有發行人股份（包括直接持股與間接持股）的董事、監事、高級管理人員作出專項承諾，自發行人股票在境外交易所上市交易之日起 1 年內不轉讓所持股份。請補充提供上述專項承諾。（適用於未在境內上市的境內公司）
	5. 按照《非上市公眾公司監管指引第 4 號——股東人數超過 200 人的未上市股份有限公司申請行政許可有關問題的審核指引》計算，發行人及其控股股東、實際控制人、重要控股子公司的股東人數是否超過 200 人；如超過，請補充提供有關申請文件並履行有關程序。其中，「重要控股子公司」的核查標準為：報告期內，營業收入、淨利潤或者資產總額達到合併報表數額 50% 以上的控股子公司，以及其他目前或未來對集團整體的經營、財務狀況有重要影響的控股子公司。（適用於未在境內上市的境內公司）
	6. 發行人是否存在內部職工直接持股、股份代持或間接持股的情形；如存在，是否違反《關於金融企業內部職工持股的通知》（財金〔2010〕97 號）等有關規定；請發行人相關高管和其他持有內部職工股超過 5 萬股的個人按照有關規定對股份轉讓鎖定期和出售限額作出專項承諾。（適用於金融企業）
	7. 是否按照《中國證券監督管理委員會關於境外上市公司非境外上市股份集中登記存管有關事宜的通知》（證監國合字〔2007〕10 號）將非境外上市股份在中國證券登記結算有限責任公司集中登記存管。（適用於擬在境外增發的 H 股公司）

資料來源：中國證監會官方網站。

1.3　H 股架構上市的最新進展

1.3.1　H 股上市公司已經成為港股市場的重要組成部分

1993 年 6 月，香港聯交所《上市規則》准許中國註冊企業來港上市。同年 7 月，青島啤酒（股份代號：00168）成為首家發行 H 股在香港上市的內地企業。2000 年以後，內地大型企業相繼來港以 H 股上市，包括四大國有銀行、中國聯通、中國石油股份等在內，形成內地企業赴港上市潮。

截至 2019 年 6 月 30 日，共有 1,197 家內地企業在香港上市，包括 274 家 H 股；內地企業市值佔香港證券市場約 68%，2019 年 1 月至 6 月的交易額佔比約 80%。其中，已上市的 A+H 股數量為 112 家，其中 90 家同時於上海證券交易所

（簡稱「上交所」）掛牌，22 家同時於深圳證券交易所（簡稱「深交所」）掛牌。

所有 274 家 H 股的總首發募集資金為 1.81 萬億港元。主要行業分佈於工業（69 家）、金融（66 家）和可選消費（25 家），合共佔所有 H 股首發募集資金近八成。香港資本市場服務內地企業的能力不斷深化並加強。

	表 2：香港聯交所的上市公司概況						
上市公司類別	數量			總市值（2019 年 6 月底）		日均成交額[11]（2019 年 1 月至 6 月）	
	合計	主板	GEM	百萬港元	佔比（%）	百萬港元	佔比（%）
內地公司	1,197	1,093	104	22,247,751.84	67.98	55,134.84	79.89
H 股	274	250	24	6,376,083.15	19.48	24,257.63	35.15
紅籌[12]	176	171	5	5,595,665.82	17.10	7,908.10	11.46
MPE[13]	747	672	75	10,276,002.88	31.40	22,969.11	33.28
香港公司	1,006	765	241	7,921,907.18	24.21	9,484.60	13.74
海外公司	179	138	41	2,556,988.79	7.81	4,393.95	6.37
合計	2,382	1,996	386	32,726,647.81	100.00	69,013.39	100.00

資料來源：香港交易所。

此外，自 2014 年「滬港通」、2016 年「深港通」相繼開通以來，香港、上海、深圳三地股票市場的互聯互通機制進一步促進中國內地與香港資本市場雙向開放和健康發展。於 2019 年 6 月底，納入「滬港通」的 326 隻港股中有 118 隻為 H 股；納入「深港通」的 481 隻港股中有 144 隻為 H 股。

1.3.2　H 股全流通試點順利實施

H 股上市公司的股本結構中，包括允許在港股市場流通交易的「外資股」，以及在上市前原法人股及國有股等「內資股」。「內資股」在境內市場公開上市交易之前就處於非流通狀態。H 股全流通則將容許把尚未公開交易的「內資股」轉為「外資股」，並在港股市場上公開交易。

11　日均成交額包括普通股和優先股。
12　此處「紅籌」僅指以大紅籌架構於香港上市的內地公司。
13　MPE 指非以 H 股架構於香港上市的內地民營公司。

2017 年 12 月 29 日公佈的《中國證監會深化境外上市制度改革開展 H 股「全流通」試點》文件顯示，將按照積極穩妥、循序漸進的原則，以「成熟一家、推出一家」的方式有序推進 H 股全流通試點。

2018 年聯想控股（03396）、中航科工（02357）和威高股份（01066）三家公司 H 股全流通試點工作全部順利完成。在試點已經成功完成的基礎上，中國證監會副主席方星海公開表示盡快全面推出 H 股全流通，條件具備的 H 股公司，都可申請實現全流通，以便利境內創新企業境外上市。

2019 年 6 月，中國證監會主席易會滿在上海陸家嘴論壇表示「將進一步加大資本市場對外開放，包括全面推進 H 股全流通改革」。隨着未來 H 股全流通試點工作的深化，以 H 股架構模式在香港聯交所上市的內地公司將必獲得更加廣闊深厚的資本市場支持。

1.3.3　新三板公司在港發行「新三板+H 股」試點

2018 年 4 月，香港交易所與營運內地新三板的全國中小企業股份轉讓系統公司簽訂合作備忘錄，允許符合香港上市條件的公司以「新三板+H 股」模式在兩地同時掛牌，即新三板企業無需事先除牌即可申請到香港聯交所掛牌交易。

根據統計，截至 2019 年 6 月 30 日已有超過 25 家新三板公司（或分拆子公司）向香港聯交所遞交上市申請；有 7 家原新三板掛牌企業已經完成香港上市，其中君實生物-B（01877）以「新三板+H 股」模式成功登陸香港資本市場。「新三板+H 股」模式的成功試點為境內外資本市場聯動服務 H 股公司提供新的思路和有益嘗試。

1.3.4　B 股與 H 股的聯動深化

B 股一般是指以人民幣標明面值、以外幣認購和買賣、在內地的上交所或深交所上市交易的外資股。截至 2019 年 6 月 30 日，上交所 B 股上市為 50 家，深交所 B 股上市為 47 家。

近年來，有不少 B 股上市公司選擇從 B 股退市，並將股份轉為 H 股形式在香港聯交所掛牌上市。2012 年 8 月，中集集團（02039）作為首家成功完成 B 股轉 H 股的公司登陸香港資本市場。在此之後，萬科企業（02202）和麗珠醫藥（01513）

也相繼從 B 股市場除牌，並於 2014 年由「A 股+B 股」公司轉為「A 股+H 股」上市公司。此外，還有多個擁有 B 股股份的上市公司選擇同時在香港聯交所掛牌，例如晨鳴紙業 (01812) 為「A 股+B 股+H 股」模式的上市公司，伊泰煤炭 (03948) 為「B 股+H 股」模式的上市公司等。

「B 股轉 H 股」、「B 股+H 股」的模式日趨成熟，標誌着香港與內地資本市場的聯動進一步深化，也為內地企業多幣種、國際化股權融資提供了更多可行之路。

1.4 香港上市涉及的專業機構

1.4.1 保薦人及合規顧問

擬申請在香港聯交所主板上市的公司，必須委任至少一名具備獨立性的香港證券及期貨事務監察委員會 (簡稱「香港證監會」) 持牌 (SFC Licensed) 保薦人以協助其處理上市申請 [14]。不論是否已經提交上市申請，保薦人一經委任或解聘，須盡快書面通知香港聯交所。自保薦人獲正式委任起兩個月，新申請人方可向香港聯交所提交上市申請。保薦人的職責分工通常包括以下方面：

- 上市的總體籌劃及協調；
- 協助公司制定融資方案；
- 進行合理的盡職調查；
- 審閱公司準備的盈利預測和現金流預測；
- 深度並密切參與編製公司的上市申請材料；
- 代表公司與監管機構聯繫，並陪同參加會議。

此外，申請人必須委任一名合規顧問，任期由上市之日起，至其公佈上市後首個完整財政年度的財務業績之日止。

1.4.2 承銷商

承銷商主要從股票的發行及銷售角度參與 IPO 過程，尤其是在發行規模較大或市況受壓時，引入不同類別的承銷商可以在一定程度上分擔銷售壓力，降低發

---

14 詳情參閱《上市規則》第 3A.02 條。

行風險。承銷商的職責分工通常包括以下方面：

- 協助公司準備路演材料；
- 安排分析師撰寫研究報告；
- 為公司尋找並鎖定潛在投資者；
- 組織路演及市場推介；
- 公開發售及國際配售；
- 必須全數發行 IPO 擬供投資者認購的新股；
- 協助公司確定發行價格；
- 為股價提供後市穩定支持。

1.4.3　律師

　　由於內地及香港兩地適用的法規體系不同，因此在申請香港上市的過程中一般需要涉及四方律師團隊，分別是發行人律師 (分別聘請境內律師 [15] 及境外律師 [16]) 及保薦人與承銷商律師 (分別聘請境內律師及境外律師)。律師團隊的職責分工通常包括以下方面：

- 起草招股書；
- 協助完成公司重組和股改；
- 協助盡職調查工作；
- 就上市要求，在盡職調查過程中對公司法律問題發表意見；
- 就公司合規事宜向保薦人與承銷商提供法律意見；
- 協助公司及保薦人回答監管機構問詢。

1.4.4　會計師

　　申請人須聘請專業會計師審核財務報告及協助公司 IPO 事務，會計師的職責通常包括以下方面：

- 準備審計報告；

15　境內律師一般是指具有中國內地法律背景的律師。
16　境外律師一般是指香港、開曼羣島、美國 (如涉及美國證券法) 以及其他公司主要海外業務分佈地區 (如適用) 的律師。

- 協助公司處理稅務事項（如果公司沒有聘請獨立的稅務顧問）；
- 審閱公司準備的盈利預測和現金流預測；
- 協助保薦人回答監管機構問詢。

此外，香港 IPO 過程中還可能涉及到如下外部機構提供的專業服務：

- 行業顧問：主要負責行業資料調研並撰寫行業分析報告等；
- 物業評估師：負責對上市申請人的物業權益或其他資產出具估值報告等；
- 專項資產評估：如礦產等特殊專項資產的評估等；
- 印刷商：主要負責招股書的校對、翻譯及印刷等；
- 公關公司：主要負責發行路演、媒體對接、上市儀式安排等。

1.5　GEM 上市條件

香港聯交所的 GEM 板塊（前稱「創業板」）是專為中小型企業而設的市場，該板塊的上市資格低於主板、但其持續責任與主板類似。2017 年 12 月 15 日，香港聯交所刊發《創業板諮詢總結》，該次《GEM 上市規則》的修訂取消了 GEM 發行人轉往主板上市的簡化轉板申請程序，再次明確 GEM 成為獨立市場為中小企服務 [17]。該次修訂以後，申請在 GEM 上市的公司應滿足基本條件如下：

- 經營業務有現金流入，前兩年營業現金流合計≥ 3,000 萬港元 [18]；
- 市值≥ 1.5 億港元；
- 最低公眾持股量（一般為 25%，如上市時市值 >100 億港元可酌情減至 15%）；
- 上市後有至少 100 名股東 [19]；
- 管理層最近兩個會計年度大致維持不變；
- 擁有權和控制權最近一個會計年度不變；
- 至少三名獨立董事，並必須佔董事會成員人數至少三分之一；
- 如為 H 股上市申請人，至少須有一名獨立非執行董事通常居於香港；
- 要求每一季度提交財務報告。

17　該次《諮詢總結》已於 2018 年 2 月 15 日生效。
18　詳情參閱《GEM 上市規則》第 11.12A 條。
19　詳情參閱《GEM 上市規則》第 11.23(2)(b) 條。

符合要求的 GEM 上市公司也可以申請轉為香港聯交所主板上市公司掛牌交易 [20]。2018 全年及 2019 上半年，分別有 10 家和 8 家 GEM 公司成功轉至主板上市。

1.6　有關在港「借殼」上市

近年，香港聯交所留意到「炒殼」、「造殼」的市場活動有所增加，這是由於市場上有通過「殼股」借殼上市的需求。為維持市場質素，香港聯交所在應用《上市規則》方面已採取更嚴格的方針，以打擊不斷演變的「殼股」活動，包括借殼上市、剝離上市發行人業務的公司行動，以及發行人營運業務水平過低等問題。2018 年 6 月 29 日，香港聯交所刊發有關借殼上市、持續上市準則及其他《上市規則》條文修訂的諮詢文件，提出多項修訂《上市規則》的建議，旨在解決市場關注有關借殼上市及「殼股」活動的問題，並就有關建議徵詢市場意見。

2019 年 7 月 26 日，香港聯交所刊發有關借殼上市及其他殼股活動等諮詢總結，諮詢中的建議按市場的回應意見作調整後實施，相關的《上市規則》條文修訂於 2019 年 10 月 1 日生效 [21]。香港聯交所亦刊發了三封新指引信，就《上市規則》經修訂條文的應用提供指引 [22]。

有關借殼上市的主要修訂包括以下分節的內容。

1.6.1　反收購交易的定義

原則為本測試

將有關指引信 [23] 原則為本測試的六個評估因素正式編入《上市規則》，主要修改了最後兩項內容：

- 交易規模；
- 目標資產品質；
- 發行人業務性質及規模；
- 主營業務出現根本轉變；

20　詳情參閱《上市規則》第 9A 章。
21　這次諮詢總結將《GEM 上市規則》一併作修訂，詳情參閱「關於《GEM 上市規則》的修訂（第六十二次）」。
22　詳情見香港交易所指引信 HKEX-GL104/105/106-19。
23　詳情見香港交易所指引信 HKEX-GL78-14。

- 控制權或實際控制權變動；
- 一連串的交易及/或安排（包括在合理接近的時間內進行（一般指 36 個月內）又或在其他方面互有關聯的收購、出售及/或控制權或實際控制權變動）。

明確測試

修改明確測試，使有關測試適用於發行人控制權變動後 36 個月內向發行人控股股東進行的非常重大的收購事項。同時限制發行人不得在控制權變動之時或其後 36 個月內建議將其全部或大部分原有業務出售或作實物配發。香港聯交所亦可限制發行人不得在實際控制權（如原則為本測試中所載）轉手之時或其後 36 個月內進行有關出售或作實物配發。

透過大規模發行證券借殼上市

將相關指引信 24 正式編入《上市規則》，禁止透過以下方式進行借殼上市：大規模發行證券換取現金，當中牽涉到又或會導致發行人控制權或實際控制權轉變，而所得資金將用作收購及/或開展規模預計遠較發行人現有主營業務龐大的新業務。

1.6.2　收緊反收購及極端交易的合規規定

極端交易

將相關指引信 25 中的「極端非常重大的收購事項」規定正式編入《上市規則》，並將此交易類別更名為「極端交易」。同時為可使用此交易類別的發行人增設合資格準則：一是發行人須有規模龐大的主要業務；二是發行人須長時間（一般不少於 36 個月）受同一人控制或實際控制，而且有關交易不會令發行人控制權或實際控制權有變。

反收購及極端交易的規定

修改《上市規則》要求反收購或極端交易的收購目標須符合《上市規則》第 8.04 及 8.05 條（或第 8.05A 或 8.05B 條）的規定，並要求擴大後的集團須符合《上市規則》第八章的所有新上市規定（第 8.05 條除外）。若反收購是由第 13.24 條發行人提出，則收購目標亦須符合《上市規則》第 8.07 條的規定。

24　詳情見香港交易所指引信 HKEX-GL84-15。
25　詳情見香港交易所指引信 HKEX-GL78-14。

這次諮詢過程中，香港聯交所共收到各界回應意見合計 121 份，普遍對解決借殼上市和殼股活動問題的各項措施表示支持。這次諮詢及對《上市規則》的修訂也將有助於香港聯交所繼續維持並提升香港市場質素及穩健持續發展。

2 企業赴港上市的申報及審核流程

2.1　香港聯交所與香港證監會的分工

香港聯交所遵循披露為本的審核理念、審核基本程序透明、審核時間表可預期。根據法例（《證券及期貨條例》），香港聯交所負責在合理切實可行的範圍內，確保香港市場是公平、有秩序及資訊透明的。

香港證監會在市場監管方面及上市規管的若干範疇扮演着領導的角色，並且在涉及企業失當行為的個案中，透過行使其法定調查及執法權力，肩負起相輔相成的角色。

香港證監會的法定責任之一是監督及監察香港聯交所履行其與上市事宜有關的職能及職責。香港聯交所與香港證監會的職員會定期開會討論與上市有關的事宜。根據雙重存檔制度，香港聯交所會將由上市申請人遞交的資料副本送交香港證監會。假如香港證監會認為有關的上市資料內所作的披露看來載有虛假或具誤導性的資料，香港證監會可以否決有關的上市申請。香港證監會亦會定期稽核香港聯交所在規管與上市有關的事宜方面的表現。

香港聯交所涉及上市事宜的職能由上市部及上市委員會執行，包括制定、發佈及執行《上市規則》。此外，所有就《上市規則》所作的修訂及會強制執行或應用範圍廣泛的政策決定，均須獲香港證監會批准。

2.2 IPO 審核流程概覽

　　企業發展到一定階段後，如有赴港 IPO 發行上市的意向，則從前期的籌劃考慮，到中期的上市申請 (A1 申請表) 審核，再到後期的上市聆訊，最後成功發行並掛牌上市交易，大致需要經歷如圖 4 所示的各個階段。

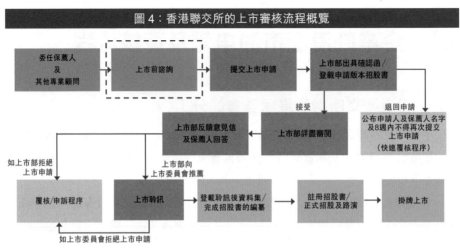

圖 4：香港聯交所的上市審核流程概覽

註：上圖僅為示意性表述，實踐操作中可能有所不同。

(1)　第一步：委任保薦人及其他專業顧問

　　委任有經驗的專業顧問團隊是新股成功上市的關鍵。專業顧問一般包括保薦人及承銷商、境內律師及境外律師團隊、會計師等。上市申請人須於提交上市申請前至少兩個月委任保薦人，並於委任後五個營業日內書面通知香港聯交所。公司應徵詢專業顧問團隊的意見，以商討公司是否適合上市、上市發行結構、上市所需的時間及費用、上市工作團隊分工及在上市後將面對的問題、挑戰與持續責任。

(2)　第二步：籌備上市流程

　　保薦人對公司進行盡職審查，並協助擬備招股章程 (此階段的招股章程擬稿亦稱「申請版本」)。招股章程須載有一切重要資料，能讓投資者作出有根據的投資決定。在準備招股章程的過程中，保薦人應幫助公司制定或優化中長期發展戰略，並完成業務和財務預測以便承銷商建立估值模型、確定募集資金用途、梳理投資故事亮點。

(3)　第三步：向上市部提交上市申請

向上市部提交上市申請（「A1」，包括申請版本）。如 A1 上市申請的資料大致完備，上市部會確認收悉，並在網站登載中英文申請版本。

同時，香港聯交所鼓勵擬申請人在正式提交申請材料之前，就非常見事項尋求非正式及保密的指引，即通常所指的「上市前諮詢」程序（Pre-A1 Submission）。此舉可有助於發行人結合自身實際情況，及早得知上市發行的建議是否符合要求，以進一步明確赴港上市的重點考慮事項。

(4)　第四步：上市部審閱申請

上市部其後會對 A1 上市申請進行詳細審核，評估公司是否符合上市資格、是否適宜上市、業務是否可持續、公司是否遵守規條以及作出充分披露。

首輪意見會於接獲申請後盡快發出，一般僅需耗時 15 個營業日左右。後續所需審核時間主要取決於公司回覆的時間及質量。

(5)　第五步：上市委員會聆訊

上市委員會審閱新上市申請，確定申請人是否適合進行首次公開招股。根據《上市規則》規定，上市委員會的成員由 28 人組成，包括最少 8 名代表投資者權益的人士，19 名以適當的比例提名的、能夠適當代表上市發行人與市場的從業人士，包括律師、會計師、企業融資顧問等高級人員，以及香港交易所集團行政總裁。上市委員會的職能主要包括 [26]：

- 上市審核職能；
- 上市審查並監督上市部工作；
- 批准新股上市申請以及規則豁免申請；
- 通過、更改或修改上市部以及上市委員會的決定；
- 提供政策諮詢，批示重要政策及《上市規則》修訂。

(6)　第六步：市場推廣及新股銷售

在此階段，香港聯交所要求公司滿足最低公眾持股及股東人數標準，但不會干預定價過程及最終定價，體現市場定價精神。

26　詳情見香港交易所官方網站。

承銷商或承銷團一般負責協助公司進行上市籌備推廣活動，包括投資者教育、潛在基石投資者溝通、分析師路演、管理層新股路演等。香港 IPO 公開發行股份的分配包括「國際配售」和「公開認購」兩個部分，並根據回撥機制及股份重新分配機制予以適當調整，通常分別佔最終總新股發行數量的 90% 和 10% 左右。

許多公司在香港 IPO 上市的同時選擇引入基石投資者。基石投資者一般是指上市申請人在首次公開招股時，將部分股份優先配售予若干投資者。對基石投資者的一般配售原則包括[27]：

- 配售必須按首次公開招股價進行；
- 所配售的首次公開招股股份須設禁售期，一般為上市日期起計至少六個月；
- 每名投資者在上市申請人的董事會內沒有代表，同時獨立於上市申請人、其關連人士及其各自的聯繫人；
- 配售安排的詳情（包括投資者身份及背景）須在上市文件內披露；
- 假如該投資者就《上市規則》第 8.24 條（《GEM 上市規則》第 11.23 條附註 2 及 3）而言會被視為公眾人士，即使有上述的至少六個月禁售期，該等股份亦屬《上市規則》第 8.08 條（《GEM 上市規則》第 11.23 條）所指的公眾持股量的一部分；
- 該類投資者除給予優先配售股份的保證之外，不得向基石投資者提供直接或間接利益。

(7) 第七步：掛牌上市及上市後交易

成功定價及分配股份予機構投資者和散戶後，公司股份便會在香港聯交所上市及進行買賣。

在公司上市之後，股票價格會因為各種各樣的原因產生波動。承銷團通常會指定某一個或多個承銷商作為「穩定市場經紀人」（Stabilisation Agent）負責在公司股價低於 IPO 的價格時買入公司股票來維護上市之後一段既定時間內股價表現的穩定。

香港 IPO 可以由擬上市公司授予承銷商「超額配售選擇權」，通常為原 IPO

---

27 詳情見香港交易所指引信 HKEX-GL51-13。

總發行股數的 15%，需在香港公開發行結束後 30 天內行使完畢；這一機制又被稱為「綠鞋機制」(Green Shoe)。穩定市場經紀人可以根據上市後股價的漲跌表現，判斷二級市場投資者對公司股票的需求情況，並自主決定是否行使「超額配售選擇權」，以平滑市場波動，實現上市公司、投資者、承銷商的多方共贏。

3 香港上市公司的監管及股權再融資

3.1　上市公司的監管

香港聯交所對已上市發行人的基本監管原則是確保發行人持續遵守《上市規則》，以保障公平、有序、有效的市場運作。其監管職能主要體現在監察、審批和指導等三個方面，例如對上市發行人的媒體報導及股價監管、處理投訴、豁免審批、交易停復牌、處理上市公司有關《上市規則》的查詢、對上市公司提供指導、市場教育等。

香港上市公司監管的核心在於落實持續披露責任，其中又包括三個種類，一是按照法規要求[28] 及時公佈須披露之內幕消息；二是避免出現虛假市場[29] 須披露的消息；三是回應交易所查詢有關上市公司證券交易 (股價/成交量) 的異常波動[30]。

其他須披露的一般事項包括但不限於以下各項：

- 財務業績/報告 (中期/年度)；
- 修訂公司組織章程大綱或章程細則；
- 董事會或監事會的人士變動；
- 更換核數師、更改財政年度；
- 更換公司秘書、股份過戶登記處、註冊地址；

---

28 詳情見《證券及期貨條例》(第 571 章) 第 XIVA 部 —《內幕消息條文》。按照定義，內幕消息是指「並非普遍為慣常 (或相當可能會) 進行該法團上市證券交易的人所知，但該等消息或資料如普遍為他們所知，則相當可能會對該等證券的價格造成重大影響」。

29 「虛假市場」一般指市場上流傳重大失實或嚴重缺漏的資料，影響市場發現正常價格的功能。

30 詳情見《上市規則》第 13.10 條。

- 股本變更；
- 股東會議通知。

3.2　上市公司的股權再融資

　　公司在香港聯交所 IPO 掛牌上市以後，將擁有多元化、高效率的再融資及股東增/減持渠道。事實上，香港上市公司及其股東在上市後的股權類再融資渠道包括後續新股增發、供股、可轉換/可交換債券、發行認股權證/期權、大宗交易等。

　　在許多年份，香港股市的股權再融資金額都超過了新上市公司 IPO 募資金額，為香港上市公司的持續發展注入強勁的資本動力。

圖 5：於香港聯交所上市的公司的 IPO 集資金額與股權再融資金額
（2001 年至 2019 年 6 月）

資料來源：香港交易所。

3.2.1　後續新股增發

- 指上市公司發行新股融資，僅在首發融資後的六個月內不能增發新股[31]；
- 基本原則是保障現有股東的優先權，發行新股須事先獲得股東批准，避免股東權益被動攤薄；
- 年度股東大會的一般性授權機制允許香港上市公司可以自由靈活地進行新

31　詳情見《上市規則》第 10.08 條。

股或股票類證券的增發（通常不超過 20% 總股本），且能夠快速發行，通常隔夜即可完成新股增發交易 [32]；

- 如果為現金募資，在一般性授權下，增發價格不得低於市場價格的 20%。

3.2.2　供股

- 指向現有股東作出供股要約，使他們可按其現時持有證券的比例認購證券；
- 公司須確保所有股東均受到公平及平等對待，以及讓股東掌握充分資料作投資/投票決定；
- 需要董事會和股東大會批准（如適用）；
- 採用供股方式上市，必須刊發公告/上市文件（如適用）。

3.2.3　可轉換/可交換債券

- 上市發行人及其股東可以發行可轉換/可交換債券；
- 初始轉換價不可低於進行發行時上市發行人股份的基準價；
- 可轉換債券（Convertible Bond）為上市公司發行及融資；
- 可交換債券（Exchangeable Bond）為股東以其持有的可流通上市公司股票（其控股的上市公司或其持有少數股權的上市公司皆可）發行，從而為股東補充資本。

3.2.4　認股權證/期權

- 只允許轉換價格大於市場價格時，以一般性授權發行認股權證。

3.2.5　大宗交易

- 股東可通過大宗交易出售持有的上市公司股票以獲取融資；
- 控股股東在 IPO 後首 6 個月內不得減持，第 7 個至第 12 個月內減持不得影響其控股地位。

32　詳情見《上市規則》第 13.36(2) 條。

第5章

香港的新股上市制度改革與突破

香港交易及結算所有限公司
首席中國經濟學家辦公室
及
環球上市服務部

摘 要

2018 年 2 月香港聯合交易所（簡稱「香港聯交所」）進行新興及創新公司諮詢（簡稱「創新公司諮詢」），建議修訂《上市規則》以拓寬原有上市機制，便利新興及創新產業公司上市。創新公司諮詢亦提議，為已在合資格交易所上市的大型創新公司增設全新的優待第二上市機制。創新公司諮詢中提出的建議廣獲市場支持，香港聯交所宣佈《主板規則》的相關修訂建議已於 2018 年 4 月 30 日生效。

本章簡潔概述《上市規則》為三類公司新增章節的上市規定條文。這三類公司為：（1）擁有「不同投票權」架構的公司；（2）生物科技公司；（3）擬作第二上市的公司。

1 「不同投票權」章節

1.1　設計理念及原則

股東投票權與股權比例相對稱的概念（常被稱為「一股一票」原則）是投資者保障的重要環節，有助於維持控股股東權益與其他股東權益一致。如果管理層表現不理想，持有發行人最大股權的股東也有能力罷免相關管理層。

香港聯交所認為「一股一票」原則仍然是賦予股東權力及使股東權益一致的最理想方法。因此，如果公司希望以「不同投票權」架構上市，則需要滿足相應的資格要求，證明其具有所需的創新及增長元素，並證明其建議的不同投票權受益人的貢獻。

在這一理念下，香港聯交所推出了全新的《上市規則》第 8A 章，允許符合條件的「不同投票權」架構公司赴港上市。截至 2019 年 6 月 30 日，已有小米集團-W（股份代號：01810）和美團點評-W（股份代號：03690）兩家公司在此章節框架下成功登陸香港資本市場。

1.2　上市資格及基本條件

- 只限新上市的申請人，已上市公司不可以申請更改為「不同投票權」架構；
- 上市時市值至少為 400 億港元；或市值至少 100 億港元及經審計的最近一個會計年度收益至少為 10 億港元；
- 必須是創新產業公司；
- 業務經營成功，具備高增長業務的紀錄並預計可持續（可用業務活動、使用者、客戶、收益、盈利或市場價值等運營數據客觀計量）；
- 受到外界認可，必須已得到至少一名「資深投資者 [1]」提供相當數額的第三方投資。

1　「資深投資者」是指經考量其淨資產或管理資產、相關投資經驗以及於相關範疇的知識及專業技能等方面的因素後，香港聯交所認為具備豐富經驗的投資者。

1.3　對創新產業的範疇的界定 [2]

申請人必須是創新產業公司，應具備多於一項的下述特點：

(1)　能證明公司成功運營有賴其核心業務應用了新科技、創新理念及/或新業務模式，使得公司有別於現有的行業競爭者。

　　申請人應詳細解釋其業務運作與業內傳統運作模式有何不同，能令其從同業中脫穎而出。若申請人業內其他公司也採用類似技術/業務模式，香港聯交所會將申請人應用技術、創新舉措及/或業務模式的實施時間與業內和其最貼近的公司作比較，判斷申請人是否業內的先行者。

(2)　研究及開發為公司貢獻一大部分的預期價值，是公司的主要活動，佔大部分開支。

　　申請人除要提供其業績紀錄期的研發開支金額（數字及佔收入/總開支百分比）外，亦應解釋研發為申請人創造了甚麼價值。就此而言，香港聯交所會看申請人的賬目是否將研發開支資本化為無形資產，作為評估有關研發活動對申請人所產生價值的指標。若大部分研發開支都沒有資本化，申請人應解釋箇中理由。

(3)　能證明公司成功運營有賴其獨有業務特點或知識產權及/或相對於有型資產總值，公司的市值/無形資產總值極高。

　　僅提供專利及商標名單並不足以證明申請人符合有關特點。申請人應詳細解釋其知識產權如何助其取得業務成功。

香港聯交所認為，「創新產業」的範疇將視乎申請人所屬行業及市場狀況，且會隨着科技、市場及行業發展而有所變化。因此某家公司符合以不同投票權架構上市的資格，不一定表示其他有類似技術、創新或業務模式的申請人也符合以不同投票權架構上市的資格。香港聯交所將根據個案事實及具體情況認定申請人能否證明其滿足相關要求，單憑表面在傳統業務中應用新科技並不足以證明有關公司具備本節所述的特點。

2　詳情見香港交易所指引信 HKEX-GL93-18。

1.4　對不同投票權受益人的限制

- 只限個人：對推動公司業務增長有重大貢獻（會就法團受益人另作諮詢）；
- 只限董事：每名不同投票權受益人在上市時及其後必須一直擔任公司董事；
- 最低持股量：上市時合共持股至少 10%；
- 禁止轉讓：自然的日落條款、容許信託及合法的稅務規劃。

1.5　額外的上市規定及股東保障措施

- 公司上市後，具有不同投票權的股份類別不得流通，也不得將股份比例增至超過上市時該等股份所佔比例；
- 公司上市後，不得提高不同投票權比例；
- 不同投票權股份的投票權不得超過普通股可就任何股東大會議案投票權的 10 倍；
- 普通股股東必須佔股東大會議案合資格投票權的至少 10%；
- 首次申請上市時，不同投票權架構受益人實際擁有的公司相關經濟利益合計不少於 10%。

以下重大事宜必須按「一股一票」的基準投票表決：

- 修訂組織章程文件；
- 委任及罷免獨立非執行董事；
- 委聘及辭退核數師（審計師）；
- 任何類別股份所附帶權利的變動；及
- 上市發行人自願清盤。

加強披露及企業管治：

- 在上市文件及公司通訊加入示警字句；
- 股份名稱有「W」的標記；
- 設立企業管治委員會，以檢視、監察及申報發行人是否有遵守不同投票權保障措施；
- 一直聘用合規顧問，就是否有遵守不同投票權保障措施及相關規則提供意見；

- 企業管治委員會成員必須全部是獨立非執行董事。

組織章程的法律效力及法律補救行動：

- 不同投票權保障措施必須納入組織章程文件；
- 不同投票權受益人必須以香港聯交所接納的方式向發行人承諾會遵守不同投票權保障措施。

2 生物科技公司章節

2.1 設計理念及原則

2018 年 4 月，香港聯交所就主板上市規則全新推出了第 18A 章「生物科技公司」章節，主要適用於未能通過《上市規則》第 8 章所列相關的「盈利測試」或「市值/收益測試」或「市值/收益/現金流量測試」、但尋求在香港聯交所主板上市的生物科技行業公司，以進一步增強香港市場服務新經濟企業的能力[3]。

在此理念的指引下，僅經過一年多的培育，香港已然成為全球第二大生物科技企業上市中心。自新章節頒佈以來至 2019 年 6 月 30 日止，共有 14 家生物科技公司在香港聯交所主板上市，總計集資 475 億港元；其中包括 8 家尚未有營業收入的生物科技公司，集資總額約 235 億港元。此外還有 5 家生物科技公司已經提交 A1 申請表，未來將進一步強化香港在生物科技行業領域的領先資本市場地位。

---

[3] 需要注意的是，如果相關生物科技公司已經符合《上市規則》第 8 章所列相關的「盈利測試」或「市值/收益測試」或「市值/收益/現金流量測試」，則不可按照第 18A 章要求申請上市。香港聯交所鼓勵發行人盡早與交易所聯繫並確認是否符合相關規定。

	股票簡稱	股票代碼	上市日期	主要營運地點
	表 1：香港聯交所依生物科技公司章節的新上市情況（2019 年 6 月底）			
1	歌禮製藥-B	01672	01/08/2018	杭州
2	百濟神州	06160	08/08/2018	北京
3	華領醫藥-B	02552	14/09/2018	上海
4	信達生物-B	01801	31/10/2018	蘇州
5	君實生物-B	01877	24/12/2018	上海
6	基石藥業-B	02616	26/02/2019	上海
7	康希諾生物-B	06185	28/03/2019	天津
8	邁博藥業-B	02181	31/05/2019	泰州

註：由於百濟神州的經營及股價表現已可通過上市規則第 8.05(3) 條之「市值/收益」測試，經香港聯交所批准，該
　　公司已於 2019 年 6 月 28 日轉為普通上市，「B」標誌已從其股票代碼中移除。

資料來源：香港交易所。

2.2　上市資格及基本條件 [4]

(1)　申請人最少有一項核心產品已通過概念開發流程

　　適用法律法規表明須經有關主管當局 [5] 根據臨床試驗（即人體試驗）數據評估及批准，方可在有關主管當局所規管的市場行銷及發售的生物科技產品，是構成生物科技公司根據《上市規則》第 18A 章申請上市的最重要基礎之一。根據產品類型的不同，香港聯交所一般會將符合以下條件的產品認定為已通過概念開發流程：

- **藥劑（小分子藥物）**——已通過第一階段臨床試驗，或先前已獲批准的產品（例如美國 FDA 505(b)(2)）至少已通過一次人體臨床試驗；同時有關主管當局不反對開展第二階段（或其後階段）的臨床試驗。
- **生物製劑**——如屬於新生物製劑產品已通過第一階段臨床試驗，或生物仿製藥至少已通過一次人體臨床試驗；同時有關主管當局不反對開展第二階段（或其後階段）臨床試驗。
- **醫療器材（包括診斷器材）**——屬於有關主管當局分類標準項下的第二級（或相等級別）或以上分類醫療器材；已至少通過　次人體臨床試驗；有關

4　詳情見香港交易所指引信 HKEX-GL92-18。

5　有關主管當局包括美國食品和藥物管理局（FDA）、中國國家食品藥品監督管理總局以及歐洲藥品管理局。其他機關會因應個別情況作考慮。

主管當局同意或不反對開展進一步臨床試驗又或開始銷售有關器材。

- **其他生物科技產品** — 需要按情況逐一個別考慮；需要證明有關生物科技產品已通過概念開發流程；應具備適合框架或客觀指標可供投資者作出知情投資決定。

(2) 申請人應主要專注於研究與開發核心產品

- 上市前最少 12 個月一直主要從事核心產品的研發；
- 上市集資主要用於研發，以將核心產品推向市場；
- 必須擁有多項與核心產品有關的長期專利、已註冊專利、專利申請及/或知識產權；
- 如申請人從事小分子藥物或生物產品研發，需證明其擁有多項潛在產品。

(3) 申請人應獲得「資深投資者」認可及投資

申請人的研發及生物科技產品應有相當程度的市場認可，應在首次公開招股至少 6 個月前，已得到至少一名資深投資者「相當數額的投資」，且直到首次公開招股時仍未撤回。

香港聯交所對於以上所述「資深投資者」定義的考量，包括其淨資產或管理資產、相關投資經驗、知識、專業技能等因素。一般而言，「資深投資者」主要包括以下四種類型：

- 專門的醫療保健或生物科技投資基金，或旗下有專門或側重於投資生物製藥領域的分支/部門的大型基金；
- 主要的製藥/醫療保健公司；
- 大型製藥公司/醫療保健公司的風險投資基金；
- 管理資產總值不少於 10 億港元的投資者、投資基金或金融機構。

香港聯交所對於以上所述「相當數額的投資」定義的考量，包括投資的性質、金額、份額、時機等因素。一般而言，「相當數額的投資」的參考指標包括以下三個等級：

- 如申請人市值在 15 億至 30 億港元之間，則投資佔申請人上市時已發行股份不少於 5%；
- 如申請人市值在 30 億至 80 億港元之間，則投資佔申請人上市時已發行股份不少於 3%；

- 如申請人市值高於 80 億港元，則投資佔申請人上市時已發行股份不少於 1%。

(4)　申請人應滿足的其他上市條件

- **市值**：上市時的市值至少達到 15 億港元；
- **業績記錄**：上市前最少兩個會計年度一直從事現有業務，管理層大致相同；
- **營運資金**：申請人應確保有充足的營運資金（包括首次公開招股的集資額），以應付上市文件刊發之日起至少 12 個月開支（包括一般、行政及營運開支及研發開支）的 125%。

2.3　向基石投資者及原股東配售的要求

根據《上市規則》第 18A 章申請上市的生物科技公司，須於上市時符合額外的公眾持股量最低要求，即在撇除基石投資者及公司的現有股東在公司首次公開發售（IPO）過程中認購的股份後的公眾持股量須不少於 3.75 億港元，方可將 IPO 股份配售給基石投資者及公司的現有股東。

2.4　特別的風險管理措施

(1)　重大業務變化

未經香港聯交所批准，根據第 18A 章上市的生物科技公司不得進行任何個別或一系列收購、出售或其他交易及安排，令其於上市申請文件中描述的主營業務活動出現根本性變化。

(2)　較短的除牌流程

當發行人出現財政困難、嚴重損害發行人繼續經營業務的能力、或導致其部分或全部業務停止運作，以及/或發行人於結算日的負債額高於其資產值時[6]，香港聯交所將發行人的證券停牌甚至除牌[7]。如果發行人在 12 個月內未能重新符合並遵守《上市規則》第 13.24 條的相關要求，香港聯交所會將發行人的證券除牌。

6　詳情參閱《上市規則》第 13.24 條。
7　停牌、除牌等相關事宜詳情參閱《上市規則》第 6.01 條。

(3) 特殊的股份標記

根據第 18A 章上市的生物科技公司的股本證券的股份名稱結尾須有「B」字作為標識。如果相關生物科技公司上市後，已滿足《上市規則》第 8 章所列相關的「盈利測試」或「市值/收益測試」或「市值/收益/現金流量測試」，經香港聯交所批准，可以轉為一般上市公司，其股份名稱結尾將不再有「B」字作為標識。

3 第二上市章節

3.1 第二上市的概念及優勢

在實踐中，許多已在海外上市的發行人也希望能夠同時在香港上市，使其股份獲得香港資本市場的流動性及估值支持。如果公司在香港作主要上市，則公司須完全遵守《上市規則》的規定。如果公司在香港作第二上市，則公司主要受其作主要上市的司法權區的法規及監管機關所規管，香港聯交所則預期公司證券主要在該海外交易所交易。在此基準上，香港聯交所豁免第二上市或尋求第二上市的海外發行人遵守《上市規則》的若干規定。

因此，對於這部分已在海外上市的發行人而言，相較於退市再上市或雙重主要上市等形式，選擇在香港作第二上市是操作性最強、難度最低的首選。尤其是已在海外上市的「中國概念股」，在香港聯交所作第二上市的主要優勢包括：

- 赴港第二上市可使用《美國公認會計原則》，毋須就三個會計年度的營業記錄額外/重新審計，且不需要提供調節表，以減低發行人在準備財務報表及相關招股書披露方面的工作量；

- 在符合條件的情況下，赴港第二上市可享有多項自動及有條件的豁免，包括豁免遵守多項於香港聯交所主要上市的持續責任（例如：有關須予公佈的交易及關連交易的規則）；部分能符合條件的海外上市發行人還可以按保密形式提交上市申請；

- 赴港第二上市可以增發新股進行融資，為公司發展資金進行補充；也可以

不增發新股，以介紹上市、老股出售、回購並再次出售等方式進行，則毋
須稀釋現有股東權益；

- 在香港聯交所上市流通的股份具有價格傳遞作用，更加貼近亞洲交易時段
  的投資者，形成全天候的股票流通體系，促進股票活躍度；

- 香港提供資本運作的長期便利性，赴港第二上市可以使發行人更加接近國
  內消費用戶市場，進一步提升品牌知名度和影響力。

傳統上，香港聯交所僅允許業務重心在大中華地區以外的海外發行人申請來
港作第二上市[8]。此限制的主要用意在於防止「監管套利」，即業務以大中華地區為
重心的公司通過先在海外交易所作主要上市，然後在香港作第二上市，以規避遵
守香港主要上市的規定。近年來，不少內地及非內地大型的新興及創新產業公司
都在美國及其他主要國際交易所作主要上市。當中的內地公司都是因為香港有「業
務重心」限制，而不是為了「監管套利」而到當地上市，因此「業務重心」限制在一
定程度上阻礙了內地公司通過第二上市來接觸香港投資者。

在此背景下，香港聯交所於 2018 年 4 月 30 日正式實施修訂後的《上市規
則》，專門增加了第 19C 章「合資格發行人第二上市」章節，吸引高增長及創新產
業公司，特別是為大中華地區申請人赴港作第二上市提供便利。

3.2　第 19C 章項下的合資格發行人在香港聯交所作第二上市須符合的附加條件及例外情況

3.2.1　合資格發行人

第 19C 章項下的「合資格發行人」分為以下三個類別：

- **「獲豁免的大中華發行人」**，即於 2017 年 12 月 15 日或之前已在「合資格
  交易所」[9] 作主要上市，而業務以大中華為重心的發行人；

- **「不獲豁免的大中華發行人」**，即 2017 年 12 月 15 日後才在合資格交易所
  作主要上市，而業務以大中華為重心的發行人；

8　詳情請參閱 2018 年 2 月香港交易所《新興及創新產業公司上市制度》諮詢文件。

9　「合資格交易所」包括紐約證券交易所、納斯達克證券市場或倫敦證券交易所主板市場（並屬於英國金融市場行為監
　管局「高級上市」分類）。

- **「非大中華發行人」**，即在合資格交易所作主要上市，而業務以大中華以外為重心的發行人。

3.2.2　申請人應屬於創新產業公司

除了一般第二上市申請的規定外，合資格發行人如要根據《上市規則》第19C章作第二上市，必須是符合「創新產業」範疇的公司。一般而言，創新產業公司具備下述兩項或以上的特點：

- 應用了有別於競爭者的新科技、新理念或新業務模式；
- 研發活動為公司貢獻大部分的預期市值，同時亦是公司的主要活動及佔去公司大部分開支；
- 成功營運有賴於獨有業務特點或知識產權；及/或
- 相對於有形資產總值，公司市值或無形資產價值極高。

何謂「創新」則視乎申請人所屬行業及市場的狀況，且會隨着科技、市場及行業發展而有變。

3.2.3　獲豁免的大中華發行人

條件及上市資格：

- 發行人須在合資格交易所作主要上市並且於至少兩個完整會計年度保持良好合規紀錄；
- 上市時的市值至少400億港元，或者上市時的市值至少100億港元及最近一個經審計會計年度的收益至少10億港元；
- 發行人須證明其當地法律、規則及規例以及其組織章程文件合起來如何達到相當於香港提供的主要股東保障水平（如有需要，公司或要修訂組織章程文件）；

  例如：若干事項（包括發行人任何類別股份所附帶權利的變動、組織章程文件的重大變動及自動清盤）須經股東以絕大多數票批准；審計師的委聘、辭退及薪酬須由多數股東或獨立於董事會以外的其他組織批准等。

例外情況：

- 與傳統的非創新產業公司並業務重心在大中華地區以外來港申請第二上市

的發行人一樣，獲豁免的大中華發行人能豁免遵守多項於香港聯交所作主要上市的持續責任（例如：有關須予公佈的交易及關連交易的規則）；

- 發行人能豁免遵守《上市規則》對於組織章程的內容及條文的有關規定；
- 發行人可保留現行的不同投票權架構及可變利益實體（VIE）結構作第二上市（如適用）；
- 豁免遵守《上市規則》對於不同投票權的保障措施（有關披露規定除外）；
- 如發行人股份的主要成交地永久轉移到香港聯交所市場 [10]，便需要遵守香港聯交所主要上市的規定，但豁免遵守《上市規則》對不同投票權保障規定（有關披露規定除外）。

3.2.4　未獲豁免的大中華發行人

條件及上市資格：

- 發行人須在合資格交易所作主要上市並且於至少兩個完整會計年度保持良好合規記錄；
- 上市時的市值至少 400 億港元，或者上市時的市值至少 100 億港元及最近一個經審計會計年度的收益至少 10 億港元；
- 發行人須證明其作主要上市的海外交易所為股東提供的保障至少相當於香港提供的保障水平；
- 發行人須遵守《上市規則》對於組織章程的內容及條文的有關規定。

例外情況：

- 與傳統的非創新產業公司並業務重心在大中華地區以外來港申請第二上市的發行人一樣，未獲豁免的大中華發行人能豁免遵守多項於香港聯交所作主要上市的持續責任（例如：有關須予公佈的交易及關連交易的規則）。

為防止「監管套利」：

- 不同投票權架構及 VIE 結構必須遵守於香港作主要上市的規定（如適用）；
- 如發行人股份的主要成交地永久轉移到香港聯交所市場，便需要遵守香港聯交所主要上市的規定。

10　倘發行人最近一個會計年度的上市股份全球成交量（包括該等股份的預託證券的成交量）總金額有 55% 或以上在香港聯交所市場進行，香港聯交所即視其上市股份交易已大部分永久轉移到香港聯交所市場。

3.2.5　非大中華發行人

條件及上市資格：

- 發行人須在合資格交易所作主要上市並且於至少兩個完整會計年度保持良好合規紀錄；

- 同股同權的發行人於上市時的市值至少 100 億港元；擁有不同投票權架構的發行人於上市時的市值至少 400 億港元，或者上市時的市值至少 100 億港元及最近一個經審計會計年度的收益至少 10 億港元；

- 發行人須證明其當地法律、規則及規例以及其組織章程文件合起來如何達到相當於香港提供的主要股東保障水平（如有需要，公司或要修訂組織章程文件）。

例外情況：

- 與傳統的非創新產業公司並業務重心在大中華地區以外來港申請第二上市的發行人一樣，非大中華發行人能豁免遵守多項於香港聯交所作主要上市的持續責任（例如：有關須予公佈的交易及關連交易的規則）；

- 發行人可保留現行的不同投票權架構及 VIE 結構作第二上市（如適用）；

- 豁免遵守《上市規則》對於不同投票權保障措施（有關披露規定除外）；

- 如發行人股份的主要成交地永久轉移到香港聯交所市場，適用的監管規定及例外情況不變。

第6章

雙重股權架構的香港實踐

香港交易及結算所有限公司
首席中國經濟學家辦公室

摘 要

從全球發展趨勢來看，隨着互聯網、高科技，以及生物醫藥研發等成長型公司不斷增加，包括美國、英國、新加坡等全球各主要國際金融市場都提供或正考慮允許不同形式的不同投票權架構，為這些公司上市提供便利。

採用「不同投票權」架構的主要顧慮在於同股不同權使得控制權和現金流權分離，可能加劇公司代理問題，削減管理層對股東的問責程度。但是另一方面，雙重股權結構有利於初創企業的長遠發展，尤其是對高投入、高不確定性、高增長的科創型企業。具體表現在：雙重股權結構有利於實現創新公司的長期價值；可作為有效的激勵機制，促使創始人向企業投入更多創新能力和人力資本；可作為防止敵意收購的有效屏障；在金融市場機構投資者過度集中的情況下，雙重股權結構在一定程度上被視為新興企業規避市場短期行為而不得不採取的自我保護措施。另外，從一些實證研究結果來看，採用雙重股權架構後，企業價值和代理成本也會得到改善。

當然，雙重股權企業是否比單一股權結構企業帶來更高的代理成本，不利於股東權益保護，在理論上和實證上依然存在着大量爭論。如何完善雙重股權結構的公司內部監督機制，確保對公司具有控股權的股東進行有效監督？各界提出多種方式：第一，對超級表決權的運用進行適度限制，包括限定表決權的差額，以及規定上市公司在設計公司架構時，盡可能明確或劃定超級投票權的應用範圍。第二，對超級投票權設置明確的退出機制和轉讓機制，包括通常所提到的「日落條款」機制，以及超級表決權股轉讓限制。第三，加強企業管治，共同使用公司內部和外部替代性約束機制。

2018 年 4 月，香港交易所提出的雙重股權架構安排，增加了相應措施，對制度的適用加以限制和約束，比如要求申請人須具備特定的特點方可以不同投票權上市。香港交易所亦將保留權利，若申請人的不同投票權架構極端不符合管治常態（例如普通股完全不附帶投票權），可以不適合上市為由拒絕該等申請。同時，香港交易所也提出了不同投票權公司上市後的詳細投資者保障措施，包括限制不同投票權權力、保障同股同權股東投票權的措施、加強企業管治和加強披露等要求。不同投票權發行人的股份名稱結尾將要加上標記「W」，與其他發行人區分開來。此外，不同投票權受益人必須擔任公司董事，使其對公司的經營承擔法律法規規定的董事責任。一旦不同投票權受益人轉讓其不同投票權股份，或他們身故或失去行為能力，又或他們不再為董事，其不同投票權即會失效，令不同投票權受限於「自然」日落條款，不會無限期存在。

通過建立適宜的上市規則，推動一批大型創新型企業湧現，將大大助力區域內新經濟核心產業的發展，實現區域經濟升級和跨越發展，這正是資本市場改革為香港經濟帶來的長遠正面影響。

1 主要國家對「同股不同權」架構的運用

「同股不同權」是指某些股東享有與其在企業中所持的經濟利益不成比例的投票權或其他相關權利,包括無投票權股份、優先投票權股份,及具有較大或獨有的董事選任權的股份等多種形式。較常見的股權架構為 A、B 股結構:A 類股份的投票權為每股一票;B 類股份的投票權則每股多票,持有這種較高投票權比例股份的股東,一般都是公司的創辦人、早年加入公司的合夥人、主要策略性投資者或公司高層。在具體運用中(以 Google、百度在美國的股權架構為例),大多數創新型公司於首次公開發行(IPO)時,附帶每股一票的 A 股在交易所上市;及附帶每股多票(最常見為每股 10 票)的 B 股發行予公司在任管理人。B 類股可按照 1:1 的比例轉換成 A 類股,而 A 類股卻不能轉換為 B 類股。

從本質上看,同股不同權架構是一種雙重股權架構的公司管治架構,通過不同投票權的股權設置,公司創始股東僅持有小比例但高投票權的股份,便可有效控制企業,使新經濟公司創始人在不受新投資者壓力的情況下追求創新,為股東謀取最高的增長及價值。從全球發展趨勢來看,隨着互聯網、高科技,以及生物醫藥研發等成長型公司不斷增加,全球各主要國際金融市場都提供考慮允許不同形式的不同投票權架構,為這些公司上市提供便利。

1.1　「同股不同權」架構在美國的演變過程及效果

美國是最早實行雙重股權架構的國家之一,同時也是雙重股權架構運行得比較好的國家。雙重股權架構起源於 1898 年 International Silver Company 發行 900 萬股優先股和 1,100 萬股無投票權的普通股,開啟了歷史上股權和投票權分離的先河。19 世紀 20 年代,雙重股權架構在美國逐步流行,1927 年至 1932 年間共 288 家公司發行了無投票權或限制投票權的股票[1]。此後近 40 年,雖然上市體系轉向一股一票的發行架構,但仍有小部分企業(如福特汽車)發行了不同投票權的股份。

1980 年代,企業收購兼併浪潮興起,交易所之間競爭加劇,極大地促使了

1　Ashton, D. C. (1994). "Revisiting Dual-Class Stock",*John's Law Review*,第 68 期,863 頁。

雙重股權架構的採納和運用。當時美國三家主要證券交易機構（紐約證券交易所〔NYSE〕、美國證券交易所〔AMEX〕和納斯達克股票交易所〔NASDAQ〕）對雙重股權制度取態不一。NYSE 對不同投票權禁止最為嚴格，要求借助股權重組實現同股不同權的公司強制退市，但 NASDAQ 對不同投票權股份發行則不設任何規定。AMEX 介於二者之間，有條件地允許兩類不同投票權的普通股上市。1984 年，NYSE 宣佈暫停對發行雙重投票權股份的公司實施摘牌措施，同時設立專門委員會重新評估其長期堅持的一股一票政策。經過數輪改革，1994 年美國證監會向美國三個證券交易所發出倡議，要求在雙重股權公司上市交易的問題上採用一致政策，得到了 NYSE、AMEX 和 NASDAQ 的相繼認可。此後，各交易所達成協議：對已經發行的股票，公司不能通過任何行動或發行來減少或限制普通股股東的投票權，但是公司在發行新股時可以引入雙重股權結構[2]。美國對雙重股權結構的政策趨於統一，這之後的四年間（1994 年至 1998 年），使用雙重股權架構的企業逐步增加，佔 IPO 的 11.9%，佔首發融資金額的 24.9%。

2000 年以後，高科技網絡公司成為主要的上市發行主體，美國雙重股權架構的運用比重呈明顯增加趨勢，特別是 2004 年 Google 上市採用的雙重股權結構，被同行競相仿效。2003 年至 2008 年金融海嘯之前，美國 681 宗 IPO 中有 64 家選擇了雙重股權架構，佔比為 9.4%，選擇雙層股權架構企業首發融資金額佔比達到 20.8%。而金融海嘯後至 2013 年，461 宗 IPO 中有 76 家選擇了雙重股權架構，佔比為 16.5%，在首發融資金額中佔比達到 34.1%[3]。（見表 1。）

表 1：1980 年代以後美國運用雙重股權結構的演變過程	
年期	使用情況
1985 年	有 180 家上市公司使用雙重股權結構，佔上市公司總數 2.8%
1988 年—1992 年	使用雙重股權結構的公司佔 IPO 比重為 5.4%
1994 年 6 月—1998 年 9 月	使用雙重股權架構的公司佔 IPO 比重為 11.9%，佔首發融資金額的 24.9%
2003 年—2008 年	使用雙重股權架構的公司佔 IPO 比重為 9.4%，佔首發融資金額的 20.8%
2008 年—2013 年	使用雙重股權架構的公司佔 IPO 比重為 16.5%，佔首發融資金額的 34.1%

資料來源：Howell, J. W. (2017). "The survival of the US dual class share structure"，*Journal of Corporate Finance*，第 44 期，440-450 頁。

2　參見蔣小敏〈美國雙層股權結構：發展與爭論〉，《證券市場導報》，2015 年 9 月。

3　Howell, J. W. (2017). "The survival of the US dual class share structure"，*Journal of Corporate Finance*，第 44 期，440-450 頁。

　　從行業分佈上來看，雙重股權架構在美國運用十分廣泛。美國上市公司 44 個行業中，只有 6 個行業沒有企業使用雙重股權架構，圖 1 顯示使用雙重股權架構的企業佔比超過 6% 的行業達到 24 個（2010 年）。機械、零售、農業等傳統行業使用雙重股權架構的企業佔比在 6%-8%，而在通訊、印刷和出版行業的分佈則分別達到了 26.58% 和 22.64%，是所有行業中最廣泛使用雙重股權架構的行業，説明使用雙重股權架構與行業特性高度相關。信息化、電子化程度越高的行業，或組織結構更需符合新經濟行業發展特點的公司，投資者對雙重股權架構的接受度越高。

圖 1：美國使用雙重股權的上市公司在不同行業中的比例（比例超過 6% 的行業）（2010 年）

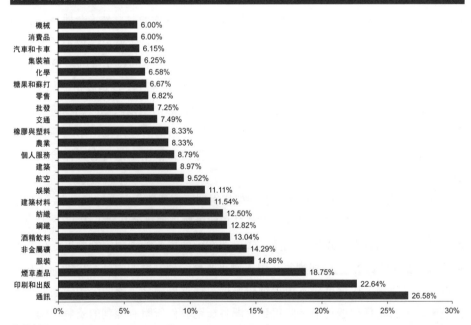

資料來源：Arugaslan, O., D. O. Cook 與 R. Kieschnick (2010). "On the decision to go public with dual class stock"，*Journal of Corporate Finance*，第 16(2) 期，170-181 頁。

1.2　「同股不同權」架構在其他國家的運用情況

　　英國在 1960 年代中期，由於機構投資者廣泛參與股票投資，並對公司管治及股東權益保護提出更多要求，雙重股權架構股票在英國發行量下降，90 年代後雙重股權架構又得到了廣泛使用。表 2 顯示，1996 年英國企業使用雙重股權架構的企業佔比達到了 23.9%。與美國略有不同的是，目前英國僅允許在部分板塊中運用雙重股權架構。英國主板市場分為高級上市（Premium Listing）和標準上市（Standard Listing）兩種形式。其中高級上市的要求較為嚴格，發行人須符合嚴謹的英國超等同規則（super-equivalent rules）進行信息披露。例如，申請人必須能夠證明自己是經營獨立業務、提供三年收入業績紀錄、有充足的營運資金及無保留意見的財務報表，給投資者更多信心。而標準上市的規則，主要運用於股份、預托證券和債券，適用歐盟最低要求（而非英國的超等同規則）。2014 年 5 月英國修訂了《上市規則》，規定同股不同權的上市架構形式只適用於標準上市。

在新加坡，雙重股權結構曾一度被抵制，根據新加坡公司法 Section 64(1) 規定，除報業公司根據《報紙和印刷出版法案》發行的管理層股票外，一股普通股應具有相應的一票表決權。2011 年，監管部門對此條例進行了修訂，取消了原先對一股只包含單一投票權的限制，允許公眾公司發行包含不同投票權的股票。2016 年，新加坡上市諮詢委員會建議新加坡交易所可接受兩級投票制股份結構（dual-class shares）的公司上市，並提出需同時推出合適保障措施以限制該結構帶來的風險。2018 年 3 月 28 日，新加坡交易所就雙重股權結構的運用推出第二輪市場諮詢，包括考慮公司商業模式、由事件觸發的日落條款機制、是否有資深投資者參股等。

從全球範圍來看，雙重或多重股權架構得到了廣泛運用，全球最大的 46 個國家股票市場中有 30 個國家有使用或曾經使用雙重股權架構的公司上市 [4]，尤其是在歐洲地區，包括芬蘭、瑞典、法國、愛爾蘭在內的主要歐洲國家都曾廣泛使用雙重股權架構（見表 2）。這些國家在引入雙重股權結構制度的同時，也增加了相應的股東保障措施，以促進資本市場健康發展。據世界經合組織的公司管治原則，投資者購入公司任何股份前，應得到有關各類別股份所附有權利的資料；倘若投票權有任何改變，必須得到受負面影響的股份持有人通過，務求所有股東均得到公平待遇。說明雙重或多重股權架構的運用在一定程度上取決於投資者的投資認知水平、信息披露完善性，以及相關約束機制。這些將在下文詳細討論。

---

4   Nenova, T. (2003). "The value of corporate voting rights and control: a cross-country analysis"，*Journal of Financial Economics*，2003 年，第 68 期，325-351 頁。

表 2：各國使用雙重股權結構的企業所佔比例		
國家	百分比	統計截止年度
瑞典	66.10%	1998
瑞士	51.20%	1999
意大利	41.40%	1996
芬蘭	37.60%	1999
愛爾蘭	28.10%	1999
英國	23.90%	1996
澳洲	23.30%	1999
德國	17.60%	1996
挪威	13.20%	1998
加拿大	10.20%	1998
美國	6.10%	2002
法國	2.60%	1996

資料來源：Howell, J. W. (2017). "The survival of the US dual class share structure"，*Journal of Corporate Finance*，第 44 期，440-450 頁。

2 「同股不同權」架構為企業帶來哪些益處？

採用「同股不同權」架構的主要顧慮在於，此種雙重股權結構導致控制權和現金流權分離，使得公司內部管理層擁有更多投票權，在管理項目時有可能不是以公司的最佳利益出發而是出於一己私利，因而違反公平原則，加劇公司代理問題（agency problem）。具體問題包括：內部管理者對私人利益的攫取、管理層過高的薪酬和津貼待遇、對公司重大決策不負責履行等問題[5]。

此外，「同股不同權」架構有可能削減股東對管理層的問責程度。由於大多數股東擁有次等投票權，即使管理層的表現不佳，股東也沒有能力撤換管理層。如

5　詳見 Howell, J. W. (2010). *The dual class stock structure in the United States*.

果出現收購建議而創辦人或管理層持反對意見，股東也可能無法接受收購建議。因此，雙重股權架構被不少投資者 (特別是機構投資者) 所詬病[6]。

但是另一方面，雙重股權結構對初創企業的長遠發展，尤其是對於高投入、高不確定性、高增長的科創型企業來說，具有明顯優勢。具體表現在：

第一，雙重股權結構有利於實現創新公司的長期價值。

面對創新行業飛速變化和外部環境的巨大不確定性，科創型企業創始人及其團隊擁有深入的專業知識和行業判斷力，能夠作出更快、更有效率的決策，對企業發展擁有很強的影響力。雙重股權結構可使得創始人通過較高比例的投票權控制公司運作，專注於提升公司的長遠收益，而不會因為短期股價波動或短期收益而改變公司發展方向，對執行商業模式和長遠戰略佈局具有重要的推動作用。

以 2014 年 2 月 Facebook 收購 WhatsApp 事件為例，Facebook 僅耗時 11 天，就斥資 190 億美元天價收購了只有 50 多名員工的 WhatsApp。包括投資者在內的大多數市場人士認為 Facebook 收購價格過高，並不看好這筆交易，以致收購翌日股價下跌，市值縮水 30 多億美元。但是，Facebook 認為 WhatsApp 在多個國家 (地區) 覆蓋率超過 90%，將為 Facebook 帶來以數十億計的活躍用戶；而且，Facebook 認為通過收購將獲取對方的優秀人才，減少潛在競爭對手，彌補 Facebook 在移動社交平台的劣勢。在一定程度上，雙重股權結構使 Facebook 管理層避免市場短視壓力，選擇了有利於公司長期發展的決策，進而可望為中小投資者提供可持續的投資利益。收購後一年內，Facebook 股票收益率超過同期 Google 和納斯達克指數，收購後近三年時間內，Facebook 股票累計收益率超過 Google 和納斯達克指數[7]。可見採用雙重股權結構應有助創新公司的商業決策運用，達致公司的長遠增值。

第二，雙重股權結構可作為有效的激勵機制，促使創始人向企業投入更多創新能力和人力資本。

6　詳見 Jarrell, G. A. 與 A. B. Poulsen (1988). "Dual-class recapitalizations as antitakeover mechanisms: The recent evidence"，*Journal of Financial Economics*，第 20 期，129-152 頁。

7　有關 Facebook 這一個案的描述，具體參見李海英、李雙海、畢曉方 (2017)〈雙重股權結構下的中小投資者利益保護——基於 Facebook 收購 WhatsApp 的案例研究〉，《中國工業經濟》，2017 年第 1 期。

　　新興企業最大的特徵是其創新性，而這在很大程度上依賴創始人的原創能力、精神感召力與趨勢洞察力。如果管理層發生變動，創始人投入的知識產權和人力資本可能無法獲得回報，這對於長週期的生物醫藥企業、需要複雜知識的新經濟企業來說，會無法完成足夠的知識積累和創新。與外部股東相比，創始人在投資目的、資源基礎、企業發展方面承擔了更大風險，因此，從本質上看，雙重股權結構更類似於一種激勵機制，通過對創始人權益的保護，幫助創始人專注投入不斷創新，增加企業的歸屬感和凝聚力，使得創始人團隊為公司創造更大價值[8]。

**　　第三，雙重股權結構可作為防止敵意收購的有效屏障。**

　　敵意收購一般指在未經目標公司董事會同意、或者未與目標公司股東談判的情況下，對目標公司作出股權收購，繼而更換公司管理層，獲取公司經營權的行為。在雙重股權結構中，集中的投票權使擁有超級投票權的股東擁有重要交易權，即使因為敵意收購而獲得了普通股股票，往往不能獲得足夠多的決策權影響公司行為，亦不足以更換管理者，使公司很難成為敵意收購的目標，這也成為家族公司傾向採用雙重股權架構的主要原因之一。

**　　第四，在金融市場機構投資者過度集中的情況下，雙重股權結構在一定程度上被視為新興企業規避市場短期行為而不得不採取的自我保護措施。**

　　研究顯示[9]，具有主要投票權的團體大多由管理他人資金的基金管理者組成。這些人的動力來自短期投資結果，從中獲得報酬，所以，機構投資者更關注短期股票價值的上漲。過度「金融化」改變了股本投資性質：長期以來認為股權資本更傾向於長期投資，會關注公司長期發展，但是由於當前壟斷股權市場的是眾多機構投資者，更注重股東的回報而不是企業長期的健康發展。為了規避這種過度「金融化」給企業帶來的負面影響，雙重股權結構成了應對機構投資者、公眾市場僅關注短期利益問題的主要出路。

8　He, L. (2008). "Do founders matter? A study of executive compensation, governance structure and firm performance"，*Journal of Business Venturing*，第 23(3) 期，257-279 頁。

9　參見 David Berger (2018). *Why Dual-Class Stock: A Brief Response to Commissioners Jackson and Stein*，Wilson Sonsini Goodrich & Rosati，2018 年 2 月 22 日。

第五，從一些實證研究結果來看，採用雙重股權架構後，企業價值和代理成本[10]也會得到改善。

比如，Dimitrov and Jain（2006）[11]使用了 176 家美國企業樣本，發現從單一股權轉向雙重股權後帶來了 23.11% 的超額回報，如果是新發行低投票權股票的企業，超額收益更大。Jordan, Liu, Wu（2014）[12]則發現雙重股權公司比單一股權公司支付更多現金股息，並且常規股息支付要多於特別股息和回購，這表明企業雖然採取了雙重股權架構，但仍考慮了外部股東利益，發放更多股息兌現事先承諾。Howell（2017）[13]對於美國雙重股權企業的存續時間進行了調研，認為雙重股權結構為眾多企業創下良好業績提供股權結構上的保障，是確保創新型企業成功存活的重要原因。石曉軍等（2017）[14]利用在美國上市的全球互聯網企業的數據，發現雙重股權架構對高科技企業創新有顯著促進作用，尤其是在發達國家外部制衡監督機制較完善時，使用雙重股權並且創始人擔任主要管理者，將會更加促進企業的創新研發投入。

---

10 代理成本是由所有權和經營權分離造成的。所有者和股東希望管理層按股東財富最大化的目標盡力經營管理企業，但由於管理層本身不是股東，或持有股份比例小，往往從自身的利益出發從事企業的日常經營管理。比如通過在職消費獲取除工資報酬外的額外收益，從而造成股東利益受損。而且，股東和管理層之間存在着較為嚴重的信息不對稱。管理層在第一線從事經營活動，掌握着企業貨幣資金的流入流出，處於相對的信息優勢，而股東則處於信息劣勢，難以判定管理層行事是否符合股東財富最大化的目標。

11 Dimitroy, V. 與 P. C. Jain (2006). "Recapitalization of one class of common stock into dual-class: Growth and long-run stock returns", *Journal of Corporate Finance*，第 12(2) 期，342-366 頁。

12 Jordan, B. D., M. H. Liu 與 Q. Wu (2014). "Corporate payout policy in dual-class firms", *Journal of Corporate Finance*，第 26 期，1-19 頁。

13 Howell, J. W. (2017). "The survival of the US dual class share structure", *Journal of Corporate Finance*，第 44 期，440-450 頁。

14 石曉軍、王鷺然 (2017)〈獨特公司治理機制對企業創新的影響——來自互聯網公司雙層股權制的全球證據〉，《經濟研究》，第 1 期，149-164 頁。

3 完善對雙重股權結構公司監督機制的主要路徑

當然，雙重股權企業是否比單一股權企業帶來更高的代理成本，不利於股東權益保護，在理論上和實證上依然存在着大量爭論，研究採用的樣本不同，企業發展週期不同，都會得出不同結果。

雙重股權架構一方面可能對創始人和管理者產生激勵效應，提升公司長遠價值，但另一方面也可能弱化公司內部治理監督機制，產生更大的信息不對稱和代理問題，從而降低公司價值。如何完善雙重股權結構公司的內部監督機制，確保對公司擁有控股權的股東進行有效監督？在具體實際運作中，各國嘗試從多個角度加以控制。以下分節論述各種有關措施。

3.1　對超級表決權的運用進行適度限制

這主要體現在以下兩方面：

一是限定表決權的差額，將具有超級表決權股份所享有的表決權限定在同等數量普通股表決權的 10 倍以內。從國際做法來看，雖然美國股市並沒有對投票權設定加以限制，但最普遍的結構安排是超級表決權普通股每股有不超過 10 倍的投票權（10：1 的表決權差異），如 Google、Facebook、百度等（表 3 列出部分在美國上市的中國內地公司使用的不同投票權倍數和架構）。其他地區，如瑞典等歐洲交易所，亦要求企業的額外投票權股份，最多 1 股擁有相當於 10 股普通股的投票權。此前新加坡交易所諮詢引入雙重股權架構時，也建議倍數投票權股份（multiple vote shares）以 10 倍投票權為限。

表 3：部分在美國上市的中國內地公司的不同投票權倍數和架構				
公司	IPO 日期	業務	股份架構	控股股東
百度 （Baidu, Inc.）	04/08/2005	網上搜索引擎	類別 A（上市）：1 票 類別 B（非上市）：10 票	創辦人持有： • 15.9% 的股權；及 • 53.5% 的投票權
邁瑞醫療國際有限公司 （Mindray Medical International Ltd.）	25/09/2006	全球醫療設備研發、製造及推廣公司	類別 A（上市）：1 票 類別 B（非上市）：5 票	全體董事及行政人員集體持有： • 28.8% 的股權；及 • 64.2% 的投票權
盛大遊戲有限公司 （Shanda Games Ltd.）	24/09/2009	網上遊戲開發商及運營商	類別 A（上市）：1 票 類別 B（非上市）：10 票	盛大網絡（Shanda International）持有： • 70.8% 的股權；及 • 96.0% 的投票權
當當網 （eCommerce China Dangdang Inc.）	07/12/2010	網上 B2C 商貿平台	類別 A（上市）：1 票 類別 B（非上市）：10 票	創辦人持有： • 35.3% 的股權；及 • 83.3% 的投票權
奇虎 360 （Qihoo 360 Technology Co. Limited）	29/03/2011	互聯網以及移動安全產品	類別 A（上市）：1 票 類別 B（非上市）：5 票	全體董事及行政人員（包括兩名共同創辦人）集體持有： • 40.4% 的股權；及 • 64.9% 的投票權
鳳凰新媒體 （Phoenix New Media Limited）	12/05/2011	媒體內容提供商	類別 A（上市）：1 票 類別 B（非上市）：1.3 票	鳳凰衛視持有： • 52.8% 的股權；及 • 59.2% 的投票權
優酷土豆網 （Youku Tudou Inc.）	優酷網與土豆網於 23/08/2012 合併	網上視頻	類別 A（上市）：1 票 類別 B（非上市）：4 票	創辦人持有： • 21.3% 的股權；及 • 51.5% 的投票權
蘭亭集勢 （LightInTheBox Holding Co., Ltd.）	06/06/2013	全球網上零售公司	單一類別股份，股份持有人在大多數事宜上每股有一票。創辦人在控制權變動方面每股有 3 票	在控制權變動方面事宜，創辦人持有 20.1% 的股權及 43.0% 的投票權
汽車之家 （Autohome Inc.）	10/12/2013	網上汽車銷售	類別 A（上市）：1 票 類別 B（非上市）：每股 1 票，但若控制人持有該公司總股權低於 51% 但高於 39.3%，則可有 51% 的投票權	澳洲電訊（Telstra）持有： • 65.4% 的股權；及 • 65.4% 的投票權
愛康國賓 （iKang Healthcare Group, Inc.）	08/04/2014	民營體檢提供者	類別 A（上市）：1 票 類別 C 非上市：15 票	創辦人持有： • 14.3% 的股權；及 • 35.9% 的投票權
京東商城 （JD.com）	21/05/2014	互聯網直銷公司	類別 A（上市）：1 票 類別 B（非上市）：20 票	創辦人持有： • 20.7% 的股權；及 • 83.7% 的投票權

資料來源：2014 年 8 月香港交易所「不同投票權架構」諮詢文件。

　　二是，公司在設計公司架構時，盡可能明確或劃定超級投票權的應用範圍。當公司事務涉及到公司或其控制人的經營管理（如敵意收購）、戰略決策（如企業文化或經營理念傳承）、國家安全與社會公共利益等重大事項時，創始人及部分股東可擁有較多的投票權。但是，如果表決事項直接關係到外部投資者的切身合法權益（如公司進行關聯交易或對外擔保等對公司財產重大處分、對其核心信息強制披露、監事或獨立董事提名等），則可以考慮將超級投票權和普通投票權的比例縮減或恢復到「一股一票」的投票機制 15，讓普通股東在公司的重大交易和關聯交易決策上有更多發言權，這樣既不會違背雙重股權結構創制的初衷，防止控制權被稀釋，也可以使股東的監督職能得以恢復。

3.2　對超級投票權設置明確的退出機制和轉讓機制

　　其中一種自動轉換機制體現在超級表決權股份的轉讓。普通股東接受投票權受限條件，主要基於對創始人的信任，包括對創始人創新實力、經營水平以及綜合能力，這些作為讓渡投票權的附屬條件。當這些創始人或者控制股東退出公司或將其股票轉讓給第三人時，對公司控制、經營的狀況發生改變，票權讓渡的附屬條件不再具備，就需要考慮恢復原狀。

　　從各國實際運用的情況來看，不同國家對超級表決權股轉讓也設有一定限制，美國規定超級表決權股一般不能流通，即使轉讓也會自動轉換為一股一票的普通股。加拿大多倫多證券交易所則要求給外部股東提供「燕尾保護」（coat-tail protection），即擁有超級表決權股權架構的公司應當確保提出要約購買超級表決權股的購買方在同等條件下提出要約購買次級表決權股，以防止內部股東將公司的控制權出售並獲得大量溢價，掠奪其他股東利益。新加坡交易所在近期的諮詢文件中，則建議在特定條件下超級投票權股須轉為普通股，例如股東辭任董事、轉讓手上股份等。

　　另一種機制則為「日落條款」機制。有研究表明，在 IPO 之後一段時間內，採用雙重股權結構的公司會產生交易溢價，但隨着公司不斷成熟發展，這種公司溢價最終會消失，說明雙重股權架構對企業帶來的成本和收益在企業不同的生命週

15　參見曹陽（2017）《中國雙層股權結構的引入與規制》，2017 年 12 月。

期是不斷變化的：在公司發展早期，基於創新能力對企業競爭力構建的重要性，保護創立者的控制權會有利於公司，是明智選擇，但是在某個時間點（這一個時間點通常被稱作「日落」（sunset）之後，雙重股權架構的有效性需要重新考慮 [16]。基於此邏輯，公司在採用雙重股權架構進行 IPO 時，有時會對其永續雙重股權結構適用的標準加以一定限制，表明管理層願意在某一天把投票權「還給」股東。

在實際運用中，「日落條款」機制並不常見，表 4 列明了近年來部分公司對「日落條款」機制的運用，即以創始人最低持股佔公司總發行股本比重為強制其回轉的下限，或以上市達到一定年限為界，將超級投票權轉換為普通投票權。

表 4：部分企業使用「日落條款」的觸發條件		
公司	IPO 年份	「日落條款」觸發條件
Groupon	2011	上市後 5 年（2016 年已轉為一股一權）
Kayak Software	2012	上市後 7 年
Yelp	2012	上市後 7 年或超級投票權股份低於 10% 普通股
Workday	2012	上市後 20 年或超級投票權股份低於 9% 普通股
Apptio	2016	上市後 7 年或超級投票權股份低於 25% 普通股
Nutanix	2016	上市後 17 年
Hamilton Lane	2017	上市後 10 年或創始人及僱員持有的投票權低於所有投票權的 25%
MuleSoft	2017	上市後 5 年或超級投票權股份低於 15% 普通股

資料來源：Council of Institutional Investors。

3.3　加強企業管治，
共同使用公司內部和外部替代性約束機制

公司內部建立防範雙重股權架構企業內部代理人問題（agency problem）的有效措施，包括強制信息披露機制，比如規定公司要明確披露其不同投票權架構及與此有關的風險。此外，也可能要求公司披露不同投票權持有人的身份，和構建控制股東的激勵和懲罰機制等。

公司外部存在約束機制也會對企業管理層具有替代性制約作用。比如 IPO 價

16　詳見 Robert J. Jackson (2018). *Perpetual Dual-Class Stock: The Case Against Corporate Royalty*，2018 年 2 月 15 日。

格方式可體現資本市場本身的懲罰機制，如果投資者預見到管理層存在內部代理人問題，企業上市價格就會有很大的折扣，以覆蓋投資者未來可能遭到的損失。另外，市場也會自發選擇企業合適的股權架構。Howell（2017）[17] 發現，美國 61 家雙重股權公司從雙重股權模式轉向一股一票模式，而股權架構變化後，資本市場作出了正向的反應，表明市場自身的約束能力發揮着決定性作用。

雙重股權結構有利有弊，一方面來自於控制權人的戰略眼光和企業家精神推動企業長遠發展，而不利之處在於對外部股東權利的削弱和對公眾投資者利益的潛在危害。通過對企業內部控制制度的完善，並在企業外部設置靈活適宜的法律架構，使得公司根據自身情況對股權結構作出最有利於公司未來發展的安排，那麼就能在一定的制約機制下充分發揮雙重股權結構的優勢。

4 「同股不同權」架構在香港的運用和討論

歷史上，香港曾有 5 家公司使用雙重股權架構，此後隨着公司私有化或退市。太古集團是目前港股市場唯一發行 B 股的上市公司 [18]。香港於 1987 年禁止採用不同投票權架構的公司在香港上市，上市架構轉為「一股一權」安排，即所有上市公司股東所持有的每一股都享有相同的投票權（voting rights），以確保投票權與股本對等，股東權益獲得平等對待。

2004 年，以美國 Google 為代表，開啟了科創企業、創新型公司以「同股不同權」架構的上市浪潮，除 Twitter 之外的美國大多數高科技公司上市都採用了雙重股權結構，相當一部分在美國上市的中國概念股也採取了類似的架構。2017 年 6 月，116 家在美國上市的內地公司中，33 家（28%）採用了不同投票權架構，市值

17　Howell, J. W. (2017). "The survival of the US dual class share structure"，*Journal of Corporate Finance*，第 44 期，440-450 頁。

18　太古 B 的投票權與太古 A 相同，但股值為 A 股的五分之一。

高達 5,610 億美元，佔所有美國上市內地公司市值的 84%，其中 18 家（55%，市值佔比 84%）為科創型公司[19]。此外，美國採用的雙重股權結構也引領了其他國家競相仿效，英國、德國、加拿大等國家已成功採納劃分多個不同上市板塊或在上市板塊內細分不同市場的方法，以引入「同股不同權」架構。

在此背景下，2014 年香港針對阿里巴巴上市訴求，曾就「同股不同權」架構進行了廣泛討論，2017 年香港交易所再次推出設立創新板的市場諮詢，為引入「同股不同權」架構探索新的可能性。2018 年 4 月發表的雙重股權架構諮詢意見總結，提出了相應措施對制度的適用加以限制和約束，比如要求申請人須具備若干特點方可以不同投票權上市。香港交易所亦將保留權利，若申請人的不同投票權架構極端不符合管治常態（例如普通股完全不附帶投票權），可以不適合上市為由拒絕該等申請。同時，香港交易所也提出了不同投票權公司上市後的詳細投資者保障措施，包括限制不同投票權權力、保障同股同權股東投票權的措施、加強企業管治和加強披露要求等。不同投票權發行人的股份名稱結尾將要加上標記「W」，與其他發行人區分開來等。此外，不同投票權受益人必須擔任公司董事，使其對公司的經營承擔法律法規規定的董事責任。一旦不同投票權受益人轉讓其不同投票權股份，或他們身故或失去行為能力，又或他們不再為董事，其不同投票權即會失效，令不同投票權限於「自然」日落條款，不會無限期存在[20]。

就是否需要引入集體訴訟權來保障投資者的議題，2015 年 6 月香港交易所在不同投票權架構諮詢及 2017 年有關建議設立創新板的諮詢中發現，只有少數回應者認為引入集體訴訟機制該是准許不同投票權的先決條件，同時市場的回應也顯示，在美國的大多數集體訴訟是涉及信息披露問題，而不是在不同投票權架構下可能發生濫用控制權的問題。實際上，研究表明香港市場與美國一樣高度重視投資者保障，只是做法不同。美國的投資者保護重點在於事後保護，投資者對於其自身權益侵犯方面較易採取訴訟並獲得賠償，司法成本較香港低。而香港對於投資者保護更側重於前端防護，其披露要求主要基於《上市規則》，盡量對可能產生的控制股東濫權的情況加以預先防範，香港證監會亦代表股東作出事後法律行動[21]。

19　有關數據見香港交易所刊於其網站的《有關建議設立創新版的框架諮詢文件》，2017 年 6 月。
20　詳見香港交易所刊於其網站的 2018 年 4 月諮詢意見總結。
21　詳見香港交易所刊於其網站的 2014 年 8 月《不同投票權架構的概念文件》。

　　另外，隨着雙重股權架構的不斷演化，有可能出現類似阿里巴巴「中國合夥人制」的非典型雙重股權架構，甚至可能加入中國特色的雙重股權架構。如果進一步強制信息披露，或仿照美國進一步引入「吹哨者計劃」（whistle-blower programmes），鼓勵內部舉報，將可能對市場產生震懾作用，使企業從內部得到淨化，防止採用雙重股權結構公司的實際控制人及管理人團隊觸碰欺詐和內幕交易的紅線。

　　從 2016 年起，香港交易所開始醞釀上市架構改革方案，最主要的長遠目的在於利用同股不同權結構，為創投產業的高增長公司、或尚未有盈利的生物醫藥公司掃清上市障礙，提供上市渠道，回應全球新經濟行業的發展趨勢。對香港來說，也可從上市集資的國際企業及內地公司中獲得龐大商機，成為鞏固香港全球金融中心地位的重要保證。「同股同權」雖然運作多年，一直被認為是香港投資者保障的重要內容，但隨着市場機制進一步完善，上市架構不應該成為企業創新、經濟增長的阻礙，而是應為制度創新提供更大的空間，由市場檢驗制度的創新性和生命力。如果眾多科創類新經濟企業，以及尚未盈利的生物技術企業無緣利用公開資本市場，既不利於風險投資為他們提供大規模的資金支持，也難以幫助中小型企業進一步和國際企業建立戰略關係，不利於創新性企業的進一步成熟和行業集羣的形成。

　　根據科創類新經濟企業在成長期不同財務特點和投資風險，設置相應的上市標準，將在最大程度扶持其發展，引導更多創業風險投資和私募股權類投資進入產業和企業內部，最終推動一批大型創新型企業的湧現，從而促進區域內新經濟核心產業的發展，實現區域經濟升級和跨越發展，這正是資本市場改革為香港經濟帶來的長遠正面影響。

第7章

同股不同權：
對投資者是好？是壞？

香港交易及結算所有限公司
首席中國經濟學家辦公室

摘 要

　　不同投票權股份一般被稱為雙重股權股份。具不同投票權架構的公司（簡稱「不同投票權公司」）會發行至少兩類股份，兩類股份的每股投票權並不相同。創辦人和管理層持有的通常是一股多票的股份類別，而發行給一般投資者的則是一股一票的股份類別。新經濟創新產業公司首次公開招股時多傾向採用不同投票權架構，務求集資發展業務的同時，創辦人可持有相對其投票權作不成比例攤薄後的股份數量，而能繼續維持對公司的控制，自主追求創新。中國內地正值新經濟轉型，這類型的公司有如雨後春筍，供應源源不絕。這些不同投票權公司可選擇於接受不同投票權公司上市的主要海外市場（如紐約、倫敦、香港、新加坡及東京）上市。

　　香港的新上市機制特別為不同投票權公司的投資者加入一系列的保障措施，與其他主要的環球金融中心相比，香港市場在這方面的要求更全面，監管規定更嚴謹。這些投資者保障措施包括對公司市值的高門檻要求、日落條款、加強企業管治及披露要求等。該等措施不單有效降低「代理成本」，還可對「委託人成本」起限制作用。委託人成本在投資者對創新產業並無專業知識（「委託人能力成本」）和投資者之間的潛在利益衝突（「委託人衝突成本」）的情況下尤其重要。當市場有適當的保障措施，公司採用不同投票權架構上市便不一定傷害投資者的利益，反而或能支持股價和營運表現，有機會為投資者帶來裨益，有利創造長遠價值。

　　事實上，國際經驗表明，投資者投資於不同投票權架構的公司看來利多於弊。首先，在 MSCI 的股票指數成份股中，具不同投票權架構的創新產業公司，其長期股價回報跑贏其他在相同地區和環球指數的成份股。第二，並沒有一致的實證顯示不同投票權架構對公司的業務表現有潛在負面影響，對在美國上市、採用不同投票權架構的中國公司所得的觀察亦如是。況且，不同投票權公司的創辦人若能不受投資者干預而按自己的專業知識實現各種創新意念，往往能為公司帶來更多價值。事實上，不同投票權公司能為投資者擴闊投資機會，切合其不同的風險承擔能力和投資需求。有鑒於此，MSCI 指數公司也推出經投票權調整的全新指數系列，應對不同投資者的喜好。

　　總括而言，接受不同投票權公司來港上市並非逐利之爭，而是要拓闊投資市場機遇。對實體經濟來說，不同投票權公司由此得到的融資支援為經濟轉型和新的增長動力帶來貢獻；對投資者來說，他們可選擇投資信譽可靠的不同投票權公司，在投資的風險與機遇之間取得平衡。

1　不同投票權公司的上市：全球證券交易所概覽

　　不同投票權股份一般被稱為雙重股權股份。具不同投票權架構的公司（「不同投票權公司」或「雙重股權公司」）會發行至少兩類股份，兩者的每股投票權並不相同 ——「A 類」股份每股僅可投一票，「B 類」股份則每股可投多於一票。持有 B 類股份的都是公司的內部人士或管理層，包括創辦人及董事，他們手上的未上市股份的投票權不合比例地高於其收益權。在此情況下，B 類股東毋須擁有同等高比例的持股亦可主導公司政策上的決定，例如提名董事會成員、股份發行及公司行動[1]。相比之下，A 類股份則是上市股份而每股僅有一票，由外部投資者持有。下文中，每股可投多於一票的股份一概簡稱「不同投票權股份」，而每股僅可投一票的股份則簡稱「普通股份」。

　　儘管不少國家當地的公司法都採用「一股一票」原則，但不少證券交易所亦容許發行人根據不同的規則及標準申請雙重股權架構股份上市[2]。這些交易所大多在歐美，亞洲則在最近才見相關例子，當中包括巴西、加拿大、丹麥、法國、芬蘭、中國香港、意大利、日本、俄羅斯、新加坡、瑞典、瑞士、英國及美國[3]。

　　最新一輪不同投票權公司的首次公開招股（「IPO」）（簡稱「不同投票權 IPO」）熱潮主要受創新科技公司所推動。科技公司一向活躍於 IPO 市場，現時越來越偏好採納雙重股權架構。科技及創新產業公司的收入及投資期結構與傳統公司不同，其對資金有強烈的需求。這類公司會有高投資風險，但通常亦會有高發展潛力；其營運初期的收入會極不穩定，但長線增長前景卻潛力巨大。這些公司的創新科技發展很大程度上依靠創辦人的遠見及魄力。這些公司 IPO 時多傾向採用不同投票權架構，務求集資發展業務的同時，創辦人可持有相對其投票權作不成比例攤薄後的股份數量，而能繼續維持對公司的控制，自主追求創新。創新產業公司的

1　公司行動包括股息、供股、股份拆細、分拆上市以及合併和收購。

2　見香港交易所研究報告〈雙重股權架構與生物科技行業的上市制度改革〉，載於香港交易所網站，2018 年 4 月 24 日。

3　部分例子引述於 Tan, A. 與 B. Robertson〈為何投資者為雙重股權股份苦惱〉（"Why investors are fretting over dual-class shares"），載於 *Bloomberg QuickTake*，2017 年 7 月 10 日。

創辦人對採用不同投票權架構的偏好，理由可能包括幫助實現公司長遠價值、推動創辦人持續對公司投入創新能力和人力資源、隔擋惡意收購的威脅，以及避免剛冒起的公司在決策上受到機構投資者的短線利益所影響等[4]。

因此，接受不同投票權公司的上市制度正好切合科技及創新產業公司的需要。根據普華永道的資料，2017 年全球共有 100 宗科技業 IPO，合共籌得 251 億美元的資金[5]。另一資料來源則顯示在美國所有 IPO 活動中，科技公司約佔 36%，而不同投票權科技公司的 IPO 的佔比在 2017 年更升至 43% 的歷史新高，2018 年則回落至 34%（見圖 1）。

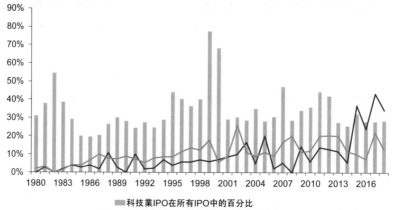

圖 1： 美國市場的科技業 IPO 佔比以及科技與非科技產業中的不同投票權 IPO 佔比（以數目計）（1980 年至 2018 年）

■ 科技業IPO在所有IPO中的百分比
── 不同投票權IPO在科技業IPO中的百分比
── 不同投票權IPO在非科技業IPO中的百分比

資料來源：Ritter, J. R.〈首次公開招股：統計更新〉（"Initial public offerings: Updated statistics"）（手稿），佛羅里達大學，2018 年 12 月 31 日（https://site.warrington.ufl.edu/ritter/files/2019/01/IPOs2018 Statistics_Dec.pdf）。

這些科技業 IPO 當中，有若干是來自內地的科技巨頭。為了將這些科技公司迎回內地上市，中國內地正考慮以不同方式，接納具不同投票權架構的新經濟公

4　見香港交易所研究報告〈雙重股權架構與生物科技行業的上市制度改革〉，載於香港交易所網站，2018 年 4 月 24 日。

5　資料來源：普華永道〈全球科技行業 IPO 回顧：2017 年第四季度及全年〉（"Global technology IPO review: Full-year and Q4 2017"），2017 年 12 月 19 日。

司上市。國務院於 2018 年 9 月澄清[6]，於海外上市且具有不同投票權架構或尚未有盈利的新經濟公司可於在岸 A 股市場發行中國存託憑證（CDR）。2018 年 11 月 5 日，國家主席習近平宣佈[7]計劃在上海證券交易所設立「科創板」，並試行註冊制；根據於 2019 年 3 月推出的相關試行措施[8]，中國科技公司及具不同投票權架構的科技公司將可在內地進行 IPO。

現時，新經濟公司可選擇於紐約、倫敦、香港、新加坡及東京等主要環球金融中心上市。這些市場因應不同投票權公司所設的上市制度的演進過程各有不同，概述如下：

(1) 紐約： 美國在 1926 年曾一度禁止雙重股權股份上市，紐約證券交易所於 1940 年正式宣佈禁令[9]。其後因面對美國其他證券交易所的競爭，紐約證券交易所於 1985 年 1 月建議放寬上市政策，允許雙重股權架構的公司上市。1994 年，美國的證券交易所就雙重股權股份實施統一的上市規定，但鑒於投資者憂慮已上市公司的現有股東可能要被迫放棄其投票權，美國各家證券交易所自那時起自願性地禁止美國公司於上市後才採納雙重股權架構（稱為「雙重股權資本重整」）。在這方面，一項法律研究[10]指出雙重股權資本重整可以為公司帶來好處，但僅限於發展初期。

現時在美國，不同投票權公司的 IPO 越來越普遍。實證顯示，美國的不同投票權 IPO 的佔比於 2017 年升至 28% 的歷史新高[11]（見圖 2）。美國之所以能吸引這些不同投票權公司，可能與其披露為本的上市制度有關，在這制度下，公司僅須遵守若干關於非傳統管治架構風險因素的披露規定[12]。

6　《國務院關於推動創新創業高質量發展打造「雙創」升級版的意見》，2018 年 9 月 26 日。

7　見〈習近平在首屆中國國際進口博覽會開幕式上的主旨演講〉，載於《新華網》，2018 年 11 月 5 日。（http://www.xinhuanet.com/world/ciie2018/jbhkms/index.htm）

8　《科創板首次公開發行股票註冊管理辦法（試行）》，中國證監會發佈，2019 年 3 月 1 日。

9　見香港交易所諮詢文件《不同投票權架構概念文件》附錄三，載於香港交易所網站，2014 年 8 月。

10　Gilson, R. J. (1987)〈評估雙重股權股份：代用品的關聯〉（"Evaluating dual class common stock: The relevance of substitutes"），載於 Virginia Law Review 第 73 期，807-844 頁。

11　資料來源：Ritter, J. R.〈首次公開招股：統計更新〉（"Initial public offerings: Updated statistics"）（手稿），佛羅里達大學，2018 年 12 月 31 日。研究所用的數據僅包括發售價為 5 美元或以上的 IPO，並不包括美國存託憑證、單位要約（unit offers）、封閉式基金、房地產投資信託基金、天然資源有限責任合夥公司、小型最大努力要約（small best-effort offers）、銀行及存貸款，以及證券價格研究中心（CRSP）未有列入的股份。

12　見美國證券交易委員會的投資者諮詢委員會〈上市公司的雙重股權架構和其他鞏固管治的架構〉（"Dual class and other entrenching governance structures in public companies"），2018 年 2 月 27 日。

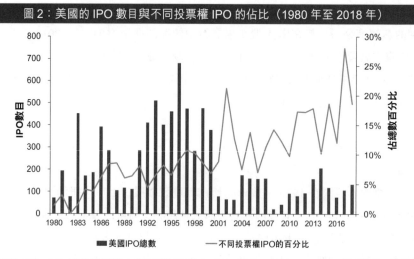

圖 2：美國的 IPO 數目與不同投票權 IPO 的佔比（1980 年至 2018 年）

資料來源：Ritter, J. R.〈首次公開招股：統計更新〉（"Initial public offerings: Updated statistics"）（手稿），佛羅里達大學，2018 年 12 月 31 日（https://site.warrington.ufl.edu/ritter/files/2019/01/IPOs2018Statistics_Dec.pdf）。

(2) 倫敦：英國市場僅接受不同投票權 IPO 根據「標準上市」的規則（即歐盟的最低要求）上市。要在倫敦證券交易所主板上市，「標準上市」的申請人僅須符合最低的歐盟協調標準，而「高級上市」的申請人則須遵守英國比歐盟最低要求更為嚴謹的「超等同」規則。「高級上市」的透明度較高，資金成本因而可能較低，不過「高級上市」的原則在 2014 年收緊了，投票權與股份持有量需成比例[13]，亦即必須一股一票。

實證顯示，2007 年英國市場上有多重投票權的上市公司僅佔 5%[14]（見圖 3）。據一篇研究論文[15] 所解釋，背後的原因有二：機構投資者的反對，以及收購規則移除了要約前的防禦措施（例如透過採用雙重股權架構），迫使公司遇到要約收購時只能採用一股一票制。

---

13 見英國金融市場行為監管局〈PS14/8：對 CP13/15 的回應 — 加強上市制度的效能〉（"PS14/8: Response to CP13/15 — Enhancing the effectiveness of the listing regime"），2015 年 8 月 3 日。

14 資料來源：Shearman & Sterling LLP〈歐盟上市公司的擁有權與控制權的相稱性：法規比較研究 — 個別國家的法規研究〉（Proportionality between Ownership and Control in EU Listed Companies: Comparative Legal Study — Legal Study for Each Jurisdiction）（由歐盟委員會委託的外部研究），2007 年 5 月 18 日。

15 Huang, F. (2017)〈全球主要金融中心的雙重股權股份〉（"Dual class shares around the top global financial centres"），載於 Journal of Business Law 第 2 期，137-154 頁。

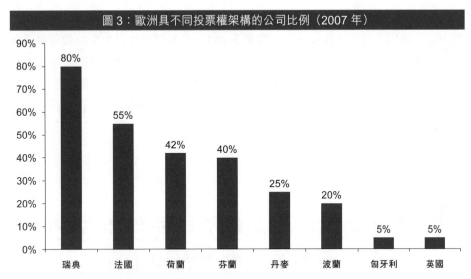

圖 3：歐洲具不同投票權架構的公司比例（2007 年）

資料來源：Shearman & Sterling LLP〈歐盟上市公司的擁有權與控制權的相稱性：法規比較研究 — 個別國家的法規研究〉（Proportionality between Ownership and Control in EU Listed Companies: Comparative Legal Study — Legal Study for Each Jurisdiction）（由歐盟委員會委託的外部研究），2007 年 5 月 18 日。

(3) 香港：香港旨在成為創新產業公司的上市中心，於 2018 年 4 月實施上市制度改革，對香港聯合交易所（聯交所）的上市規則作出修訂，接受具不同投票權架構的創新產業公司及未能符合上市財務資格的生物科技公司於主板上市，並為合資格的海外上市公司提供新的第二上市渠道。有關改革修訂了於 1989 年制訂推出的相關上市規則，允許偏離一股一票原則，使具不同投票權公司也可申請上市。自上市制度改革實施後，首六個內即有兩家具不同投票權架構的內地科技巨頭進行 IPO 上市。

(4) 新加坡：新加坡市場把對雙重股權架構的接納範圍從報業公司擴大至其他公司。除了報業公司因為新加坡《報紙與印刷出版法》的規定須發行兩類股份（管理層股東的投票權為一般股東的 200 倍）之外，新加坡交易所以往一直禁止其他公司作雙重股權股份上市。後來新加坡於 2014 年 10 月修訂《公司法》，容許公眾公司發行無投票權或多重投票權的不同類別股份。新加坡交易所其後改革其上市制度，於 2018 年 6 月 26 日起允許創新產業的雙重股權股份上市。

(5) 東京：東京證券交易所（東京交易所）於 2008 年修訂其上市規則 [16]、放寬相關規管，開始按個別情況接受不同投票權 IPO，惟繼續禁止已上市的公司發行新類別股份。2014 年 3 月，東京交易所正式迎來首宗不同投票權 IPO —— 生化人公司（Cyberdyne）（可穿戴機械人開發商）。在這宗 IPO 後，東京交易所在 2014 年 7 月再修訂其上市規則，對雙重股權股份的上市新增兩項規定：包括 (i) 採用雙重股權架構的必要性及適當性，以及 (ii) 日落條款。

2 香港為不同投票權公司而設的新上市制度：適當的投資者保障措施

市場對投資創新產業公司的需求，推動了香港市場上市機制的改革，以接受不同投票權公司的普通股在港上市（不同投票權公司的上市股票於下文稱為「W 股」）。經過市場諮詢並獲得持份者有力支持下，香港交易及結算所有限公司（簡稱「香港交易所」）於其主板的《上市規則》新增適用於不同投票權公司上市的章節，自 2018 年 4 月 30 日起生效。在諮詢過程中，回應者提出對投資不同投票權公司的潛在風險有所顧慮。以下分節將討論此等顧慮，以及香港的新上市機制如何引入適當的保障措施應對有關風險。

2.1　創辦管理人是不同投票權公司成功的關鍵

一般所見，創新產業公司的創辦人顯然都是最有能力發展及管理其公司的人。採用不同投票權架構就是要確保創辦人對公司有絕對控制權。不同投票權架構可以是雙刃劍，公司日後的發展全繫於創辦人身上。若創辦人目光遠大，又能一直作出正確的商業決定，就是公司和投資者之福。在這情況下，不同投票權架構的

---

16 見〈2008 年財政年度的上市系統改善〉（"Listing system improvement FY2008"），載於東京交易所網站，2008 年 5 月 27 日。

確保障了創業精神，亦即保障了股價。反之，若創辦人能力不高又或不慎作出錯誤決定，不同投票權之設就可能成為公司隱患，導致代理人問題。

股東與管理人之間的「代理人問題」(或稱「潛在利益錯配」) 是投資者關注的一大問題，指的情況是股東追求的是股票價值的最大化，而負責實際營運的管理人追求的則是自身薪酬和福利的最大化。不同投票權公司的管理人都是持股相對不多但擁有大多數的控制權。即使管理人未能按股東利益行事，其他股東也不能通過投票表決影響公司決策。因此，市場應設置恰當的投資者保障措施，以制約不同投票權公司的代理人成本風險，否則，代理人問題可能損及投資者利益。

不過，不同投票權架構不一定要犧牲投資者利益，方可惠及管理人。一方面，投資者須承擔代理人成本 (具體上即上述的代理人利益衝突成本)，另一方面，學術文獻 [17] 提出「委託人成本」理論，來解釋雙重股權架構如何能降低「代理人能力成本」、「委託人能力成本」和「委託人利益衝突成本」：

- **代理人能力成本**：源自管理層犯下非欺詐性錯誤的成本。成本高低各有不同。如管理人有才智，能不受偏見影響並把握最新資訊，他們可能會較少犯錯。對於不同投票權公司而言，管理人離職的風險較低，因此他們可能更勇於認錯，這或有助避免隱瞞過錯而令長遠增長受損。
- **委託人能力成本**：因投資者專業知識不足而犯錯所產生的成本。如管理層 (代理人) 已具備所需專業知識，投資者 (委託人) 可「坐享其成」，成本將會較低。不同投票權公司的管理人通常都能證明其能力為公司長遠增長作出貢獻 (例如往績顯示曾成功推行有遠見的項目)。
- **委託人利益衝突成本**：源自投資者之間的利益衝突的成本。假如公司的多方委託人 (投資者) 有利益衝突時 (例如維權投資者着重短線表現，而退休基金則注重長線表現)，成本預期會較高。對於不同投票權公司而言，外界投資者影響力較少，他們之間的利益衝突多數不會影響商業決策。

換句話說，投資者應同時考慮代理人成本及委託人成本，即控制權的總成本。若不同投票權架構能降低控制權的總成本，投資者便可受惠。這要視乎上市機制是否有適當的保障措施讓投資者免受代理人成本影響。

---

17 Goshen, Z. 與 R. Squire. (2017)〈委託成本：一個企業法律和管治的新理論〉("Principal costs: A new theory for corporate law and governance")，載於 *Columbia Law Review* 第 117 期，767-796 頁。

2.2 機構投資者呼籲為不同投票權架構設立投資者保障

在沒有適當保障措施的情況下，世界各地機構投資者都會要求 W 股在價格上有折讓。歐盟委員會曾向世界各地 445 名合共管理資產超過 4.9 萬億歐元（歐洲投資者佔 13%）的機構投資者進行調查 [18]，當中八成均認為採納加強控制權機制的公司，其股價應有折讓。歐盟委員會表示，世上並無一種能應對各種加強控制權機制的保障方案，但應該要求公司對加強控制權機制（例如雙重股權架構）的存在及其影響提升透明度。

2000 年代初期的歐洲曾見雙重股權股份合併個案（每股多於一票的股份轉換為一股一票的股份）增加的趨勢 [19]。同一研究亦顯示，法律不應強制雙重股權股份公司轉為單一股權類別的一股一票公司（以下稱為「一股一票公司」）[20]。

在美國，證券交易委員會（SEC）檢討上市制度後，於 2017 年 3 月建議加強雙重股權架構及其他強化治理架構的披露規定 [21]。美國 SEC 專員 Robert Jackson 於 2018 年 2 月認同這些架構有其優點之餘，亦倡議對永久雙重股權架構加設日落條款 [22]。

在加拿大，加拿大良好治理聯合會（CCGG）[23] 認同採用雙重股權架構有利有弊，表示「鼓勵加拿大企業精神很重要，不希望影響到加拿大企業家將公司上市」[24]。由於加拿大的上市制度及法規二十多年來不曾改變，CCGG 於 2013 年針對新上市的雙重股權公司發佈了七項保障投資者利益的最佳實踐 [25]，範圍包括董事選

---

18 Institutional Shareholder Services、Shearman & Sterling LLP 與 European Corporate Governance Institute《歐盟相稱性原則報告》（"Report on the Proportionality Principle in the European Union"）（由歐盟委員會委託的外部研究），2007 年。

19 Pajuste, A (2005)〈雙重股權股份合併的決定因素和後果〉（"Determinants and consequences of the unification of dual-class shares"），《歐洲央行工作報告》第 465 號，2005 年 3 月。

20 另見 Lauterbach, B. 與 A. Pajuste (2015)〈雙重股權股份的自願性統一對長期價值的影響〉（"The long-term valuation effects of voluntary dual class share unifications"），載於 Journal of Corporate Finance 第 31 期，171-185 頁。

21 見美國 SEC 投資者諮詢委員會〈上市公司的雙重股權架構和其他鞏固管治的架構〉（"Dual class and other entrenching governance structures in public companies"），2018 年 2 月 27 日。

22 見〈永久雙重股權的股份：反對企業皇朝的例子〉（"Perpetual dual-class stock: The case against corporate royalty"），美國 SEC 專員的演說，2018 年 2 月 15 日。

23 CCGG（Canadian Coalition for Good Governance）是加拿大的企業管治組織，代表加拿大機構股東爭取企業改革。

24 見 CCGG〈雙重股權股份政策〉（"Dual class share policy"），2013 年 9 月。

25 同上。

舉、設立投票權比率上限、聯票 [26] 及日落條款。

　　亞洲方面，市場專業人士追求一股一票的原則，若然不同投票權架構是合法的，亦會要求公司充分披露這些架構。CFA 協會 [27] 曾於 2018 年 4 月就雙重股權股份對 454 名會員進行調查。調查結果與 CCGG 的觀點相若，都認為不同投票權架構有利有弊 [28]。為雙重股權架構公司而設的上市制度的公認好處包括增加交易所的吸引力、吸引科技及其他創新產業的公司上市，而弊端則有小股東保障不足及所有權和控制權不合比例。該項調查總結呼籲交易所設置適當的保障措施，包括強制性的企業管治措施、有時間限制的日落條款、投票權比率限制、聯票條款，以及每股多於一票的股份一旦轉讓便須轉為一股一票股份等。

2.3　香港過去並無為不同投票權架構制定投資者保障措施

　　在香港，上市公司有兩類股權股份並非新鮮事。1972 年至 1973 年間共有七家公司先後發行「B」股 [29] 上市，其面值較「A」股為低，但每股投票權與 A 股相同。這些公司背後的用意不一，包括集資購買房地產或擴充業務，以及吸引更多不同類型的投資者 [30]。這七家公司中六家 [31] 已被收購或已私有化，現只有一家公司 ── 太古集團 ── 仍有「B」股在香港上市。

　　當時的上市規則並沒有為發行這類股份的特定章節。發行「B」股時，公司原本的上市股份變成「A」股，「B」股則發行予原有股東 [32]。「B」股的每股投票權與「A」股相同 ，但面值較低（僅為「A」股的五分之一或十分之一），派息亦較低。控股股東可出售所持有的「A」股，並以較低價購入相同數目的「B」股來保持相同的投票權，以騰出更多現金抵抗收購行動 [33]。就太古集團而言，所發行「B」股的每股面

26　聯票條款是法定條款，容許次級股東平等參與任何以獲取多重投票權股份的正式競投。

27　CFA 協會是全球投資界專業人士的業界組織。

28　見 CFA 協會〈雙重股權股份及其所需的保障〉（"Dual class shares and the need for safeguards"），2018 年 4 月。

29　此等「B」股有別於美國附不同投票權的 D 類股份的一般做法（例如 D 類股份通常不上市）。

30　見香港交易所諮詢文件《不同投票權架構的概念文件》第二章，載於香港交易所網站，2014 年 8 月。

31　包括會德豐集團五家公司 ── 會德豐有限公司、會德豐船務國際有限公司、香港置業信託有限公司、聯邦地產有限公司及連卡佛有限公司。另一家有「B」股的公司是 Local Property and Printing Company Limited。資料來源：香港交易所 2014 年 8 月的《不同投票權架構的概念文件》。

32　例如會德豐於 1972 年將普通股重設為 A 股並向現有股東發行 B 股（擁有 A 股面值的十分之一），股東每持有兩股 A 股可購買一股 B 股（資料來源：馮邦彥（2017）《香港企業併購經典（增訂版）》，三聯書店（香港）有限公司。

33　見周顯（2013）《財技密碼（下篇）》第 26.3 節，香港財經移動出版。

值為每股「A」股面值的五分之一（0.12 港元相對 0.60 港元），派息亦按此等面值支付（即使股票須訂面值的規定已於 2014 年香港修例後被廢除）。若控制權受挑戰，「B」股價格將相對「A」股出現溢價。然而於 2000 年至 2018 年間，在無人爭奪控制權的情況下，太古「B」股經調整派息後的成交價較低，平均較「A」股折讓12%（見圖 4），主要因為其流通量不高 [34]。

圖 4：太古集團「A」股與「B」股價格比較（2000 年 1 月至 2018 年 12 月）

註：基於太古「B」股的派息是 A 股的五分之一，故「B」股股價乘以 5 以作比較。
資料來源：彭博。

　　1987 年，多間上市公司曾建議發行「B」股，但考慮到當時香港及海外經紀的強烈反對，聯交所及當時的證券監管機構決定禁止新的「B」股上市。同年，政府就「B」股事宜委託進行的檢討結果表示「在特殊情況下，這類型的股票有繼續存在的合理需要」[35]。其後，限制「B」股上市的規定於 1989 年 12 月正式寫入主板《上市規則》第 8.11 條。自此以後，香港再沒有「B」股新上市。

34　同上。
35　資料來源：公司註冊處《公司法改革務常務委員會第三次中期報告：B 股》，1987 年 7 月。

2.4　香港新上市制度下為不同投票權架構設有適當的投資者保障措施

香港的新上市制度旨在於投資者保障與發行人利益之間取得平衡，使香港有更大靈活性提供更多投資機會予不同風險胃納的投資者，又可讓創新產業公司毋須攤薄創辦人的控制權而進行直接融資。要達到這些目的，唯一方法就是在上市制度中設置適當的保障措施。

與舊制不同，不同投票權股份在香港上市須遵循嚴格規定。過去「B」股上市並無特定規定，但在新制度下，聯交所主板《上市規則》新增不同投票權公司上市的章節，訂明一系列保障措施應對投資者可能關注的問題。這些措施包括更高的上市門檻、持股及投票權上限、日落條款、轉讓股份須符合的條件，以及加強企業管治及披露規定。在這些方面，香港為不同投票權架構而設的新上市制度所提供的保障措施，要比其他主要環球金融中心的交易所所提供的更多（見附錄）。這些保障措施有如下特色。

在香港作新上市的不同投票權公司須符合的市值要求是全球主要交易所中要求最高者。市值最低要有 400 億港元（如收入有 10 億元或以上，最低市值要求為 100 億港元）。高市值意味公司已有一定合理規模，不是高風險的初創企業。環顧全球市場，MSCI 旗下的全球指數便有若干科技巨頭，而當中採用雙重股權架構的公司的市值通常都遠高於同業。按 2017 年 9 月 1 日的統計數字，MSCI 所有國家世界指數（ACWI）有 253 隻 W 股，佔指數總市值的 11.2%[36]，當中又以資訊科技行業的佔比最大（23.2%）。採用不同投票權架構的科技公司的平均市值最高，達 720 億美元，相比指數內所有 W 股的平均市值僅為 190 億美元（見表 1）。

表 1：MSCI ACWI 內 W 股的平均市值（2017 年 9 月）			
行業	W 股的數目	行業總市值 （百萬美元）	每隻股票平均市值 （百萬美元）
資訊科技	24	1,740,097	72,504
醫療保健	11	358,682	32,607

36　資料來源：《股票指數應否納入不同投票權架構公司的股票》（*Should Equity Indexes Include Stocks of Companies with Share Classes having Unequal Voting Rights*），是就《MSCI 有關處理不同投票權架構的諮詢》的討論文件，2018 年 1 月。

（續）

表 1：MSCI ACWI 內 W 股的平均市值（2017 年 9 月）			
行業	W 股的數目	行業總市值（百萬美元）	每隻股票平均市值（百萬美元）
金融	44	722,027	16,410
非必需消費品	61	870,424	14,269
必需消費品	29	388,716	13,404
房地產	6	70,459	11,743
工業	30	306,094	10,203
電訊服務	10	90,228	9,023
原料	19	136,032	7,160
能源	10	71,168	7,117
公用事業	9	28,126	3,125
MSCI ACWI 中所有 W 股	253	4,782,053	18,901

資料來源：Melas, D.〈把聚光燈照在 Spotify：為何不同投票權的股票跑贏大市？〉("Putting the Spotlight on Spotify: Why have stocks with unequal voting rights outperformed?"），載於 MSCI 博客，2018 年 4 月 3 日。

　　為限制代理人利益衝突成本，香港市場要求不同投票權股份的實益擁有人（「不同投票權受益人」）符合若干規定。不同投票權受益人只限於身份是董事以及在促進業務增長方面有重大責任的自然人。不同投票權受益人還受制於以下其他規定：

- **已上市公司不得發行不同投票權股份。**公司上市後，不同投票權受益人不得增加不同投票權的比例。換言之，普通股股東（「同股同權股東」）的投票權不會被進一步不成比例地攤薄。同股同權股東的投票權任何時候均須至少佔總投票權的 10%。

- **不同投票權股份的投票權比率設置上限。**在香港，不同投票權股份與普通股的投票權比率上限（或最高投票權差異）為 10 比 1，與許多有雙重股權股份的市場的做法一致。10 對 1 的比率上限有其歷史原因，就是當年（1985年）紐約交易所轄下一個小組委員會曾提出放寬一股一票原則的規則建議時，這便是其中的一項建議，只是紐約交易所最後並無對投票權差異設置任何上限。一個極端的個案是 Snap，於 2017 年 3 月在美國進行 IPO 時只公開發行無投票權的股份。就產品性質而言，無投票權的股份就類似定息產品，是一種回報與公司表現掛鈎、但不影響公司決策的金融產品[37]。

37　資料來源：CFA 協會〈雙重股權股份：好處、壞處和危險〉("Dual-class shares: The good, the bad, and the ugly"），2018 年 8 月。

　　香港亦引入不同投票權架構自然日落條款以減少潛在的代理人利益衝突成本。日落條款為創辦人提供合理時間，使其可投資於高風險項目創造長線價值，而要求不同投票權股份其後轉為普通股亦屬合理。在香港的制度下，不同投票權股份的不同投票權不設時間限制，而是遵從自然日落條款，即在受益人身故、或不再是董事、或不再被視為有行為能力可履行董事職責、或不再被視為符合董事的規定時，又或在不同投票權股份被轉讓之後，原有的不同投票權便不再有效。換言之，該等人士的繼承人不能繼承超級投票權。美國市場並沒有要求雙重股權架構有日落條款，但美國 SEC 專員 Robert Jackson 提醒市場：「永久」的雙重股權股份存在風險，因永久雙重股權架構永久的存在，不僅惠及目光遠大的創辦人，也會惠及目光不一定遠大的繼承人 [38]。Robert Jackson 特別指出，永久雙重股權股份的「相對估值」[39] 的中位數經過一段時間之後均見下降，但設有日落條款的，其相對估值的中位數在長線（上市後 7 年或以上）會略有改善（見圖 5）。

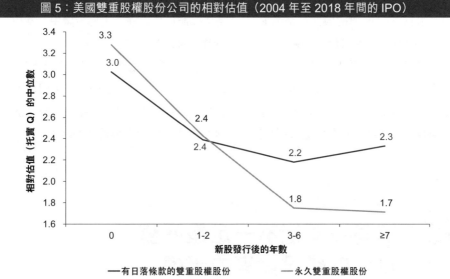

圖 5：美國雙重股權股份公司的相對估值（2004 年至 2018 年間的 IPO）

資料來源：〈永久雙重股權股份：反對企業皇朝的例子〉（"Perpetual dual-class stock: The case against corporate royalty"），美國 SEC 專員的演説，2018 年 2 月 15 日。

38　資料來源：〈永久雙重股權股份：反對企業皇朝的例子〉（"Perpetual dual-class stock: The case against corporate royalty"），美國 SEC 專員的演説，2018 年 2 月 15 日。

39　公司的「相對估值」是以托賓 Q 比率計算。托賓 Q 比率是公司市值與公司資產重置成本之比率。

香港的不同投票權公司須接受強化的企業管治措施約束，有關措施可減少代理人能力成本。在香港，不同投票權受益人的多重投票權並不適用於一些須按一股一票表決的決議，包括修改組織章程文件、委任及罷免獨立非執行董事或核數師、更改任何股份類別所附帶的權利及上市發行人自願清盤等。不同投票權公司必須擁有企業管治委員會及常設合規顧問。這些保障措施須納入公司的組織章程文件，不同投票權受益人須向公司保證遵守這些保障措施。新加坡也有類似的措施。日本的一家不同投票權公司（生化人公司 Cyberdyne）的組織章程文件規定，在要約收購中，如果要約人持有不少於 75% 的已發行股份，不同投票權股份便須轉換為普通股，意即在這情況下將需按一股一票處理。同樣，在英國，一股一票亦適用於持有不同投票權公司的 75% 股權的收購要約。美國現時反而並無任何針對不同投票權公司的強化企業管治措施。

香港加強披露規定，有助保障投資者權益。新上市制度規定發行人必須在上市文件和通訊中加入警告字眼，指明公司是不同投票權發行人，投資者投資時必須審慎行事。W 股的股票名稱加上「W」字樣標記以作區別。主要金融中心當中，只有香港和新加坡採用這些獨特的股票名稱標記提醒投資者。

總括而言，香港市場的投資者保障措施較其他主要環球金融中心更為完備。一方面，不同投票權架構可減少委託人成本，而另一方面，這些措施可同時有效限制投資者的總體控制權成本（包括代理人成本）。在此制度下，投資者投資於具不同投票權架構的股票以追求更高潛在回報時，亦更可能獲得較佳保障。

3 不同投票權不一定是惡魔，也可以是投資者的天使

有為投資不同投票權公司設置適當的投資者權益保障措施是市場所需求，然而這並不代表投資者要對不同投票權架構敬而遠之。事實上，實證顯示不同投票權架構可以是投資者的天使，有利公司股價及營運表現，長遠有助創造價值。所

以，若上市制度已有針對不同投票權公司的適當投資者保障措施，投資者便可在享有更豐富投資選擇和保障權益之間取得平衡。以下分節作詳細論述。

3.1　股價雖然折讓，但回報可能很出色

基於不同投票權架構有潛在的代理人問題，一般投資者通常都預期這些公司的普通股股價會較其不同投票權股票的股價有所折讓，以補償第 2 節所述管理層盤踞及公司資源被徵用的潛在風險。實證研究顯示附有次級投票權的上市股票都有價格折讓。在美國，研究 [40] 發現在 1990 年至 1998 年間，253 家雙重股權公司在 IPO 後的五年內，市盈率平均有 17% 的折讓。另一研究 [41] 估計，在 1990 年至 2000 年間，分別在 39 個國家的 393 批控股股權股份在私下協議轉讓時，控制權的價值相對公司普通股權的股票價值平均有高出 10% 至 14% 的溢價。還有一項研究發現，不同投票權股份相對其普通股的股價在不同市場中有不同程度的估計溢價（見圖 6）。

40 Smart, S., R. Thirumalaib 與 C. Zutter (2008)〈一個投票權包含甚麼？雙重股權股份對首次公開招股的公司價值的短期和長期影響〉（"What's in a vote? The short- and long-run impact of dual-class equity on IPO firm values"），載於 *Journal of Accounting and Economics* 第 45 期第 1 號，94-115 頁。

41 Dyck, A. 與 L. Zingales (2004)〈控制權的私人利益：國際比較〉（"Private benefits of control: An international comparison"），載於 *Journal of Finance* 第 59 期，537-600 頁。

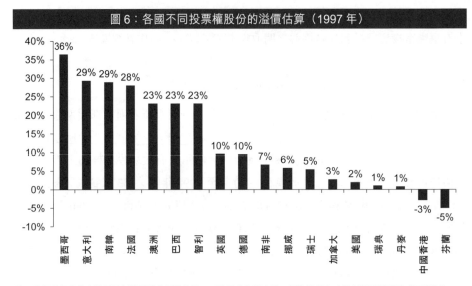

圖 6：各國不同投票權股份的溢價估算（1997 年）

註：股價溢價的數字按具控制權票數的價值估計，而這是按每股多於一票的股份與有限制投票權股份（經調整為
　　投票權的 50%）的股價計算再按公司市值調整。

資料來源：Nenova, T. (2003)〈企業投票權與控制的價值〉（"The value of corporate voting rights and
　　　　　control"），載於 *Journal of Financial Economics* 第 68 期，325-351 頁。

　　然而，一項研究 [42] 發現，在全球及區域市場中，MSCI 指數成分股中 W 股的
長線平均回報均超過指數中的其他成份股。按 2017 年 9 月的數字，MSCI 的全球
基準指數 ACWI 共有 2,493 隻成份股，當中 243 隻為不同投票權公司的 W 股 [43]。
2007 年 11 月至 2017 年 8 月期間，MSCI ACWI 的 W 股成份股的年化總回報率為
7.2% 左右，比起指數整體回報的 4.5% 及一股一票成份股回報的約 4.2% 均要高。
MSCI ACWI 的北美、歐洲和新興市場區域指數中的 W 股成份股表現，亦跑贏指
數整體回報及一股一票成份股的回報。W 股在不同市場的超額回報 [44] 各有不同。
（見圖 7。）

42　Melas, D.〈把聚光燈照在 Spotify：為何不同投票權的股票跑贏大市？〉（"Putting the Spotlight on Spotify: Why
　　have stocks with unequal voting rights outperformed?"），載於 MSCI 博客，2018 年 4 月 3 日。

43　MSCI ACWI 可包含不同投票權公司多於一類的股份，例如 Google 母公司 Alphabet 的 A 類股份（一股一票）及 C
　　類股份（無投票權）均包括在指數內，而有多重投票權的非上市 B 類股份則不包括在內。

44　超額回報的定義為股價與基準指數之間的回報差異。

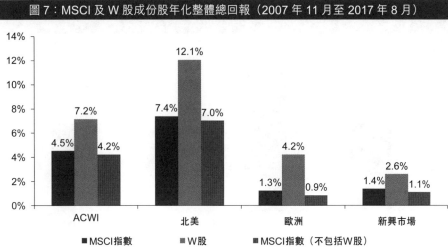

圖 7：MSCI 及 W 股成份股年化整體總回報（2007 年 11 月至 2017 年 8 月）

資料來源：Melas, D.〈把聚光燈照在 Spotify：為何不同投票權的股票跑贏大市？〉（"Putting the Spotlight on Spotify: Why have stocks with unequal voting rights outperformed?"），載於 MSCI 博客，2018 年 4 月 3 日。

　　同一項研究又將 MSCI 指數中 W 股的超額回報拆解分析，將之分為源自不同風險因素的回報（見圖 8）。行業因素帶來正回報（MSCI 新興市場指數除外），這與過去十年科技業表現強勁吻合。其他主要風險因素（如貨幣、國家及市場因素）對跑贏大市的貢獻卻未見一致。相反，股票本身的特有風險在超額回報中的貢獻程度較高（歐洲除外），該等風險可能包括創辦人對創新產業的遠見及其選擇採用不同投票權架構。同一項研究還發現，與一股一票股份相比，W 股成份股具有市值相對大、盈利能力較高但較不穩定、資產增長率較高但槓桿較低及股息較低的特徵，這印證了不同投票權公司具遠見的創辦人會傾向投資於高風險項目，以盡量提升公司的長遠價值。

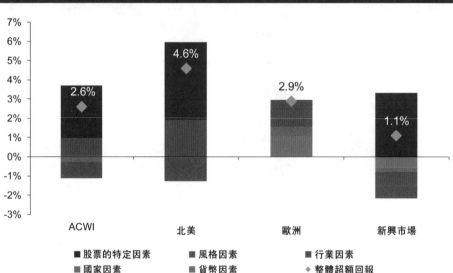

圖 8：各類風險因素對 MSCI 指數中 W 股整體超額回報的貢獻率
（2007 年 11 月至 2017 年 8 月）

資料來源：Melas, D.〈把聚光燈照在 Spotify ：為何不同投票權的股票跑贏大市？〉（"Putting the Spotlight on Spotify: Why have stocks with unequal voting rights outperformed?"），載於 MSCI 博客，2018 年 4 月 3 日。

　　部分實證研究亦發現 W 股有超額回報。一項專題研究 [45]（以下簡稱「D&J 研究」）對 1978 年至 1998 年間 178 家美國上市公司的雙重股權資本重整進行研究（「雙重股權資本重整」是指一股一票公司重整為不同投票權公司並將其一類有限投票權或無投票權的普通股上市），估算到買入並持有該等公司既有的一股一票股份在其宣佈進行資本重整後首年及第四年，與沒進行雙重股權資本重整的配對樣本比較下所獲得的超額回報率分別為 5% 及 23%。另一項實證研究 [46] 發現美國 2018 年上半年的 W 股 IPO 回報較市場平均為高 —— 於研究期間，美國 W 股 IPO 的平均回報率（自發售日期起計）高於所有 IPO 的平回報率（見圖 9）。行業方面，科技、媒體及電訊（TMT）行業的 W 股的平均回報率優於業內所有 IPO 的平均回報率；

45　Dimitrov, V. 與 P. C. Jain (2006)〈從單一類別股權資本重整成雙重股權股份：增長與長期股價表現〉"Recapitalisation of one class of common stock into dual-class: Growth and long-run stock returns"），載於 *Journal of Corporate Finance* 第 12 期，342-366 頁。

46　Klausner, D.〈雙重股權股份首次公開招股正在冒起：科技獨角獸乘着最新趨勢〉（"Dual class IPOs are on the rise: Tech unicorns jump on board this new trend"），載於普華永道 Deals Blog，2018 年 7 月 18 日。

而當中的 TMT 行業的 IPO 平均回報率於 2018 年上半年亦是所有行業中最高者，顯示創新科技產業公司的 W 股相當受投資者歡迎。

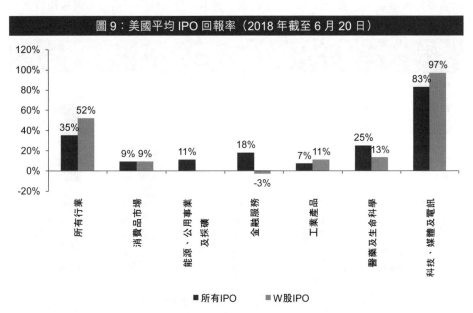

圖 9：美國平均 IPO 回報率（2018 年截至 6 月 20 日）

■ 所有IPO　　■ W股IPO

註：某隻股票的「IPO 回報率」是指該股於 2018 年 6 月 20 日的股價相對於當初發售價的回報率。研究數據涵蓋 2018 年 1 月 1 日至 2018 年 6 月 20 日期間所有美國 IPO（不包括「特別併購上市」（SPAC）的公司）。

資料來源：Klausner, D.〈雙重股權股份首次公開招股正在冒起：科技獨角獸乘着最新趨勢〉（"Dual class IPOs are on the rise: Tech unicorns jump on board this new trend"），載於普華永道 Deals Blog，2018 年 7 月 18 日。

3.2　不同投票權公司業務表現跟一股一票公司相若，甚至更佳

上文第 3.1 節討論的 W 股價格折讓意味着 W 股的市盈率相對於實際價值而言大多會較低，公司採用不同投票權架構進行股權融資的成本因而可能會受到影響。這或會影響到不同投票權公司的槓桿率及資金成本以至盈利，繼而損害投資者的投資回報。然而，實證顯示事實未必如此。

第 3.1 節引述的 D&J 研究結果顯示美國不同投票權公司的收入增長及盈利能力優於其屬於非不同投票權公司的競爭對手——美國不同投票權公司在銷售、資產及營運收益方面的增長較業內規模相近的競爭對手為高。尤值得注意的是，不

同投票權公司在宣佈進行雙重股權資本重整後四年內的營運收益上升了 72%，而競爭對手則只上升了 50%（見圖 10）。盈利能力方面，不同投票權公司與競爭對手的資產回報率並無差異，但資本重整後數年間不同投票權公司的股本回報卻較高。

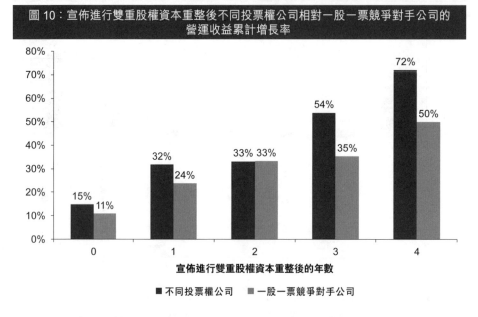

圖 10：宣佈進行雙重股權資本重整後不同投票權公司相對一股一票競爭對手公司的營運收益累計增長率

註：「0 年」指宣佈進行雙重股權資本重整之年。
資料來源：D&J 研究。

在加拿大，不同投票權架構亦未有被發現對公司的業務表現有任何負面影響。一項實證研究[47]分析了 1996 年至 2005 年間所有加拿大上市公司 10,366 項公司年度觀察，研究不同投票權架構對盈利能力（資產回報率）、公司估值（托賓 Q 比率）及收購溢價（累積超額回報率）的影響。該研究中的加拿大不同投票權架構公司的控股股東大多仍為創辦人的家族成員，在研究期間仍於公司擁有集中股權。因此，公司資源若被徵用，對該等股東亦會受財富效應的影響。研究結果並未發現由不同投票權架構造成的顯著正面或負面影響。

---

47 Jog, V., P. Zhu 與 S. Dutta (2010)〈受限制投票權的股權結構對公司價值和表現的影響〉（"Impact of restricted voting share structure on firm value and performance"），載於 *Corporate Governance: An International Review* 第 18 期，415-437 頁。

3.3　有利於創造價值

　　除價格折讓及業務表現外，投資者亦會關注到不同投票權公司的長線估值，始終公司管理層行事未必與投資者看重為股東創造價值的利益一致。事實上，有些外部股東（例如維權投資者）多數也看短線，會向所投資的公司的業務決定施加壓力。若是一股一票公司，維權投資者及某些類型的機構投資者（例如對沖基金或其他注重短期回報的資產管理公司）可利用投票表決去左右業務決定，務求改善短期股價表現（例如派高息而非再投資），又或停止投資可能會創造長遠價值的高風險項目。相反，在採用不同投票權架構的創新產業公司，具遠見的創辦人在這方面所受的干擾會較少，相對較能夠自主投資高風險項目，儘管短期回報可能極不穩定，但期望長遠而言能建立豐厚價值。

　　具遠見的創辦人與投資者之間可能產生的矛盾通常源自信息不對稱。投資者大多欠缺對創新產業的專業知識，創辦人則一般有較豐富的相關知識或獨到的遠見去創造長期優厚回報，並與投資者按比例分享這些回報。因此，部分長期投資者願意在影響力受限的情況下，投資於不同投票權公司。雖說外部股東對不同投票權公司業務決定的影響力相對不高，但不同投票權架構可減低公司因缺乏專業知識的投資者投票而被迫作出不當業務決定的機會（所謂「委託人能力成本」）。

　　確實有研究發現，不同投票權公司的管理層可利用其專業知識創造額外價值，從而降低委託人能力成本。其中一項[48]研究觀察了一股一票類股份在被收購時的溢價。研究對象是 142 家於 1978 年至 1998 年間在美國進行雙重股權資本重整的公司，其中有 48 家公司在資本重整中將內部持股比例大幅下降（「淨賣出」公司），但內部人士所持股權仍佔全部股份的 39% 左右。這些淨賣出公司在股權分散後，規避風險的傾向反而減輕。不同投票權架構使這些公司可推行風險較高但能提升價值的項目。事實上，這些公司在增加資本開支及槓桿率之後反令資產回報大幅上升。整個研究樣本中有 67 家公司其後被收購，當中包括同時有兩類別股權股份作上市買賣的公司。有雙重股權股份上市的公司當中，一股一票股份的收購溢價（定義為據報收購價格與宣佈收購前四天的股價之百分比差異）的平均值為

48　Bauguess, S. W., M. B. Slovin 與 M. E. Sushka, (2012)〈大股東分散投資，企業承受風險以及轉變為不同投票權的影響〉（"Large shareholder diversification, corporate risk taking, and the benefits of changing to differential voting rights"），載於 *Journal of Banking and Finance* 第 36 期，1244-1253 頁。

53.4%，高於不同投票權股份的平均值（39.7%）以及只得單一類別股份上市的對比指標樣本公司的股份平均值（36.1%）。這些研究結果減輕了人們對代理人問題在創造價值方面的潛在影響之顧慮。

　　不同投票權架構亦可避免因為長期與短期投資者之間的潛在衝突而產生的委託人利益衝突成本。有些短期投資者（如維權投資者）或傾向於對其投資的公司行使控制權，但研究發現維權投資者的行為不一定能提升公司價值。根據 McKinsey 的一篇文章 [49]，針對美國公司進行的維權投資者運動的次數從 2010 年的 108 次增至 2016 年的 280 次。文章特別指出，TMT 產業的維權投資者大多是缺乏相關產業專業知識的一般人士。維權投資者雖有意創造價值，但其進行的活動卻不一定能提高對股東的整體回報率（「整體回報率」是指按公司營運表現、股票市值及市場對公司表現及財務槓桿所預期的轉變而作出的整體量標 [50]）。超額的股東整體回報率（相對於相應的標準普爾產業指數的股東整體回報率）在維權投資者運動後首三年內可為正數或負數（見圖 11）。

49　Cyriac, J., S. Otto 與 D. Wells〈用維權投資者方式思考的好處〉（"The benefits of thinking like an activist investor"），載於 McKinsey 網站，2017 年 6 月。

50　請參閱 Deelder, B., M. H. Goedhart 與 A. Agrawal〈一個更好的方式去明白股東整體回報率〉（"A better way to understand TRS"），載於 McKinsey《策略及企業金融》（*Strategy and Corporate Finance*）網誌，2008 年 7 月。

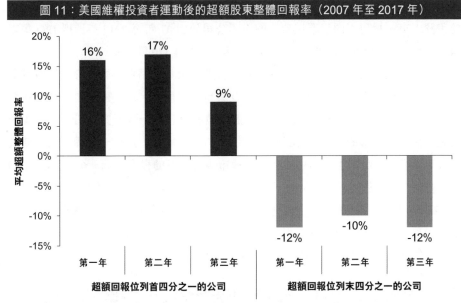

圖 11：美國維權投資者運動後的超額股東整體回報率（2007 年至 2017 年）

註：研究樣本為 151 家市值過百億美元及收入超過 10 億美元、並且有 2017 年 3 月 15 日之前三年整體股東回報資料的美國上市公司，於 2007 年以來的 252 次獨有的積極／維權投資者活動。股票超額回報率為對比標準普爾相應產業指數計算。

資料來源：Cyriac, J., S. Otto 與 D. Wells〈用維權投資者方式思考的好處〉（"The benefits of thinking like an activist investor"），載於 McKinsey 網站，2017 年 6 月。

3.4　於美國上市的中國不同投票權公司表現的實證

在 2018 年底，於美國的證券交易所上市的中國公司共有 220 家，當中不同投票權公司佔了一定的比例，其中包括 28 家於 2018 年進行 IPO 上市的公司[51]。2018 年第一季的九宗 IPO 中有六宗（佔總數約 67%）來自不同投票權公司[52]。香港交易所 2014 年有關不同投票權架構的概念文件[53] 中提到，於美國上市的中國公司中約有總數的 30% 採用不同投票權架構；按 IPO 集資的累積金額計算，佔比數字更達總額的 48%（1998 年 1 月至 2014 年 5 月）。該文件亦指出，截至 2014 年 5 月，於美

51　資料來源：Ritter, J. R.〈首次公開招股：統計更新〉（"Initial public offerings: Updated statistics"）（手稿），佛羅里達大學，2018 年 12 月 31 日（https://site.warrington.ufl.edu/ritter/files/2019/01/IPOs2018Statistics_Dec.pdf）。

52　資料來源：Hu, K.〈中國公司在 2018 年大量流入美國首次公開招股市場〉（"Chinese companies flooded into the U.S. IPO market in 2018"），載於雅虎財經網站，2018 年 12 月 29 日。

53　香港交易所諮詢文件《不同投票權架構的概念文件》，載於香港交易所網站，2014 年 8 月。

國上市的中國不同投票權公司中有 70% 來自資訊科技產業。另一資料來源則表示於美國上市、採用不同投票權架構的中國公司於 2007 年至 2017 年透過 IPO 集得340 億美元[54]，相對於同期美國所有 IPO 集資額的 4,420 億美元[55]。

　　基於公司文化或有不同，中國的不同投票權公司的情況與國際經驗比較可會是相同或有異？部分實證顯示，就股價及業務表現而言，於美國上市的中國 W 股表現沒有跑輸一股一票公司的股份（「一股一票股份」）。一項研究[56]分析了在美國的證券交易所上市的 33 家中國雙重股權公司及 88 家在中國單一股權公司。就價格表現而言，於美國上市的中國 W 股在熊市表現下滑的程度較低——W 股在 IPO上市後第二年超額回報的中位數下跌 13%，一股一票股份的中位數則下跌 38%（見圖 12a）。不同投票權公司及一股一票公司的市盈率及托賓 Q 比率並無重大差異。業務表現方面，不同投票權公司及一股一票公司在 IPO 上市後第二年的資產回報率的中位數並無統計學上的顯著差異（見圖 12b）。

圖 12：美國上市的中國公司股價及業務表現的比較（不同投票權公司與一股一票公司）

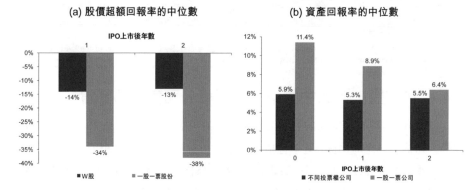

(a) 股價超額回報率的中位數 　　　(b) 資產回報率的中位數

註：超額回報率定義為特定時間內的實際回報率與市場回報率之間的差異。

資料來源：Abdullah, J. Zhou 與 M. H. Shah (2017)〈不成比例的投票權對公司的市場表現的影響：在美國上市的中國公司的實證〉（"Effect of disproportional voting rights on firm's market performance: Evidence from Chinese firms cross-listed on US exchanges"），載於 *International Journal of Financial Studies* 第 5 期，1-11 頁。

54　資料來源：Robertson, B. 與 A. Tan〈雙重股權股份〉（"Dual-class shares"），載於彭博網站，2018 年 5 月 3 日。

55　資料來源：國際證券交易所聯會。

56　Abdullah, J. Zhou 與 M. H. Shah, (2017)〈不成比例的投票權對公司的市場表現的影響：在美國上市的中國公司的實證〉（"Effect of disproportional voting rights on firm's market performance: Evidence from Chinese firms cross-listed on US exchanges"），載於 *International Journal of Financial Studies* 第 5 期，1-11 頁。

3.5　擴闊投資機會

儘管有代理人問題的隱憂，但不同投票權公司可能會有高營運效率，以及有機會為投資者帶來良好的回報（見上文）。對於這些公司，一些規避風險的投資者可能會敬而遠之，但也有些投資者受潛在升幅吸引而願意投資。不同投票權公司的上市有助迎合不同的投資者喜好。對新經濟創新產業公司有興趣的投資者也許願意持有相對不同投票權股份較少投票權但現金流權益相同的股份。事實上，新經濟創新產業公司的 W 股可以帶來的增值機會，即使對經驗豐富的投資者而言也可能有一定程度的吸引力。Snap 的無投票權股份便是個極端例子 —— 該公司 2017年 3 月在美國進行 IPO，超額認購逾 10 倍 [57]。

為切合不同投資者的需求，全球指數供應商 MSCI 已重新將 W 股納入其指數。在 2017 年 3 月 Snap 的無投票權股份 IPO 後，倡導一股一票制的美國機構投資者委員會（CII）[58] 向指數供應商施壓，要他們剔除指數中的 W 股，MSCI 因而曾一度暫停納入新的 W 股。為重新考慮有關情況，MSCI 向市場參與者就 W 股作了詳細諮詢。MSCI 於 2018 年 10 月公佈了諮詢總結，支持持份者「在投資者觀點與股本證券產品的全面代表性之間尋找適當的平衡」[59]。根據諮詢總結，MSCI 在2019 年 3 月起取消對將其全球基準指數再加入 W 股的限制（現有指數編纂方法不變 [60]），並同時推出經投票權調整的新指數系列 [61]。有關做法與美國 SEC 專員對不同投票權公司投資價值的看法一致 ——「如果各大指數均禁止納入任何雙重股權公司，投資大眾可能就會錯失參與市場上走在創新最前沿的公司的業務增長機會」[62]。加拿大的情況也類似，加拿大的不同投票權公司大多為主要行業中最大的

57　資料來源：路透社〈Snap 在期待已久的首次公開招股的定價超出預期〉（"Snap tops expectations in pricing of long-awaited IPO"），2017 年 3 月 1 日。

58　CII（Council of Institutional Investors）是美國退休基金的業界組織。

59　見〈MSCI 全球可投資市場指數維持不變，同時 MSCI 將推出反映投資者在不同投票權架構方面的偏好的全新指數系列〉（"MSCI will retain the MSCI Global Investable Market Indexes unchanged and launch a new index series reflecting the preferences of investors on unequal voting structures"），MSCI 於 2018 年 10 月 30 日發佈的聲明。

60　見 MSCI〈不同投票架構諮詢結果的問答〉（"Q&A on the results of the unequal voting structures consultation"），2018 年 10 月。

61　見 MSCI〈MSCI 經投票權調整指數系列的編纂方法〉（"MSCI voting rights-adjusted indices methodology"），2019 年 3 月。

62　資料來源：〈永久雙重股權的股份：反對企業皇朝的例子〉（"Perpetual dual-class stock: The case against corporate royalty"），美國 SEC 專員的演講，2018 年 2 月 15 日。

公司及僱主[63]，所以難以在投資組合剔除這些公司。

4 總結

世界各地主要金融中心的股票市場紛紛進行制度改革，接受不同投票權公司申請上市。背後的主要動力源自發行人和投資者兩邊的需求均不斷增加。發行人方面，那些具遠見的創新產業公司（包括來自中國內地的公司）創辦人，尋求上市時都傾向採用不同投票權的架構，務求保留其對業務決策的控制權。投資者方面，投資新經濟產業的增長潛力愈來愈吸引，甚或足以抵銷持有投票權受限的股份的壞處和不同投票權架構所涉及的風險。香港便是全球已進行上市機制改革、允許不同投票權公司上市的主要市場之一。

香港於 2018 年 4 月推出其新的上市機制，納入了比其他主要市場更為全面的投資者保障措施。這些措施不單有效消減不同投票權架構可能涉及的代理人問題，還可遏抑因投資者缺乏相關專業知識和相互之間存在利益衝突所連帶引致的委託人成本。因此，新上市機構可說不單有利發行人亦有益於投資者。

從國際經驗所見，投資於不同投票權架構的公司對投資者來說可會是利多於弊。淨得益包括潛在的股價超額回報和締造長遠價值的機會。不同投票權公司的上市令市場有更多投資選擇，迎合投資者不同的需要和喜好。只要對潛在風險進行徹底評估，不同投票權公司的股票也可成為投資者趨之若鶩的選擇。

63　資料來源：Allarire, Y.〈雙重股權股份的例子〉（"The case for dual class of shares"），載於 Institute for Governance of Private and Public Organisations 網站，2018 年 12 月 20 日。

附錄

主要金融中心的證券交易所有關不同投票權公司上市的投資者保障措施

保障措施	紐約證券交易所	倫敦證券交易所（根據「標準上市」規則）	東京證券交易所	新加坡交易所	香港交易所
最低市值要求	4,000 萬美元	700,000 英鎊（約 540,000 美元）	第一板的要求為 250 億日圓（約 2.2 億美元）	3 億新加坡元（約 2.14 億美元）	市值 400 億港元（約 51 億美元）或市值 100 億港元（約 13 億美元）及收入 10 億港元（約 1.27 億美元）
新發行人限制	有	無	有	無	有
創辦人或其他不同投票權受益人須持有最低持股量佔總股本的百分比	無限制	無限制	無限制	未有規定	10%
最大投票權差異	無	無	無	有（10:1）	有（10:1）
特定行業限制	無	無	無	有	僅限「創新產業」公司
日落條款	無	無	自然日落條款	自然日落條款	自然日落條款
創辦人退休/喪失能力/去世後自動轉為一股一票股份	否	否	是	否	是
股份轉讓時自動轉為一股一票股份	否	否	是	否	是
強化企業管治措施	無	有	有	有	有
股票名稱特別標記	無	無	無	有	有

資料來源：各證券交易所網站上有關上市準則的資料；CFA 協會〈雙重股權股份：好處，壞處和危險〉（"Dual-class shares: The good, the bad, and the ugly"），2018 年 8 月；及 Toshima, K. (2014)〈Cyberdyne 的雙重股權股份 IPO〉（"Cyberdyne's dual-class IPO"），載於 *International Financial Law Review* 第 33 期，43-45 頁。

詞彙

「一股一票」	一股一票制（或比例原則）是指每股享有同等投票權，亦享有同等份數的現金流權利（例如股息）。
「一股一票公司」	只發行單一類別、遵循一股一票制的股份的公司。
「一股一票股份」	公司所發行的、遵循一股一票制的股份類別（發行公司可能有單類別或多類別股份）。
「不同投票權」	有別於一股一票制的股權架構，某類股份每股可享有多份投票權。
「不同投票權受益人」	每股可享有多份投票權的股份（有別於一股一票股份）的持有人。
「不同投票權公司」	至少有兩類股份、其中一類是不同投票權股份的公司。
「不同投票權股份」	有別於一股一票股份，每股可享有多份投票權的股份。
「W股」	不同投票權公司在交易所上市的非「不同投票權股份」的股份類別；在香港，此等股份是一股一票股份。

第8章

海外債市支持新經濟公司發展

邱志明

花旗環球金融亞洲有限公司資本市場部董事總經理、
亞洲債務發行部聯席主管

蔣國榮

花旗環球金融亞洲有限公司董事總經理、
中國企業與投資銀行部主席兼主管

方寶榮

花旗環球金融亞洲有限公司中國投資銀行部董事總經理

茅駿翔

花旗環球金融亞洲有限公司資本市場部董事總經理

摘 要

　　近十年來，隨着中國經濟的迅速發展和金融開放政策的落地，中國企業的國際化程度顯著提高，而新經濟企業作為其中走在最前沿的不可忽視的貢獻者，逐漸成長為國際資本市場的關注焦點。在此進程中，作為規模最大的國際資本市場，海外債券市場扮演了十分重要的角色。本章將通過簡述債券市場趨勢與分析發行實例來敍述中資發行人如何在十年間逐漸取得國際投資者的認可，並展望中國新經濟企業如何利用海外債券市場助力其進一步的發展。

1 中資發行人海外債券市場的發展歷程 (2009 - 2019 年)

　　隨着中國經濟的迅速發展和中國企業國際化、拓展海外業務的需求不斷增長，海外債券市場融資逐漸成為中國企業的新潮流。中資企業海外債券的發行規模從 2010 年起開始直線上升，並逐漸成為國際債券資本市場的主要組成部分，總發行量從 2009 年的 33 億美元，於 2017 年達到 2,172 億美元的歷史最高點，為 2009 年的 65 倍；儘管 2018 年發行量較 2017 年有所回落，但仍然保持在 1,755 億美元的歷史高位，2019 年截至 7 月中資企業的發行量已達到 1,367 億美元，市場預計有望超過 2017 年的最高點（見圖 1）。在亞太債券市場上，中資發行人已然成為貢獻發行量的中堅力量（見圖 2）。

圖 1：中國企業海外債券的發行規模和數量（2009 年至 2019 年 7 月）

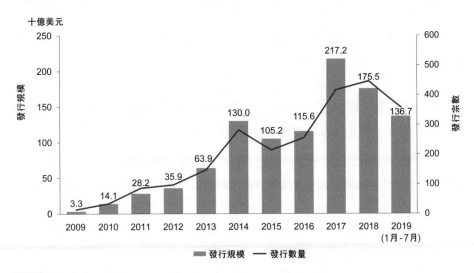

資料來源：Dealogic。

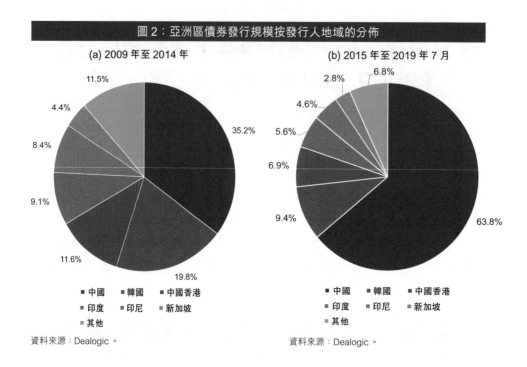

圖 2：亞洲區債券發行規模按發行人地域的分佈

(a) 2009 年至 2014 年

(b) 2015 年至 2019 年 7 月

資料來源：Dealogic。

伴隨着發行量的迅速增長，發行的多樣性不斷湧現。從發行人種類到發行債券的品種和特點均呈多樣化發展。

(1) 發行人種類不斷擴展，包含不同行業領域，由少數核心央企拓展到各個領域的國企、民企、金融機構、省市級發行人

中資債券市場上起初主要的發行人為「三桶油」（中石油、中海油、中石化）等大型央企，逐漸拓展到大型金融機構和房地產公司等傳統行業的企業（見圖3），而近幾年，依託信息技術和高端製造業為發展動力的新經濟企業取得了顯著的發展。隨着國際業務需求不斷增長，新經濟企業也成為中資企業發行的中堅力量，逐漸成長為海外債券市場關注的亮點。

回顧截至 2019 年 7 月的中資海外債券發行，總規模在 50 億美元以上的交易共有 7 筆，而新經濟企業已佔據 3 筆，最大規模的記錄由阿里巴巴在 2014 年首次發行 80 億美元創造，可見新經濟行業的發行人已成為海外債券市場不容忽視的存在。2011 年以前，中國新經濟企業仍處在發展前期，在國際債券上也未有任何公開發行。2011 年底，騰訊作為第一位公開亮相的新經濟企業發行人，完成了 5 年期 6 億美元的固息美元債券發行。8 年間，騰訊通過 6 次公開發行，持續向投資者更新企業發展及信用亮點，其發行規模明顯增長，發行品種逐漸多樣。在 2014 年騰訊第一次實現了總值 25 億美元的雙年期發行。到 2018 年，騰訊又進一次擴大了發行規模，成功完成了 50 億美元 4 個不同年期品種的債券發行，並在其中首次加入了浮息債券，滿足了投資者在加息通道中的資產配置需求，受到了以美國投資者為代表的追捧，實現了超額認購 9.6 倍。而到 2019 年，騰訊已成長為國際市場上備受關注的成熟發行人，成功定價了 60 億美元債券發行 —— 這也是其規模最大的美元債券發行 —— 分為 5 年浮息及 5/7/10/30 年固息五個系列，取得了 270 億美元的訂單。在這 8 年間，其發行人評級也從 Baa1/BBB+/-- 逐級提升至 A1/A+/A+，再次反映出公司的成長與市場的認可。

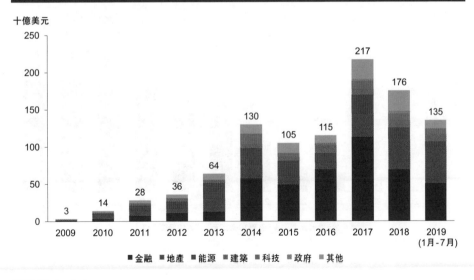

圖 3：中國企業海外債券發行規模按發行人行業的分佈（2009 年至 2019 年 7 月）

資料來源：Dealogic。

(2) 發行規模不斷提升

2009 年前，來自中資的發行主要集中在主權和政府機構的發行，企業債券的發行規模很少超過 5 億美元，但隨着海外投資者對中資發行人的認可不斷加深，高至 20 億/30 億等值及以上的發行便不再罕見 (見圖 4)。尤其在近五年，隨着新經濟企業逐漸實現人口與政策紅利，資質較好的發行人在國際上均有十分顯著的發行 (表 1)。阿里巴巴集團在 2014 年首次發行便通過 6 個年期固息/浮息雙品種的發行達到了 80 億美元，2017 年再次以長期限為主發行了 5.5/10/20/30/40 共 5 個年期的固息美元債券，貢獻出 70 億美元的發行量。騰訊自 2011 年首次於海外債券市場亮相後，相繼發行了另外 5 筆美元債券，從 5 年期 6 億美元的單系列發行，演變到 5 年至 30 年期共計 60 億美元的大規模交易。

圖 4：中資海外債券發行數量按發行規模的分佈（2009 年至 2019 年 7 月）

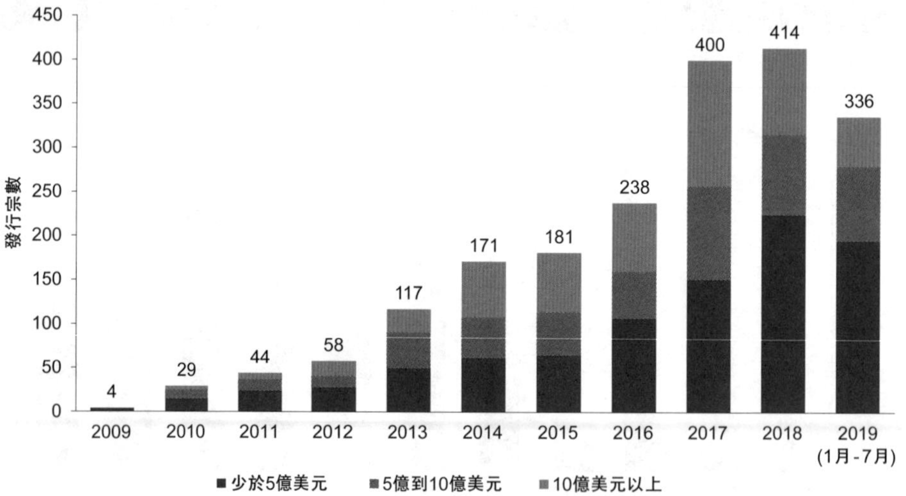

資料來源：Dealogic。

表 1：規模在 15 億美元或以上的中國企業海外債券發行（截至 2019 年 7 月）			
定價日	發行人	發行人評級 （穆迪/標普/惠譽）	交易總額（億美元）
20/11/2014	阿里巴巴	A1/A+/A+	80
29/11/2017	阿里巴巴	A1/A+/A+	70
03/04/2019	騰訊控股	A1/A+/A+	60
11/01/2018	騰訊控股	A2/A+/A+	50
22/04/2014	騰訊控股	A3/A-/--	25
29/04/2016	華為	--/--/--	20
03/02/2015	騰訊控股	A3/A-/--	20
25/01/2018	清華紫光	--/--/--	19
22/03/2018	百度	A3/--/A	15
28/06/2017	百度	A3/--/A	15
14/02/2017	華為	--/--/--	15
29/04/2014	聯想	--/--/--	15

資料來源：Dealogic。

(3)　發行年期組合更加多元化

　　起初中資海外債券發行的年期一般為較為單一的 5/10 年期固息品種，隨着發行人需求更加多元化，發行的年期也逐漸拓展為短至 365 天、長至 20/30/40 年浮固息結合或永續的多種選擇（見圖 5）。對於投資級發行人，多年期組合發行更逐漸成為其傾向的選擇。例如：阿里巴巴於 2014 年發行的 80 億美元債券中，其年期組合包括了 3 年浮息以及 3/5/7/10/20 年固息；在 2017 年發行時，發行年期更是拓展到長至 40 年；騰訊在 2019 年發行的 60 億元美元債券包含了 5 年浮息和 5/7/10/30 年固息的年息組合；聯想在 2017 年成功發行永續債券，並成功地在會計上計為權益。

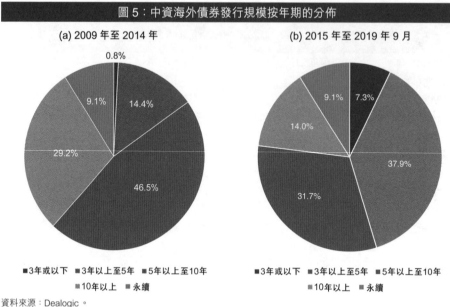

圖 5：中資海外債券發行規模按年期的分佈

(a) 2009 年至 2014 年　　　　(b) 2015 年至 2019 年 9 月

資料來源：Dealogic。

(4)　發行的品種結構也更加複雜

　　例如聯想集團（Lenovo）於 2017 年成功發行了總值 10 億美元的永續債券，成功地在會計上被確認為權益，穩定其負債水平；公司又於 2018 年完成了對 2019 年到期和 2022 年到期的總計 20 億美元債券的收購要約，並同時在其中期票據計劃下發行了 7.5 億美元新債券對收購要約進行再融資，有效地進行了負債管理，降低了短期內的再融資風險，有效延長了公司的債務到期組合，並利用當時有利的市場條件減少了公司的利息開支。

(5)　發行債券幣種更加多元化

　　發行的貨幣也從單一的美元發展到包含歐元、英鎊、離岸人民幣、港元、澳元、日圓、新加坡元等多種貨幣組合。雖然總體來看在歷年的發行中，美元債券發行仍居主導，每年平均佔比 87%（見圖 6），但非美元幣種例如歐元、離岸人民幣等，也在不同的市場條件下貢獻了顯著的發行量（見圖 7）。同時這些市場上較低的絕對收益率成本也為發行人提供了更多樣的融資渠道，特別是在特定市場上具有對應貨幣收入的發行人，可以形成天然的對沖。

圖 6：中資海外債券年均發行量按貨幣種類的分佈（2009 年至 2019 年 7 月）

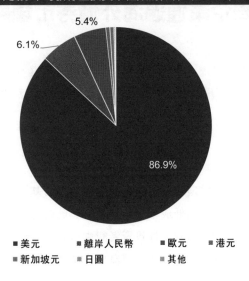

5.4%

6.1%

86.9%

- 美元
- 離岸人民幣
- 歐元
- 港元
- 新加坡元
- 日圓
- 其他

資料來源：Dealogic。

圖 7：中資海外債券中離岸人民幣債券及歐元債券發行量的變化（2009 年至 2019 年 7 月）

十億美元

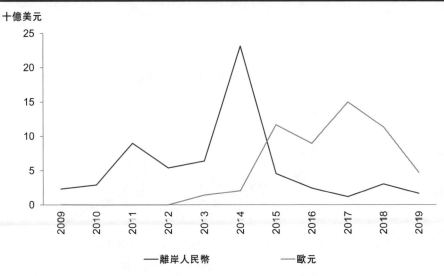

── 離岸人民幣　　── 歐元

資料來源：Dealogic。

2　新經濟企業通過海外債券市場融資的優勢

　　近十年的海外債券發展，對中資發行人拓寬融資渠道、降低再融資風險、優化資本結構以及進一步實現國際化均有顯著成效，具體表現為以下幾個方面。

(1)　發行成本低

　　從宏觀經濟的角度來看，2009 年全球金融危機成為了一個關鍵的轉折點。近十年來，在全球寬鬆的貨幣環境下，美元等國際貨幣的基準利率處於歷史低位，這也使企業能夠以前所未有的低成本發行債券。與此同時，全球高流動性的市場環境、對優質資產的追逐，以及境外投資者對中資企業越來越高的認可度，也推動中資企業海外債券的息差不斷收窄，這為中資企業海外債券發行提供了極為理想的市場條件（見圖 8）。

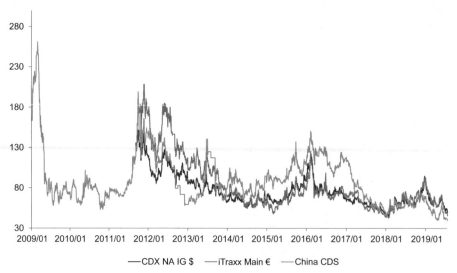

圖 8：全球信用息差指數的每日走勢（2009 年 1 月至 2019 年 7 月）

—— CDX NA IG $　　—— iTraxx Main €　　—— China CDS

資料來源：Dealogic。

(2)　發行規模大、年期長

相較境內債券市場（一般主要由銀行投資人主導），海外債券市場的投資者更加多元化，覆蓋全球性投資者類型更多，包括高質量的大型基金、對沖基金、銀行、保險公司、養老基金等，令資金池深度更大。這些全球性大型基金、保險公司和養老基金一般對較長期限債券發行更有興趣，因此發行人可以在海外債券市場上取得更大規模、更長期限的發行。例如阿里巴巴在 2017 年的發行中成功定價了 10 億美元的 40 年期債券，其中大多數投資者來自歐洲和美國，全球性基金和保險公司均有顯著貢獻量。

(3)　募集資金用途更加靈活

相對境內市場，海外債券發行對發行人的限制相對較低，無論是在發行募集資金的用途，還是發行窗口的選擇均給予企業更大的靈活性，可以更有效的滿足中資發行人在國際化進程中拓展海外業務的需求。以阿里巴巴為例，公司管理層在投資者會議中表示，境外美元債券融資所得將有助於其在實體零售及物流資產方面的境外收購，助力其在全球範圍的產業佈局。再如順豐 2018 年首次發行了美元債券，融資所得有效地支持了其在海外收購相關固定資產的需求。

(4)　發行執行流程靈活高效

海外債券發行的執行流程非常高效，一旦完成首次發行，建立起融資平台後，再次發行的流程非常簡單，時間也可以大大縮短。同時，海外債券發行的窗口也十分靈活，有利於公司開闢適應不同市場狀況的多樣化融資渠道，與國內市場融資形成良好補充，減少境內依賴，以緩解單一渠道所帶來的周轉風險。

(5)　海外投資者對新經濟企業認可度更高

海外債券市場投資者基礎非常廣泛且一般對高科技、互聯網、新經濟等行業的理解程度更高，認購興趣更濃。很多歐美投資者具有專門的團隊負責新經濟不同行業企業的研究，更加具有針對性。例如近期完成首次美元債券發行的微博，作為中國社交平台的獨角獸，其發行的稀缺性成功吸引了大量國際投資者的踴躍參與。

(6) 有效提升企業管理水平和透明度

在公開募集資金的過程中，通過發行文件的籌備、獲得國際評級、與國際投資者深入交流等，發行人在企業管理水平和業務透明度方面均得到了有效的提升——成功的發行即是對發行人在國際市場上的一次強而有力的推介，對進一步發展國際業務、海外投融資、建立可靠的國際形象均有重要幫助。而對於在市場上較活躍的多次發行人而言，每一次發行都是一次極好的機會，向國際投資者更新業務狀態及未來發展規劃，從而及時將新的信用亮點轉化為更低的融資成本。從阿里巴巴在 2017 年第二次發行可看出，公司通過和投資者的有效溝通，最終定價了 70 億美元的多年期債券，所有年期定價均低於二級市場交易水平（即全面實現負的「新發行溢價」），重新定義了阿里巴巴的信用曲線，反映出投資者對阿里巴巴信用資質強化的充分認可。

3 新經濟企業海外債券發行的具體案例 [1]

鑒於新經濟企業相對於國企一般具有更先進的技術、更高的業務競爭力、更好的財務表現以及透明度等特點，他們對境外資本市場上的投資者擁有更大的吸引力。特別是發達市場投資者，普遍對信息技術行業和高新科技產業擁有更深入的了解和更高的敏感度，投資興趣和接受能力更高。而在債券資本市場上，目前更多中資的發行仍集中在傳統行業，因此，新經濟企業對海外投資者具有相當的稀缺性價值。這些因素都給新經濟企業提供了極具吸引力的發行融資機會。

海外債券市場為優秀的新經濟行業發行人提供了難得的發展機遇，不僅有利於新經濟企業獲取更低成本、更大規模、更長期限、更多幣種的融資，而且可與境內市場融資形成良性補充，助力企業進一步開拓海外市場、實施全球戰略。市場業已見證如阿里巴巴、騰訊、百度、京東、微博等互聯網巨頭，和聯想、舜宇

---

1 本小節案例分析中的交易信息均為公開渠道可獲得的信息。

光學、順豐速運等新興行業領袖在海外成功發行債券的例子。以下我們特選取三例供讀者一覽。

3.1 阿里巴巴海外債券發行

作為中國信息技術和高新科技領域的先驅企業，2014 年 11 月 20 日，阿里巴巴集團成功完成了在國際債券市場上的首次發行，發行了 80 億美元 3/5/7/10/20 年 6 個年期組合的債券。這次發行創造了多項紀錄，引起了市場的廣泛關注 —— 是當時全球最大規模的首次債券發行，也是亞洲歷史上最大規模和亞洲歷史上最多期限組合的債券發行。此前，阿里巴巴剛剛在 2014 年 9 月 19 日成功於紐交所完成其首次股權公開債券，這也是當時紐交所上市的全球最大的 IPO 發行，引起了巨大的反響。憑藉全球投資者對其充分的認可，阿里巴巴隨後又成功取得了來自全球三大評級機構 (穆迪、標普和惠譽) 的 A1/A+/A+ 首次評級，與中國主權評級一致，這也是首次中國民營企業取得與國家主權同樣的評級，為債券發行做好了充分的準備。

在此基礎上，阿里巴巴啟動了其首次國際債券發行，由於剛剛完成 IPO 項目，因此債券文件準備的過程也大大地縮短，在不到 6 週的時間內即完成了項目的準備工作，隨後向市場宣佈交易。交易宣佈後，阿里巴巴選派兩支團隊，在全球五個金融中心 (香港、新加坡、倫敦、波士頓、紐約) 展開了為期三天的投資者推介，通過一對一會議和電話會議的形式與 300 多家投資者進行了溝通，對公司的經營業績、行業地位和信用故事進行了詳細的介紹，得到了投資者的積極回響。簿記建檔當日，認購倍數達到 6.9 倍，在當時是除了 Verizon 的 490 億美元債券發行之外最大規模的訂單需求，在當時全球美元債券發行中名列第二，在所有亞洲發行人中居首位。

交易也憑藉強勁的訂單勢能，最終取得了極低的發行成本，其中 5 年期和 10 年期的發行息差均為當時來自中國發行人的最低水平，直逼美國藍籌科技巨頭亞馬遜、思科和甲骨文公司債券的發行水平，顯示出全球投資者對阿里巴巴信用的強烈認可。更值得注意的是，該債券發行的最終配售中相當一部分給了美國投資者，顯示出美國投資者對阿里巴巴此類信息技術和高科技行業公司濃厚的投資興趣和高敏感度，也體現了海外投資者對優質中國新經濟企業的認可和支持。阿里巴巴成功地利用了國際債券市場有效地拓展了投資者基礎，在全球範圍內完成宣傳推介，進一步提升了企業形象，並以創紀錄的低成本取得了大規模長年期的融資，多元化了國際融資渠道，有效地優化了資本結構，為公司未來海外發展打下了牢固的基礎。

時隔三年之後，2017 年 11 月，隨着阿里巴巴的海外擴張以及國際業務的進一步深化，公司再一次憑藉極低的發行成本完成了 70 億美元的債券發行，這也是其 2014 年首發之後最大規模的亞洲企業發行。更值得注意的是，這次發行中公司進一步在年期上取得了突破，實現了 5.5/10/20/30/40 年期的組合，將發行年期進一步拉長，其中 40 年期發行為過去 20 年裏亞洲發行人中首次。交易更是改寫了阿里巴巴的信用曲線，所有的年期定價均取得了負的「新發行溢價」（即定價低於二級市場），債券定價比 2014 年阿里巴巴的首筆發行低至少 20 基點，使其與全球大型科技公司的信用息差進一步收窄。這筆交易突出地展現了國際債券市場的支持為阿里巴巴在降低融資成本和資本結構優化上帶來的巨大優勢。

3.2 舜宇光學海外債券發行

在香港上市的舜宇光學，作為科技製造業的翹楚，在 2018 年 1 月成功在海外債券市場上完成了 6 億美元 5 年期 S 條例下的首次債券發行，交易取得巨大的成功，實現了當時亞洲除日本外同區間評級（BBB）企業發行人首次發行的最低息差。舜宇光學憑藉其在高端光學器材製造行業的領先地位、極低的財務槓桿，以及盈利能力，成功取得了穆迪授予的 Baa2 的投資級評級，在同類企業裏表現出眾，不僅成功地為債券發行打下了扎實的基礎，更為其同行業的公司樹立了重要的標杆。與此同時，公司把握住了極為理想的市場窗口。2018 年初，公司在股票資本市場上獲得投資者大力追捧，股價在之前的 12 個月上漲近 170%，公司充分抓住了這一理想時機，進入債券市場進行發行。交易宣佈後，公司在香港、新加坡和倫敦等金融中心分別與超過 160 名投資者進行了會面；路演期間，多場都有 20 多家投資者參與（相對一般為 3 至 4 位投資者）。儘管定價時宏觀環境略有波動，但交易仍然得到了投資者的支持，簿記訂單峰值高達 91 億美元，實現 15.2 倍超額認購，是當時市場上亞洲發行人美元債券中最高的超額認購倍數之一，強烈的訂單簿使交易實現了發行價格較初始指引達 35 個基點的收窄。定價後二級市場交易於首天進一步收窄 10 多個基點，反映市場需求。首發美元債交易幫助公司實現了極低的融資成本，以及融資渠道的多元化，鞏固了公司在國際資本市場上的地位，優化了公司的資本結構。

3.3　聯想集團收購要約及新債券發行

作為香港的上市公司，聯想集團境外業務非常廣泛，收入和資產均以美元計價，因此，公司自然選擇通過美元債券市場進行融資，有效對沖匯率風險。2014年，聯想集團業績大幅上升，公司成功在海外市場完成了 15 億美元的 5 年期美元債券發行。儘管公司沒有採用國際信用評級，但仍然取得了投資者的充分認可，創造了當時亞洲單個年期僅 S 條例下高級債券發行的最大規模記錄，訂單總量超過 80 億美元，實現 5.3 倍超額認購，為公司的業務佈局和戰略發展提供了充足的資金儲備。此後，公司在業務上展開了對 Motorola 和 IBM x86 服務器的併購，但經過一系列整合，業績表現並未如預想中成功，償債壓力有一定上升。因此，聯想集團在 2018 年決定充分利用海外債券市場的靈活性，以及產品的多樣性，以主動債務管理的形式對即將在 2019 年到期的債券進行回購，並通過新發行 5 年期債券的形式進行再融資，調整公司資本結構，有效地延長了公司債務的到期年限，降低了短期內的再融資風險。這次收購要約非常成功，聯想最後收到了 10.2 億美元的投資者賣回指令，也成功完成了 7.5 億美元的 5 年期債券新發行，募集資金用於債券回購。這一成功的債務管理操作，協助了聯想在穩定總債務水平的前提下，有效延展了平均債券到期期限，並降低了聯想的再融資成本。

4　總結

綜上所述，對於新經濟企業而言，海外債券市場提供了極具吸引力的機會，諸多企業成功發行的經驗值得借鑒。海外債券發行不僅有利於新經濟企業獲取更大規模、更長期限、更多幣種的融資，而且可與境內市場融資形成良性補充，助力企業進一步開拓海外市場、實施全球戰略。海外投資者對中國信息技術和高科技行業的領先企業充滿興趣,但整體而言中國科技板塊債券發行的體量和頻率相對有限,當前稀缺價值顯著。我們深信伴隨中國新經濟行業的持續成長,海外債券市場將見證更多優質中資企業的活躍參與。當然企業應根據自身實際充分考慮潛在的風險和挑戰,並向有專業經驗的金融機構和法律顧問等進行諮詢、充分商討。

註:花旗環球金融亞洲有限公司資本市場部董事張瑾、資本市場部副總裁倪一卜、中國投資銀行部副總裁張希、資本市場部分析員吳天昊、資本市場部分析員祁顥清與中國投資銀行部分析員韓天玥也參與了本文的寫作。

第9章

新經濟公司的海外併購與融資安排

吳衛軍

德勤中國副主席兼金融服務業領導合夥人

許思濤

德勤中國首席經濟學家兼德勤研究負責人及合夥人

余雲

德勤中國風險諮詢兼司庫服務合夥人

周穎

德勤中國稅務服務兼全球基礎設施業務中心合夥人

摘 要

　　近年來，中國新經濟公司在全球經濟體系的數字化變革中表現亮眼，未來其全球化活動將越來越活躍。在推進全球化的各種方式中，海外併購扮演着重要的角色，可幫助新經濟公司快速開拓海外市場、吸收國外先進技術。為籌措足夠的併購資金，新經濟公司往往需要融資支持，但許多新經濟公司尚處於發展初期，難以獲得傳統的銀行信貸，因此對它們來說，海外融資是更方便的渠道。而香港作為國際金融中心，能利用其制度、渠道優勢為內地新經濟公司提供吸納海外資金的平台，包括吸引國際銀行支持，在香港上市或發債等，助力其全球化發展。

1 中國新經濟公司海外併購的趨勢和特點

1.1. 海外併購整體放緩

2008 年金融危機爆發，低估值國際資產成為中國企業海外「抄底」對象，中國企業海外併購進入快速增長階段。直至近年來，政策法律限制逐漸趨嚴，海外併購逐漸回歸理性。2016 年 11 月開始，中國政府陸續出台政策加強外匯管制、嚴控資本外流[1]，加強了對大型海外併購交易的外匯審查，導致 2017 年開始中國企業海外併購案例和交易金額大幅下滑，並且放緩趨勢仍然持續。根據商務部數據，2019 上半年中國企業的海外併購交易總額僅 169.5 億美元，與 2018 年上半年同期相比下降 72%。

1.2 新經濟公司海外併購依然活躍

儘管併購趨勢整體放緩，但「一帶一路」倡議仍為中國企業海外併購提供了新的機遇。特別是目前以互聯網、知識經濟、高新技術為代表，以滿足消費者需求為核心的新經濟形態迅速發展，相應的以「獨角獸」為代表的中國新經濟公司海外併購廣受關注。

新經濟企業的商業模式和產品以技術創新和應用為基礎，在發展早期競爭激烈，企業需要投入大量資金和資源佔領市場，因而通過併購獲得海外技術和市場成為理想的戰略佈局方式，實踐表明，大量新經濟企業在短短幾年內高速發展，成長為「獨角獸」。在 2018 年至 2019 年上半年，中國企業大型海外併購中，新經濟公司表現搶眼，如清華紫光集團以 26 億美元收購全球最大的智慧安全晶片組件廠商 Linxens SA，及社交媒體平台 YY 歡聚時代以 14.53 億美元全資收購新加坡視頻社交平台 BIGO。

1　2016 年 11 月，國家外匯管理局（簡稱「外管局」）要求資本項下 500 萬美元或以上資金的匯出須報外管局審批；2017 年 1 月，外管局發佈了《進一步推進外匯管理改革完善真實合規性審核的通知》；2017 年 7 月，商務部發佈《關於修改〈外商投資企業設立及變更備案管理暫行辦法〉的決定》。

1.3　新經濟公司海外併購的特點

1.3.1　併購主體：民營企業活躍、多元化股東合作形式增多

　　傳統併購主體國有企業在「一帶一路」建設上扮演着「領頭羊」和「主力軍」的角色，其優勢集中於能源、基建等大型傳統行業。而新經濟公司代表創新力和競爭力，其優勢表現為數字技術、通訊、媒體、互聯網、生物醫藥、智能製造和文旅等行業領域，民營企業和以風險投資基金與私募股權基金為代表的財務投資者，是新經濟公司海外併購的主要形態，如華為、阿里巴巴、復星等大型企業集團的併購案例均表現亮眼。而且，為了減少風險，主體併購企業通常會尋找其他合作夥伴成立合資公司實施併購，因此採取多元化的股東合作模式會越來越多。

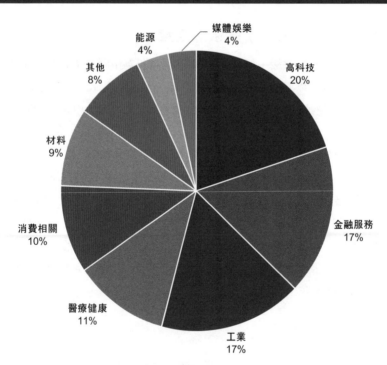

圖 1：進行海外併購的中國公司按行業的分佈（2018 年）

能源 4%
媒體娛樂 4%
其他 8%
高科技 20%
材料 9%
消費相關 10%
金融服務 17%
醫療健康 11%
工業 17%

資料來源：湯森路透、德勤研究。

1.3.2　目標行業：集中於高技術含量和高附加值的新興產業

在國內經濟轉型升級趨勢的帶動下，中國企業的海外併購更加趨向具有高技術含量和高附加值的新興產業，如互聯網、TMT[2]、新零售[3]、生物醫藥、智能製造等。湯森路透的數據顯示，2018 年，中國新經濟企業海外併購交易較多的行業為高科技類，佔比 20%，其次醫療健康及消費領域比例也較高。最近的代表收購包括：騰訊以 86 億美元收購芬蘭手機遊戲開發商 Supercell 公司 84.3% 的股份；上海萊士血液公司宣佈以 54 億美元收購全球血液檢測領域龍頭企業 GDS。

1.3.3　目標地區域：歐美優勢下降，亞太地區比重提高

一直以來，中國企業海外併購的首選區域為歐美，但自 2017 年開始，對歐美企業的併購數量大幅下降，亞太地區的併購佔據半壁江山[4]，東南亞企業的併購增長尤為明顯。這一主要併購區域的轉變主要源自兩方面的因素：一是歐美各國設置直接投資及併購的限制、貿易保護主義情緒升溫。如 2018 年 6 月，美國總統特朗普宣佈推進《外國投資風險評估現代化法案》，歐盟也於 2017 年 9 月對外公佈《歐盟外資審查框架法案》草案並在 2019 年 3 月正式通過，這些法案均嚴格限制高科技領域外來投資，使歐美不再是中國企業首選的投資目標國家。二是「一帶一路」倡議使中資企業加大了在東盟等地的佈局，例如近年中國電商的海外併購尤其青睞東南亞，阿里巴巴曾先後於 2017 年和 2018 年收購印尼網購公司 Tokopedia 及新加坡電商平台 Lazada。

1.3.4　併購目的：獲得關鍵能力和開拓海外市場

一方面，併購可直接獲得目標企業的關鍵能力與無形資產，如研發能力、商標、商譽、技術、管理、銷售渠道等，還可通過一定跨領域併購來實現企業業務組合的優化。新經濟公司發展依賴研發人才及高科技投入支撐，因此，新經濟公司常將擁有眾多專利或前沿性技術的海外公司作為兼併對象，以期獲得關鍵技術

2　TMT 為科技、媒體及通訊的簡稱。

3　新零售：即個人、企業以互聯網為依託，通過運用大數據、人工智能等技術手段，對商品的生產、流通與銷售過程進行升級改造，進而重塑業態結構與生態圈，對線上服務、線下體驗以及現代物流進行深度融合的零售新模式。

4　資料來源：德勤《開放新征程 德勤 2018 中國企業海外投資運營指南》。

方面的快速突破和研發能力的提升。如 2017 年 4 月，楚天科技併購了全球領先的固體製劑製藥裝備產品供應商 Romaco 集團，大大增加了楚天的產量及工廠規模。

另一方面，跨國併購並不需要很長建設期，成為想要加快市場擴張速度，在短時間內進入目標市場的跨國公司的首選方式。而且當國內市場接近飽和、缺乏增長空間時，新經濟公司可以通過併購來開拓國際市場。海外併購、跨國經營，可以幫助公司更快取得未開發的海外市場的優先權。如 2018 年 1 月，滴滴收購巴西本土最大的出行平台「99」，正是出於拓展巴西乃至拉美出行市場的考慮。

2 新經濟公司海外併購面臨的挑戰和問題

2.1 風險和監管是最主要的挑戰

大量實例證明，事前對風險準備不足、事中對風險不善應對、事後對風險不予總結改進是許多企業海外併購失敗的主要原因。當前「一帶一路」背景下，風險管理的思路更應得到全面更新，相關國家大多情況複雜，單純進行一時一地一方面的風險分析遠遠不夠，而是需要考慮多方面因素，包括政策變化風險、經濟動盪風險、法律差異風險、利率匯率風險、財務風險，以及文化差異風險等。

國內外監管要求和執法力度日益嚴格，特別是歐美對外國投資審核趨嚴，企業也越來越意識到事前了解監管環境、開展合規經營的重要性。

2.2 問題貫穿於併購全生命週期

受多方面因素影響，中國企業海外併購活動並不一帆風順。以海外併購的全生命週期來看，問題存在於併購前、併購中和併購後整合三個階段。

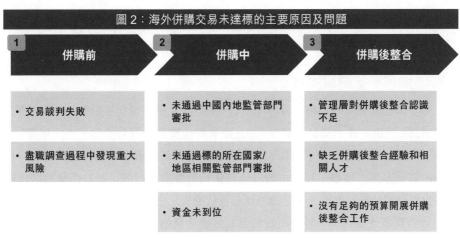

資料來源：德勤 2017 年 8 月發佈的《中國企業海外併購及整合後現狀調查》、德勤研究。

2.2.1　併購全週期中的問題

德勤的一項調查結果[5]顯示：在併購前，交易談判失敗和盡職調查過程中發現重大風險從而終止交易是導致交易未能完成的兩大主要原因，在未完成的併購案例中，這兩項因素分別佔 24% 和 22%。中國投資者往往不擅於合理規劃項目進度，或對賣方所需材料的具體要求等理解不夠，可能因程序性原因而非報價因素導致項目流產。

在併購中，未能通過國內外監管部門審批和資金不到位是主要的交易終止因素。首先，中國企業對外投資須獲得國家商務部、國家發展和改革委員會（簡稱「發改委」）、外匯管理局等政府部門批准，較長審批週期和嚴格的審批流程可能導致錯過最佳併購時機；其次，出於增加對本國技術控制權的考慮，併購標的所在國也會對跨國併購進行嚴格審查，特別是注重知識產權保護的高科技企業海外併購所在國家／地區的相關監管；另一個難題是新經濟公司併購資金不到位或合作夥伴中途退出導致交易流產。

在併購後的整合過程中，許多整合未取得理想效果，影響了預期交易價值的實現。原因來自多方面：一是管理層認識不足，認為沒有必要進行併購後整合；

5　德勤《併購活躍，整合滯後——中國企業海外併購及併購後整合現狀調查》，載於德勤網站之文章頁。

二是缺乏併購後整合經驗和相關人才；三是部分投資類機構缺乏足夠預算開展相關工作。

2.2.2　亟需解決融資困境

以上問題部分可借助第三方專業服務機構（涉及服務如交易談判、盡職調查、監管審批）解決，或企業自身發起（如注重併購後整合、戰略優化、人才培育）。然而，資金困境是更為棘手的問題，原因在於許多新經濟公司處在發展初期，可抵押的資產不足、經營不穩定，難以達到銀行信貸准入門檻，而且內地的多層次資本市場仍處於發展階段，企業得不到相應的融資、貸款支持。

面對這類資金困境，香港可以成為新經濟公司的海外併購融資平台，為新經濟公司破解資金難題。隨着香港與內地互聯互通機制的搭建，內地新經濟公司可以更加方便地通過香港這一平台尋求海外融資，而香港可以利用其制度、渠道、人才優勢服務於新經濟公司的海外併購。香港的相關服務提供者會為企業搭建融資平台，幫助企業設計多維度融資渠道，助力其對接相關金融以及資產管理機構，解決境外運營資金需求，降低融資成本。

3　香港作為海外併購平台的優勢和活用方式分析

香港多年位列全球金融中心指數（Global Financial Centres Index）[6] 第三位，是亞洲地區最大的國際金融中心，也是全球最大的離岸人民幣市場，自然成為了內地多種經濟類型的公司積極應對海外併購風險和挑戰的平台。香港不僅與內地聯繫緊密，在跨境融資和制度環境等方面也具有明顯優勢。

6　「全球金融中心指數」是全球最具權威的國際金融中心排名指數，由英國智庫 Z/Yen 集團和中國（深圳）綜合開發研究院共同編制。

3.1　與內地的緊密聯繫

香港與深圳市毗鄰，在粵港澳大灣區中處於相對核心的位置。對於計劃開拓海外市場的內地企業來說，香港的區位具有排他性，而香港在語言和文化方面的優勢則不言而喻。

中國改革開放以後，香港與內地的經濟交流日益增多，兩地貿易總額在1979-1997年間擴大了65倍[7]。在頻繁的貿易往來中，香港作為全球自由貿易港口的優勢逐漸被內地企業挖掘。回歸以後，兩地在《內地與香港關於建立更緊密經貿關係的安排》(CEPA) 框架下展開了更緊密的經濟合作。香港國際金融中心優勢得到突顯，更多的內地企業開始赴港融資，在香港上市的 H 股數量從1997年1月的26家增長到了2018年年底的267家[8]。

近年來，香港與內地逐漸通過互聯互通機制發展成為一個共同資本市場。「滬港通」和「深港通」先後於2014年和2016年啟動，雙向股票成交金額快速增長。2017年上線的「債券通」也在兩年內匯集了全球上千家境外機構投資者[9]。

互聯互通模式進一步加強了香港與內地的聯繫，為內地投資者開啟了更多的全球投資機會，在一定程度上促進了內地企業的海外投資和併購。

3.2　香港的融資優勢

基於香港與內地的緊密聯繫，香港在內地新經濟企業進行海外併購的過程中，具有很多明顯的優勢，其中最重要的是能夠為內地企業提供充足多元、低成本以及便利化的融資。

我們在實際業務過程中觀察到，內地企業在進行海外併購時普遍傾向於銀行貸款的融資方式，希望能夠得到國際大型商業銀行的貸款支持，以便按時完成併購業務交割。然而，根據中國銀行保險監督管理委員會 (簡稱「銀保監會」) 的不完全統計，在內地銀行業貸款餘額中，民營企業只佔25%[10]。大多數為民營的新經濟

7　資料來源：香港政府統計處。

8　資料來源：中國證券監督管理委員會。

9　資料來源：新華社 (http://hm.people.com.cn/n1/2019/0704/c42272-31213073.html)。

10　資料來源：銀保監會主席郭樹清答記者問實錄，載於銀保監會網站 (http://www.cbirc.gov.cn/chinese/home/docView/7F7EDCDDD5A04396A00E8E23F8E2E813.html)。

企業，其順利通過貸款融資的難度可想而知。香港的銀行業能夠更好地滿足新經濟公司的併購需求，截至 2019 年 7 月，香港共有接受認可的銀行機構 194 家，其中有全球 100 大銀行中的 70 家在香港設立的分支機構[11]。內地企業在香港能夠獲得的貸款中，內地民營企業所佔的比例已經超過 40%（見圖 3）。

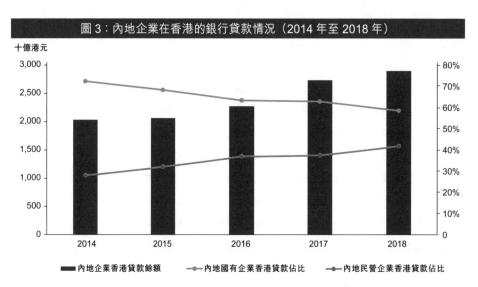

圖 3：內地企業在香港的銀行貸款情況（2014 年至 2018 年）

十億港元

■ 內地企業香港貸款餘額　　—●— 內地國有企業香港貸款佔比　　—●— 內地民營企業香港貸款佔比

資料來源：香港金融管理局《貨幣與金融穩定情況半年度報告》（2014 至 2019 年度）。

　　除了銀行貸款，香港在股權融資和債券融資方面同樣具備優勢。根據 Refinitiv 的統計數據，香港交易及結算所（香港交易所）因改革上市制度，於 2018 年創下了八年來最高的 IPO 籌資金額。該年的新上市公司有 208 家，IPO 集資總額約 2,866 億港元，其中 95% 由內地企業貢獻[12]。在實施新的上市制度後，包括內地小米、美團等獨角獸企業在內的 28 家新經濟公司赴港上市[13]。在債券發行方面，香港也提供了更加充足的資金量。根據世界交易所聯會（World Federation of Exchanges）的統計，2018 年香港交易所債券市場的募資金額遠超深圳證券交易所（見圖 4）。

11　資料來源：香港金融管理局（https://www.hkma.gov.hk/gb_chi/key-functions/banking-stability/banking-policy-and-supervision/three-tier-banking-system.shtml）。

12　資料來源：《德勤中國內地及香港 IPO 市場 2018 年回顧與 2019 年前景展望報告》。

13　資料來源：https://tech.sina.com.cn/roll/2019-01-05/doc-ihqfskcn4174725.shtml。

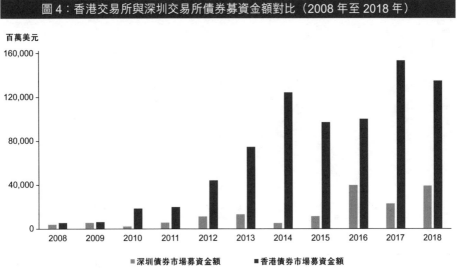

圖 4：香港交易所與深圳交易所債券募資金額對比（2008 年至 2018 年）

百萬美元

■深圳債券市場募資金額　　■香港債券市場募資金額

資料來源：世界交易所聯會、Wind。

　　此外，香港的聯繫匯率制度，決定了香港具備更低的融資成本和風險。香港的聯繫匯率制度通過香港貨幣發行局機制來運行。港元匯率通過自動利率調節機制穩定在 7.8 港幣兌換 1 美元，即香港的利率是完全市場化的。聯繫匯率制度維持了香港金融市場的穩定，完全市場化的利率則讓企業享受到了低廉的融資成本（見圖 5）。

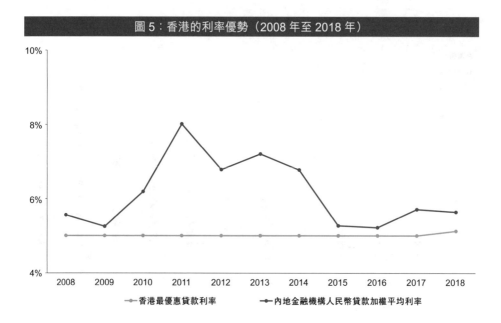

圖 5：香港的利率優勢（2008 年至 2018 年）

── 香港最優惠貸款利率　　　── 內地金融機構人民幣貸款加權平均利率

註：內地機構利率為相應年度第四季度的取值。

資料來源：香港金融管理局、中國人民銀行、Wind。

3.3　香港的制度環境優勢

　　內地對於商業銀行併購貸款以及外匯資金流動都有嚴格的監管限制，這突顯了香港的制度環境優勢。2015 年 2 月，中國銀行監督管理委員會（現為銀保監會）修訂了《商業銀行併購貸款風險管理指引》，表示將通過嚴格的統計、匯總和分析制度對併購貸款進行嚴格的監管。2016 年後，國家外匯管理局等多個部門聯合對內地企業境外投資實行限制。香港則能夠在這些方面為內地企業提供更大的靈活性和便利。

　　此外，在「一國兩制」基本國策的保障之下，香港在回歸之後依然保持了原有的普通法法律體系。在跨境併購的實務操作過程中，大多數交易文件都採用了普通法的框架。而一旦發生法律爭端，在香港做出的判決在全球其他司法轄區內具有效力。此外，香港擁有大量的法律、會計、投資銀行以及諮詢等專業服務機構和相應的高端人才，這提高了香港從事跨國併購的支持能力。

最後，香港的金融基礎設施建設也為內地企業的融資提供了極大的便利性。香港金融管理局在支付系統（如即時支付結算系統（RTGS））、債券交收系統（如債券工具中央結算系統（CMU））以及聯網系統方面進行了大量的投資，使得香港的金融基建日益完善。

事實上，隨着人民幣國際化的推進和中國企業「走出去」的增加，香港也面臨着亞洲其他金融市場的挑戰。經濟自由度高的新加坡在外匯、期貨以及其他金融衍生產品的交易上具有很高的深度和廣度，財政與外匯儲備管理亦穩健，逐漸受到內地投資者的青睞。位於中東地區的阿布達比也正在積極構建高科技數字化的金融產業鏈條，已經與發改委等多家內地機構展開合作，共建內地企業「一帶一路」區域的投融資中心。但短期內，香港的優勢更明顯。

3.4　跨境併購過程中活用香港平台的介紹

大多數內地企業都希望在海外併購的過程中，充分利用香港的各種優勢。常見的一種方式是內地企業在香港註冊一家平台公司或是通過在港上市的子公司，使用香港靈活的金融工具和資本市場完成融資活動。但在最近的實踐中，為了保持融資的效率，越來越多的公司選擇在香港成立財資中心，充分利用香港的資金優勢。

3.4.1　利用香港財資中心助力全球資金配置整合

以一家大型綜合交易金融服務集團為例，該企業的業務在全球範圍內分佈廣泛，覆蓋福費廷（forfaiting，出口信貸的一種，即應收賬款買斷業務）、國際保理、資產管理、場外交易等業務。隨着其業務在全球多個金融中心的擴張，若以境內資金直投的方式則難以為繼，且境內外資金分散管理極大程度上降低了資金使用效率，故亟需設置離岸財資中心運營其經營性資金和交易性資金，通過資金錯配管理、頭寸管理、主動外幣管理等方式實現有效資金運營及交易。

3.4.2　財資中心方案與效益

香港的專業機構團隊（以下簡稱「專業團隊」）幫助企業搭建香港財資中心，

設立投融資平台、離岸資金集中平台及離岸現金池、資金風險智慧平台，助力企業完成全球資金和資源配置的整合和優化。

- **海外財資中心選址**：根據企業的現狀與需求，專業團隊與企業溝通後決定建立海外財資中心，規劃其投資、融資、資金集中管理、資金風險管控等職能。同時對香港和新加坡等金融市場在資本市場活躍度、金融監管、離岸人民幣結算等作對比分析，最終選址在香港。
- **設計投資架構，優化海外再投資佈局**：專業團隊協助企業以香港財資中心作為境外業務控股公司，重新進行境外公司股權調整，優化投資與紅利分配；同時，利用香港相對寬鬆的政策監管環境，優化投資組合。
- **設計融資方案並輔助實踐**：以香港財資中心作為融資平台，專業團隊幫助企業設計多維度融資渠道、對接相關金融以及資產管理機構、解決境外運營資金需求，降低融資成本。
- **構建資金池實現資金實物和虛擬集中管理**：利用香港相對寬鬆的匯率政策和貨幣政策，選擇合作銀行建立跨境多幣種現金池，最大限度實現全球範圍內的資金集中。

4 結論

在全球地緣政治不確定性增強、貿易保護主義抬頭以及各國監管和法律趨向嚴格的情況下，內地企業的海外併購會受到影響。然而，新經濟公司的海外併購將保持活躍這一看法得到若干樂觀積極因素的支持。

首先，全球經濟體系正在進行數字化革新，數字經濟領域將是新一輪產業變革的先驅者。大多數中國的新經濟類型公司已在該領域處於全球領先的位置。為了保持自身的優勢，他們將通過直接投資或跨國併購積極地向海外市場拓展業務。

其次，中國內地正在積極通過供給側結構性改革、消費市場升級、營商環境改善和更高水平的開放來實現「高品質發展」。這些措施保證了中國經濟的活力，並將對新經濟公司的發展起到積極作用。

另外，中美貿易摩擦懸而未決，使得向北美地區尋求技術領域併購的新經濟企業的熱情可能會受到打擊，但在「一帶一路」倡議下，未來在歐洲市場、以色列、印度以及東南亞地區的併購活動會逐漸活躍起來。

香港應該充分地抓住這一機遇，發揮好自己「亞洲最大國際金融中心」的角色，與內地實現經濟雙贏的發展。這需要香港自身進行積極的改革，充分利用自己在人民幣離岸市場的優勢，抓住大灣區發展的機遇，完善與內地的互聯互通機制，做大做強香港的「國際金融中心」優勢，避免被邊緣化的風險。

註：感謝德勤研究總監陳嵐、德勤中國風險諮詢兼司庫服務總監黃強發、德勤公共政策研究經理趙昱和德勤金融行業研究經理周菲的參與。

互聯互通成為海內外資產跨境配置的新平台

第10章

MSCI 納入中國 A 股的歷程和影響

魏震

MSCI

中國研究主管

摘 要

自 2018 年 MSCI ACWI 全球指數和新興市場指數開始部分納入中國 A 股，隨後在 2019 年 A 股權重得到進一步增加，這大大促進了全球投資者重新審視股票投資組合中的 A 股配置。

2019 年 2 月，MSCI 宣佈將在 2019 年底前，將中國 A 股在其旗艦指數中的納入因子從 5%上調到 20%，這說明 A 股市場准入水平的提升得到國際投資者的認可。截至 2019 年 5 月 30 日，按自由流通調整市值計算，中國股票飯佔 MSCI 新興市場指數的權重是 31%。隨着 A 股權重繼續提升，長期投資者會更加重視在其投資政策中與中國配置相關的決策。

1　聚焦中國：納入 MSCI 的歷程

中國的股票市場是伴隨着改革開放的春風而孕育發展起來的。從 1990 年初創建兩地交易所和證監會開始，經過二十多年對市場制度的發展，當前滬深兩地 A 股的總市值已經躍居全世界第二位。

儘管成績斐然，但 A 股市場從成立至今卻經常為人所詬病。單從投資者的角度，有一些因素嚴重制約中國股市進一步健康發展並且不利於市場積極發揮作用。例如，A 股市場長期以個人投資者為主、整體投資理念不成熟、投機氛圍較濃等。自 2015 年匯改以來，人民幣從近年單邊升值轉向雙向波動所帶來的資本外流壓力至今仍然存在。

國際機構投資者從真正意義上開始關注和投資中國境內股市是從 2002 年引入合格境外機構投資者（QFII）制度之後才發生的。金融市場的開放在期初相對較慢。2013 年，當 MSCI 初次考慮將中國 A 股加入其主流指數的時候，全球投資者反應並不熱烈。儘管 MSCI 中國指數（包括除 A 股外的中國股票和海外中概股）已經是 MSCI 新興市場指數中最大市值的國家板塊，但熟悉中國國內股票市場的投資者並不多。總體上，由於投資渠道不足，全球投資者對中國國內市場並不熟悉。

中國已經小心地對全球機構投資者打開了國內 A 股市場的大門：QFII 和人民幣合格境外機構投資者（RQFII）[1]。雖然這些制度吸引了一些前期資金，但仍不能滿足機構投資者進入市場的最低要求。

因此，將中國納入指數的過程遠非一帆風順。在 2014 年至 2016 年的 MSCI 的數次市場諮詢中，投資者提出了若干關於市場可投資程度的問題。除了合格投資者制度下的資本流動限制和不平等的市場限制外，投資者還存在其他一些疑慮，例如不確定的資本利得稅（於 2014 年諮詢中提出），滬深港通制度[2] 早期的受益人歸屬問題（於 2015 年諮詢中提出），大範圍的股票主動停牌（於 2016 年諮詢中提出），以及發行金融產品的預先審批限制（於 2016 年諮詢中提出）。儘管存在很多

1　QFII 制度發佈於 2002 年，允許外國投資者直接投資於中國資本市場。RQFII 制度發佈於 2011 年，允許將在大陸之外的合規地點募集的人民幣投資於國內證券市場。

2　滬深港通制度允許國際和中國大陸投資者在本地的交易所買賣對方市場的股票。

困難，但中國政策制定者和監管部門在解決投資者的擔憂方面取得了進展。最終，當滬深港通制度在 2016 年 12 月擴大至深圳交易的股票時，全球投資者希望的突破 —— 向所有投資者開放的市場出現了。

自從 2017 年 6 月 MSCI 宣佈將中國 A 股部分納入其一系列重要指數，我們發現，由於市場投資渠道改進，全球投資者投資中國股票的興趣也逐步高漲：

- 特別獨立賬戶（Special Segregated Accounts，簡稱 SPSA[3]）的開立可以用來推測國際投資者的準備和參與 A 股交易，該類型賬戶已經從 2016 年 6 月的少於 1,700 個上升至 2019 年 5 月的超過 8,000 個。

- 2019 年 2 月，通過滬深港通下北向通完成的組合交易清算金額達到人民幣 9,100 億（約 1,320 億美元）[4]。

- 滬深股通投資者持有包括已經納入 MSCI 新興市場指數的超過 1,800 隻中國 A 股股票。滬深股通正在快速成為投資 A 股的主要渠道。

中國監管層為減少交易限制做出多方面努力。例如，2018 年，中國人民銀行（PBoC）將股票北向通的日度限額擴大到 4 倍至 520 億人民幣（約 83 億美元），香港金融管理局（HKMA）推出了一系列措施來保證離岸人民幣市場（CNH）的充足流動性。

3　SPSA 的定義請見香港交易所網站（https://www.hkex.com.hk/Services/Settlement-and-Depository/Special-Segregated-Account-Services?sc_lang=en）；有關數據亦見香港交易所網站。

4　資料來源：香港交易所網站。

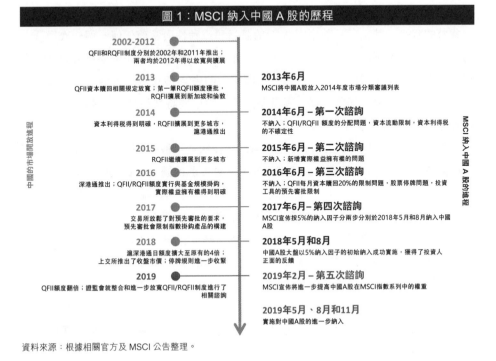

圖 1：MSCI 納入中國 A 股的歷程

中國的市場開放進程

2002-2012
QFII和RQFII制度分別於2002年和2011年推出；兩者均於2012年得以放寬與擴展

2013
QFII資本贖回相關規定放寬；第一筆RQFII額度獲批，RQFII擴展到新加坡和倫敦

2014
資本利得稅得到明確，RQFII擴展到更多城市，滬港通推出

2015
RQFII繼續擴展到更多城市

2016
深港通推出；QFII/RQFII額度實行與基金規模掛鈎，實際權益擁有權得到明確

2017
交易所放鬆了對預先審批的要求，預先審批僅限指數掛鈎產品的構建

2018
滬港通日額度擴大至原有的4倍；上交所推出了收盤市價；停牌規則進一步收緊

2019
QFII額度翻倍；證監會就整合和進一步放寬QFII/RQFII制度進行了相關諮詢

MSCI 納入中國 A 股的進程

2013年6月
MSCI將中國A股放入2014年度市場分類審議列表

2014年6月 – 第一次諮詢
不納入；QFII/RQFII 額度的分配問題，資本流動限制，資本利得稅的不確定性

2015年6月 – 第二次諮詢
不納入；新增實際權益擁有權的問題

2016年6月 – 第三次諮詢
不納入；QFII每月資本贖回20%的限制問題，股票停牌問題，投資工具的預先審批限制

2017年6月 – 第四次諮詢
MSCI宣佈按5%的納入因子分兩步分別於2018年5月和8月納入中國A股

2018年5月和8月
中國A股大盤以5%納入因子的初始納入成功實施，獲得了投資人正面的反饋

2019年2月 – 第五次諮詢
MSCI宣佈將進一步提高中國A股在MSCI指數系列中的權重

2019年5月、8月和11月
實施對中國A股的進一步納入

資料來源：根據相關官方及 MSCI 公告整理。

2 對中國金融市場和資產管理行業的潛在影響

　　歷史上，國際投資者在某個市場被納入 MSCI 新興市場指數之後，會增持該市場的股票[5]。

　　在考慮 A 股獲部分納入 MSCI 新興市場指數對中國金融市場可能產生的各種影響的時候，我們還應考慮到一些特殊情況。其一，人民幣已經加入了國際貨幣

5　Michelle Liu, "Bringing Their 'A' Game"，Lazard 資產管理公司，2019 年 2 月。

基金組織（IMF）特別提款權（SDR）的貨幣籃子，其國際使用率正逐步接近發達國家貨幣的水平；其二，目前中國 A 股的體量和流動性與當前任何一個初次獲納入的新興市場都不可同日而語；其三，當前外資對 A 股的持股比例要遠低於歷史上初始納入其他國家時的比例；其四，中國國內除去 A 股市場之外的可投資機會非常多，如債券市場、期貨市場、私募基金等。從這些特殊性來看，中國 A 股市場國際化道路是獨樹一幟的。

中國 A 股的總市值已然位居世界第二，僅次於美股（根據 MSCI 股票數據庫計算得到）。若以可作投資的流通市值計，中國 A 股的體量事實上還落後於日本、英國這些發達國家。隨着中國股市改革的深化，股市自由流通率的提升有助於提高機構投資者參與率，並且為資產管理機構的發展壯大開拓空間。

從屬於國際資產管理機構的現有新興市場基金管理人已經諳熟國際化的投資流程。他們的優勢是對國際資產擁有者投資流程的深入理解和參與，以及對標準化組合管理與風險評估工具的使用經驗。他們的投資理念也深受國際資產擁有者的影響，並且在組合管理過程中通常能及時準確地反映資產委託方的投資價值觀。例如，這種價值觀的體現之一可能是對所投資企業在環境影響、社會責任或者公司治理（ESG）方面的特殊要求。

然而，具備多年投資 A 股經驗並持良好紀錄的新興市場基金管理人並不多見。這種 A 股組合管理經驗的「稀缺性」給中國本地的基金管理人創造了良機。中國資產管理人的優勢是對中國經濟和行業周期的深入了解，對內地政策、行業動態和公司信息的近距離接觸，以及多年管理 A 股組合的實戰經驗。

3　對投資過程的潛在影響

對於使用 MSCI 新興市場指數作為策略基準的國際機構投資者，如何應對 A 股的納入取決於該投資者是否已經有 A 股的敞口。對於尚沒有 A 股敞口的全球機構投資者，他們可以考慮對於新興市場及中國的策略配置。這種重新評估或許可以更精確地反映他們的長期風險和收益偏好。

　　已經在基準外配置了中國 A 股的全球機構投資者，則會面對這部分持倉與基準偏離的問題。此外，他們可能會考慮將原來的 A 股專門配置，保留或者劃轉給常規的新興市場或全球市場的投資經理。在這個過程中，中國 A 股的專業投資經理可能具有優勢，一方面他們具有更長的業績記錄，另一方面新興市場和全球投資經理可能並未對管理中國 A 股做好準備。

　　而對於 A 股專業投資經理的挑戰，是證明他們的投資過程與投資人的信條是一致的，同時還要證明業績的穩定性和持續性。雖然常規的投資經理可能在投資中國 A 股方面經驗不足，但他們可以更好的追隨資產所有人的信條和基準，穩定地在長時間內創造超額收益。

　　無論是否擁有 A 股的敞口，全球機構投資者可能希望審視他們目前新興市場和全球股票經理是否已經準備好應對將 A 股納入到他們的基準之中。這可以幫助了解經理們是否對一些具體問題具有清晰的認識，如不同投資路徑的優缺點、風險和收益的來源（包括區別板塊組合）、宏觀影響因素、A 股和 H 股的折溢價、風格因子和其他中國 A 股的特殊驅動因素。

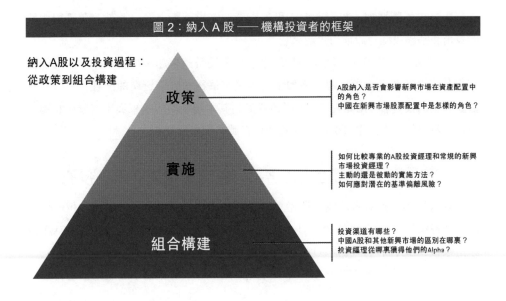

圖 2：納入 A 股 —— 機構投資者的框架

納入A股以及投資過程：
從政策到組合構建

政策

A股納入是否會影響新興市場在資產配置中的角色？
中國在新興市場股票配置中是怎樣的角色？

實施

如何比較專業的A股投資經理和常規的新興市場投資經理？
主動的還是被動的實施方法？
如何應對潛在的基準偏離風險？

組合構建

投資渠道有哪些？
中國A股和其他新興市場的區別在哪裏？
投資經理從哪裏獲得他們的Alpha？

4 權衡中國在股票組合中的配置

投資者對股票資產進行戰略配置的原因各不相同。許多投資者希望獲得股票投資組合現金流隨着績效增長所帶來的長期溢價；有些是希望用股票的回報來彌補未來負債；還有一些則是希望滿足預定的儲備貨幣組合預期，或是提高資產組合整體的流動性和透明度，又或是為其他資產類別分散風險等等。

回報導向型的投資者通常會關注股票所在市場的經濟規模變化。根據 IMF 的最新預測，到 2024 年，中國將佔全球本地生產總值（GDP）的 19%，很快就會接近美國對全球 GDP 22% 的貢獻[6]。

按購買力平價，中國的經濟規模早在 2014 年就已經超越美國，並且預計在 2024 年將達到美國的 1.5 倍[7]。世界經濟格局的這種巨大轉變，對那些期望從相關的經濟成長中獲取投資收益的投資者對股票資產的配置方式，將會產生重大影響。

中國經濟增長的另一個直接影響體現在中國與世界上許多國家的跨境貿易和投資聯繫正不斷增加。投資者對人民幣資產的需求不斷上升。此外，各國對中國的技術和基礎設施建設能力也越來越認可。

舉例來說，繼美元、歐元和日圓之後，人民幣越來越多地被用作新的儲備貨幣，各國央行和貨幣當局都已增加了對人民幣債務和股權資產的配置。有一些負債驅動型的投資者（比如養老金）開始意識到其本國經濟的未來與中國緊密相連，並且重新評估對人民幣計價的金融資產進行戰略配置的意義。

4.1 全球和新興市場投資者是否已對上述變化做出反應？

數據顯示，投資者才剛剛開始重新考慮中國在其股票配置的戰略地位。這樣的反應特別是發生在 2017 年 6 月，當 MSCI 宣佈將中國 A 股納入 MSCI ACWI 全球指數和 MSCI 新興市場指數以及其他相關指數之後。

而且，相對於 MSCI ACWI 全球指數而言，全球基金在中國股票上的持倉依

6　國際貨幣基金組織（IMF）《世界經濟展望報告》，2019 年 4 月。
7　同上。

然偏低。相比之下，由於中國自 2007 年以來一直是 MSCI 新興市場指數中權重最大的板塊，中國在新興市場策略中的地位會高出許多。2018 年，新興市場共同基金投資組合的四分之一配置在中國股票市場，但這仍少於中資股在 MSCI 新興市場指數中的權重[8]。

無論是全球型還是新興市場型基金，其投資組合中中國股票的權重都偏低。這種現狀可能是基於投資者對中國的各種擔憂，例如中國的宏觀經濟風險、對中國市場缺乏認知，以及在此市場可用的風險管理工具不足等。因此，隨着中國資本市場和金融基礎設施的發展，全球基金和新興市場基金投資者都可能會重新權衡其在資產組合中對中國股票的配置。

4.2　如何「切割」全球股票市場？

許多機構投資者都習慣於用市值加權指數來指導資產配置決策。市值加權指數旨在反映市場中可供投資標的的總體表現，是資產配置決策提供中性出發點。

儘管中國目前已佔全球經濟總量以及全球企業盈利產生地的 16%，但中國市場僅佔全球市值加權指數的 4%[9]。

這種權重差異主要反映了兩個大方面問題：其一、欠發達資本市場上的上市公司的自由流通股份通常較少；其二、中國 A 股尚未全部被納入全球指數中。

不過，中國目前在 MSCI 新興市場指數中的權重並沒有明顯低於其經濟總量或者作為企業盈利來源地的佔比。

4.3　投資信念

在典型的資產配置框架下，投資者在投資於特定資產或資產類別時通常會考慮兩個目標：其一是提高投資組合的預期回報，另一方面是提升投資組合的多樣化。換句話說，投資者的選擇基本上由其回報目標和風險承受能力驅動。不過實際情況下，償還負債、准入約束和監管限制等因素也可能影響其決策。

沒有人能夠預測未來，因此提高預期回報可能需要堅信當前投資機遇的前景。

8　根據康橋匯世（Cambridge Associates）的數據，主動型新興市場機構基金對中國的配置較新興市場共同基金稍高，但仍低於中國在 MSCI 新興市場指數的權重。

9　數據截至 2019 年 2 月 28 日，根據 MSCI ACWI 指數和 MSCI 證券經濟暴露度數據庫計算得到。

然而，機構投資者長期以來一直認為新興市場股票是其全球股票投資組合中的重要成分，而中國目前已經成為了新興市場中最大的組成部分。

儘管中國在 MSCI 新興市場指數中的權重已經很大，但對於一些尋求在早期加大中國 A 股敞口的投資者而言，他們認為雖然中國國內股市目前相對受限，但有朝一日終將會對國際投資者敞開大門，屆時中國 A 股也將成為他們資產組合的重要組成部分。此前，這些投資者通過 QFII 或 RQFII 制度對中國進行了「基準之外」的專業化投資。

假設所有其他因素保持不變，就算中國 A 股全部納入 MSCI 指數，這些投資者的中國配置思路也不一定會改變。也就是說，這些投資者可能繼續相對於其基準指數高配中國，並採取適合他們的策略。

另一方面，基準敏感型投資者可能會對「基準之外」的任何投資採取更保守的方法。這些投資者也許會繼續謹慎地接觸新的機會，同時加強對相關市場的了解。

5 政策型基準指數設計框架

考慮在股票資產中進行中國配置時，國際投資者通常要麼是選擇對中國採取專業化配置策略，要麼就會跟隨基準指數（如 MSCI 新興市場指數）中的中國權重來進行配置。

看好中國長期前景的投資者通常會選擇設立專門的中國配置策略，從而彌補中國在其基準中權重較低的影響。而對標廣泛基準的投資者則更傾向於跟蹤相關指數的變化。這些不同的方法反映了投資者在投資中國股票時的不同經驗水平和市場看法。隨着 A 股進一步獲納入基準指數，投資者的方法也可能會發生變化。

對中國配置的考量分很多層次，包括政策型基準考量、戰略資產配置考量，或是投資組合實施層面的考量。由於政策型基準選擇對整體投資過程的影響最大，我們着重對這個問題進行以下的具體討論。事實上，大多數投資者在資產配置和投資組合實施方面通常比較靈活。

有些投資者認為中國在全球股票市場中的規模和重要性都會不斷增加，不過也有些投資者持有不同的觀點。因此他們在進行全球股票配置時對中國採取的投資策略不盡相同。

對中國持中性或相對積極看法，但並不急於採取行動的投資者，可以選擇沿用目前的廣泛基準，其對中國股票的配置敞口隨着基準指數對納入 A 股的實施相應增加。理論上，因為投資者希望追蹤基準（對中國的配置相對於其基準指數保持中性或接近中性），只要基準指數中的 A 股權重增加，這些投資者對中國的配置也會相應增加。

對中國經濟增長的長期前景持樂觀態度，或看好中國股市的投資者，則可能希望迅速提高其中國敞口。這意味着他們的中國敞口可能高於其全球股票基準中的權重。取決於投資者對偏離基準的接受程度，他們可能會選擇不同的投資路線：

- 希望盡量向基準靠攏的投資者，可能會採取「綜合」配置的思路，對標增加了中國股票權重的特定指數進行配置；
- 對基準敏感度較低的投資者，則可以專門針對中國資產（甚至是僅對 A 股）進行配置，增加其中國相關的敞口。

不太確定增加中國 A 股權重對現有新興市場投資組合影響的投資者，可以考慮推遲是否增加 A 股投資比重的決定。

也有些投資者可能對中國在全球經濟比重的增長或中國股市的長期發展前景沒那麼樂觀。在這種情況下，他們可能更願意採取將中國從現在基準中「分割」出來的解決方案。

圖 3：政策基準配置時可能的思考過程

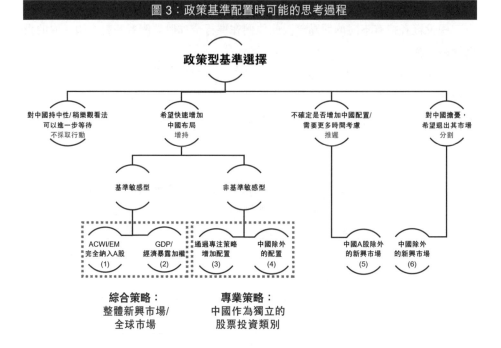

5.1　採取綜合還是專業的中國投資策略？

　　投資者最終採用的投資策略很大程度上取決於他們的目標和投資限制。對於採取甚麼策略並沒有「一刀切」的標準答案。

　　如果採取綜合的配置方案，投資者可以通過在更廣泛的股票投資組合中微調中國的權重來保持現有投資流程和治理結構的連續性。如果採用專業的配置方法，投資者則可以使用「增持」的方式分步調整其中國配置，更靈活地反映他們對中國的看法。然而，將中國獨立為新的股票類別可能意味着投資流程發生變化，投資組合的治理結構也可能會受到影響。

表 1：綜合與專業的中國配置方法比較		
綜合策略	專業策略（中國成為新的股票投資類別）	
中國權重中性或調高的特定新興市場/ACWI 指數	增持策略：專門針對中國策略配置	分割策略：中國和新興市場（中國除外）兩部分
優點 • 與現有投資流程和治理結構的連續性 • 對經理人的選股或量化策略的影響較小 • 無需在研究中國市場單獨配置上投入額外的資源	• 在政策型基準和戰術資產配置中進行中國配置的靈活性 • 可在弱有效市場享受持續風險溢價 • 對現有的新興市場/ACWI 指數基準沒有影響	• 基準配置的更高精確度和戰術資產配置決策的靈活性 • 可在弱有效市場享受持續風險溢價
缺點 • 戰術資產配置靈活性較低 • 缺乏能夠在中國權重顯著提高的情況下管理全球股票組合的經理人	• 可能需要改變投資流程和治理結構 • 需要在中國投入專門的資源 • 可能需要重新部署現有資產，為新的中國策略提供資金	• 可能對現有的投資流程和治理結構造成干擾 • 缺乏能夠管理中國除外新興市場股票的經理人 • 對中國市場相關的選擇時風險更高

註：以 MSCI 指數作基準指數為例。

5.2　MSCI 中國相關指數選項

考慮專業中國配置策略的投資標的時，投資者通常會選用 MSCI 中國指數。已部分納入 MSCI 指數的中國 A 股成分股也已包含在 MSCI 中國 A 股指數、MSCI 中國全市場指數，以及其他相關的 MSCI 指數中。

5.3　投資機會集的選擇：比較容量、廣度和因子溢價

尋找政策型基準時，投資者通常需要詳細評估所選基準對更廣義的資產配置決策的影響。具體而言，投資者會評估：

- **資產類別可投資性**：基準指數的可投資市值和流動性是多少？
- **市場廣度和阿爾法機會**：基準指數包括多少可投資證券？潛在的阿爾法（"alpha"，即相對於基準指數的超額回報）和因子溢價機會有多大？
- **投資結構可用性**：相關機會有多少種投資結構？有多大？新結構有甚麼趨勢？

換句話說，除了投資理念和資產配置的優點外，確定政策配置基準的過程中還需要考慮能夠吸納一定規模投資的各類投資產品和結構的現狀和未來發展情況。

MSCI 中國全市場指數表現出了中國整體市場（包括內地 A 股和海外上市中資股）的廣泛性和多樣性。另一方面，包括中國在內的綜合性新興市場基準則提供了更多的投資可能性。在選擇基準指數時，投資者也應考慮這些因素。

我們發現，相對於其他主流指數，MSCI 中國 A 股指數和 MSCI 中國全市場指數的成分股收益的分散系數（即表現好的股票比表現差的股票之間的收益落差，體現市場廣度的指標）較大，其次是 MSCI 新興市場指數和 MSCI 亞洲（日本除外）指數。在過去十年中，美國和其他發達市場的成分股收益的分散系數則低於新興及亞洲（日本除外）市場。

儘管不代表未來表現，但基於因子的量化策略，過去在中國股市中的表現比世界其他市場更好 [10]。此外，整合 ESG 因子對於中國、新興市場和亞洲（日本除外）指數歷史回報率的提升要高於發達市場 [11]。

6 結語：中國和未來的股票資產配置

中國不斷增長的經濟實力和市場准入條件的改善，為全球投資者帶來了重大的投資機遇和挑戰。鑒於 MSCI 已經將中國 A 股部分納入了 MSCI 新興市場指數和 MSCI ACWI 指數，了解中國對全球股票配置的重要性可以幫助投資者做出更穩健的投資策略。

投資者的投資經歷、目標和限制因素各不相同，因此無法一次性解決所有投資者的問題。採用更具戰略性的股票配置方法的投資者可能希望跳出短期市場和宏觀經濟波動，來對其中國股票的戰略配置進行評估。評估過程中，投資者可能希望關注以下兩個要點：（1）中國配置的權重大小，這包括在全球和新興市場策略中的配置；（2）如何在政策基準指數中實踐這一配置。

10　在 2018 年 6 月 27 日的博客文章 "Can your investment strategy work with China A shares?" 中，我們解釋了中國 A 股市場上，因子策略收益更高的潛在原因，詳見 MSCI 官方網頁。

11　基於 MSCI ESG 領先者指數從 2008 年 12 月到 2019 年 4 月的年化超額收益。

對於第一個要點，投資者可能需要評估在每種情景下如何才能滿足自己的投資目標，匹配其對全球股票機會集的分類看法，並與其投資信念無縫接合。

至於第二點，投資者可能要在兩種配置類型之間做出選擇，要麼是將中國放在基準指數的股票機會集中綜合考量，要麼是對中國股票採取專業的配置方法，將其視為獨立的股票類別。如果是後者，投資者可能根據實際情況作出不同的最終決策：譬如決定從新興市場投資組合中將中國板塊「分割」出去，或是採取整合後的「中國全市場」策略，又或是專門對中國 A 股形成特定的策略。在決定採取的方法時，投資者可能需要納入考量其投資信念，潛在配置方案對整體投資組合的影響，以及相關投資結構的可用性。

無論投資者做出何種選擇，理解中國在其股票資產配置中的地位尤為重要。

第11章

滬深港互聯互通上市公司
可提供的「一帶一路」投資機遇

香港交易及結算所有限公司
首席中國經濟學家辦公室

摘 要

　　中國國家主席習近平於 2013 年首次提出「一帶一路」倡議，並於 2015 年發表有關願景及行動方案。「一帶一路」倡議預期有利於促進沿線各國經濟繁榮與區域經濟合作、加強不同文明交流互鑒和相互學習，以及促進世界和平與發展。逾百國家透過不同合作方式參與「一帶一路」倡議，當中超過 60 個國家位於「一帶一路」沿線。中國提出「一帶一路」倡議後的五年來，經貿合作、投資及服務等方面都取得重大進展。

　　「一帶一路」倡議為全球各行各業的企業帶來無盡商機，讓他們於「一帶一路」沿線國家的投資爭取潛在豐厚回報。為了協助資本市場參與者深入了解相關投資機遇，中華證券交易服務有限公司（中華交易服務）進行了有關「一帶一路」的研究調查，以揭示滬深港上市公司在「一帶一路」進行業務活動的趨勢。調查結果顯示：（1）上市公司的「一帶一路」參與度持續提升，「一帶一路」業務穩步增長；（2）上市公司參與「一帶一路」業務越趨積極，尤其於工程合同及保險承保業務方面增幅顯著；（3）內地公司已在非金融業務中佔主導地位，其在金融業務的份額亦不斷增加；（4）公司參與「一帶一路」業務所涉及的行業越見多元化；及（5）「一帶一路」業務所覆蓋的地域亦越來越廣。

　　市場上亦陸續出現多隻「一帶一路」指數，追蹤箇中商機。當中，中華交易服務一帶一路指數是首隻兼唯一一隻追蹤「一帶一路」投資的中港跨境指標，指數成份股為可透過內地與香港股票市場交易互聯互通機制試點計劃（即「滬港通」及「深港通」）買賣的上市股票。市場大可以此指數為基礎開發投資工具（如交易所買賣基金），利便投資者跨境捕捉「一帶一路」投資機遇。

1 「一帶一路」倡議

1.1　啟動及重要意義

　　中國國家主席習近平 2013 年出訪亞洲國家期間，首次提出了「一帶一路」倡議。「一帶一路」涵蓋「絲綢之路經濟帶」和「21 世紀海上絲綢之路」。絲綢之路經濟帶貫穿中亞、西亞、中東至歐洲，延伸至南亞和東南亞。海上絲綢之路則貫穿東南亞、大洋洲和北非。同年稍後的一個場合中，中國總理李克強強調，有需要鋪就面向東南亞國家聯盟（東盟）[1] 的海上絲綢之路。 2015 年 3 月，中國政府發佈了關於「一帶一路」願景和行動的基礎文件（「一帶一路」行動計劃）[2]，指出加快建設「一帶一路」有利於促進沿線各國經濟繁榮與區域經濟合作、加強不同文明交流互鑒，促進世界和平與發展。按此行動計劃，「一帶一路」旨在：促進經濟要素有序自由流動、資源高效配置和市場深度融合；推動沿線各國實現經濟政策協調，開展更大範圍、更高水平、更深層次的區域合作；並共同打造開放、包容、均衡、普惠的區域經濟合作架構（見第 1.2 節）。

　　據悉「一帶一路」共涉及 100 多個國家，其中超過 60 個為「一帶一路」沿線國家（「一帶一路」國家）[3]，其餘還有許多通過加入亞洲基礎建設投資銀行（亞投行）[4]、與中國共同興建交通基礎設施又或以其他合作方式參與進來[5]。在「一帶一路」國家，一系列如公路和鐵路的基建項目已啟動並在進行中，中國無論在融資和工程方面都一直積極參與其中。在 2017 年 5 月的「一帶一路」國際合作論壇上，中國國家主席習近平重申中國對促進全球化和自由貿易的決心，承諾向「一帶一路」國

1　東盟由汶萊、柬埔寨、印度尼西亞、老撾、馬來西亞、緬甸、菲律賓、新加坡、泰國和越南共十個國家組成。

2　《推動共建絲綢之路經濟帶和 21 世紀海上絲綢之路的願景與行動》，中華人民共和國國家發展和改革委員會、外交部和商務部在國務院批准下共同發佈，2015 年 3 月。

3　中國國際貿易研究中心 2015 年 8 月公佈的《「一帶一路」沿線國家產業合作報告》列出將會參與倡議的 65 個「一帶一路」沿線國家。

4　總部設於北京的亞投行是多邊發展銀行，由中國推動成立支持「一帶一路」的發展，旨在提高亞洲以至世界各地的社會和經濟效益。截至 2018 年 10 月 8 日，亞投行擁有 44 個地區成員、24 個非地區成員和 19 個來自世界各地的潛在會員（來源：亞投行網站，2018 年 12 月 12 日資料）。

5　馮氏集團利豐研究中心 2016 年 5 月的報告《The Belt and Road Initiative: 65 Countries and Beyond》，識別出 48 個這樣的國家。

家投資 1,240 億美元。

許多「一帶一路」國家，不管是在公營還是私營市場，都欠缺自行發展基礎設施的財政能力。現時在中國「一帶一路」項目的支持下，這些國家將可受惠，得以加強其基礎設施建設、吸引更多外來投資以至經貿活動等。中國本身已承諾進一步開放市場，加強國際合作。

按已知的「一帶一路」國家計算，其人口合計約佔全球總數的 65%，國內生產總值 (GDP) 佔全球的 40%，能源資源佔全球四分之三，商品和服務貿易亦佔全球四分之一[6]。「一帶一路」預計將為上述地區的人民和企業創造豐厚的就業和貿易機會。

1.2 多邊合作重點及舉措

根據「一帶一路」行動計劃，「一帶一路」堅持和平共處的原則[7]; 堅持開放合作、和諧包容、遵循市場運作，謀求各方共同利益。合作或「聯通」的五大要點概述如下：

- **政策溝通** —— 加強政府間合作，積極構建多層次政府間宏觀政策溝通交流機制，達成合作新共識；

- **設施聯通** —— 逐步形成連接亞洲各次區域以及亞歐非之間的基礎設施網絡；

- **貿易暢通** —— 透過投資及貿易合作，改善投資以及貿易便利化問題，消除投資和貿易壁壘，構建區域內和各國良好的營商環境；

- **資金融通** —— 深化金融合作，推進亞洲貨幣穩定體系、投融資體系和信用體系建設；

- **民心相通** —— 傳承和弘揚絲綢之路友好合作精神，廣泛開展文化交流、學術往來、人才交流合作、媒體合作、青年和婦女交往、志願者服務等。

「一帶一路」倡議自推出以來便得到「一帶一路」國家的熱烈積極反響。截至

6　資料來源：〈China says it's building the New Silk Road. Here are five things to know ahead of a key summit〉，載於《時代雜誌》(*Time*)，2017 年 5 月 12 日；新加坡政府網站 (https://www.iesingapore.gov.sg)。

7　和平共處五項原則是：尊重各國主權和領土完整、互不侵犯、互不干涉內政、平等互利、和平共處。

2018 年 8 月止的五年間，中國與 103 個國家、地區和國際組織共簽署了 118 項發展「一帶一路」的合作協議，先後與 13 個「一帶一路」國家簽署或升級了 5 項自由貿易協定。倡議的範圍由此從歐亞大陸延伸到非洲、拉丁美洲及加勒比地區以至南太平洋地區 [8]。

　　除基礎設施項目外，「一帶一路」的商機也涉及貿易投資以及金融和專業服務。在金融方面，有機構估計未來 15 年「一帶一路」項目需要的融資金額高達 5 至 6 萬億美元 [9]。據報導，多個「一帶一路」國家已與中國合作發展基礎設施項目，這包括斯里蘭卡、馬來西亞、菲律賓、巴基斯坦、比利時、埃塞俄比亞、肯尼亞、柬埔寨及莫桑比克等 [10]。過去五年來中國與「一帶一路」國家的貿易總額已超過 5 萬億美元，年均增長率為 1.1%，並在這些國家建立了 82 個海外經貿合作區，累計投資額達 289 億美元 [11]。英國、新加坡和中國香港等國家和地區也抓緊機會為「一帶一路」提供金融及專業服務，包括銀行及保險、法律、工程及城市規劃等。

　　據報導，在「一帶一路」的推動之下，中國企業增加了在「一帶一路」國家的經營和投資，與「一帶一路」商業夥伴共創雙贏局面。從事建築、港口運營、貨運/航運及物流、能源、貿易和金融服務等行業的企業均從中受益 [12]。以貿易及貨運業為例，「一帶一路」倡議下建立的鐵路及海運航線以及港口設施 [13] 大大縮短了貨物運輸時間。2018 年截至 10 月止，中國企業於 55 個「一帶一路」國家新增 119 億美元投資，同比增長 6.4% [14]，另 2018 年（截至 7 月）中國企業與「一帶一路」國家新簽訂了 571.1 億美元的建設合同 [15]。此外，中資銀行機構也在「一帶一路」融資中發揮着重要作用。例如據報導，中國銀行新加坡分行為「一帶一路」項目的融資先

8　資料來源：《新華社》文章〈China sees trade, investment growth with B&R countries〉，2018 年 8 月 27 日，及《新華社》文章〈Belt and Road Initiative in 5 years〉，均載於中國國務院新聞辦公室網站，2018 年 8 月 27 日。

9　滙豐估計為 6 萬億美元，普華永道估計為 5 萬億美元（資料來源：〈BRI helps companies build global bridges of connectivity〉，載於《今日中國》（*China Daily*），2018 年 11 月 14 日）。

10　資料來源：不同媒體報導。

11　資料來源：《新華社》文章〈B&R countries eye production capacity cooperation with China〉，載於中國國務院新聞辦公室網站，2018 年 10 月 22 日。

12　據不同媒體報導。

13　其中包括中新鐵路 - 海運公路、鄭州 - 列日（比利時）貨運列車路線、蘭州 - 伊斯蘭堡（巴基斯坦）鐵路和公路貨運服務、中歐陸海快速路線和中歐貨運鐵路計劃等。

14　資料來源：〈前 10 個月我國企業對「一帶一路」沿線國家新增投資 119 億美元〉，載於中國國務院新聞辦公室網站，2018 年 11 月 15 日。

15　資料來源：《新華社》文章〈Belt and Road Initiative in 5 years〉，載於中國國務院新聞辦公室網站，2018 年 8 月 27 日。

後籌集了 160 億美元的資金 [16]。

因此，「一帶一路」為內地、香港乃至世界各地的企業創造了無限商機，讓他們於「一帶一路」國家的投資爭取潛在豐厚回報。

2 中華交易服務的「一帶一路」研究調查

有見「一帶一路」對全球商業及投資市場的重大意義，香港交易及結算所有限公司（香港交易所）與上海證券交易所（上交所）及深圳證券交易所（深交所）的合資公司 —— 中華證券交易服務有限公司（中華交易服務）—— 啟動了關於「一帶一路」的研究調查，收集上市公司參與「一帶一路」的業務範圍及性質的數據及資料。該調查定期進行，希望可緊貼資本市場在「一帶一路」不斷發展下的格局變化。調查結果發揮以下作用，預期會有助資本市場參與者深入了解相關機遇：

(1) 揭示「一帶一路」經濟活動的現狀特徵和發展趨勢；

(2) 展示上市公司把握「一帶一路」商機時就產品和服務市場採納的不同商業模式；再由此

(3) 展現滬深港上市公司參與「一帶一路」為資本市場創造的投資機會。

該調查的目標回應公司涵蓋上交所、深交所及香港交易所旗下股票市場互聯互通 [17] 的上市公司。首次調查於 2017 年進行（2017 年調查），收集了 2014 年至 2016 年財政年度的數據；最近一次調查於 2018 年進行（2018 年調查），收集了 2017 年財政年度的數據。2017 年調查及 2018 年調查分別覆蓋的目標樣本有 1,931 及 1,939 家上市公司，回應率分別為 67%（1,286 份有效回應）及 66%（1,274 份有效回應）。

---

16 資料來源：《新華社》文章〈China, Singapore explore more potentials for BRI cooperation〉，載於中國國務院新聞辦公室網站，2018 年 11 月 12 日。

17 此機制涵蓋「上海與香港股票市場交易互聯互通機制」（滬港通）及「深圳與香港股票市場交易互聯互通機制」（深港通）。

　　至於有關上市公司參與「一帶一路」的計量指標，則是參照「一帶一路」國家的五大關鍵聯通領域（見上文第 1.2 節）來確定。該五大關鍵領域中有三個是可量化的，分別是：「設施聯通」、「貿易暢通」及「資金融通」。最後定出以下六個「一帶一路」商業參與度指標：

(1) **商品/服務銷售金額** —— 向「一帶一路」國家提供貿易出口、於當地銷售商品及/或提供服務等所獲取的收入金額；

(2) **商品/服務採購金額** —— 從「一帶一路」國家進行貿易進口、從當地採購貨品及/或接受當地所提供的服務等，以作自家生產或轉售用途的採購金額；

(3) **新簽工程項目合同金額** —— 於「一帶一路」國家從事基建設計及/或建設、土木工程及設備製造等所涉及的新簽工程項目合同總金額；

(4) **新增投資金額** —— 在「一帶一路」國家進行的資本開支及/或股權投資的新增投資總額；

(5) **融資/貸款金額（僅限金融機構）** —— 在「一帶一路」國家進行直接融資、債券發行及/或貸款業務所涉及的金額；

(6) **保險承保金額（僅限金融機構）** —— 在「一帶一路」國家提供客戶保險服務及承保業務所涉及的金額。

　　調查向受訪者提供「一帶一路」國家的參考名單[18]，請他們報告其於上述六個指標（按其適用者）中所涉及的金額，並提供他們參與「一帶一路」相關業務的任何其他信息。由於「一帶一路」沒有明確的地理邊界，調查中這份「一帶一路」國家名單會按年擴張[19]。

　　以下第 3 節將根據研究調查的結果，展示受訪公司在基建、貿易及投資以及金融服務等關鍵領域的「一帶一路」參與趨勢。

18　有關 2018 年調查的「一帶一路」國家參考名單，請參見附錄一。香港為中國特別行政區，未有列入名單。

19　與 2017 年調查相比，2018 年調查增添了 9 個國家：東北亞的韓國；大洋洲的新西蘭；非洲的埃塞俄比亞、馬達加斯加、摩洛哥及南非；東歐的奧地利；美洲的巴拿馬及特立尼達和多巴哥。

3 上市公司的「一帶一路」參與情況[20]

3.1 「一帶一路」參與情況總覽

「一帶一路」參與現況及未來計劃

2018 年調查（2017 財政年度數據）中，大部分受訪公司（52%）均有參與「一帶一路」業務（下稱「一帶一路」公司），結果亦顯示參與度自 2014 年至 2017 年財政年度漸次上升（2014 年、2015 年及 2016 年分別為 42%、44% 及 46%）。2018 年調查中的「一帶一路」公司有 60% 擬在未來 3 年增加其「一帶一路」業務，33% 擬維持該等業務。不涉「一帶一路」業務的公司中，29% 擬在未來 3 年會參與「一帶一路」業務。整體而言，94% 受訪公司擬增加或維持於「一帶一路」業務的參與度。（見圖 1。）

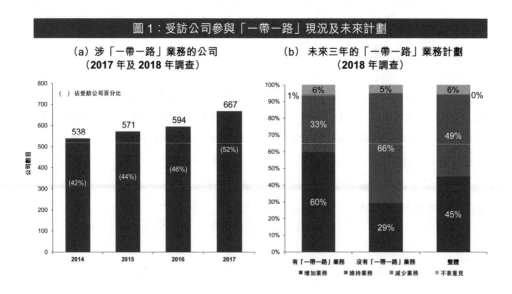

圖 1：受訪公司參與「一帶一路」現況及未來計劃

（a）涉「一帶一路」業務的公司
（2017 年及 2018 年調查）

（b）未來三年的「一帶一路」業務計劃
（2018 年調查）

---

20 詳情請參閱 2018 年 12 月 11 日中華交易服務刊發的 2018 年調查報告（https://www.ces.com）。

「一帶一路」公司的類別及行業（2018年調查）

回應 2018 年調查的「一帶一路」公司主要為內地公司[21]（98%）—— 36% 為國有企業（國企）[22]，62% 為非國企。行業分佈上，工業類別的公司佔最大比重（28%），其次為原材料（18%）、可選消費（15%）及信息技術（14%）。上述四大行業的公司合計佔有回應 2018 年調查的全部「一帶一路」公司樣本的 75%。（見圖 2。）

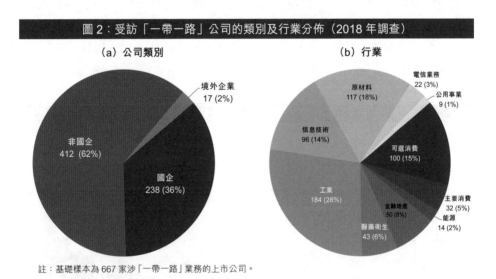

圖 2：受訪「一帶一路」公司的類別及行業分佈（2018 年調查）

（a）公司類別　　　　　（b）行業

註：基礎樣本為 667 家涉「一帶一路」業務的上市公司。

3.2　「一帶一路」業務的增長

根據 2017 年調查及 2018 年調查（涵蓋受訪公司 2014 年至 2017 年財政年度）的結果，不同類別的「一帶一路」業務均呈明顯增長趨勢[23] —— 2014 年至 2017 年錄得正數的複合年增長率。計及新增國家的業務後，錄得最大複合年增長率的「一

21　「內地公司」指符合以下任何一項準則的公司：（1）實體在中國內地註冊成立；（2）公司的營運中心在中國內地；及（3）公司收入至少 50% 來自中國內地。

22　國企涵蓋（1）由國有資產監督管理委員會（國資委）地方分局或地方政府又或其他地方國企作為實際控制人的「地方國企」，及（2）國務院屬下國資委或財政部作為實際控制人的「中央國企」。

23　雖然兩項調查的受訪樣本數目不一，但實際數目（1,286 家及 1,274 家）及回應比率（67% 及 66%）均很接近，因此其結果可作比較。

帶一路」業務類別為新簽工程合同（63%），然後是保險承保（14%）及融資/貸款（10%）。即使不包括新增國家，新簽工程合同的複合年增長率仍處於 60% 的高水平，保險承保及融資/貸款則分別為 10% 及 9%。（見圖 3。）

以金額計（包括新增國家），受訪「一帶一路」公司於 2017 年的保險承保及融資/貸款金額分別達人民幣 40,366 億元及人民幣 25,461 億元，當中保險承保金額更較 2016 年大幅增長 30%。相對來說，2017 年新增的「一帶一路」投資只有人民幣 942 億元。「一帶一路」採購及銷售金額於 2017 年幾乎相同，分別為人民幣 11,137 億元及人民幣 11,109 億元。「一帶一路」新簽工程合同金額則為人民幣 6,806 億元，較三年前的人民幣 1,584 億元增加逾三倍，較 2016 年的人民幣 3,277 億元亦增加一倍。

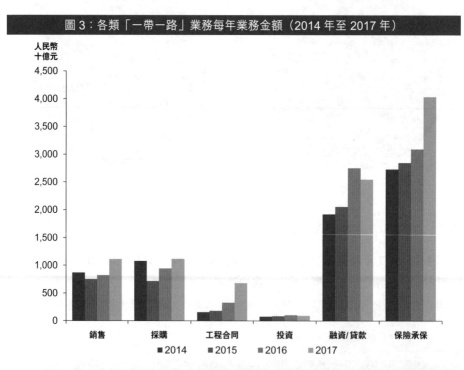

圖 3：各類「一帶一路」業務每年業務金額（2014 年至 2017 年）

包括/不包括新增國家	複合年增長率（2014 — 2017）					
	銷售	採購	工程合同	投資	融資/貸款	保險承保
包括	9%	1%	63%	7%	10%	14%
不包括	5%	-1%	60%	6%	9%	10%

3.3　按上市公司類別及業務分析「一帶一路」參與情況

按公司類別分析

　　受訪公司中，內地公司（國企及非國企）佔「一帶一路」非金融業務最大比重 —— 分別佔 2017 年銷售、採購及新簽工程合同的 83%、93% 和 100%，佔新增投資 76%（見圖 4）。在銷售、採購及新簽工程合同這三個業務類別，國企較非國企更為重要（2017 年所佔比重分別為 65%、82% 及 97%）。另一方面，非國企在新增投資方面所佔的比重亦快速增長，由 2014 年的 6% 增至 2017 年的 39%。

　　金融業務類別一直由境外企業主導，但內地公司正逐步趕上。境外企業 2017 年於融資/貸款及保險承保的比重分別為 67% 及 79%，低於 2014 年的 98% 及 88%。（見圖 4。）

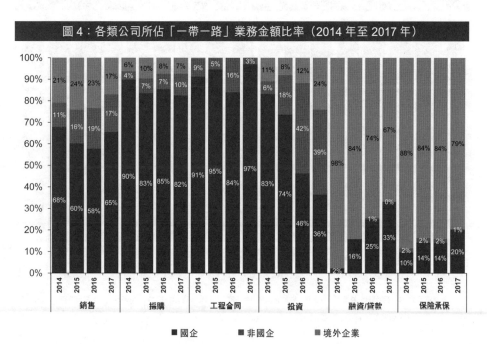

圖 4：各類公司所佔「一帶一路」業務金額比率（2014 年至 2017 年）

■ 國企　　■ 非國企　　■ 境外企業

註：由於四捨五入的關係，百分比的總和未必相等於 100%。

按行業分析

　　能源類公司僅佔 2018 年調查中受訪「一帶一路」公司數目的 2%，但卻主導「一帶一路」採購及銷售業務，2017 年佔這兩類業務最大比重（分別為 68% 及 33%，已較 2014 年的 85% 及 52% 下跌）。在全部四個「一帶一路」非金融業務類別中，工業類公司的佔比皆有所上升 ── 於新工程合同的佔比由 2014 年的 76% 升至 2017 年的 99%，銷售的佔比由 2014 年的 13% 升至 2017 年的 25%，採購的佔比由 2014 年的 6% 升至 2017 年的 11%，新增投資的佔比由 2014 年的 8% 升至 2017 年的 17%。在新增投資及銷售業務類別中，金融類公司的佔比明顯相對較高（2017 年分別佔 23% 及 12%）。要注意的是，基於業務性質的關係，只有金融類公司提供「一帶一路」融資/貸款及保險承保服務。（見圖 5。）

圖 5：各行業所佔「一帶一路」業務金額比率（2014 年至 2017 年）

■ 可選消費　■ 主要消費　■ 能源　■ 金融　■ 醫藥衛生　■ 工業　■ 信息技術　■ 原材料　■ 房地產　■ 電信業務　■ 公用事業

註：由於四捨五入的關係，百分比的總和未必相等於 100%。

3.4　「一帶一路」業務的區域及國家分析

圖 6 顯示受訪上市公司 [24] 2016 年及 2017 年「一帶一路」業務金額的區域分佈及相關比重。2018 年調查（收集 2017 年財政年度的數據）多加了五個區域合共九個國家 [25]，非洲、大洋洲及美洲都是 2017 年調查（收集 2014 年至 2016 年財政年度的數據）所未有涵蓋的新增區域。

受訪「一帶一路」公司的「一帶一路」業務比較集中於東南亞地區 ── 2017 年，東南亞的銷售佔比為 59%，採購佔比為 42%，新簽工程合同為 50%，新增投資為 55%，融資/貸款為 45%，而保險承保的佔比尤其高，達 78%。中東地區於採購及融資/貸款業務的佔比很大（2017 年分別佔 31% 及 36%），南亞於新簽工程合同及銷售業務的比重亦相對較高，2017 年於新簽工程合同的佔比為 18%（與中東相若），於銷售的佔比為 11%（中東：9%）。

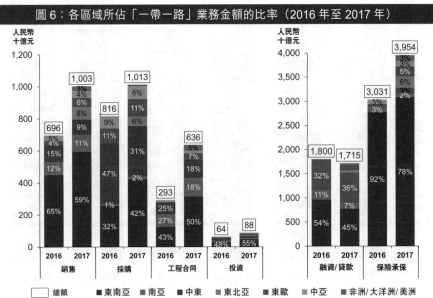

圖 6：各區域所佔「一帶一路」業務金額的比率（2016 年至 2017 年）

註：基礎樣本為 512 家（2017 年調查）及 657 家（2018 年調查）有提供其「一帶一路」業務區域分佈的公司。
　　由於四捨五入的關係，百分比的總和未必相等於 100%。

24　2017 年調查及 2018 年調查分別有 512 及 657 家公司回答有關國家分佈的問題。這批受訪公司的「一帶一路」業務於 2016 年及 2017 年財政年度的合計金額分別為人民幣 67,000 億元及人民幣 84,090 億元，分別佔這兩年受訪「一帶一路」公司整體樣本中「一帶一路」業務總額的 83% 及 88%。

25　2018 年調查的「一帶一路」國家名單見附錄一，有關變動見上文註 19。

　　「一帶一路」業務的主要地區東南亞於 2017 年錄得新簽工程合同最大年度增長（以金額計為 153%），採購、投資及銷售業務的增長也不俗（分別為 64%、58% 和 32%）。反之，中東於 2017 年「一帶一路」業務在採購及銷售方面全年分別下跌 17% 及 11%，但新簽工程合同仍有 51% 的年度增長。東北亞於 2017 年「一帶一路」業務的增長（2017 年整體佔比為 5%，2016 年則為 0%）主要源自新列入國家名單的韓國。

　　雖然 2017 年於每個業務類別的首三大區域的佔比合計均約達八成或以上，但集中度已較 2016 年下降，而且 2017 年受訪公司的「一帶一路」業務所覆蓋的地域分佈亦見越廣。（見圖 7。）

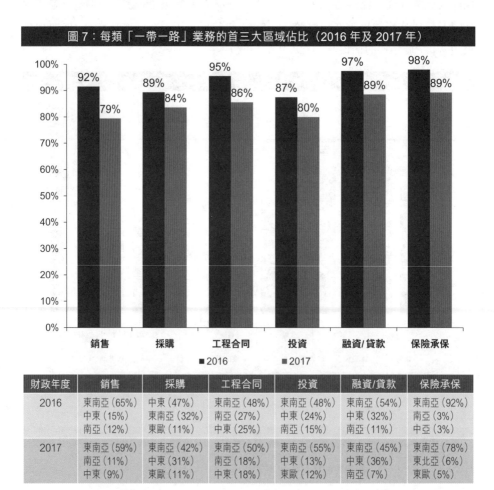

圖 7：每類「一帶一路」業務的首三大區域佔比（2016 年及 2017 年）

財政年度	銷售	採購	工程合同	投資	融資/貸款	保險承保
2016	東南亞（65%） 中東（15%） 南亞（12%）	中東（47%） 東南亞（32%） 東歐（11%）	東南亞（48%） 南亞（27%） 中東（25%）	東南亞（48%） 中東（24%） 南亞（15%）	東南亞（54%） 中東（32%） 南亞（11%）	東南亞（92%） 南亞（3%） 中亞（3%）
2017	東南亞（59%） 南亞（11%） 中東（9%）	東南亞（42%） 中東（31%） 東歐（11%）	東南亞（50%） 南亞（18%） 中東（18%）	東南亞（55%） 中東（13%） 東歐（12%）	東南亞（45%） 中東（36%） 南亞（7%）	東南亞（78%） 東北亞（6%） 東歐（5%）

　　圖 8 顯示 2017 年財政年度受訪「一帶一路」公司每類「一帶一路」業務中佔比最高的首五個國家。新加坡在六個「一帶一路」業務類別中的五個均踞首位，馬來西亞則在新簽工程合同業務中排首位。

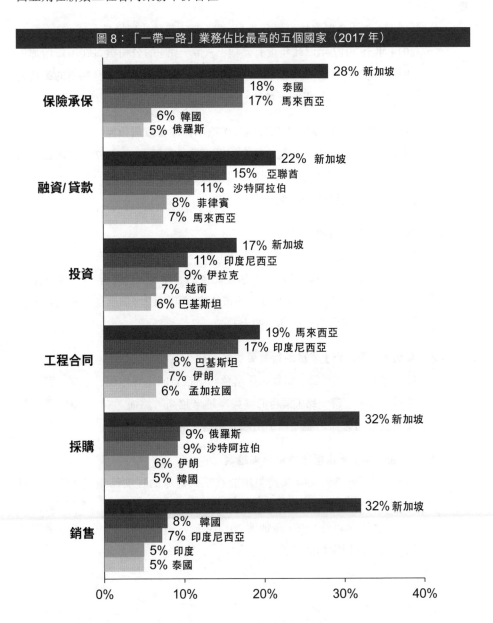

圖 8：「一帶一路」業務佔比最高的五個國家（2017 年）

保險承保
- 28% 新加坡
- 18% 泰國
- 17% 馬來西亞
- 6% 韓國
- 5% 俄羅斯

融資/貸款
- 22% 新加坡
- 15% 亞聯酋
- 11% 沙特阿拉伯
- 8% 菲律賓
- 7% 馬來西亞

投資
- 17% 新加坡
- 11% 印度尼西亞
- 9% 伊拉克
- 7% 越南
- 6% 巴基斯坦

工程合同
- 19% 馬來西亞
- 17% 印度尼西亞
- 8% 巴基斯坦
- 7% 伊朗
- 6% 孟加拉國

採購
- 32% 新加坡
- 9% 俄羅斯
- 9% 沙特阿拉伯
- 6% 伊朗
- 5% 韓國

銷售
- 32% 新加坡
- 8% 韓國
- 7% 印度尼西亞
- 5% 印度
- 5% 泰國

0%　10%　20%　30%　40%

3.5 調查結果的主要趨勢

總括而言，調查結果顯示受訪上市公司於「一帶一路」的參與情況有以下主要趨勢：

(1) 「一帶一路」參與度持續提升及「一帶一路」業務持續增長

2014 年至 2017 年財政年度有參與「一帶一路」的公司數目在不斷增加，佔比由前數年的少數增至 2017 年的多數。調查中不同業務類別的業務金額在這數年間都錄得顯著增幅。

(2) 企業參與「一帶一路」工程合同和保險承保業務愈趨積極

調查的業務類別中，新簽工程合同及保險承保業務於 2017 年的增長特別高（與 2016 年比較增長分別超過 100% 及 30%），反映「一帶一路」國家的基礎設施發展持續吃重。

(3) 內地公司參與度顯著重要

「一帶一路」的銷售、採購、新簽工程合同和新增投資等業務均由內地公司為主導。一直以境外企業佔大比重的金融業務（融資/貸款及保險承保），內地公司（主要是國企）的市佔率亦正不斷增加及越來越吃重。非國企於新增投資的佔比亦明顯上升。

(4) 參與「一帶一路」業務的行業更加多元化

能源業於「一帶一路」銷售、採購及新增投資的主導地位已下降，其他行業於此等「一帶一路」業務的參與率越來越高。然而，在「一帶一路」新簽工程合同方面仍繼續以工業類公司為主導。

(5) 「一帶一路」業務覆蓋地域越來越廣

這數年來「一帶一路」業務集中於首三大區域的程度已漸次下降。不同「一帶一路」業務類別佔比最大的首三個國家都有所不同。調查所評估的國家（大中華以外）中，新加坡 2017 年於六個評估的「一帶一路」業務類別中的五個類別均排名首位。

4　中華一帶一路指數及其他「一帶一路」指數

2017 年 6 月，香港交易所委託中華交易服務開發一帶一路指數，以回應市場對「一帶一路」指標的需求。「中華一帶一路指數」由此產生。

中華一帶一路指數是離岸市場首隻與「一帶一路」相關的跨境指數，旨在反映積極參與「一帶一路」倡議的公司的股價表現。

中華一帶一路指數成份股篩選自滬深港互聯互通機制下可供北向或南向交易的股票（不包括上交所或深交所的 ST 或 ST* 股票 [26]），參考基準為按照「一帶一路」研究調查中取得該等公司於「一帶一路」六大業務範疇的參與度評估 [27]。那即是說，中華一帶一路指數所包括的都是在滬深交易所或香港交易所上市的公司。指數的組成反映「一帶一路」倡議在滬深港互聯互通上市公司有參與的各個經濟領域中的發展情況。換言之，該指數是跟蹤與「一帶一路」業務相關的股票表現的指標。

目前除中華一帶一路指數外，內地主要有三隻以從事「一帶一路」主題相關行業的公司的 A 股作為成份股的在岸市場「一帶一路」指數：

(1) **中證一帶一路主題指數**（下稱「中證一帶一路指數」）[28] —— 2015 年 2 月 16 日由中證指數有限公司（中證）推出；

(2) **上證一帶一路主題指數**（下稱「上證一帶一路指數」）[29] —— 2015 年 6 月 24 日由中證為上交所推出；及

(3) **國證新絲路主題指數**（下稱「國證新絲路指數」）[30] —— 2015 年 1 月 8 日由深圳證券信息有限公司推出。

26　ST 或 ST* 指「特別處理」，提示投資者買賣該等股票具有特別風險。

27　中華一帶一路指數編制方法見附錄二。

28　指數成份股是從事五大「一帶一路」相關行業中最具代表性的上交所或深交所上市 A 股，該五大相關行業是：基礎建設、交通運輸、高端裝備、電力通信以及資源開發。（資料來源：中證網站，2018 年 12 月 10 日資料。）

29　指數成份股是同樣從事中證一帶一路指數所涉的五大「一帶一路」相關行業中最具代表性的上交所上市 A 股。（資料來源：中證網站，2018 年 12 月 10 日資料。）

30　指數成份股是公司在陝西、甘肅、寧夏、青海及新疆五省（均為「一帶一路」沿線經濟區）註冊成立的上交所或深交所上市公司的 A 股。（資料來源：上證網站，2018 年 12 月 10 日資料。）

除此之外，國際上亦至少有一隻「一帶一路」主題的相關指數 ── **MSCI
Global China Infrastructure Exposure Index（下稱「MSCI 一帶一路指數」）**。指數
所追蹤的，是發達市場、新興市場及前沿市場中，從事某些指定「一帶一路」相關
行業類別而其大部分收入與中國基礎設施發展掛鈎的上市公司。於 2018 年 11 月
底，指數成份股公司所屬國家中中國所佔比重最高（43.5%），其次分別為新加坡
（10.2%）、馬來西亞（7.2%）、俄羅斯（6.6%）及泰國（6.1%）[31]。

圖 9 所示為這五個「一帶一路」指數的行業分佈。相較中證、上證及 MSCI 等
一帶一路指數，中華一帶一路指數的行業分佈較廣。國證新絲路指數的行業分佈
廣度與中華一帶一路指數相若，但其選擇的成份股是按公司註冊地劃分，而不是
依據公司實質的「一帶一路」業務關係劃分。

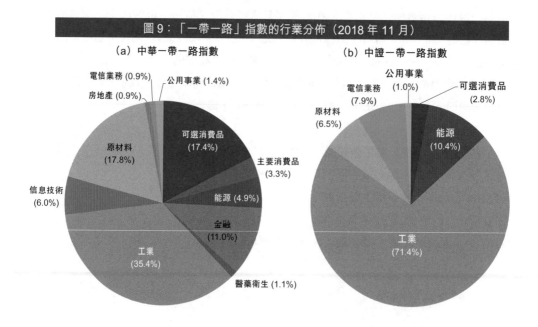

圖 9：「一帶一路」指數的行業分佈（2018 年 11 月）

（a）中華一帶一路指數

（b）中證一帶一路指數

31　資料來源：MSCI 網站所載的 2018 年 11 月指數資料表（2018 年 12 月 10 日網絡資料）。

(續)

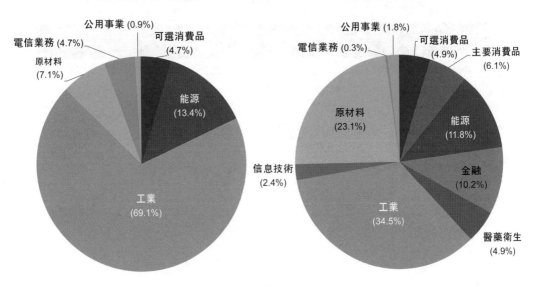

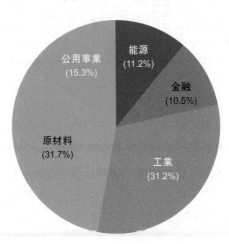

圖 9：「一帶一路」指數的行業分佈（2018 年 11 月）

（c）上證一帶一路指數　　　　（d）國證新絲路指數

（e）MSCI 一帶一路指數

註：由於四捨五入的關係，百分比的總和未必相等於 100%。

資料來源：中華一帶一路指數資料源自中華交易服務；其他指數資料來自相關指數公司網站的指數資料表。

圖 10 所示，是以滬深港上市公司為成份股的四隻「一帶一路」指數於 2015 年 7 月至 2018 年 10 月 [32] 期間的相對表現。

圖 10：「一帶一路」指數的每日收市指數（2015 年 6 月 30 日至 2018 年 10 月 31 日）（2015 年 6 月 30 日重訂基數）

註：重訂基數日期（2015 年 6 月 30 日）為中華一帶一路指數的基準日。表中所有指數的基準點於此重訂基準日設定為 2000 點。
資料來源：Wind。

市場能提供各有專屬範疇的「一帶一路」指數，投資者於「一帶一路」相關股票的投資就有相應的對照基準。此外，市場亦可開發以這些指數為基準的交易所買賣基金等主題投資工具，利便散戶及機構投資者參與。特別是，中華一帶一路指數是唯一一隻以滬深港互聯互通上市公司為成份股的跨境「一帶一路」指數。以此指數為基礎的投資工具有助投資者跨境捕捉「一帶一路」投資機遇。

---

32 MSCI 一帶一路指數沒有 2017 年 8 月之前的數據，故圖中並未包括該指數。

5 總結

　　「一帶一路」倡議是推動多邊經濟合作及發展的重要動力，可為全球各行各業的企業帶來無盡商機，讓他們於「一帶一路」沿線國家的投資爭取潛在豐厚回報。中華交易服務就「一帶一路」進行的研究調查披露了滬深港上市公司參與「一帶一路」業務活動的趨勢，有助資本市場參與者更深入了解相關投資機遇。相關的「一帶一路」指數正好針對追蹤這些投資機遇，當中的中華交易服務一帶一路指數更是首隻兼唯一一隻追蹤「一帶一路」投資的中港跨境指標。市場大可開發以此指數為基準的投資工具如交易所買賣基金，利便投資者跨境捕捉「一帶一路」的投資機遇。

附錄一

2018年「一帶一路」調查所涉及的「一帶一路」國家名單

區域	國家
非洲	埃塞俄比亞、馬達加斯加、摩洛哥、南非
美洲	巴拿馬、特立尼達和多巴哥
大洋洲	新西蘭
東歐	阿爾巴尼亞、奧地利、白俄羅斯、波斯尼亞與黑塞哥維那、保加利亞、克羅地亞、捷克、愛沙尼亞、匈牙利、拉脫維亞、立陶宛、馬其頓、摩爾多瓦、黑山、波蘭、羅馬尼亞、俄羅斯、塞爾維亞、斯洛伐克、斯洛文尼亞、土耳其、烏克蘭
中亞	阿富汗、哈薩克斯坦、吉爾吉斯斯坦、塔吉克斯坦、土庫曼斯坦、烏茲別克斯坦
中東	亞美尼亞、阿塞拜疆、巴林、埃及、格魯吉亞、伊朗、伊拉克、以色列、約旦、科威特、黎巴嫩、阿曼、巴勒斯坦、卡塔爾、沙特阿拉伯、敍利亞、阿拉伯聯合酋長國、也門
東北亞	蒙古國、韓國
南亞	孟加拉國、不丹、印度、馬爾代夫、尼泊爾、巴基斯坦、斯里蘭卡
東南亞	汶萊、柬埔寨、印度尼西亞、老撾、馬來西亞、緬甸、菲律賓、新加坡、泰國、東帝汶、越南

資料來源：中華交易服務 2018 年「一帶一路」調查報告。

附錄二

中華一帶一路指數編制方案

特徵	詳情
成份股選取範疇	• 「滬股通」、「深股通」及「港股通」的「北向交易」和「南向交易」的合資格股票。 • 不包括滬深市場上市的 ST 及 ST* 股票。
成份股篩選準則	• **盈利甄選**：剔除最近年度報告每股收益（EPS）是負數的公司 • **參與度甄選**： (1) 每家公司將根據與「一帶一路」相關的 6 項參與度指標 —— 銷售、採購、新簽工程合同、新增投資、融資/貸款、保險承保的業務金額進行排名； (2) 每項指標排名前 1% 的公司將立即成為指數成份股 *，而其餘公司將分別於 6 項指標中，將相應的業務金額轉化成一個分數，該分數為每項指標的 t 分佈累積分佈函數（CDF）的累積概率； (3) 未進入指數的公司將以其 6 項指標中最高的分數進行排名。排名最前的公司將成為指數成份股，最終得出 100 隻成份股。 * 若所選公司為兩地同時上市的「AH股公司」，選擇近一年日均成交額較高的股票作為成份股。
成份股數目	100 隻
權重計算方式	採用等權重加權
定期審核	每年一次
再平衡	半年一次
收市指數（計算貨幣）	人民幣
基值	2000
基日	2015 年 6 月 30 日
指數發佈	每日一次，於收市後發佈

資料來源：中華交易服務。

第12章

投資亞洲生物科技市場的指標：
中華交易服務香港生物科技指數

香港交易及結算所有限公司

首席中國經濟學家辦公室

及

中華交易服務有限公司

摘 要

　　全球經濟的發展、人口總量的增長和社會老齡化程度的不斷升高，為生物醫藥產業發展帶來新的增長動力。與傳統行業相比，生物醫藥企業呈現出高投入、高產出、高風險、高技術密集型的特點。股權融資（而非債務融資）是助力生物醫藥企業成長的重要融資方式。不同性質資金可以根據生物醫藥類企業的特點分階段介入，為生物醫藥行業不同時期的發展提供具規模的資金支援。近年來，國際主要的證券市場逐步建立了適用於生物科技企業上市的證券市場規則，有效引導創業風險投資和私募股權合理佈局，完成「投入一產出」的資本循環，進一步優化資源配置。

　　內地生物醫藥產業發展欣欣向榮，很大程度上得益於政策紅利不斷釋放、資本投入日漸活躍、行業加速整合等眾多利好因素。2018 年 4 月，香港交易所修改《上市規則》，為未有收益或盈利的生物科技公司開闢新的上市途徑，為生物科技這一門最具活力和未來發展空間的行業營造良好的投融資環境。

　　目前全球各主要資本市場針對生物醫藥板塊已推出相關指數及金融產品，方便投資者參與這個市場。最具代表性的是「納斯達克生物科技指數」和「標普生物技術精選行業指數」，以及基於此等指數開發的一系列交易所買賣基金（ETF）產品。「中華交易服務香港生物科技指數」（簡稱「中華香港生物科技指數」）立足亞洲市場，將基於香港的新上市規則上市的生物科技公司以及發展相對成熟的、在港上市的生物醫藥企業都納入指數中，方便投資者綜合觀察香港市場生物醫藥板塊的表現，可作為行業配置的標竿。

　　中華香港生物科技指數的推出，將進一步促進香港資本市場的多樣化，其中涵蓋的成份股公司專注於利用生物技術研發人類疾病的新型治療方式，代表了亞洲生物科技未來的發展方向和趨勢，對中國形成生物醫藥核心領域的創新競爭力具有重要意義。

1 資本市場成為生物科技醫藥行業發展的助推器

1.1　近年來生物科技醫藥行業獲得長足發展

　　全球經濟的發展、人口總量的增長和社會老齡化程度的不斷升高，為醫藥產業的發展帶來新的增長動力，全球醫藥市場呈現出遠高於經濟增長的強勁增長速度。遺傳學、分子生物學和生物化學等領域不斷出現重大技術突破，在技術進步與需求增長的雙向驅動下，生物醫藥產業迎來發展的黃金時期。根據 EvaluatePharma 2017 年醫藥市場報告，未來 5 年，全球處方藥市場將以 6.5% 的複合年增長率穩步增長，2022 年可達 1.06 萬億美元。其中生物製品行業的發展將最為迅速，至 2022 年在銷售排名前 100 名的藥物產品中將有 52% 為生物科技製品，生物醫藥行業佔整個醫藥市場份額將從 2008 年的 17% 增長到 2022 年的 30%，達到 3,260 億美元的規模 [1]。

　　生物醫藥產業飛速增長，逐步成為經濟發展的新引擎。以美國為代表的生物醫藥大國，生物醫藥行業總產值佔國內生產總值（GDP）的比重已經達到 17% 左右，湧現了波士頓、洛杉磯等生物醫藥聚集區；日本、德國等發達國家的健康醫療產業增加值佔 GDP 的比重也超過了 10%，成為社會和經濟發展的重要引擎 [2]。中國醫藥產業亦增長強勁，2011-2016 年，內地藥品市場規模持續上升，複合年增長率達到 15%[3]。根據 2016 年內地推出的《「健康中國 2030」規劃綱要》，內地將建設「健康中國」提升為國家戰略，預計 2030 年內地健康服務業總規模將達到 16 萬億元人民幣 [4]。可以預見，隨着現代生物技術將在疾病治療領域廣泛應用，生物醫藥產業將逐步成為健康醫療行業中發展最快、技術含量最高的子分類產業，一批在生物醫藥領域創新突出的企業將逐步湧現，成為國民經濟增長及行業創新的新動力。

1　資料來源：EvaluatePharma《World Preview 2017》，2017 年 6 月。
2　資料來源：〈發展健康產業是引領我國經濟增長的重要動力〉，載於《新華網》，2017 年 7 月 20 日。
3　資料來源：麥肯錫《中國醫院藥品報告：深度洞察》，2017 年 8 月。
4　參見中國國務院《健康中國 2030 規劃綱要》，2016 年 10 月。

1.2 股權融資為生物醫藥企業發展注入必需的金融資源

第一，股權融資（而非債務融資）是生物醫藥企業成長期的重要融資方式。

與傳統行業相比，生物醫藥企業呈現出高投入、高產出、高風險、高技術密集型的特點。從生產週期來看，新藥推出前須經過多個階段臨床試驗，分別對藥物的安全性、治療作用、風險和不良反應四個層級作驗證，並獲得監管機構批准才可上市。美國生物技術創新組織（BIO）發佈的新藥臨床試驗通過率報告[5] 顯示，一款候選藥物由一期臨床試驗到最終獲得美國 FDA[6] 批准並上市的概率僅為 10%，大部分藥物無法通過二期向三期臨床試驗的過渡階段，導致整個研發過程失敗。因此，根據海外一些較具規模的生物醫藥企業發展經驗，投資生物醫藥企業時間漫長、代價昂貴且充滿風險，平均融資規模 2.5 億美元、產品推出時間為 8-10 年左右，年度虧損平均 3,000 萬美元[7]。

藥物研發的巨大不確定性使得生物藥物行業成為進入壁壘極高的行業，也決定了風險投資和股權資本運作將是滿足生物醫藥企業財務需求的最主要途徑。不同性質資金可以根據生物醫藥類企業的特點分階段介入，為生物醫藥行業不同時期的發展提供具規模的資金支援。近年來資本市場為生物醫藥行業上市設置新的上市機制，有效引導創業風險投資和私募股權合理佈局，完成「投入—產出」的資本循環，進一步優化資源配置。

第二，隨着藥物研發競爭日益加劇，全球生物醫藥行業的市場集中度進一步提升，全球主要的醫藥公司需要從資本市場獲得充盈的資本投入，持續保持強勢地位。

例如輝瑞製藥公司於 2000 年收購瓦納蘭百特製藥公司，2003 年又收購法瑪西亞製藥公司；2004 年安萬特製藥公司與賽諾菲製藥公司合併。2017 年，全球生物醫藥領域併購事件超過 400 宗，金額超過 1,800 億美元，主要為大型企業併購中小企業獲得新興技術，拓展市場[8]。通過風險融資和股市融資等方式，企業可利用資

5　BIO《Clinical Development Success Rates 2006-2015》，2016 年 5 月。

6　美國食品藥物管理局（U.S. Food and Drug Administration）是由美國國會即聯邦政府授權、專門從事食品與藥物管理的最高執法機關。

7　資料來源：中國外商投資企業協會藥品研製和開發行業委員會等《推動臨床研究體系設計與實施，深化醫藥創新生態系統構建》，2017 年 12 月。

8　資料來源：火石創造編《中國生物醫藥產業發展藍皮書 2017》，2018 年 8 月。

本優勢實現產品產業化，促進企業間的研發創新能力傳導，獲得加速成長的動力。
表 1 羅列全球主要的製藥企業及其業務。

表 1：全球排名前十名製藥企業及其主要的生物製藥產品			
公司	2017 年製藥業務收入（億美元）	研發支出（億美元）	主營業務
輝瑞	525.40	76.57	以研究為基礎的製藥公司，致力於廣泛治療領域的健康、預防和治療疾病，提供心臟病學、腫瘤學和免疫學等方面的藥物
羅氏	443.68	103.92	以研發為重點的醫療保健公司，開發、生產和提供創新的療法以及診斷儀器和檢測。醫藥產品涵蓋腫瘤學、糖尿病、眼科、神經科學、免疫學、傳染病等領域
賽諾菲	366.63	66.97	法國製藥公司，提供的處方藥和非處方藥主要用於中樞神經系統、心血管、腫瘤學、糖尿病、內科醫學及疫苗領域
強生	362.56	105.54	主要提供丙肝（HCV）、愛滋病（HIV/AIDS）、消化系統疾病等方面的藥物
默沙東	353.90	100.00	以研究為基礎的製藥公司，提供腫瘤學、生育、神經退行性疾病和內分泌學方面的藥物
諾華	330.00	89.72	包括多個治療領域的創新生物療法和藥物，如腫瘤學、心臟病學、神經科學、免疫學、眼科護理、仿製藥和生物仿製藥等
艾伯維	282.16	49.82	以研究為基礎的生物製藥公司，開發和銷售的藥物主要集中在治療風濕病、腸胃病、皮膚病學、腫瘤學、病毒學、神經系統疾病、代謝相關疾病等
吉利德	256.62	33.74	以研究為基礎的生物製藥公司，專注於病毒學、肝病學、血液學、腫瘤學、心血管病、炎症、呼吸道疾病等領域
葛蘭素史克	240.38	62.35	以研究為基礎的製藥公司，提供心血管疾病、婦科、糖尿病、抗感染、皮膚病、呼吸系統疾病方面的藥物
安進	228.49	35.62	總部位於美國加利福尼亞州的製藥公司，專注於發現和開發創新的生物製劑和小分子藥物

資料來源：《2018 年全球製藥企業 TOP10》，Igeahub。

　　第三，目前腫瘤免疫治療是生物技術在醫藥行業的主要運用領域之一，這部分創新生物醫藥企業大多通過上市等方式獲得資金，支援與腫瘤免疫治療有關的藥物開發。

免疫療法的代表藥物是 PD-1 及 PD-L1 抗體。全球共有 5 種抗體藥物，包括百時美施貴寶的 O 藥（Opdivo）、默沙東公司的 K 藥（Keytruda）。O 藥的全球收入從 2014 年的 2,000 萬美元增加至 2017 年的 57.53 億美元，複合年增長率 560%；K 藥全球收入從 2014 年的 5,500 萬美元增加至 2017 年的 38.09 億美元，複合年增長率超過 310%[9]。兩者成為生物醫藥行業中最為暢銷的藥物，成績既吸引了生物醫藥企業不斷加大在這一領域的研發，也為生物醫藥企業上市後的盈利提供重要保障。生物科技企業與資本市場的有效對接，將促進更多癌症治療藥物的不斷湧現。

1.3　全球資本市場改革推動生物醫藥行業進一步發展

近年來，國際主要的證券市場逐步建立了適用於生物科技企業上市的證券市場規則，吸引了眾多被傳統市場拒諸門外的生物醫藥企業。美國的納斯達克市場是全球生物醫藥企業上市的主要市場，上市制度靈活，企業可根據自身盈利、淨資產規模、總資產規模、市值規模等指標，尋求不同層級的上市板塊。2012 年 4 月，美國頒佈 JOBS 法案（Jumpstart Our Business Startups Act，初創期企業推動法案），大力扶持創業和創新型中小企業的融資和上市，尤其是對於生物科技企業上市融資提供了強而有力的支援，有利於企業再融資，支援創新型生物醫藥產品的開發和成果產業化。截止 2019 年 2 月，約有 745 家大型健康類企業陸續登陸納斯達克市場，其中生物醫藥企業約 171 家，總市值達到 4,907 億美元[10]，中國也有多家生物醫藥企業赴美上市（見表 2）。

表 2：近年來在納斯達克上市的部分中國生物醫藥企業		
公司	上市年份	企業概況
百濟神州	2016	專注於癌症治療的創新型分子標靶藥和腫瘤免疫藥物，2016 年 2 月在納斯達克上市
和黃中國醫藥	2016	專注於發現和開發腫瘤及自身免疫性疾病的創新療法，公司曾於 2006 年 5 月在英國倫敦證券交易所 AIM 市場成功上市。2016 年 3 月在納斯達克作二次上市
再鼎醫藥	2017	主要從事腫瘤、自身免疫、傳染性疾病藥物研發。2017 年 9 月在納斯達克上市

資料來源：根據公開資料整理。

9　資料來源：Frost & Sullivan。
10　資料來源：納斯達克交易所網站，數據時點為 2019 年 2 月 25 日。

其他國家的證券交易所也不斷修改其主板上市制度，推動生物醫藥產業發展。從 1993 年起，英國倫敦證券交易所（LSE）對其主板進行多次制度改革，1995 年推出另類投資市場（AIM），促進英國生物科技產業迅速發展。2014 年，英國生物醫藥企業募集的風險投資數量和金額在全歐洲位列第一，2016 年，LSE 的主板和 AIM 板塊共有 11 家和醫療大型健康相關的公司成功掛牌。其中，生物醫療領域迎來了 ConvaTec（康維德）上市，融資 14.65 億英鎊，成為歐洲醫療板塊近 20 年來最大一筆首次公開發行（IPO）的股份 [11]。

德國法蘭克福證券交易所為了解決具有高成長性的科技型企業的融資難題，曾在主板市場之外建立了「新市場」（Neuer Markt），2003 年又嘗試引進新的股票市場劃分 [12]，採用不同的透明度標準，2005 年建立了針對中小板塊公司的初級市場板塊。這些制度創新不僅促進了上市生物技術企業的研發和企業發展，也帶動了德國生物產業整體飛速發展。目前在歐洲國家當中，德國在新藥研究與開發方面居歐洲第一，生產的藥物佔歐洲市場份額的 40% 以上 [13]。2018 年 1 月，法蘭克福證券交易所迎來首家中國生物科技型企業上市 [14]。

2　亞洲生物醫藥行業獲得新的增長動力

2.1　內地生物醫藥行業的政策紅利和行業整合

內地生物醫藥產業發展欣欣向榮，很大程度上得益於政策紅利的不斷釋放。「十三五」、「健康中國 2030」等國家戰略加速落實，藥監會及各監管部門加快改革步伐，新版醫保目錄、一致性評價、優先評審、創新藥加快審批等重要政策改

11　資料來源：Beyond Laboratory。
12　包括將市場分割為兩個獨立的板塊（Prime Standard、General Standard）和構建新的行業指數，只有高級市場（Prime Standard）發行人才能獲納入法蘭克福證券交易所各項指數；市值大、交易額高的大企業被納入 DAX 指數；傳統工業領域的中小企業被納入 MDAX 和 SDAX 指數；科技行業的中小企業則被納入 TecDAX 指數。
13　資料來源：〈2017-2022 年中國醫藥工業行業市場行情動態與投資戰略研究報告〉，載於《智研諮詢》網站。
14　該企業為貝羅尼集團有限公司。

革為生物醫藥行業帶來全新的發展機遇。國家對自主創新鼓勵力度不斷加大,從「十一五」時期的以仿製為主,到「十二五」時期的自主創新,再到現在「十三五」國家科技創新規劃,中國生物醫藥企業獲得長足發展,為建設健康大國培育了重要的創新支撐能力。

內地生物醫藥行業的資本投入日漸活躍,行業亦加速整合。目前,內地已經初步形成以長三角、珠三角、環渤海為中心的生物醫藥產業集羣,逐步形成產業集聚效應。其中,深圳的生物醫藥產業規模於 2016 年達到 2,000 多億元人民幣,出現了華大基因、邁瑞、北科生物等主要的創新型企業[15]。2018 年前十個月,內地醫療保健行業完成併購交易量 357 宗,總規模超過 1,200 億元人民幣,併購規模同比增加了近 20%(見表 3 的例子);其中,生物醫藥的併購交易數量佔醫療保健全行業的五分之三。另外,在醫療保健行業細分領域投資中,生物製藥投資事件超過 240 宗,佔比 49%[16],特別是針對腫瘤的治療及藥物、基因診療產品及服務、植入介入醫療產品等加大了整合力度,進一步提高行業創新型生物醫藥產品的研發能力。

表 3:近年部分內地生物醫藥企業的兼併收購活動			
年份	收購方	被收購方	金額 / 股權	領域
2014	微創醫療	Wright	2.9 億美元	關節植入物
2014	普華和順	天新福	8 億元人民幣/100%	骨科植入器械
2015	安科生物	中德美聯	4.5 億元人民幣/100%	基因檢測
2015	上海醫療	中谷生物	2.7 億元人民幣/67.5%	抗腫瘤藥物中間體
2016	南京新百	齊魯幹細胞	34 億元人民幣/76%	造血幹細胞
2016	百花村	華威醫藥	19.54 億元人民幣/100%	抗腫瘤藥物研發
2017	三胞集團	凡迪生物	6.8 億元人民幣	基因檢測
2018	華西股份、海辰藥業	NMS	3.69 億美元	抗腫瘤藥物研發
2018	華東醫藥	英國 Sinclair	1.69 億英鎊/100%	生物技術

資料來源:《中國生物醫藥產業發展藍皮書》,2017;GBI SOURCE 數據庫。

---

15 資料來源:〈深圳生物產業規模超 2000 億元〉,《中國經濟日報》,2017 年 3 月 31 日。
16 資料來源:中國製藥網。

2.2　香港資本市場對接區內生物醫藥企業的發展需求

2018 年 4 月，香港交易所修改《上市規則》，為未有收益或盈利的生物科技公司開闢了新的上市途徑。自新規則實施後，成功進入資本市場的既包括還沒有營業收入而又符合上市新規則的公司，也有根據原有上市規則上市的有收入和盈利的公司。

2018 年，香港上市新規則認可中國國家食品藥品監督管理總局（China Food and Drug Administration，簡稱 CFDA）為有資格評判生物科技產品開發階段的監管機構，與美國的 FDA 及歐洲藥物管理局（European Medicine Agency，簡稱 EMA）具有同樣認可的效力。香港上市新規則對中國藥物標準的認可，有助於中國標準在國際市場的運用和推廣。

香港市場上市平台所提供的「退出管道」有利於吸引更多創投資金流向高風險、高潛在回報的生物科技領域。相較其他海外市場投資者，香港市場投資者對國內相關法規與市場狀況會有更好的理解，有助評估內地生物醫藥企業的風險；內地投資者也可通過互聯互通機制購買於香港上市的生物科技公司的股票，為生物技術企業營造良好的投資者基礎和投融資環境。

3 資本市場開發的生物科技指數及其相關產品

3.1　美國資本市場開發的主要生物科技指數

生物醫藥企業作為最具活力也最具未來發展空間的一個子行業，備受市場矚目。為方便投資者參與這個市場，各主要資本市場推出了相關指數及金融產品。其中最具代表性的有：標準普爾（標普）於 2006 年 1 月推出的「標普生物技術精選行業指數」（S&P Biotechnology Select Industry Index），其成份股共有 119 隻，總市值

70.78 億美元 [17]；納斯達克早於 1993 年推出的「納斯達克生物科技指數」(NASDAQ Biotechnology Index，簡稱 NBI)，其成份股共有 221 隻 [18]，已經成為涵蓋納斯達克上市企業中生物科技和基因製藥分類的一個重要行業指數。過去十年間，這兩個指數的總回報率是標普 500 指數的 2 至 3 倍（見圖 1 及表 4）[19]。

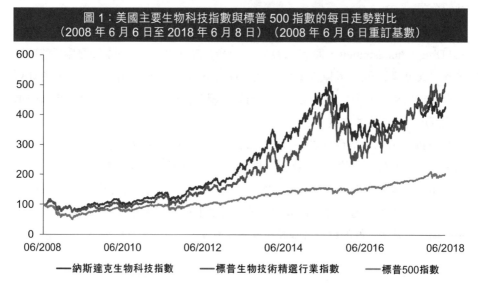

圖 1：美國主要生物科技指數與標普 500 指數的每日走勢對比 (2008 年 6 月 6 日至 2018 年 6 月 8 日)（2008 年 6 月 6 日重訂基數）

——納斯達克生物科技指數　　——標普生物技術精選行業指數　　——標普500指數

資料來源：彭博。

表 4：美國主要生物科技指數及標普 500 指數的表現對比

指數	3 年回報率	5 年回報率	10 年回報率	市盈率（倍）
納斯達克生物科技指數	34.75%	85.17%	325.96%	48.03
標普生物技術精選行業指數	110.81%	162.50%	398.97%	-13.02
標普 500 指數	42.56%	69.10%	104.24%	20.28

資料來源：彭博。回報率數據時點為 2018 年 6 月 6 日，市盈率數據時點為 2019 年 1 月 29 日。

17　資料來源：標準普爾網站，數據時點為 2019 年 1 月 31 日。
18　資料來源：納斯達克網站，數據時點為 2019 年 2 月 25 日。
19　過去的表現並非未來表現的指標。

3.2　基於生物科技指數開發的 ETF 產品

　　美國已經形成全球最大的生物科技交易所買賣基金（Exchange-Traded Fund，簡稱 ETF）市場，共有 18 隻該類 ETF，總資產管理規模約為 190 億美元（見表 5）。這些 ETF 產品的標的成份股均在美國上市，追蹤的大多是美國三大生物醫藥指數：納斯達克生物科技指數、標普生物技術精選行業指數和紐約證券交易所的生物技術指數。其中，資產管理規模最大的 ETF 是 iShares NASDAQ Biotechnology ETF，追蹤標的為納斯達克生物科技指數，資產管理規模為 96 億美元。此外，還有 6 個反向和槓桿式生物技術 ETF，進一步豐富了美國的生物科技指數產品線。在歐洲市場也有兩隻生物科技 ETF 產品，分別在倫敦和法蘭克福上市。當中較大的一隻 ETF 由 Invesco 推出，追蹤納斯達克生物技術指數，資產管理規模為 4.9 億美元。

　　內地市場有 6 隻生物科技指數基金，還沒有生物科技 ETF。這些基金中有 4 隻是跟蹤 A 股的生物科技指數，另外 2 隻跟蹤美國的生物科技指數。在 6 隻基金中，其中 3 隻是結構性基金。目前為止，內地沒有生物科技指數期貨或其他相關衍生品交易。（見表 6。）

表 5：於美國和歐洲市場上市的主要生物科技指數 ETF 產品					
股份代碼	產品名稱	產品性質	發行人	追蹤指數	資產總值 *（百萬美元）
美國上市					
IBB	iShares NASDAQ Biotechnology ETF	傳統型	BlackRock iShares	NASDAQ Biotechnology Index	9,563.22
XBI	SPDR S&P Biotech ETF	傳統型	State Street Global Advisor	S&P Biotechnology Select Industry Index	5,386.64
FBT	First Trust Amex Biotechnology Index ETF	傳統型	First Trust	NYSE Arca Biotechnology Index	1,728.72
BBH	VanEck Vectors Biotech ETF	傳統型	VanEck	MVIS US Listed Biotech 25 Index	430.79
LABU	Direxion Daily S&P Biotech Bull 3x Shares	槓桿式（300%）	Direxion	S&P Biotechnology Select Industry Index	411.66
BIB	ProShares Ultra NASDAQ Biotechnology ETF	槓桿式（200%）	ProShares	NASDAQ Biotechnology Index	363.94

（續）

表 5：於美國和歐洲市場上市的主要生物科技指數 ETF 產品					
股份代碼	產品名稱	產品性質	發行人	追蹤指數	資產總值 *（百萬美元）
PBE	Invesco Dynamic Biotechnology & Genome ETF	傳統型	Invesco	Dynamic Biotechnology & Genome Intellidex Index	280.12
ARKG	ARK Genomic Revolution Multi-Sector ETF	傳統型	ARK Investment Management	無	226.11
SBIO	ALPS Medical Breakthroughs ETF	傳統型	ALPS	Poliwogg Medical Breakthroughs Index	224.71
BBC	Virtus LifeSci Biotech Clinical Trials ETF	傳統型	Virtus	LifeSci Biotechnology Clinical Trials Index	79.18
LABD	Direxion Daily S&P Biotech Bear 3x Shares	反向式（-300%）	Direxion	S&P Biotechnology Select Industry Index	77.47
CNCR	Loncar Cancer Immunotherapy ETF	傳統型	Exchange Traded Concepts	Loncar Cancer Immunotherapy Index	59.67
BTEC	Principal Healthcare Innovators Index ETF	傳統型	Principal Financial Group	NASDAQ U.S. Health Care Innovators Index	54.62
UBIO	ProShares UltraPro NASDAQ Biotechnology ETF	槓桿式（300%）	ProShares	NASDAQ Biotechnology Index	37.64
BBP	Virtus LifeSci Biotech Products ETF	傳統型	Virtus	LifeSci Biotechnology Clinical Trials Index	34.36
BIS	ProShares UltraShort NASDAQ Biotechnology ETF	反向式（-200%）	ProShares	NASDAQ Biotechnology Index	28.97
IEIH	iShares Evolved U.S. Innovative Healthcare ETF	傳統型	BlackRock iShares	無	5.15
ZBIO	ProShares UltraPro Short NASDAQ Biotechnology ETF	反向式（-300%）	ProShares	NASDAQ Biotechnology Index	2.99
歐洲上市					
SBIO	Invesco NASDAQ Biotech UCITS ETF	傳統型	Invesco	NASDAQ Biotechnology Index	490.65
A2DWAW	iShares NASDAQ US Biotechnology UCITS ETF	傳統型	BlackRock iShares	NASDAQ Biotechnology Index	14.36

* 2018 年 7 月 10 日之數據。

資料來源：彭博及各發行人網頁。

表 6：中國市場的生物科技指數基金產品		
產品名稱	追蹤指數	資產總值 * （百萬人民幣）
申萬菱信中證申萬醫藥生物指數 分級證券投資基金	中證申萬醫藥生物指數	853.53
招商國證生物醫藥指數 分級證券投資基金	國證生物醫藥指數	394.56
廣發納斯達克生物科技指數型發起式 證券投資基金	納斯達克生物科技指數	359.15
易方達生物科技指數分級 證券投資基金	中證萬得生物科技指數	210.00
易方達標普生物科技指數 證券投資基金	標普生物技術精選行業指數	78.50
匯添富中證生物科技主題指數型發起式 證券投資基金	中證生物科技主題指數	77.50

* 2018 年 7 月 10 日之數據。

資料來源：Wind 及各發行人網頁。

3.3　立足亞洲市場的中華交易服務香港生物科技指數

相較於美國和歐洲資本市場，目前亞洲區內生物醫藥產業還處於初步發展階段。一方面，區域內缺乏大型的生物製藥公司，國際排名前十名的醫藥公司大都集中在歐美地區，如羅氏、諾華等（見上文表 1），也缺乏類似英國劍橋桑格研究院等國際級的生物技術研發中心，將生物醫藥研究轉化為商業用途。這不僅限制了亞洲區內生物科技研發集羣的形成，也不利於知識與人才匯聚，投資者和分析師對製藥公司也缺乏足夠的評估經驗和專業知識。無疑，香港上市新規則將吸引更多優秀的生物技術企業聚集於此，為區內生物醫藥企業發展注入新動力。

資本市場對於追蹤、反映生物科技行業發展指標的需求也會越來越強烈。生物醫藥企業的產品大多處於研發階段，股價變動受臨床試驗進展的影響較大，股價震盪是常態。因此，市場需要為此板塊設定一個行業指標，使得配置該板塊時可分散投資個股的風險。

為此，滬、港、深三家證券交易所的合資公司中華證券交易服務有限公司（簡稱「中華交易服務」）於 2018 年 11 月 14 日正式推出了「中華交易服務香港生物科技指數」（簡稱「中華香港生物科技指數」，代碼為 CESHKB），作為投資香港生物科技股份的基準指標。以下介紹該指數的編制與表現情況。

（1）成份股遴選方式

該指數以 2014 年 12 月 12 日為基日，以 2000 點為基點，樣本空間為香港交易所主板第一或第二上市的生物科技公司，於 2019 年 4 月 11 日的成份股數目為 17 隻。

入選中華香港生物科技指數的成份股必須符合以下條件：公司主營業務來自生物科技行業、或根據香港聯合交易所《上市規則》第十八 A 章申請上市的生物科技股、股票過去一年日均市值不少於 15 億港元。

由於希望盡量納入更多合適的生物科技股，於指數成份股少於 20 隻的期間，設有快速納入機制，即在股份上市第十個交易日結束後，便可將其納入指數。在篩選新上市股份作為成份股時，日均市值會按新股上市第四個交易日至審核截止日以來計算。

（2）指數計算方式

中華香港生物科技指數的價格指數是以成份股的流通市值加權計算，流通市值越大的公司，對指數的影響性就越大。成分股的股價變化，能夠直接反映在指數變動上。當成份股價格上漲，其市值也將隨之增加，那麼該成份股的權重也就越大。相反，當成份股價格下跌，其市值隨之減少，那麼該成份股的權重也就相應減少。具體計算方法如下：

計算公式

$$報告期指數 = \frac{報告期股本數的調整市值}{除數} \times 2000$$

其中，調整市值 $= \sum ($股價 $\times$ 調整股本數 $\times$ 權重調整因子$)$

除數的確定

除數的初始值為 2014 年 12 月 12 日成份股的總市值。為保證指數的連續性，當成份股名單發生變化、或成份股的股本結構發生變化、或成份股的市值出現非交易因素的變動時，將根據指數維護規則修正原除數。

$$\frac{修正前的調整市值}{原除數} = \frac{修正後的調整市值}{新除數}$$

其中，修正後的調整市值 = 修正前的調整市值 + / − 新增 / 減調整市值。

股價的確定

各樣本股的計算價位（X）根據以下原則確定[20]：

若當日沒有成交，則 X＝開盤參考價；若當日有成交，則 X＝最新成交價

調整股本數

調整股本數＝總股本 X 加權比例

加權比例為根據自由流通股本所佔總股本的比例（即自由流通比例[21]）賦予總股本一定的加權比例。為反映市場中實際流通股份的變動情況，在計算指數時，剔除了上市公司股本中的限售股份，以及由於戰略持股或其他原因導致的基本不流通股份，剩下的股本稱為自由流通股本，也即自由流通量。各樣本股採用分級靠檔的方法，即根據自由流通比例計算一定的加權比例，以確保計算指數的股本保持相對穩定。

權重調整

在計算中華香港生物科技指數時，權重調整因子介於 0 和 1 之間，以使樣本股權重不超過 10%。然而，當樣本股數目少於 20 隻時，樣本股權重上限變為 15%。當樣本股數目少於 8 隻時，樣本股權重上限變為 25%。當樣本股數目降至 5 隻以下時，指數將採用等權重加權計算，即每隻樣本股在指數中的權重相同。權重調整因子隨樣本股定期調整而調整，調整時間與指數樣本定期調整實施時間相同。在下一個定期調整日前，權重調整因子一般固定不變[22]。

（3）指數表現和前景

按 2019 年 3 月 6 日的數據，中華香港生物科技指數涵蓋的成份股中，藥明生物比重最高，為 16.30%，信達生物佔比達到 13.46%，中國生物製藥、三生製藥和金斯瑞生物科技的權重均超過 10%[23]。指數既包含了基於新上市規則的生物科技公司，也包含了行業中發展相對成熟的公司，可讓投資者綜合觀察板塊表現，相關的指數產品亦方便投資者分散個股風險。

20　樣本股的即時成交股價來自香港交易所通過各種管道發佈的行情信息。即時指數計算時間覆蓋香港交易所的交易時段。

21　自由流通比例＝自由流通量／總股本。

22　指數的具體計算方式請參見指數計算和維護細則（Calculation and Maintenance Methodology）（https://www.cesc.com/en/Index/Hong-Kong-Overseas/Ces-Hk-Biotech.html）。

23　資料來源：中華交易服務。

　　中華香港生物科技指數的年化波動率與納斯達克生物科技指數相近，較納斯達克綜合指數為高，其年化收益率達 42.41%（過去三年），遠遠超過恒生指數及納斯達克生物科技指數的同期表現（見圖 2 及表 7）。

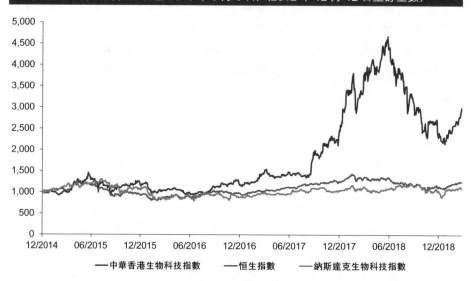

圖 2：中華香港生物科技指數與恒生指數及納斯達克生物科技指數的走勢對比（2014 年 12 月 12 日至 2019 年 3 月 6 日）（2012 年 12 月 12 日重訂基數）

——中華香港生物科技指數　　——恒生指數　　——納斯達克生物科技指數

資料來源：彭博。

表 7：中華香港生物科技指數與恒生指數及納斯達克生物科技指數的表現比較

指數	年化回報率			年化波動率			經風險調整後的回報率 *		
	2019 年初至 3 月 6 日	1 年期	3 年期	2019 年初至 3 月 6 日	1 年期	3 年期	2019 年至 3 月 6 日	1 年期	3 年期
中華香港生物科技指數	30.00%	-10.03%	42.41%	28.52%	41.84%	35.43%	1.05	-0.24	1.20
恒生指數	12.35%	-2.84%	12.90%	15.48%	18.64%	15.97%	0.80	-0.15	0.81
納斯達克生物科技指數	13.51%	-0.93%	7.85%	24.09%	24.69%	31.10%	0.56	-0.04	0.25

* 經風險調整後的回報率是將年化回報率除以年化波動率所計算出的比率。

資料來源：根據彭博提供的每日收市指數計算，截至 2019 年 3 月 6 日。

4 結語

長遠來看,中國生物科技行業已迎來發展的春天。香港交易所根據生物醫藥企業成長期獨有的財務特點(上市前後長期無盈利或收入),以及其風險特性,建立適宜的上市規則,引導更多創業風險投資和私募股權類投資進入,將有助於推動一批創新型生物科技企業的湧現,從而促進亞洲區內核心產業的發展,實現區域經濟升級。

中華香港生物科技指數的推出,將進一步促進香港資本市場的多樣化,其中涵蓋的成份股公司專注於利用生物技術研發人類疾病的新型治療方式,代表了亞洲生物科技未來的發展方向和趨勢,對中國形成生物醫藥核心領域的創新競爭力具有重要意義。

附錄

中華交易服務香港生物科技指數成份股的基本情況

股份代號	公司名稱	上市日期	2018 年資產回報率(%)	2018 年股本回報率(%)	2018 年每股盈利年增長率(%)	2018 年每股盈利(港元)	2018 年收入增長率(%)	2018 年負債與股東權益比率(%)
775	長江生命科技	16/07/2002	2.50	5.92	1.86	0.03	11.50	108.98
1035	BBI 生命科技	30/12/2014	8.24	10.82	22.88	0.17	25.78	0.86
1061	億勝生物科技	27/06/2001	17.71	27.70	36.13	0.41	30.78	5.92
1177	中國生物製藥	29/09/2000	25.59	47.09	274.67	0.87	40.96	9.09
1530	三生製藥	11/06/2015	9.26	15.95	35.14	0.59	22.75	36.99
1548	金斯瑞生物科技	30/12/2015	2.99	6.09	-22.37	0.09	51.34	2.13
1672	歌禮製藥	01/08/2018	-0.32	-0.36	86.54	-0.01	212.63	0.00
1801	信達生物	31/10/2018	-134.69	-512.60	-926.38	-20.44	-48.88	18.88
1877	君實生物	24/12/2018	-26.16	-28.56	-123.29	-1.41	-18.64	0.00
2269	藥明生物	13/06/2017	8.86	10.49	116.67	0.62	56.56	0.00
2359	藥明康得	13/12/2018	12.83	18.81	70.23	2.64	23.85	0.85
2552	華領醫藥	14/09/2018	-415.05	-1602.61	-1221.06	-11.94	274.07	0.00
2616	基石藥業	26/02/2019	-158.92	-370.68	不適用	-12.28	不適用	0.00
6118	奧星生命科技	07/11/2014	0.01	0.02	-175.00	0.00	49.30	4.16
6160	百濟神州	08/08/2018	-41.03	-56.00	-416.67	-7.29	-22.17	2.82
6185	康希諾生物	28/03/2019	-16.64	-24.92	不適用	不適用	不適用	29.86
6826	昊海生物科技	30/04/2015	9.74	12.17	11.16	3.07	14.94	0.96

註:成份股名單為 2019 年 4 月 11 日之名單。

資料來源:成份股名單源自中華交易服務網站;資料源自彭博及 Wind,取於 2019 年 4 月 4 日。

第13章

內地及香港證券市場
對大手交易需求漸殷

香港交易及結算所有限公司
首席中國經濟學家辦公室

摘 要

對股本證券市場而言，大手交易（又稱「大宗交易」）機制的設置有其重要性，可將大額交易所帶來的價格影響盡量減輕，亦避免向交易發出不必要的信號，有助經紀按最佳條件執行交易。這種機制能有效地讓有需要人士只需進行一次性交易來大手買賣股份，而成交價格和時間都比在競價市場交易有更高的確定性。環顧全球，大手交易機制各有不同，包括場外自行議定成交（即協議交易）後向交易所申報、獨立的場外交易平台（黑池），又或專設的場內大手交易機制。雖然大手交易機制各有不同，但各大證券交易所協議交易的增長顯示其大手交易有增長趨勢。

資產管理行業的增長是大手交易需求的主要推手。對主動型資產管理人來說，在場內顯示買賣盤來進行大額交易的價格影響會侵蝕其主動型回報。對被動型資產管理人而言，在指數調整成份股時所需的股份交易涉及的成本也不能忽視，甚至比收取的管理費還要高許多。再者，申贖基金單位有時亦涉及在短時間內進行大額交易。其他需要大手交易的原因包括：合併收購、股份回購以及私募股權投資和大股東交易等。此外，股本證券衍生產品和結構性產品的發展或會帶動市場對大手交易的需求，以提升結算時股份交收的效率。另外，有實證研究顯示，協議交易（所謂「樓上市場」）有效地提供額外的流動性和降低價格影響，尤其是對場內競價市場流動性低的股份而言。

在內地市場，上海和深圳兩家證券交易所均專設「大宗交易」平台。但與環球主要市場不同的是，內地 A 股的大手交易以賣方驅動的交易為主導 —— 大多是大股東在禁售期結束後減持名下的大額持股，較少是資產管理業務的需求而衍生的交易。這也許是為何內地協議交易的平均每宗成交金額一直遠高於主要股票市場相對的金額的原因。不過，「滬股通」和「深股通」的開通讓國際投資者更廣泛地參與 A 股市場，估計情況將會改變。再者，A 股獲納入國際指數，加上「滬股通」和「深股通」淨買入的每日額度擴大至四倍，預料能推動投資 A 股的離岸機構投資者其資產管理業務進一步增長。由此可見，隨着國際機構投資者提高參與度，進行 A 股大手交易的需求也將會上升。

在香港，透過交易所的非自動對盤交易機制，交易所參與者可向交易所申報經雙方經紀自行議定、內部兩邊客交易，又或於黑池進行的大手交易。協議交易的成交於 2018 年達到高峰。鑒於平均每宗成交金額相對偏高，香港的協議交易相信大都是大手交易。大手交易與日俱增，某程度上是香港的資產及財富管理業務持續增長所帶動。再者，內地投資者愈來愈偏好進行多元資產配置，包括買入投資港股的互惠基金。內地基金透過「港股通」進入香港市場將推高大手交易的潛在需求。

由於跨境大手交易機制的潛在需求日高，而內地與香港之間的跨境交易現時大都經

「滬港通」及「深港通」進行，市場值得研究可否進一步優化內地與香港市場的互聯互通，在適當的安排和可控環境下，引進大手交易的安排，切合投資者的需求。屆時，跨境市場的流動性將可進一步提高，投資者也可以較佳的條件執行交易。

1 大手交易機制對股票交易的重要性

1.1 大手交易是甚麼？

大手交易（又稱「大宗交易」）是指在一宗交易中大額買入或賣出證券。大手交易通常不在公開競價市場進行，以減低交易成本或價格影響（即為了預防買入時價格被推高或賣出時價格被推低，而避免向公開市場披露大額買賣盤），從而獲得「最佳執行」價。有大手交易機制的證券交易所，都對大手交易的股數或交易金額設置不同的門檻，而大手交易的模式亦不盡相同。

大手交易通常在場外執行，包括經紀之間磋商成交後向證券交易所申報交易（即協議交易），以及在其他交易場所進行的非透明交易（黑池交易）。投資者在場外進行交易的好處在於交易前隱藏身分，其他投資者不能在交易所的買賣輪候隊伍中知道有大額買賣盤，這樣就能避免大額買賣盤在執行前的價格走向對落盤者不利。此外，大額買賣盤還可在場外與一些沒有在交易所下盤的「保留」或「隱藏」流動性作配對。

常見的大手交易方式有以下兩種：

- **協議交易**：亦稱「樓上市場」，經紀自行代客戶私下商議大額買賣盤，成交後再向證券交易所申報作中央結算，以減低對手方風險。這些交易主要受流動性帶動，目的是要減低執行成本。
- **黑池大手交易**：黑池大手交易日益普及。不過，雖說投資者身份可以隱藏，也不怕交易金額龐大而發出不必要訊號影響價格，但投資者也會擔心資訊不對稱[1]而令黑池交易與場內交易的價格可能有偏差，另外也怕被黑池營運者超前交易。所以，黑池交易應有妥善監管及投資者保障的配套。

場內（又稱「樓下市場」）大手交易的價格影響通常頗高，因為在競價市場下顯示的買賣盤可能會有訊號效應，吸引市場進行更多同一方向的交易。曾有實證研

---

1 為了限制黑池交易與場內交易之間的價格差異，美國監管當局在 2005 年採納了多項法規（例如 Regulation NMS Rule 611），確保買賣盤是以最佳顯示價格或更佳價格執行。

究[2]估計過主要市場的場內大手交易不對稱的價格影響：在 1999 年至 2002 年間，大手買入的價格較公開市場價（沽盤報價）高約 0.16% 至 0.34%，而大手賣出的價格則較公開市場價（買盤報價）低 0.32% 至 0.42%。有人或會透過程式買賣，以小額交易去執行大手買賣盤（例如採用成交量加權平均價（VWAP）或時間加權平均價（TWAP）的買賣策略），但可能會令執行成本增添不確定性。若可設法避免資訊外洩問題（例如推出隱藏買賣盤類型），說不定會有更多人在場內進行大手交易。

這些大手交易執行方式有助減低價格影響和交易價格的不確定性。

1.2　全球市場對大手交易的相關需求

資產管理行業的增長是買方機構對大手交易有更大需求的主因之一。隨着投資基金的資產管理規模不斷增長，買賣的總交易金額亦水漲船高。根據羅兵咸永道的資料，資產及財富管理業的全球資產管理規模在 2016 年為 84.9 萬億美元，預計於所有地區都會有所增長，到 2020 年及 2025 年時將分別達至 111.2 萬億美元及 145.4 萬億美元（見圖 1）。中國內地是增長最快的市場之一，其資產管理規模在 2016 年為 48 萬億元人民幣（7.4 萬億美元），預計於 2022 年會倍增至約 90 萬億元人民幣（14 萬億美元）[3]。香港的資產及財富管理業的資產管理規模在 2018 年達至 23.9 萬億港元（3.1 萬億美元）[4]。

2　Frino, A.、D. Gerace 與 A. Lepone（2007）〈大手交易與連帶價格影響：兩種不對稱現象的國際例證〉（"Block trades and associated price impact: International evidence on the two asysmmetries"），載於《歐洲財務管理學會 2007 年周年大會論文集》（Papers for European Financial Management Association 2007 Annual Meetings）。

3　資料來源：Oliver Wyman（2018）〈中國的全球資產管理人：乘上改革浪潮〉（"Global asset managers in China: Riding the waves of reform"），載於 Oliver Wyman 的網站，2018 年 3 月。

4　資料來源：證券及期貨事務監察委員會（證監會）調查報告〈2018 年資產及財富管理活動調查〉，載於證監會網站，2019 年 7 月。（下稱「證監會 2018 年資產管理調查」。）

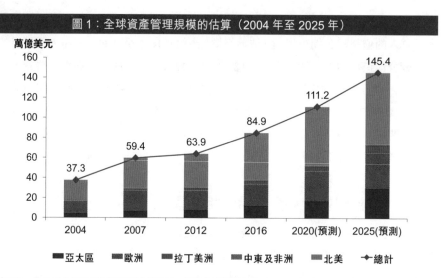

圖 1：全球資產管理規模的估算（2004 年至 2025 年）

萬億美元

145.4

2004　37.3
2007　59.4
2012　63.9
2016　84.9
2020（預測）　111.2
2025（預測）　145.4

■ 亞太區　■ 歐洲　■ 拉丁美洲　■ 中東及非洲　■ 北美　◆ 總計

資料來源：羅兵咸永道〈資產及財富管理革命：擁抱急速轉變〉（"Asset & wealth management revolution: Embracing exponential change"），載於羅兵咸永道網站，2017 年 10 月 30 日。

無論是主動型還是被動型的資產管理，大手交易機制都是減低價格影響的重要措施。若交易所沒有大手交易機制，執行交易的成本可能會大增。

- **就主動型資產管理而言**，資產管理人會採用各式投資策略，力求跑贏基準指數，取得主動回報（「alpha」回報）。主動型投資策略的基金在 2016 年佔全球資產管理規模的 71%[5]。一項實證研究[6]以澳洲 26 名主動型股票管理人為對象，發現往返交易（一買一賣）的價格影響能對其主動回報有顯著影響（0.27% 對比 0.92%）。此外，基金在市況低迷時需出售大量證券以應付贖回需要，可能會有重大贖回壓力。這説明了為何有些基金會採用「波動定價」，讓基金經理可調整基金的每股資產淨值，把估計贖回及認購所涉及的交易成本轉嫁投資者[7]。

5　資料來源：羅兵咸永道〈資產及財富管理革命：擁抱急速轉變〉（"Asset & wealth management revolution: Embracing exponential change"），載於羅兵咸永道網站，2017 年 10 月 30 日。（下稱「羅兵咸永道 2017 年文章」）。

6　Gallagher D. R. 與 A. Looi（2003）〈審視澳洲主動型股票管理人的市場影響成本〉（"An examination of the market impact cost of active Australian equity managers"），載於《SIRCA 研究報告》（*SIRCA Research Report*）。

7　見 Arnold and Porter Kaye Scholer LLP（2016）〈美國證券交易委員會採納新規：基金流動性、匯報與披露以及「波動定價」〉（"SEC adopts new rules: Fund liquidity, reporting and disclosure and 'swing pricing'"），載於 Lexology.com，2016 年 12 月 6 日。

- **就被動型資產管理而言**，資產管理人純粹追蹤基準指數的表現。被動型投資策略的基金在 2016 年佔全球資產管理規模的 17%[8]。一份研究報告[9]提到，追蹤指數的基金的市值在 2000 年佔標普 500 指數（美國大型股指數）股份市值逾 10%，同時佔羅素 2000 指數（美國小型股指數）約 6%。同一研究發現在美國市場上，與指數基金管理年費的約 10 個點子相比，被動型基金的指數調整活動其市場影響可以非常大（高達 168 個點子）（見表1）。若沒有大手交易機制，在收市前最後數分鐘或指數調整交易日的收市競價時段期間，有關的價格影響會更大。在交易所買賣基金（ETF）申贖基金單位方面，則與互惠基金相類似，可能要在短時間內買賣大額證券作套戥，以減低 ETF 的溢價／折讓，又或變賣手上證券以應付贖回需要[10]。

表 1：美國指數調整時的市場影響估計		
指數	增減涉及市值佔指數總市值的比率	增／減的異常回報	對回報率的影響
標普 500 指數	5.0%	5%	25 個點子
標普 1500 指數	3.4%	7%	25 個點子
羅素 2000 指數	16.8%	10%	168 個點子
羅素 3000 指數	1.8%	10%	18 個點子
納斯達克 100 指數	5.6%	3%	17 個點子
道瓊斯全市場指數	1.8%	0%	0 個點子

資料來源：Quinn 及 Wang 2003 年論文。

　　除了只持長倉的資產管理外，有些成熟的專業投資者也有興趣投資股本結構性產品（例如股票掛鈎票據）或衍生產品。這些產品大多透過實物交收作結算。例如，投資者投資長倉股票掛鈎票據時，若結算日當天正股股價跌低於協議行使價的氣墊水平，投資者須以協議行使價買入股份。為投資者進行結算的交易金額有時可以很高，特別是涉及只為少數高淨值人士而設的產品時。若投資者需為股票

8　資料來源：羅兵咸永道 2017 年文章。

9　Quinn, J. 與 F. Wang（2003）〈增減對股票指數回報的影響〉（"The impact of adds and deletes on the returns of stock indexes"），載於加洲大學柏克萊分校 Hass 學院金融工程碩士《應用金融項目》。（下稱「Quinn 及 Wang 2003 年論文」）。

10　資料來源：Dickson, J. M. 與 J. J. Rowley（2014）〈ETF 交易最佳典範：七大守則〉（"Best practices for ETF trading: Seven rules of the road"），載於 Vanguard 研究，2014 年 9 月。

掛鈎票據的結算而在場內買入大額股份，市場可能會接收到錯誤訊息，增加股價波動。在此情況下，投資者以隱藏身份的方式進行大手交易，可為結算股票掛鈎票據提高交易價的確定性，有利投資者以至整體市場。這亦能推動市場開發更多不同的結構性產品及衍生產品，應對資產管理的需求。

除資產管理外，公司行動亦能助長市場對大手交易機制的需求。公司可透過大手買入進行回購，以支持股價。有研究 [11] 提到透過私下商議回購比在場內進行回購優勝，因為公司可在短時間內透過大手交易回購大量股份，省卻要與大量股東接觸周旋的行政開支。

另一個對大手交易機制的需求來自公司併購。收購人要取得目標公司大部分股份控制權，其中一個方法是經商議後大手買入，一般都是以溢價成交 [12]。按一份研究報告 [13] 解釋，在 1987 至 2002 年間的 756 宗商議大手交易中，併購後有最高管理層變動的大手交易有較高溢價，這溢價當中包括了管理人及擁有人的得益。該報告亦指出買入大股東持股時可參考大手交易的交易價，這顯示出大手交易對機構投資者之間的價格發現十分重要。

此外，私募股權和大股東都會受惠於大手交易。私募股權保薦人（或財務保薦人）收購公司時，他們的目標是最終能把項目通過轉售或公開發售予潛在投資者。在買賣股份時，若買賣雙方需要成交價有較高確定性和較短的執行交易時間，大手交易正是其中一種可行的方法。根據 Dealogic 的數據，美國與保薦人相關的大手交易由 2005 年至 2010 年間的 286 億美元升至 2011 年至 2016 年間的 1,708 億美元 [14]。亞太區亦有類似情況：在 2001 年當騰訊尚在創立初期，納斯帕斯（Naspers）便持有其 33% 權益，其後一次性以較當時市價折讓 7.8% 的價格在香港大手出售共

---

11 參見 Atkins, P. 與 P. Korff（2013）〈股份回購的問題〉（"Questions surrounding share repurchases"），載於《哈佛法學院企業管治及財務監管論壇》（*Harvard Law School Forum on Corporate Governance and Financial Regulation*），2013 年 3 月 14 日。

12 參見 Bittlingmayer, G.（1999）〈為企業控制（包括收購）服務的市場〉（"The market for corporate control (including takeovers)"），載於《法律與經濟百科全書第三冊—合約監管》（*Encyclopedia of Law and Economics, Vol. III — The Regulation of Contracts*），Boudewijn Bouckaert 及 Gerritt De Geest、Edward Elgar 以及根特大學。

13 Hwang, J. H.（2004）〈控制權的私人利益誰屬—擁有人還是管理人？〉（"Whose private benefits of control — Owners or managers?"），尚未發表的論文底稿，印第安納大學，布盧明頓。

14 資料來源：〈美國股票資本市場大手交易市佔率日高〉（"Building up US ECM a block at a time"），載於 Dealogic 網站，2017 年 10 月 17 日。

值 98 億港元的騰訊股份 (佔股權的 2%) [15]，是亞太區有史以來最大宗的大手交易 [16]。
回想 2015 年牛市時，騰訊部分主要股東也在香港市場大手買賣股份，折讓不過是
3% 至 4% 左右，少於一般約 10% 的水平 [17]。

　　總而言之，全球市場買賣雙方的機構投資者都會大手買賣股份，既可盡量減
低價格影響，亦可掌握到交易的時機及執行交易的成本，並且避免影響場內 (樓下)
市場的其他投資者。下一分節會詳述。

1.3　國際經驗說明大手交易能補足場內市場

　　一些投資者關注到公平性的問題，例如樓上市場究竟是在增加還是減少樓下
市場的流動性，以及經紀會否趕在客戶前頭先自行交易 (就如黑池交易一樣)。這
方面國際經驗或可帶來啟示。

　　實證顯示，大手交易的樓上市場能彌補服務小額買賣盤的樓下市場的不足，
而絕非其替代品。儘管有人憂慮市場分化會引致買賣盤在兩個市場的替代問題，
不過也有指樓上市場能造就更多買賣盤，提高流動性及成交 —— 文獻 [18] 指出，有
些樓上市場的交易員想以較低成本大手買賣股票 (受流動性帶動)，但不想讓樓下
市場知道其買賣盤的詳情或數據。這類交易就不會在樓下市場執行，而只能在樓
上市場進行。

　　以泛歐交易所為例，大手交易的方式包括透過商議完成交易、使用成交量加
權平均價交易及行使期權及衍生產品。一項研究 [19] 分析過 2007 年 1 月至 3 月的數
據，發現與其他股份相比，樓上市場交易活躍的股份的流動性更高 (買賣差價較
小)，在樓下市場的交易量也較大。研究亦發現，樓下市場若流動性不高 (買賣差

15　資料來源：〈納斯帕斯減持　騰訊再跌 220 億元〉("Tencent drops a further \$22 billion after Naspers trims stake")，載於彭博網站，2018 年 3 月 23 日。

16　資料來源：同上。

17　資料來源：〈DEALTALK：香港旺市　大手交易大增；中資銀行股沽售接二連三〉("DEALTALK: Hong Kong stock surge spurs block trade boom; China bank deals seen")，載於路透社網站，2016 年 4 月 21 日。

18　見 Burdett, K. 與 M. O'Hara (1987)〈構建模塊：大手交易入門〉("Building blocks: An introduction to block trading")，載於《銀行及金融期刊》(Journal of Banking and Finance)，第 11 期，193-212 頁；及 Seppi, D. J. (1990)〈均衡大手交易與對稱資訊〉("Equilibrium block trading and asymmetric information")，載於《金融期刊》(Journal of Finance)，第 45 期 (1990 年 3 月)，73-94 頁。

19　見 Lefebvre, J. (2010)〈泛歐交易所的大手交易及市場流通量〉("Block trades and market liquidity on Euronext Paris")，載於 EUROFIDAI — AFF 2010 年 12 月巴黎金融會議工作文件 (Working Paper, Paris December 2010 Finance Meeting EUROFIDAI — AFF)。

價大）亦不會驅使投資者轉到樓上市場落盤。

另一個例子是多倫多證券交易所。這家交易所為樓上市場的買賣盤專設獨立系統，但交易完成後立即向市場公佈。樓上市場買賣盤的價格起碼要與樓下市場同一時間所提供的價格一樣優越。一項研究[20]探究了 1997 年 6 月向該所申報的 5,840 宗（約佔交易量的 55.5%）樓上交易的數據，發現樓上市場並沒有把樓下市場「吞噬或佔其便宜」。相反，樓上市場中四分之一的大手交易使樓下市場的限價盤價格更理想。研究亦發現，當樓下市場的買賣差價較大，或與之對盤的買賣限價盤深度較淺時，有關交易更有可能在樓上市場執行。

證據顯示，在樓上市場進行的大手交易受流動性推動，可引出隱藏的流動性。一項研究[21]探究了在 1997 年 4 月至 1998 年 3 月期間的 92,170 宗法國股票的大手交易，發現其中的 67%（相對於紐約證券交易所（紐約交易所）的 20% 至 30%）都在樓上市場進行。研究結果顯示，隱藏的流動性可引發買賣盤，而樓上市場的實際執行成本只是樓下市場已顯示及隱藏流動性的 35%。而相類的另一項研究[22]分析了紐約交易所的 21,077 宗大手交易，亦佐證了在樓上市場進行的大手交易受流動性推動，這些交易並不會在沒有樓上市場的情況下出現。

樓上市場的經紀不只擔任促成大手交易的中介人，還在維持對投資者的公平性方面擔當重要角色。一項研究[23]審視了澳洲證券交易所在 1993 年至 1998 年的 69,449 宗大手交易，當中沒有證據證明樓上市場對樓下市場產生較大市場影響又或令其買賣差價擴大。此外，該研究還建議調低或取消 100 萬澳元的大手交易門檻，以推動小型股的大手交易，讓更多大手交易人士得享較低執行成本的好處。該研究還強調樓上市場的經紀在避免超前投資者交易及物色交易對手方方面可發揮重要作用。為了避免超前投資者交易，儘管樓上經紀都得悉對手方的身份，但他們都

20 Smith, B. F.、D. A. S. Turnbull 與 R. W. White（2001）〈自營及代理交易的樓上市場：負面信息與價格影響的分析〉（"Upstairs market for principal and agency trades: Analysis of adverse information and price effects"），載於《金融期刊》（Journal of Finance），第 56 冊第 5 期，1723-1746 頁。

21 見 Bessembiner, H. 與 K. Venkataraman（2004）〈電子證券交易所可需要樓上市場？〉（"Does an electronic stock exchange need an upstairs market?"），載於《金融經濟期刊》（Journal of Financial Economics），第 73 冊第 1 期，3-36 頁。

22 見 Madhavan, A. 與 M. Cheng（1997）〈尋找流通量：樓上及樓下市場的大手交易〉（"In search of liquidity: Block trades in the upstairs and downstairs markets"），載於《金融研究評論》（The Review of Financial Studies），第 10 冊第一期，175-203 頁。

23 Fong, K. Y. L.、A. Madhavan、P. L. Swan 與 F. D. Foster（2004）〈樓上、樓下：樓上市場有損樓下市場？〉（"Upstairs, downstairs: Does the upstairs market harm the downstairs?"），尚未發表的論文底稿，新南威爾斯大學。

願意放棄自營交易可以獲得許多短線資訊的好處，以求保住與客戶的長期關係，收取較高的代理交易費用 [24]。至於物色交易對手方，一篇論文 [25] 引述了巴黎證券交易所限價盤市場內的大額買賣盤被轉往倫敦的交易商市場這個例子，其中部分原因便是成本較低，以及倫敦的交易商市場深度大，可相對容易地找到交易對手方。

這些研究結果以實際例子證明了樓上（場外）市場有助彌補樓下（場內）市場的不足。

1.4　樓上市場大手交易在環球市場的增長趨勢

近年許多主要交易所的樓上市場協議交易的成交量都有所增長。國際證券交易所聯會（WFE）關於個別交易所的協議交易（經商議後的交易及申報的交易 [26]）的數據，可作為樓上市場大手交易量的參考指標。數據顯示，歐美及亞洲主要交易所的協議交易在 2016 年至 2018 年間的每日平均成交額，比在 2013 年至 2015 年的為高（見圖 2）。按協議交易佔總成交額的比例看，歐美交易所協議交易的佔比高於亞洲。在美國，協議交易的增長主要源於每宗成交的平均金額有所上升（協議交易的平均每宗成交金額由 2013 年至 2015 年的 9,473 美元，增至 2016 年至 2018 年的 13,830 美元）。在歐洲，協議交易的增長可能源自交易所與其他交易場所之間的交易量重組。（有關歐洲場外其他交易場所在監管變動下的近期發展，見附錄一。）至於香港，這兩段期間內的協議交易成交額相對平穩。值得注意的是，深圳證券交易所（深交所）的協議交易在 2016 年至 2018 年間的交易金額，比在 2013 年至 2015 年間的水平大增約 3.1 倍，而上海證券交易所（上交所）的協議交易金額在這期間亦上升了 30%。

24　見 Aitken, M. J.、G. Garvey 與 P. L. Swan（1995）〈經紀如何在競爭劇烈的證券市場促成長期客戶交易〉（"How brokers facilitate trade for long-term clients in competitive securities markets"），載於《商業期刊》（*The Journal of Business*），第 68(1) 期，1-33 頁。

25　De Jong, F.、T. Nijman 與 A. Roell（1995）〈法國股票的交易成本：巴黎證券交易所與倫敦證券交易所自動報價國際系統的比較〉（"A comparison of the cost of trading French stocks on the Paris Bourse and on SEAQ International"），載於《歐洲經濟評論》（*European Economic Review*），第 39 期，1277-2301 頁。

26　經商議後的交易是指買賣雙方在協議交易的價格及數量後，經由交易所（直接或間接）管理的系統確認進行的交易。交易所的系統會自動檢查交易是否符合交易所規則，包括最常見的檢查 — 檢查交易價格是否與電子買賣盤紀錄的價格一致。申報的交易是指經報告系統匯報的交易，只需其中一個對手方提供交易資料，交易報告系統會應申報交易的交易員要求而提供信息發佈服務。若須強制申報，另一邊的對手方亦可使用此系統。請注意：交易所有其不同的交易及申報規則，所以有關數據或不能完全直接比較。

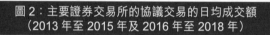

圖 2：主要證券交易所的協議交易的日均成交額
（2013 年至 2015 年及 2016 年至 2018 年）

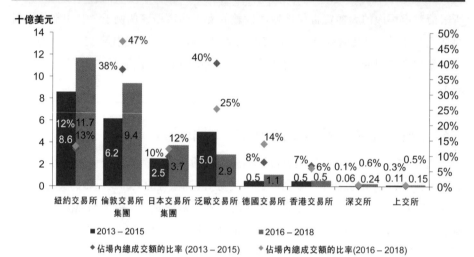

註：數字或不單包括配售或樓上大手交易，亦有可能包括其他類型的協議交易。因此，基於交易所各有不同的申報規則和計算方法，有關數據或不能完全直接比較。

資料來源：按 WFE 網站的數據計算。

　　從表 2 可見，表內所列的主要交易所（包括香港交易及結算所有限公司（香港交易所））的協議交易的平均每宗成交金額均較場內的電子交易大。2016 年至 2018 年間，上交所與深交所的協議交易的平均每宗成交金額約為 500 萬美元至 600 萬美元，遠高於其他主要證券交易所，更是其本身場內電子交易的逾 1,800 倍。上交所與深交所的協議交易的平均每宗成交金額如此龐大，或隱含了市場對 A 股的大手交易有潛在殷切需求。然而，在內地市場進行的大手交易只佔場內交易總額的極小份額（不足 1%），相對於其他交易所，例如香港交易所的 6%、紐約交易所的 13% 及倫敦證券交易所（倫敦交易所）集團的 47% 等（2016 年至 2018 年數字，見圖 2）。第 2 節將分析其背後原因及內地大手交易活動的特徵。

表 2：主要交易所按不同方式執行的交易的平均每宗成交金額（2016 年至 2018 年）		
交易所	協議交易（美元）	場內的電子交易（美元）	比率（倍）
上交所	6,105,183	3,007	2,030.4
深交所	5,262,434	2,905	1,811.6
泛歐交易所	1,582,698	8,669	182.6
倫敦交易所集團	74,354	7,241	10.3
德國交易所	64,358	11,192	5.8
香港交易所	17,400	7,634	2.3
紐約交易所	13,830	10,530	1.3

註：日本證券交易所集團因沒有相關交易宗數的數據而不包括在內。

資料來源：按 WFE 網站的數據計算。

2　中國內地的大宗交易

2.1　中國內地 A 股的大宗交易機制

中國內地 A 股的大宗交易（在香港稱為「大手交易」）須按兩家交易所的交易規則進行[27]，交易規模及價格均有限制。A 股大宗交易的最低交易限額為 300,000 股或人民幣 200 萬元。國有 A 股的主要股東在進行大宗交易時須遵守額外規定，包括若股份銷售額超過根據持股比例計算的界線，須事先取得批准才可進行，以及成交價格須在某個最低水平（過去 30 個交易日的成交量加權平均收市價或最新經審核年報的資產淨值的較高者）[28]。

A 股市價並不包括大宗交易的成交價。大宗交易價格不會用於計算指數及衍生產品（如指數期貨）的結算價，但日終收市計算的成交數字會包括大宗交易的成交數字。

27　有關內地大宗交易機制的資料來自以下版本的交易規則：《上海證券交易所交易規則》，上交所發佈，2018 年 8 月 6 日；《深圳證券交易所交易規則》，深交所發佈，2016 年 9 月 30 日。

28　資料來源：《上市公司國有股權監督管理辦法》，國務院國有資產監督管理委員會發佈，2018 年 5 月 16 日。

上交所及深交所提供大宗交易的專設平台。上交所及深交所的大宗交易平台僅供在岸機構 (交易所會員或合格投資者[29]) 使用。合格境外機構投資者 (QFII) 及人民幣合格境外機構投資者 (RQFII) 可透過有交易所會員身分的在岸經紀參與大宗交易[30]。參與機構可通過大宗交易平台提交買賣盤或申報交易。大宗交易的資料 (包括股份代號及名稱、交易的價格及成交量以及買賣雙方的名稱) 將於交易日結束時在相關證券交易所的網站上公佈。

上交所 / 深交所允許的買賣盤 / 交易申報類型為：

- **意向申報**顯示大手購入或售出的意向。此類買賣盤在上交所的提交時間為上午 9 時 30 分至 11 時 30 分及下午 1 時至 3 時 30 分；深交所的提交時間為上午 9 時 15 分至 11 時 30 分及下午 1 時至 3 時 30 分。

- **定價申報**為將價格定為收市價或成交量加權平均成交價[31]的買賣盤。此類買賣盤的提交時間為上交所的下午 3 時至 3 時 30 分，以及深交所的下午 3 時 05 分至 3 時 30 分。買賣盤會按時間優先作配對。

- **成交申報**為協議大宗交易的申報。買賣雙方均須申報大宗交易。上交所及深交所的股票成交申報可於下午 3 時至 3 時 30 分輸入[32]。

2.2 大宗交易活動

儘管 A 股的大宗交易可扣減 30% 的交易所證券交易經手費[33]，但 2012 年至 2018 年協議交易 (大部分為大宗交易) 的年度成交總額僅佔市場總額的 0.1% 至 0.6% (見圖 2)。當中成因包括機構投資者參與度相對偏低，加上流動性受到大宗交易機制及相關法規的限制等。

---

29 合格投資者包括社會保障基金、退休基金及慈善基金、向中國證券投資基金業協會登記備案的投資計劃、私募基金經理及符合中國證監會規定的合格投資者資格的投資者。詳情請參閱《私募投資基金監督管理暫行辦法》，中國證券監督管理委員會 (中國證監會) 發佈，2014 年 8 月 21 日。

30 見《上海證券交易所合格境外機構投資者和人民幣合格境外機構投資者證券交易實施細則》，上交所發佈，2014 年 3 月 19 日；《深圳證券交易所合格境外機構投資者和人民幣合格境外機構投資者證券交易實施細則》，深交所發佈，2014 年 4 月 25 日。

31 深交所接受成交量加權平均成交價的定價申報，但上交所目前暫不接受 (見《關於修訂《上海證券交易所交易規則》的通知》，上交所發佈，2018 年 8 月 6 日)。

32 上交所接受輸入成交申報的時間 (上午 9 時 30 分至 11 時 30 分、下午 1 時至 3 時及下午 4 時至 5 時) 目前暫不可使用。見《關於修訂《上海證券交易所交易規則》的通知》，上交所發佈，2018 年 8 月 6 日。

33 見上交所及深交所網站有關交易費的詳情。

　　首先，儘管內地市場基本上允許所有在岸機構投資者直接使用大宗交易平台，而 QFII 及 RQFII 亦可間接使用該平台，但大部分離岸機構投資者都沒有 QFII/RQFII 的資格，投資 A 股只能透過「滬股通」及「深股通」進行。此外，在岸的機構投資者數目還少，機構投資者的交易總額僅佔市場交易總額約 10% 至 20%[34]，低於其他主要市場 (例如香港的數字是 55%)[35]，所以，來自這些在岸機構投資者按資產管理需求而進行的大宗交易成交量會相對較低，更遑論是 QFII/RQFII。由於很多離岸機構未能使用大宗交易平台，進一步限制了在平台上的交易對手數量，繼而影響到大宗交易的時間性及定價。第二，據離岸市場參與者反映，平台上的流動性受限制亦是因為大宗交易在申報及獲披露前，其相關的款項及股份都鎖定在大宗交易平台上。第三，國有股份的大宗交易所受的限制更多 (見上文第 2.1 節)。這些都是導致中國內地大宗交易市場佔比相對不高的原因。

　　儘管如此，在內地證券交易所進行的大宗交易近年有上升的趨勢。在 2015 年股票指數創下新高後，2016 年的協議交易成交額亦升至 7,360 億元人民幣的高位 (見圖 3)。

圖 3：中國內地協議交易全年成交額與滬深 300 指數每日收市點數 (2012 年至 2018 年)

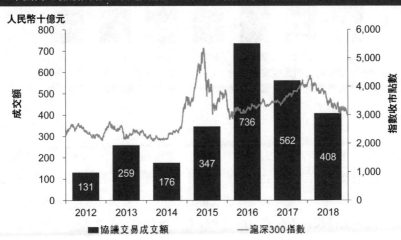

資料來源：協議交易成交額來自 WFE；股指資料來自彭博。

34　資料來源：〈A 股與美股對比：投資者結構及交易特徵〉，載於《新浪網》，2019 年 6 月 9 日。
35　資料來源：香港交易所調查報告〈現貨市場交易研究調查 2018〉，載於香港交易所網站，2019 年 7 月。

　　大宗交易是限售股解禁後進行買賣的主要方式之一 [36]。這些股份變為流通股後，大股東即可於市場上向投資者出售持股。因此，A 股的大宗交易大部分由賣方驅動，因而可能會對價格波動造成壓力。這些由賣方驅動的大宗交易，成交價格通常低於市價。據觀察，以折讓價成交的大宗交易股份，近年佔大宗交易總成交額約 74% 至 89%（見圖 4）。

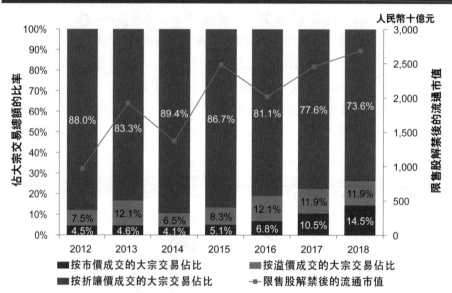

圖 4：中國內地大宗交易的分佈與限售股解禁後的流通市值（2012 年至 2018 年）

資料來源：eastmoney.com。

　　為鼓勵較平衡的交易行為及透明市場，中國證券監督管理委員會（簡稱「中國證監會」）於 2015 年 7 月、2016 年 1 月及 2017 年 5 月向上市公司大股東及高級管理層，以及「定增基金」[37] 發佈減持股份的指引 [38]。根據 2017 年的最新指引，上述權益人大手減持股份須受若干限制。持有上市公司百分之五以上權益或當初透

36　資料來源：〈減持新規限售股解禁期至　私募積極介入大宗交易業務〉，載於《中國基金報》，2018 年 6 月 4 日。

37　「定增基金」是指專門購買定向增發股份的惠основ金。

38　分別為《中證監公告 [2015]18 號 —— 上市公司大股東及董事、監事、高級管理人員 6 個月內不得減持》、《中證監公告 [2016]1 號 —— 上市公司大股東、董監高減持股份的若干規定》與《中證監公告 [2017]9 號 —— 上市公司大股東、董監高減持股份的若干規定》，中國證監會發佈。

過私人配售購入股份者，在限售股解禁後 12 個月內沽售的股數不得超過其持股的
50%。同時，在任意連續 90 日內透過大宗交易轉移的股份不得超過公司總股本數
目的百分之二，而買方購入股份後六個月內亦不得賣出 [39]。鑒於以上種種限制，按
折讓價成交的大宗交易（作為賣方驅動的大宗交易的指標）的佔比自 2016 年起逐
漸減少至 2018 年的 74%，是 2012 年以來的最低。

　　此外，某些股票互惠基金是靠大宗交易機制來應付贖回要求。這當中包括
2015 年 2 月推出的定增基金 [40]。定向增發的股票通常由上市公司以低於市價出售，
並有一年或三年的禁售期 [41]。禁售期屆滿後，基金可透過公開市場或大宗交易出售
其持股 [42]。於 2018 年底，內地共有 53 隻定增基金，涉及管理資產 280 億元人民幣
（見圖 5）。

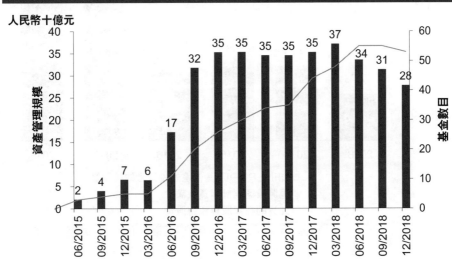

圖 5：中國內地定增基金季末資產管理規模及數目（2015 年第二季至 2018 年第四季）

資料來源：Wind。

39　《上海證券交易所上市公司股東及董事、監事、高級管理人員減持股份實施細則》，上交所發佈，2017 年 5 月 27
　　日；《深圳證券交易所上市公司股東及董事、監事、高級管理人員減持股份實施細則》，深交所發佈，2017 年 5
　　月 27 日。
40　資料來源：〈轉型 LOF，國投瑞利告別定增，該走還是留？〉，載於《每日經濟新聞》，2016 年 8 月 8 日。
41　資料來源：〈定增市場的投資方法及收益來源解析〉，載於《格隆匯》網站，2017 年 11 月 30 日。
42　資料來源：〈定增基金真會玩：大宗交易成減持新「梯子」〉，載於《經濟觀察報》，2017 年 9 月 16 日。

2.3 透過「滬股通」及「深股通」進行 A 股大宗交易的潛在需求

透過「滬股通」及「深股通」(合稱「滬深股通」)，A 股的投資者基礎擴闊至包括比 QFII 和 RQFII 更廣泛的全球機構 (與個人) 投資者。「滬深股通」於 2018 年成為最受歡迎的 A 股交易渠道 (見圖 7)。這些投資者對 A 股大宗交易或潛藏着很大的需求。

圖 6：年末外資於內地在岸股票的投資總額 (2013 年至 2018 年)

註：QFII、RQFII 及其他境外投資者的持股量為估算數字，按外資的在岸股票總持股量減去「滬深股通」的累計淨買入額計算。

資料來源：「滬深股通」累計淨買入額源自香港交易所；外資的在岸股票總持股量源自 Wind。

統計數據顯示，越來越多機構投資者參與「滬深股通」── 特別獨立戶口 (SPSA)[43] 的數目由 2017 年 6 月的約 1,700 個增至 2018 年 9 月的 6,363 個[44]，更在

---

43 特別獨立戶口是機構投資者 (以投資者號碼) 開設的戶口，以符合透過「滬深股通」進行交易的前端監控規定。一個特別獨立戶口可執行最多 20 個交易所參與者的買賣盤。交易所參與者應事先確認其基金經理客戶有權將不同特別獨立戶口的買賣盤併合計算及決定於個別戶口的股份分配，並須確定其符合適用法律、規則及規例，且並無不當挪用客戶資產。(請參閱於《常問問題》(問題 1.48.1)，載於香港交易所網站，2018 年 9 月 3 日。)

44 參閱〈有關進一步增加 MSCI 指數 A 股權重的諮詢總結〉("Conclusion of consultation on further weight increase of China A shares in the MSCI indexes")，MSCI 網站指數諮詢，2018 年 12 月。

2019 年 7 月增至逾 8,200 個 [45]。其中的原因包括「滬深股通」每日淨買入的額度，自 2018 年 5 月由 130 億元人民幣增至四倍的 520 億元人民幣，隨後是 A 股獲納入各主要全球指數。目前，透過「滬深股通」進行的大宗交易安排僅適用於某些非交易過戶 [46]。機構投資者若因其他目的需要在「滬深股通」執行大手買賣盤，唯有將之拆分為較小的買賣盤，才可在交易所買賣。這可從「滬深股通」的交易與內地交易所場內電子交易的數據看出 —— 兩者有相若的平均每宗成交金額（約 20,000 元人民幣）（分別見圖 7 與表 2）。

圖 7：「滬深股通」每月成交額及平均每宗成交金額（2016 年 1 月至 2019 年 6 月）

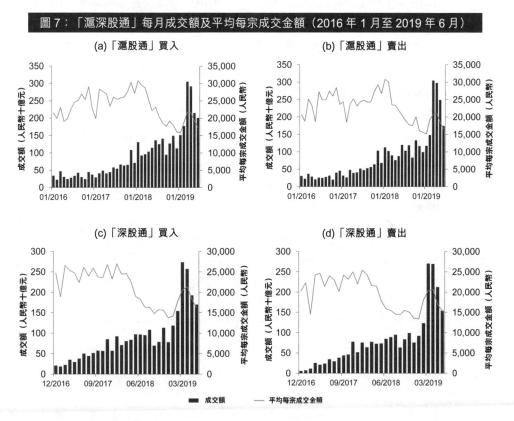

註：「深港通」於 2016 年 12 月推出。
資料來源：香港交易所。

45　資料來源：香港交易所。

46　這包括股票借貸、糾正錯誤交易、交易後由基金經理將股份併合買賣盤分配至不同戶口的基金或子基金等。請參閱「滬深股通」的《市場參與者資料文件》第 3.37 節「非交易過戶」（Non-trade Transfer），載於香港交易所網站，2019 年 2 月 15 日。

　　隨着 A 股獲納入全球股票指數，帶動更多離岸被動型投資基金投資於 A 股，預期透過「滬深股通」買賣的 A 股成交額將進一步增長。A 股獲納入各全球指數的成份股籃子中的比重預期會進一步擴張，以及相關的 ETF 會有所增長，那麼全球投資者對 A 股大宗交易的需求在指數重整或申贖基金單位時將會增加：

- **MSCI** 自 2018 年 6 月開始至 2018 年 8 月分兩個階段將 A 股納入其全球指數系列（包括新興市場指數），最初的部分納入因子為 5%（按 2018 年 5 月權重計約 0.8%）。納入因子將逐步增至 100%（按權重計約為 16.2%）[47]。2019 年 2 月，MSCI 決定由 2019 年至 2020 年分階段將大型及中型股的納入因子由 5% 增至 20%[48]（見圖 8a）。按照 2018 年 6 月追蹤 MSCI 新興市場指數的基金資產管理規模約 18,000 億美元[49] 計算，A 股在完全獲納入後所吸引的投資額將由一開始約 200 億美元躍升至約 3,000 億美元。

- **富時羅素** 於 2019 年 6 月開始了將 A 股納入富時全球股票指數系列（包括新興市場指數）的第一階段，分三期進行，直至 2020 年 3 月（見圖 8b），預期一開始可吸引指數追蹤基金投資於 A 股的金額會達到約 100 億美元[50]。

- **標普道瓊斯** 自 2019 年 9 月開始以 25% 的加權因子將 A 股納入其全球基準指數[51]。

47　資料來源：Chia, C. P.〈中國的時代來臨〉（"The world comes to China"），載於 MSCI 網站的網誌，2018 年 5 月 23 日。

48　資料來源：〈有關進一步增加 MSCI 指數 A 股權重的諮詢總結〉（"Conclusion of consultation on further weight increase of China A shares in the MSCI indexes"），MSCI 網站的指數諮詢版頁，2019 年 2 月。

49　資料來源：〈MSCI — 股票指數領袖〉（"MSCI — A leader In equity indexes"），MSCI 網站，2019 年。

50　資料來源：〈富時羅素將中國 A 股提升至新興市場級別〉（"FTSE Russell promotes China A shares to emerging market status"），富時羅素新聞稿，2018 年 9 月 26 日。

51　資料來源：〈標普道瓊斯指數 2018 年國家分類諮詢結果〉（"S&P Dow Jones Indices' 2018 country classification consultation results"），載於標普道瓊斯指數網站的公告，2018 年 12 月 5 日。

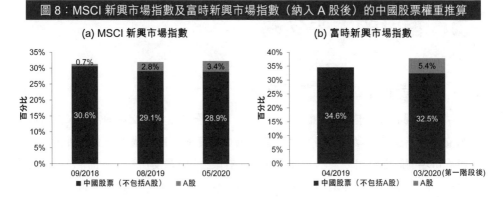

圖 8：MSCI 新興市場指數及富時新興市場指數（納入 A 股後）的中國股票權重推算

(a) MSCI 新興市場指數

(b) 富時新興市場指數

註：MSCI 新興市場指數的權重按 2019 年 1 月 22 日的股價計算。富時新興市場（大型及中型股）指數 2020 年
　　3 月第一階段三期結束後的權重按 2019 年 6 月 21 日的股價計算。

資料來源：〈有關進一步增加 MSCI 指數 A 股權重的諮詢總結〉（"Conclusion of consultation on further weight
　　　　　increase of China A shares in the MSCI indexes"），載於 MSCI 網站的指數諮詢版頁，2019 年 2 月；
　　　　　〈納入中國 A 股 ── 七大要點〉（"China A shares inclusion ── Seven key points"），載於富時羅素網
　　　　　站的網誌，2019 年 6 月 24 日。

　　結構性產品的普及化亦會增加對「滬深股通」大宗交易的需求。一項研究[52] 指出，儘管監管規定在 2007 年至 2008 年的全球金融危機後已有所收緊，但亞洲零售結構性產品的銷售額仍逐漸上升。該研究指出，主要亞洲市場（包括中國內地、中國香港、日本、南韓及新加坡）的零售結構性產品銷售額截至 2015 年年底已超過 7,500 億美元，與 2012 年數字比較的複合年增長率為 4%。以 A 股為掛鈎資產的離岸股票結構性產品及衍生產品的預期增長，會帶動大宗交易的需求增加，用以提升有關產品的結算效率（見第 1.2 節）。

　　總括而言，隨着香港的 A 股機構投資者及其 A 股所佔的資產配置增加，預期 A 股大宗交易的潛在需求亦會增加。

52　Wu, A. 與 C. Pitts（2017）〈亞洲結構性產品〉（"Asian structured products"），載於《研究基金會簡介》（*Research Foundation Briefs*），第 3 冊第 6 期，2017 年 8 月。

3 香港的大手交易

3.1　香港非自動對盤交易機制

香港市場設有「非自動對盤交易」機制，可切合港股大手交易的需要。非自動對盤交易是指交易所參與者自行協議達成的交易，而不用經過香港交易所旗下營運證券市場的附屬公司——香港聯合交易所有限公司（聯交所）——的「領航星交易平台—證券市場」(OTP-C) 的公開競價及對盤環節。非自動對盤交易須向聯交所申報交易細節。非自動對盤交易分為兩類：一類是非兩邊客交易（標記為"M"），另一類是兩邊客交易（標記為"X"，有時亦稱「交叉盤買賣」）。M 類交易是經兩名不同交易所參與者執行的交易，X 類交易則是同一名交易所參與者代表不同投資者進行的交易。兩邊客交易亦涵蓋在香港進行的黑池交易或另類平台交易。在香港，大手交易可按 M 類或 X 類交易的形式進行。

香港的非自動對盤交易對每筆交易的金額及交易時間的安排均甚靈活，但對交叉盤買賣的價格有所限制。非自動對盤交易對單筆交易並無明確的最低數量或金額的要求（例如買賣一股也可）。不過，於開市前時段及持續交易時段，如單筆大手交易涉及股數超過 3,000 手 [53]，就會因為超過了 OTP-C 自動對盤機制能接受的買賣盤最大股份數量，而只可按非自動對盤交易形式進行。非自動對盤交易可以隨時進行，包括開市前時段、持續交易時段、收市競價時段，以及收市後交易時段，但須遵守有關非自動對盤交易申報時間的交易所規則（見下段）。在持續交易時段內，屬交叉盤買賣的非自動對盤交易須遵守價格上下限（24 個價位）。換言之，價格必須處於上日收市價、當日截至交易時最低買入價及最低賣出價之下的 24 個價位、與上日收市價、當日截至交易時最高買入價與最高賣出價之上的 24 個價位之間 [54]。

在交易透明度方面，於香港市場進行非自動對盤交易要及時申報交易。如非自動對盤交易並非在黑池進行，交易所參與者須在交易日（T 日）進行交易後 15 分

53　「手」指每隻股票的買賣單位，為固定的股數。各隻股票可有不同的每手股數。
54　見聯交所的《交易所規則》第五章〈交易〉。

鐘內申報，如交易是於收市後進行的，則須在下一個交易日（T+1 日）上午 9 時 45 分或之前申報。在黑池進行的非自動對盤交易 [55]，交易所參與者則須在執行交易後一分鐘內申報。若為兩邊客的非自動對盤交易（即交叉盤買賣），包括在黑池進行的交易，交易所參與者須申報及檢視交易詳情。若為非兩邊客的非自動對盤交易，只須賣方交易所參與者進行申報，而買方交易所參與者須在申報日的交易時段內檢視所申報交易的資料。每名交易所參與者均須於交易日結束後檢視其買賣兩邊的交易日誌。

　　個別股票的市價及指數在計算時均不會包括非自動對盤交易的價格，但計算成交數字時則計及非自動對盤交易的成交。就恒生綜合大型股、中型股及小型股指數的成份股，以及有 A 股在內地交易所上市的 H 股而言，收市價於收市競價時段透過競價機制釐定。其他股票的收市價 [56] 則按持續交 時段最後一分鐘內五個按盤價（每個相隔 15 秒）的中位數計算，直至 2019 年 10 月 8 日起有關股票也會被納入收市競價機制為止。

　　上交所、深交所與香港交易所各自對於大手交易的交易安排載於附錄二。

3.2　港股的大手交易活動

　　WFE 有關協議交易的數據可用作在香港進行的大手交易的參考指標。有關數據所反映的非自動對盤交易中 [57]，大部分是大手交易。恒生指數（恒指）在 2018 年創新高時，該年的協議交易成交額亦創出歷史新高（見圖 9）。協議交易的成交量與恒指的表現趨勢似乎很相似，也在某程度上反映了在香港市場進行的大手交易一般在股市上揚時更為活躍，而在股市下挫時則較不活躍。這有別於內地的大宗交易以賣方驅動的交易佔大多數。

55　根據證監會報告（《關於香港另類交易平台的主題檢視報告》，2019 年 4 月 9 日發佈），在香港營運的黑池共有 16 個，主要是經紀行或銀行的內部交易系統。黑池內同一交易所參與者的買盤和賣盤會作內部配對（即內部化），之後才傳送至交易所市場。香港交易所的數據顯示，2016 年 10 月至 2017 年 9 月期間，交易所參與者營運的黑池成文量約在 150 億港元至 250 億港元之間（佔市場總成交的 1.0% 至 1.7%）。監管機構意識到黑池交易牽涉到投資者保障問題，2015 年 12 月起禁止散戶於黑池交易。

56　由 2019 年 10 月 8 日起，收市競價交易時段覆蓋的證券將擴大至所有股本證券（包括預託證券、投資公司、優先股及合訂證券）和基金（包括 ETF 及房地產投資信託基金）。資料來源：《香港交易所將擴大收市競價交易時段產品覆蓋範圍至所有股本證券及基金》，香港交易所新聞稿，2019 年 7 月 5 日。

57　由於並無成額限制，香港的非自動對盤交易不僅涵蓋增發以及樓上大宗交易，亦包括涉及數額非常小的交易——碎股交易（交易數量不足一手）及特別買賣單位的交易（交易量不等於一手的倍數）。香港交易所的數據表示，股份增發（大手交易的主要類型）約佔市場總成交額的 4%，而所有協議交易約佔市場總成交額的 6%。

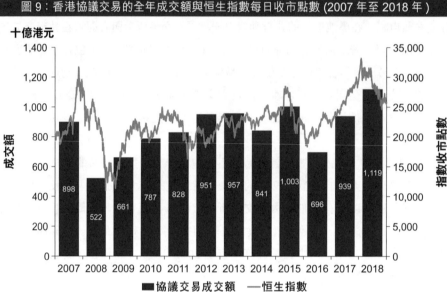

圖 9：香港協議交易的全年成交額與恒生指數每日收市點數 (2007 年至 2018 年)

資料來源：協議交易成交額來自 WFE；指數數據來自彭博。

　　為着如上文第 1 節所討論的目的，香港股票大手交易的需求隨着香港資產及財富管理業務的逐步增長 (資產規模在 2018 年年底攀至 239,550 億港元 [58]) 而大幅提升。這些業務包括資產管理以及資金管理顧問業務，資產規模在 2018 年達到 164,470 億港元。當中的機構投資者包括主動型及被動型資產管理人，他們會按上文 1.2 節所述的目的而進行股票的大手交易。涉及大手交易的香港上市股票包括 H 股，而「H 股全流通」試點計劃將會增加其流動性。中國證監會於 2017 年 12 月宣佈推行試點計劃 [59]，通過將入選 H 股公司的內地股東所持有的非流通股轉換為 H 股，實現 H 股的「全面流通」。隨後，三家中國公司於 2018 年獲批參與 H 股全流通計劃。進行「H 股全流通」後，H 股公司大股東及高級管理層持有的非流通股即可在香港市場買賣。鑒於試點計劃運行順暢，中國證監會與相關部門正計劃擴大「H 股全流通」的範圍 [60]。此舉或會增加市場對原本由內地大股東持有的 H 股作大手

58　資料來源：證監會 2018 年資產管理調查。

59　〈中國證監會深化境外上市制度改革〉，中國證監會新聞稿，2017 年 12 月 29 日。

60　資料來源：〈證監會有關負責人就進一步擴大資本市場對外開放答記者問〉，載於中國證監會網站，2019 年 6 月 14 日。

買入或賣出的需求。

3.3　經「港股通」進行大手交易的潛在需求

　　隨着內地與香港市場互聯互通的程度加深，涉及香港上市股票的跨境資產管理業務預期將持續增長。就如「滬深股通」，港股通淨買入的每日額度在 2018 年 5 月由 105 億元人民幣大增三倍至 420 億元人民幣（「滬港通」及「深港通」的額度相同）。經「港股通」淨買入香港股票的累計金額一直保持增長 — 截至 2019 年 6 月底，「滬港通」下的「港股通」淨買入達 6,571 億港元，「深港通」下的「港股通」淨買入達 2,231 億港元（見圖 10）。「滬港通」及「深港通」下的「港股通」的日均成交額（包括買入及賣出交易）於 2019 年上半年分別約為 71 億港元及 40 億港元。這佔香港股市總成交額（單邊計）約 5% 至 6%（見圖 11）。

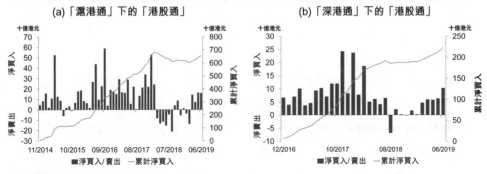

圖 10：經「港股通」進行的港股淨買入 / 賣出（2014 年 11 月至 2019 年 6 月）

(a)「滬港通」下的「港股通」

(b)「深港通」下的「港股通」

資料來源：香港交易所。

圖 11：「港股通」日均成交額（2014 年至 2019 年 6 月）

註：由 2014 年 11 月 17 日「滬港通」開通起計。「深港通」於 2016 年 12 月 5 日開通後亦計算在內。
資料來源：香港交易所。

　　「港股通」交投的增長，部分歸因於內地互惠基金的參與度日高，其中包括合格境內機構投資者（QDII）基金以及通過「港股通」投資港股的基金（下稱「內地通基金」）[61]。這些內地通基金數目於 2018 年年底錄得 375 個的新高，投資於港股的總資產淨值為 2,190 億元人民幣（見圖 12）。主動型管理模式以及申贖基金單位都可能涉及短時間內進行大額交易，增加對大手交易的需求（見 1.2 節）。由於現時「港股通」無法進行大手交易，這些互惠基金會透過程式交易的方法進行多筆小額交易來完成大額買賣盤。這從「港股通」的平均交易與香港交易所場內電子交易的數據中可見一斑 — 兩者的平均每宗成交金額相若（約 60,000 港元）（分別見圖 13 及見 1.4 節表 2）。

61　見〈滬港深基金業績報告：QDII 基金發展受限　滬港深基金崛起〉，載於《金融界》網站，2018 年 1 月 19 日。

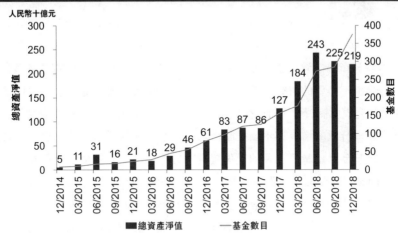

圖 12：內地通基金 —— 季末總資產淨值及基金數目（2014 年第四季至 2018 年第四季）

註：僅包括有匯報季度總資產淨值的基金。有部分基金發行量不止一批。

資料來源：Wind。

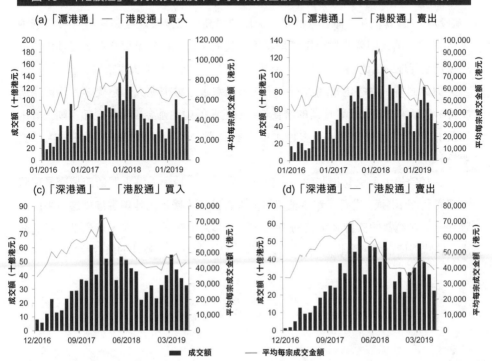

圖 13：「港股通」每月成交額及平均每宗成交金額（2016 年 1 月至 2019 年 6 月）

(a)「滬港通」—「港股通」買入

(b)「滬港通」—「港股通」賣出

(c)「深港通」—「港股通」買入

(d)「深港通」—「港股通」賣出

成交額　　平均每宗成交金額

註：「深港通」於 2016 年 12 月推出。

資料來源：香港交易所。

4 總結

　　股本證券市場設置大手交易機制可盡量減輕大額交易所帶來的價格影響、避免向市場發出不必要的信號。這令需要買入或沽出大手股份的人士，可以進行一次性交易就完成所需買賣，而成交價格和時間的確定性都比在競價市場交易更高。環球資產管理行業的增長亦不斷推動市場對大手交易的需求，這主要是來自指數調整和申贖基金單位時所涉及的股份買賣；另外，為配合結構性產品和股本證券衍生產品的發展而要提高交收效率等的因素，亦推動進行大手交易的需求。

　　不論在中國內地或香港，證券交易所都有特別機制，利便大手交易又或申報協議交易（當中包括大手交易）。在內地，大手交易以賣盤驅動的交易為主導，當中主要是大股東在禁售期結束後減持名下的大額持股。這與在包括香港在內的成熟市場進行的大手交易多由資產管理活動為誘因的情況有所不同。不過，內地市場逐漸開放，從國際投資者能通過「滬股通」和「深股通」進一步參與A股市場，「滬股通」和「深股通」淨買入的每日額度的提高，以及環球被動式的A股投資隨着A股成功納入國際指數而不斷增長等，就能看出。這些進展全都會促進內地市場為資產管理的目的而對大手交易有更大的需求，就如其他國際市場一樣。

　　與此同時，在香港市場進行大手交易的需求也愈來愈大，這受多方面因素所支持，包括：香港資產管理業的增長，投資港股的內地基金數目與金額不斷上升，以及H股全流通或會促成更多原本由內地H股公司大股東持有的股份的大手買賣等。

　　由於跨境大手交易機制的潛在需求日高，而內地與香港之間的跨境交易現時大都經「滬深港通」進行，市場值得研究可否進一步優化內地與香港兩邊市場的互聯互通，在適當的安排及可控的環境下，引進大手交易的安排，切合投資者的需求。屆時跨境市場流動性將可進一步提高，投資者也可以較佳的條件執行交易。

附錄一

歐洲另類交易場所的近期發展

　　歐洲的證券交易所都有推出大手交易配套設施，以爭取黑池流動性。在這類設施進行的交易活動增幅顯著，主要是受監管規則的發展所影響 —— 當局推出了第二版的金融工具市場指令（MiFID II）。在歐洲，黑池交易只要符合四大豁免條件[62] 其中一項（包括屬協議交易、參考價交易、在買賣盤管理設施進行的交易以及大額交易），即可豁免遵守交易前透明度的規定（見表 A1）。

表 A1：MiFID 就交易場所的交易前透明度豁免規定	
豁免類型	說明
協議交易豁免（NTW）	這豁免讓交易雙方可在掛賬冊以外自行議定交易，前提是交易價格是處於成交價加權平均價（VWAP）的範圍內（即不超出交易 VWAP 的上下限），又或交易價格須符合現行市價以外的條件。
參考價豁免（RPW）	這豁免讓於外部參照的買賣盤可按（例如）主要交易場所的買入及賣出價的中間點對盤。
買賣盤管理設施豁免（OMF）	這豁免讓有關買賣盤（例如冰山指令或止蝕指令等）在成交及向市場披露前，可在交易場所的買賣盤管理系統內繼續保留而毋須披露。
大額交易（LIS）	這豁免讓按任何價格或機制進行的大手交易都不用遵守交易前透明度規定。交易金額須超過一定規模（按股份規模而異，最高為大型股的 500,000 歐元）。

資料來源：歐洲證券及市場管理局。

　　自 2018 年起，MiFID II 禁止使用經紀交叉網絡系統（佔歐洲股本證券交易的 4% 至 5%）[63]，同時引進成交量雙重上限（DVC）[64]，以限制按參考價豁免（RPW）及協議交易豁免（NTW）進行的黑池交易 —— 佔有關股份在每個交易場所的總成交量不多於 4%，以及佔該股份於所有歐盟交易場所的總成交量不多於 8%。一旦超

62　資料來源：〈交易前透明度規定的豁免〉（"Waivers from Pre-trade Transparency"），載於歐洲證券及市場管理局（European Securities and Markets Authority，簡稱 ESMA）網站，2016 年 6 月 20 日。

63　資料來源：Puaar, A.、J. Schack 與 A. Kemmsies〈讓那裏有光：Rosenblatt 每月黑池流動性報告 —— 歐洲版〉（"Let there be light: Rosenblatt's Monthly Dark Liquidity Tracker — European Edition"），Rosenblatt 證券市場的架構分析，2018 年 9 月 21 日。

64　見〈Double volume cap mechanism〉版頁，載於 ESMA 網站，於 2019 年 8 月 2 日瀏覽資料。

越上限，相關交易場所必須停止該股份的黑池交易六個月。大額交易豁免 (LIS) 對黑池交易而言愈顯重要，而隨着證券日均成交量與日俱增，大額買賣盤交易門檻（最低交易金額）亦逐步提高。

　　眼見部分銀行設立系統化的內部撮合器 (SI) 填補市場的空隙 [65]，歐洲的證券交易所也紛紛推出大額交易的黑池配套設施，當中包括 2016 年 9 月倫敦交易所與 Plato Partnership 合作推出的 Turquoise Plato（適用於歐洲及美國股票），以及 2017 年 7 月 Euronext 夥拍 AX Trading 推出的 Euronext Block（適用於歐洲股票）。經這些設施進行的大手交易，近兩年的單日成交都先後刷新紀錄——Turquoise Plato Block Discovery（Turquoise Plato 用於配對大手交易指示或附條件指令的平台）於 2019 年 1 月 31 日錄得 6.53 億歐元 [66]，Euronext Block 於 2018 年 8 月 1 日錄得 2,400 萬歐元 [67]。

65　見 Hadfield, W.〈交易員在歐洲新規實行時會留意的兩個字〉（"These are the two words traders will be paying attention when Europe's New Rules kick in"），載於彭博網站，2017 年 12 月 18 日。

66　資料來源：〈Turquoise Plato Block Discovery〉版頁，載於倫敦交易所集團網站，於 2019 年 7 月 31 日瀏覽資料。

67　資料來源：Taylor, V.〈Euronext Block 繼續增長：記錄新高交易和改善計劃〉（"Euronext Block continues to grow: Record trades and planned improvements"），載於 Linkedin 網站的貼文，2018 年 9 月 4 日。

附錄二

內地與香港的大手交易安排

特徵	上海證券交易所	深圳證券交易所	香港交易所
執行方式	專設平台處理 交易意向申報、自行商議及固定價格競價		自行商議 （無專設平台）
數量限制	300,000 股或 200 萬元人民幣		無
交易時段	意向申報： 09:30-11:30; 13:00-15:30 成交申報： 15:00-15:30 固定價格申報： 15:00-15:30	協議大宗交易（意向申報）： 09:15-11:30; 13:00-15:30 協議大宗交易確認（成交申報）： 15:00-15:30 盤後定價大宗交易： 15:05-15:30	隨時
交易價格限制	上日收市價的 ±10% （ST 及 *ST 股票：不超過上日收市價的 ±5%；若 A 股上日收市價低於 0.1 元人民幣，則限為 0.01 元人民幣）	上日收市價的 ±10% （ST 及 *ST 股票：不超過上日收市價的 ±5%）	該交易日持續交易時段 （截至執行交易時） 全日高位之上 24 個價位與全日低位之下 24 個價位之間
對指數計算及衍生產品結算的影響	計算股票指數價格或股本衍生產品結算價時不予計算大手交易的價格，但大手交易會計入證券成交內		
申報規定	協議大手交易：買賣雙方均須申報交易 （按上列交易時段提交指令完成申報）		協議交易： 僅賣方需要申報交易 （交叉盤買賣則由該交易所參與者申報） 屬非黑池交易：須於 T 日交易後 15 分鐘內申報（收市後交易時段的交易須在 T+1 日上午 9 時 45 分或之前申報） 屬黑池交易：須於 T 日交易後 1 分鐘內申報

註：附有 ST 或 ST* 標記的股票為需要「特別處理」的股票，是對投資者的風險警示。

資料來源：香港交易所網站及聯交所的《交易所規則》第五章；《上海證券交易所交易規則》，上交所發佈，2018
　　　　年 8 月 6 日以及《深圳證券交易所交易規則》，深交所發佈，2016 年 9 月 30 日。

第14章

香港 ETF 市場作為通向環球投資的門戶

香港交易及結算所有限公司
首席中國經濟學家辦公室

摘 要

交易所買賣基金（ETF）在全球市場愈來愈受歡迎。ETF 產品因為比傳統互惠基金更具成本效益、更方便和透明度更高，所以全球 ETF 產品的數目和資產管理規模均迅速增長。隨之而來的是，ETF 產品愈見創新和多元化，定息 ETF 也愈來愈普及。在歐美的 ETF 市場繼續增長之際，亞太區市場 ETF 的增長勢頭亦不遑多讓。香港一直是亞太地區領先的 ETF 市場之一，而內地則在投資者愈來愈接受被動式投資的情況下，近年其 ETF 市場亦見大幅增長，並成為亞太區第二大 ETF 市場。

至於被動式投資，內地投資者現時只可投資內地上市的 ETF，當中大部分是追蹤內地各式各樣指數的股本證券 ETF。另外，貨幣市場 ETF 也因收益率比存款息率高而受到歡迎。除了 ETF，上市開放式基金（LOF，交易買賣像 ETF）帶來主動型回報，是個方便投資者的選擇。不過，內地追蹤環球股市指數的 ETF 和 LOF 於一級市場贖回時，收取現金的時間往往滯後。再者，內地上市的基金只以人民幣買賣。因此，可以預見內地投資者為了作全球資產配置和涉足國際貨幣投資，對參與香港的 ETF 市場有潛在需求。

香港的 ETF 市場可說是通向環球投資的門戶，當中提供涵蓋環球市場多元化資產類別的產品，還有可支持市場流動性的成熟機構投資者基礎。透過這個市場，投資者可投資於全球大部分股票市場，包括港股、內地 A 股、亞太區及其他海外地區的股票等；還有世界各地市場的定息及貨幣，以及大宗商品等資產類別。此外，這個市場中不少環球指數 ETF 均可選擇多種交易貨幣（包括港元、人民幣及/或美元），切合環球投資者的各式需要。香港的 ETF 投資者還可享受市場深厚流動性、稅款偏低以及相關風險管理工具配套齊備等好處。另一方面，對發行商來說，香港的 ETF 市場也別具吸引力，包括相對便利市場的監管環境、廣闊的環球機構投資者基礎，加上有效的莊家機制，可確保二級市場的流動性，並且有助一級市場的套戥活動。再者，在香港發行 ETF 還享有地利 —— 發行商可以「在亞洲市場買賣亞洲產品」，為旗下亞洲產品在標的物相關市場對沖風險。

香港在擔當亞洲的 ETF 發行及交易樞紐這個角色上具備優勢。香港的 ETF 市場除了能夠滿足環球投資者的需要外，還能切合內地投資者的潛在需求。目前，內地的 ETF 市場的機構投資者參與度偏低，不足以支持一級市場交易，而且內地的 ETF 市場所提供的環球資產類別有限。相反，香港的 ETF 市場則不論在產品或投資者方面都呈多樣化，對投資 ETF 興趣日增的內地投資者來說有一定吸引力。另一方面，內地的 ETF 市場追蹤許多不同的內地資產，香港市場的環球投資者亦當感興趣。若然內地與香港的 ETF 市場能夠聯通起來[1]，將可令兩邊 ETF 的一級和二級市場投資者結構變得平衡，ETF 產品選擇也更多樣化。所以，若內地與香港的市場互聯互通可再進一步，擴大至包括 ETF 的範疇，對兩邊市場的相互增長發展定必有利。

1　有待監管當局批准。

1 交易所買賣基金吸引全球投資者

交易所買賣基金（ETF）是追蹤相關資產表現的開放式基金[2]，買賣方式一如股票。全球首隻 ETF 在 1990 年於加拿大推出，追蹤基準股票指數。之後，美國（1993 年）[3]、日本（1995 年）、香港市場（1999 年）及歐洲地區（2000 年）亦相繼推出 ETF，環球 ETF 市場隨之大幅增長。全球市場上市買賣的 ETF 數目與日俱增，相關資產已由股票擴展至債券、大宗商品及一籃子資產等其他資產類別。亞洲市場近年的增長速度尤為顯著。

香港一直是亞洲領先的 ETF 市場之一，在 1999 年已經推出追蹤恒生指數（恒指）的盈富基金，成為亞太區（日本除外）內最先發行及買賣 ETF 的市場。香港亦是全球首個地區推出以內地相關資產或人民幣資產為標的的 ETF，包括首隻 A 股 ETF（2004 年推出）[4]（追蹤富時中國 A50 指數的合成 ETF[5]）、首隻實物資產 A 股 ETF（2012 年推出）（透過人民幣合格境外機構投資者（RQFII）計劃追蹤 A 股指數並設港幣及人民幣雙櫃台交易）[6]、亞洲首隻離岸人民幣債券 ETF（2013 年 6 月推出）及首隻追蹤中國內地在岸債券市場的 ETF（2014 年 2 月推出）。

現在香港 ETF 市場切合全球投資者所需，已發展成為匯聚環球市場多元化資產，且兼容多種交易貨幣的交易平台。這個市場可作為區內的 ETF 發行和交易樞紐，迎合亞太區以至全球（尤其是中國內地）投資者與日俱增的市場需求。

ETF 成為愈來愈受歡迎的資產類別也不只見於香港，全球市場同樣也出現這個現象。何以 ETF 在全球和區內市場都如此受追捧？產品創新如何令 ETF 變得普

2　開放式基金是可申購及贖回基金單位的投資基金。投資者可直接向基金經理買入基金單位。

3　美國首隻 ETF（在紐約證券交易所（紐約交易所）買賣）於 1993 年面世，追蹤標普 500 指數。在此之前，美國證券交易所及費城證券交易所曾於 1989 年 5 月推出「指數參與股份」（有如每股有長短倉的期貨），讓投資者可在股票市場買賣股權，而毋需購買個別正股。但美國監管當局興訟，聯邦法院裁定有關股份實際上是期貨合約（即衍生產品），因此不可在證券市場買賣。（資料來源：Foucher, I. 及 K. Gray（2014）〈交易所買賣基金：損益和風險的演化〉（"Exchange-traded funds: Evolution of benefits, vulnerabilities and risks"），載於《加拿大銀行金融體系評論》（Bank of Canada's Financial System Review），2014 年 12 月號，37-46 頁。

4　見 Dennison, T.（2018）〈國際投資與打破國家界限的重要性〉（"International investing and the importance of breaking the country box"），載於《突破投資框架：明瞭不同資產類別及策略》（Invest outside the box: Understanding different asset classes and strategies），Palgrave Macmillan 出版，228-278 頁。

5　合成 ETF 是以衍生工具及掉期複製相關指數表現的 ETF，有別於實物資產 ETF 以相關實物證券進行複製。

6　資料來源：〈香港交易所首隻 RQFII A 股 ETF 即將登場〉，香港交易所新聞稿，2012 年 7 月 16 日。

及？以下嘗試逐一探討。

1.1　為何 ETF 比互惠基金受歡迎？

在 ETF 還未出現前，互惠基金自上世紀 20 年代面世以來一直是投資基金類中廣受歡迎者[7]。但到近年，市場對 ETF 的需求增長已超越互惠基金，這從 2018 年底市場調整期間的資金流向可見一斑。據報導，投資者在 2018 年 12 月 19 日止的一週內贖回互惠基金的金額達 562 億美元，是 2008 年 10 月以來的新高，但在同一星期內卻錄得 252 億美元流入 ETF[8]。

跟傳統互惠基金不同，ETF 除可作申購和贖回外，還可在證券交易所二級市場買賣，所以流動性較高。在交易所交易時段內於二級市場買賣 ETF 十分方便，而且 ETF 莊家有責任提供買賣報價，確保二級市場的流動性。ETF 可於二級市場買賣的單位數目，視乎參與經紀[9]與 ETF 發行商於一級市場申購或贖回的 ETF 單位而定。一級與二級市場之間按供求進行的套戥活動，有助減低 ETF 價格相對於其資產淨值[10]的溢價或折讓。

至於對發行商而言，在成本方面，美國 ETF 的平均開支比率[11]在 2016 年為 0.23%，對比追蹤指數的互惠基金的 0.73% 及主動型互惠基金的 1.45%[12]。由此可見，投資者投資 ETF 的基金管理費會比投資互惠基金便宜。對某些環球投資者而言，投資 ETF 亦可享較低稅金的好處，例如在美國投資，可享有愛爾蘭的可轉讓證券集體投資計劃（Undertakings Collective Investment in Transferable Securities

---

7　資料來源：Divakaran, S.、S. Buxton、St. Giles、M. Valentine 與 S. Atamuratova（2015）〈發展中市場的互惠基金：迎難而上〉（"Mutual funds in developing markets: Addressing challenges to growth"），世界銀行工作文件（World Bank＇s Working Paper）第 100976 號。

8　資料來源：Stein, C.〈基金投資者撤資 560 億美元　2008 年以來最大金額〉（"Fund investors pull $56 billion in biggest exit since 2008"），載於《彭博新聞》，2018 年 12 月 27 日。

9　「參與經紀」（有時稱為「授權參與者」）是在一級市場直接向 ETF 發行商提出申購或贖回 ETF 單位指示的機構投資者。有些參與經紀同時也是莊家。

10　ETF（每個單位）的資產淨值，代表着該 ETF 所持有的所有相關證券（例如股票或債券及現金）的市值，減去任何負債（例如管理費及行政開支），再除以已發行的基金單位數目計算得出。見〈ETF 定價及估值〉（"ETF pricing and valuations"），載於貝萊德網站，2019 年 8 月 7 日。

11　基金的開支比率是將基金總營運成本除以特定期間之內（通常為一年）所管理的基金資產總值的平均數計算得出。

12　資料來源：〈ETF 對比互惠基金：成本比較〉（"ETFs vs. mutual funds: Cost comparison"），載於富達網站，2017 年 6 月。

（UCITS））ETF 的 15% 預扣稅優惠，但互惠基金則沒有此優惠 [13]。

在透明度方面，ETF 的資料通常在其網站登載，包括適時披露（多數每天更新）整個組合的資料。

1.2　全球與區內 ETF 市場概覽

全球 ETF 的數目在 2018 年底達到 6,310 隻的歷史新高（2017 年底為 5,707 隻），其資產管理規模達到 4.66 萬億美元（2017 年底為 4.69 萬億美元）（見圖 1）。2009 至 2018 年的資產管理規模的複合年增長率約為 21%。

圖 1：全球 ETF 的資產管理規模及數目（2004 年至 2018 年）

資料來源：〈ETFGI：全球上市 ETF 及 ETP 2018 年 12 月份淨流入總額 762.4 億美元，史上第二高〉（"ETFGI reports ETFs and ETPs listed globally gather net inflows of US$76.24 billion during December 2018, 2nd Highest on record"），載於 etfgi.com 的 ETFGI 新聞稿，2019 年 1 月 9 日。

ETF 變得普及，主要受益於追蹤指數表現的被動型投資不斷增長。被動型投資在 2018 年估計約佔開放式基金四分之一，預計到 2020 年年底會升至約 31%（見

13　資料來源：〈2018 年的 ETF：主動型 ETF 的機遇與障礙〉（"ETFs 2018: Opportunities and obstacles for active ETFs"），載於安永網站，2018 年 12 月。UCITS ETF 是在歐洲市場登記的 ETF，可售予全球投資者。這些 ETF 受 UCITS（歐洲市場的統一監管框架）所規管。

圖 2）。而另一資料來源提到，2018 年 11 月底美國股票基金中，主動型與被動型股票基金的佔比為十年來最接近，分別約為 52% 和 48%[14]。主要原因是被動型投資的成本較低[15]，固長遠表現較為優勝 —— 截至 2018 年的十年間，美國逾八成主動型互惠基金的表現均遜於各自的基準指數[16]。

圖 2：全球市場開放式基金 —— 主動型與被動型基金的佔比（2011 年至 2020 年）

註：2016 年後的數字並非實際統計數字，是安永的估計數字。

資料來源：〈全球 ETF 研究 2017 —— 市場重塑 投資者為本〉（"Global ETF Research 2017 — Reshaping around the investor"），載於安永網站，2017 年。

　　ETF 所涉及的相關資產類別中，以股票佔絕大多數。在 2018 年年底，交易所買賣產品（ETP）[17]（主要是 ETF）中，股本證券佔資產管理規模總額約 77%，金額達 3.7 萬億美元（見圖 3）。當中又以美國股票佔最多，資產管理規模在 2018 年年底達到 2.1 萬億美元。定息 ETP 方面，不論在資產管理規模的佔比及金額上於近

14　資料來源：Stoin, C.〈由主動型轉向被動型：2019 年達臨界點〉（"Shift from active to passive approaches tipping point in 2019"），載於《彭博新聞》，2018 年 12 月 31 日。

15　成本包括管理費、買賣差價、行政費、佣金、市場影響及稅項（如適用）。

16　資料來源：Rowley Jr. J. J.、D. J. Walker 與 C. Zhu：〈投資低成本指數基金的理由〉（"The case for low-cost index-fund investing"），載於 Vanguard Research，2019 年 4 月。

17　ETP 包含 ETF 及其他 ETP，當中包括交易所買賣工具（ETV）、交易所買賣票據（ETN）及憑證。根據 ETFGI 的數據，ETF 的資產管理規模於 ETP 的資產管理規模中的佔比，在 2004 年至 2018 年期間的每年年底介乎 89% 至 98% 之間。

年均穩步上揚,在 2018 年年底約佔所有 ETP 的資產管理規模總額的 18%,金額
為 9,000 億美元。大宗商品及其他則約佔所有 ETP 的資產管理規模總額的 4%,金
額約為 2,000 億美元。

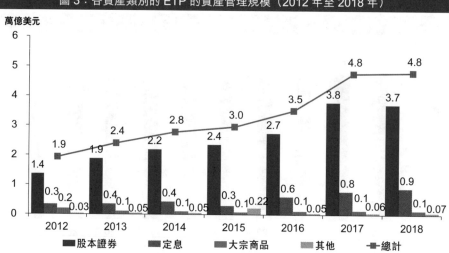

圖 3:各資產類別的 ETP 的資產管理規模 (2012 年至 2018 年)

註:「其他」包括貨幣、貨幣市場及其他類別 (例如波幅)。

資料來源:〈貝萊德環球 ETP 形勢:行業焦點〉("Blackrock Global ETP landscape: Industry highlights"),
2012 年至 2018 年各年的 12 月號,載於貝萊德網站。

儘管美國 ETF 市場是 ETF 市場中最大者 (見圖 4),但 ETP 的資產管理規模
在 2012 年至 2018 年六年期間的複合年增長率卻以亞太區的 26.1% 為最高,美國
和歐洲市場同期的增長率分別只為 16.5% 及 13.4%。

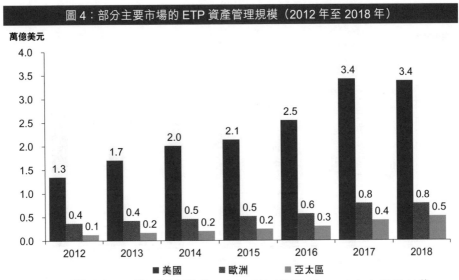

圖 4：部分主要市場的 ETP 資產管理規模（2012 年至 2018 年）

資料來源：〈貝萊德環球 ETP 形勢：行業焦點〉（"Blackrock Global ETP landscape: Industry highlights"），2012 年至 2018 年各年的 12 月號，載於貝萊德網站。

　　就二級市場的交易而言，全球交易所的 ETF 年度總成交額一直上揚，在 2018 年創出 24.1 萬億美元的新高（見圖 5）。其中以美洲的交易所的成交額佔主導，在 2012 年至 2018 年六年期間的複合年增長率為 13.9%，當中主要受美國的交易所帶動。不過，亞太區的 ETF 總成交額自 2015 年起已超越歐洲、非洲及中東的合計數字，在 2018 年創出 1.7 萬億美元的新高，在 2012 年至 2018 年六年期間的複合年增長率達到 31.2%。

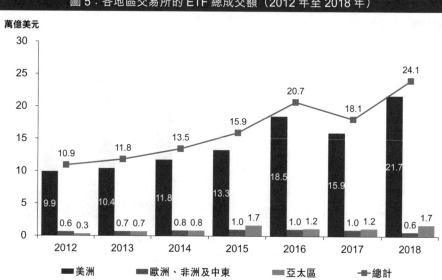

圖 5：各地區交易所的 ETF 總成交額（2012 年至 2018 年）

萬億美元

資料來源：國際證券交易所聯會網站（WFE）的統計數字。

在增長最快的亞太區，按資產管理規模計算，日本的 ETF 市場最大，在 2017 年年底佔亞太區總額的 62%，而中國內地與香港則各佔區內總額的 9%[18]。另一資料來源提到，日本市場的 ETF 資產管理規模在 2018 年 6 月底達 3,090 億美元，屬亞太區之冠，其次為中國內地（370 億美元）、韓國（360 億美元）及香港市場（340 億美元）[19]。

儘管 2018 年股市疲弱，但亞太區股票 ETP 的資產管理規模仍繼續上升（見圖 6）。定息 ETP 的資產管理規模在 2018 年上升了 1.5 倍，而大宗商品 ETP 的資產管理規模則相對平穩。其他資產類別的資產管理規模增長主要來自貨幣市場 ETF。除了資產管理業的增長趨勢外，政府政策亦促進了多個亞太區 ETF 市場的

18 資料來源：〈亞太區 ETF：尋找臨界點〉（"APAC ETF: Finding the tipping point"），Broadridge 的簡報，載於新加坡投資管理協會午餐講座系列—— 亞太區 ETF 機遇：新增長前沿（Investment Management Association of Singapore (IMAS) Lunchtime Talk Series — APAC ETF Opportunities: The Next Growth Frontier），2018 年 2 月 13 日。

19 資料來源：〈中國 ETF 起飛？〉（"Are China's ETFs taking off?"），載於 Fund Selector Asia 網站，2018 年 7 月 12 日。

增長。例子包括日本中央銀行的資產購買計劃 [20]、臺灣地區保險公司紛紛購買債券 ETF 以規避有關外國投資的新規 [21]，以及市場在澳洲禁止「衝突報酬架構」後而尋求具成本效益的投資等 [22]。

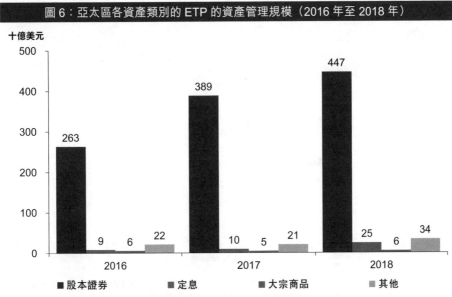

圖 6：亞太區各資產類別的 ETP 的資產管理規模（2016 年至 2018 年）

十億美元

■ 股本證券　■ 定息　■ 大宗商品　■ 其他

資料來源：〈貝萊德環球 ETP 形勢：行業焦點〉（"Blackrock Global ETP landscape: Industry highlights"），2016 年至 2018 年各年的 12 月號，載於貝萊德網站。

1.3　最新的產品發展支持 ETF 市場增長

為了滿足機構及個人投資者不斷演變的需求，產品必須不斷創新。在這方面，ETF 可以填補在有效邊界上不同資產配置的投資選擇。例如，香港市場便提供了

20　ETF 是日本央行資產購買計劃的政策工具之一。見 Petrov, A.（道富環球顧問）〈貨幣政策中的 ETF：日本央行個案研究〉（"ETFs in Monetary Policy — Case Study: Bank of Japan"），2017 年 11 月。

21　臺灣地區保險公司購買 ETF，以規避外幣投資上限的新規。見 Herrero, A. C. 及 G. Ng〈臺灣：大數據顯示市場對新型海外資產 —— 債券 ETF —— 反應正面〉（"Taiwan: Big data show positive sentiment for a new type of overseas assets — Bond ETFs"），載於 Natixis 網站，2019 年 3 月 12 日。

22　澳洲監管當局在 2013 年 7 月起禁止關於財務顧問的「衝突報酬架構」（包括佣金），促使更多人追捧 ETF 等具成本效益的投資。見《監管指引 246 —— 衝突報酬》（Regulatory Guide 246 — Conflicted remuneration），澳洲證券投資委員會發佈，2013 年 3 月。

不同風險回報系數的 ETF（見圖 7）。這些不同的投資選擇一方面吸引新投資者，另一方面亦可為現有投資者提供更多分散投資的選擇。

圖 7：香港 ETF 三年回報對三年波幅的分佈圖（2016 年至 2018 年）

註：此樣本只涵蓋 2016 年 1 月至 2018 年 12 月期間在香港上市的 ETF。累計回報是期內經調整派息後的價格回報，波幅則按三年內最高與最低價的差額除以 2018 年年底的收市價計算得出。

資料來源：Wind。

近年全球 ETF 市場的產品頻頻推陳出新，擴大了各地的投資機遇：

- **智慧型投資策略（Smart beta）ETF：**追蹤指數表現，按若干宏觀或風格因素調整投資組合成分。這些因素包括規模、價值、動力及質素（基礎指標的組成部分）等。2013 年至 2017 年間，全球新發行的智慧型投資策略 ETF 數目不斷上升，其資產管理規模由 2012 年的 2,800 億美元增至 2017 年的 9,990 億美元，複合年增長率達 29%[23]。另一資料來源顯示，2018 年年底全球共有 1,298 隻智慧型投資策略 ETP，資產管理規模合共 6,180 億

23　資料來源：〈智慧型投資策略的五年走勢及展望〉（"Five-year trends and outlook for smart beta"），載於富時羅素網站，2018 年 7 月 5 日。

美元 [24]。

- **主動型 ETF**：追蹤標的基準指數，但容許偏離基準指數對行業分配的變動或調整的時間或相關指數本身，以產生更高的潛在回報。截至 2018 年 11 月底，全球有 773 隻主動型 ETF，其資產管理規模創下 1,094 億美元的新高，並預期在 2020 年增至 2,170 億美元 [25]。

- **人工智能崛起**：美國在 2013 年推出首隻追蹤機械人及人工智能股票的全球指數 ETF。該 ETF 截至 2019 年 7 月 9 日的三年累計年度回報為 16.7%，勝過標普 500 總回報指數的 14.1%[26]。結果，不少主要 ETF 管理公司自 2016 年起發行類似的 ETF[27]。而首隻人工智能操作的 ETF 根據人工智能系統對市場數據的解讀，追蹤未來 12 個月很可能跑贏大市的 30 至 70 隻美國市場股票或房地產投資信託基金。該隻 ETF 於 2017 年 10 月 18 日在美國推出，上市首年上升了 11.8%，高於標普 500 指數同期 8.1% 的升幅，據報還比 87% 的主動型基金優勝 [28]。該 ETF 發行商在 2018 年 6 月在美國推出另一隻 ETF，涵蓋環球市場 80 至 250 隻股票，背後支援的也是同一個人工智能系統。

- **槓桿及反向 (L&I) 產品**：以期貨或掉期放大相關指數的回報率或提供與相關指數相反的回報。全球 L&I 產品的數目由 2013 年年底的 608 隻增至 2018 年 6 月底的 901 隻，其資產管理規模於同期內亦由 532 億美元上升至 799 億美元 [29]。

- **環境、社會及管治 (ESG) 投資**：支持個別經濟體的可持續發展，但不同

24 資料來源：〈ETFGI：全球上市的智慧型投資策略 ETF 及 ETP 於 2018 年 12 月份淨流入 124.2 億美元〉("ETFGI reports Smart Beta ETFs and ETPs listed globally gather net inflows of US$12.42 billion during December 2018")，載於 etfgi.com 的 ETFGI 新聞稿，2019 年 1 月 30 日。

25 資料來源：《主動型 ETF —— 亞洲 ETF 創新的下一步》("Active ETFs — The next step in Asia's ETF innovation")，載於 Risk.net 網站，2019 年 2 月 12 日。

26 該隻美國上市的機械人及人工智能股票 ETF（代號：ROBO）與標普 500 總回報指數的累計三年回報數字見 Marketwatch.com，文中所列是 2019 年 7 月 9 日的數字。

27 資料來源：Lim, J.〈機械人及人工智能股票 ETF：鹿死誰手？〉("Robotics and A.I. ETFs: Which will win the battle?")，載於輝立證券的 POEMS 網站，2018 年 7 月 20 日。

28 資料來源：Ponczek, S.〈機械人 ETF 叫專業基金望塵莫及　小型股起飛〉("Robot ETF leaves pros in dust, scoring wind on small-cap fliers")，載於《彭博新聞》，2018 年 10 月 19 日。

29 資料來源：〈ETFGI：2018 年 6 月底全球上市槓桿及反向 ETF 及 ETP 涉及投資資產 799 億元〉("ETFGI reports that assets invested in leveraged and inverse ETFs and ETPs listed globally reached $79.9 billion at the end of June 2018")，載於 etfgi.com 的 ETFGI 新聞稿，2018 年 7 月 31 日。

資產管理人對 ESG 評估有不同的定義或框架。2018 年的一項調查發現，約 99% 的大中華區機構投資者都認同 ESG 因素十分重要 [30]。ESG ETF 的數目由 2013 年年底的 47 隻增至 2018 年年底的 208 隻，其資產管理規模於同期內亦由 50 億美元上升至 230 億美元 [31]。

- **定息 ETF：** 定息 ETP 的資產管理規模由 2017 年年底的 7,800 億美元升至 2018 年年底的 8,820 億美元 [32]。當中涉及美國國債的 ETP 的資產管理規模由 550 億美元升至 1,300 億美元。同期，追蹤新興市場債務（不包括公司債券）的 ETP 的資產管理規模則由 510 億美元升至 580 億美元，這可能受益於離岸投資者通過「債券通」對內地債券的投資，推動中國債券獲納入全球債券指數而帶來的增長。定息 ETF 的增長可能意味着投資者對債務工具的興趣日濃。一項有關全球機構投資者的調查顯示，六成受訪者視債券 ETF 為另一個定息工具的選擇 [33]。投資者對債券 ETF 的興趣增加，其中一個原因可能是投資債券 ETF 的資金效率比直接購入相關債券為高 [34]。

- **ETF 跨境上市：** ETF 跨境上市（或相互認可）可擴大個別市場投資者可投資的資產類別。在歐洲，ETF 通常同時在多家歐洲交易所上市，利便投資者在自家市場的交易所買賣 ETF 及結算交易。在 2017 年年底，於歐洲上市的 ETP 有 2,260 隻，其於 27 家交易所的相應上市數目合共 7,278 宗（美國有 2,116 隻 ETP 於 4 家交易所上市）[35]。在香港，24% 的 ETF 來自其他市場的跨境上市產品（2017 年 6 月底數字）[36]。

發展 ETF 產品其中一個主因，是迎合機構投資者分散投資的需要。根據 2017

30　資料來源：〈準備起飛：大中華 ETF 機遇〉（"Ready to grow: ETF opportunities in Greater China"），載於布朗兄弟哈里曼（Brown Brothers Harriman）網站，2018 年 4 月。

31　資料來源：Wilson, D.〈ESG ETF：中國、香港均須假外求〉（"China, Hong Kong must look offshore for ESG ETFs"），載於 Fund Selector Asia 網站，2019 年 1 月 31 日。

32　資料來源：〈貝萊德環球 ETP 形勢：行業焦點〉（"Blackrock Global ETP landscape: Industry highlights"），2017 年 12 月及 2018 年 12 月號，載於貝萊德網站。

33　資料來源：McCollum, A.〈大戶為求債券市場流動性而轉投 ETF 懷抱〉（"Institutions turn to ETFs for bond market liquidity"），載於 Greenwich Associates 網站，2018 年 9 月 18 日。

34　見〈機構增加涉足於亞洲 ETF 市場〉（"Growing an institutional footprint in Asia's ETF market"），載於 Risk.net，2018 年 11 月 19 日。

35　資料來源：Crigger, L.〈歐洲 ETF 交投新生機〉（"New light shines on Europe ETF trading"），載於 ETF.com，2018 年 4 月 18 日。

36　資料來源：證券及期貨事務監察委員會（證監會）研究論文〈香港交易所買賣基金市場及相關課題〉，載於證監會網站，2018 年 1 月 16 日。（簡稱〈證監會 2018 年 ETF 論文〉）。

年一項調查，51% 的全球機構（相對於 39% 的亞洲機構）將（依據核心及戰術策略進行的）資產配置評為購買 ETF 的最重要考量因素 [37]。調查發現，ETF 的其他用途包括：風險管理或對沖、重新調整/過渡性配置、過渡期內的中期性超額回報部署、為減低閒置現金而進行現金股票化、為提供流動性而作為「流動性袖子」（liquidity sleeves）及風險管理（例如透過 L&I 產品）。

2　內地及香港 ETF 市場發展可互為補足

亞洲 ETF 市場預期將會持續增長。內地及香港 ETF 市場的增長會並駕齊驅，可收互為補足之效，並預期會因兩地市場進一步互聯互通而增速發展。一項研究預測，由 2017 年至 2025 年，內地 ETF 於亞太區 ETF 的資產管理規模佔比將由 9% 升至 19%、香港 ETF 的相應佔比將由 9% 升至 18%，而日本 ETF 相應的佔比卻將由 62% 跌至 40%[38]。該項研究亦預期，亞太區市場於 2018 年至 2025 年從 ETF 新投資的淨流入金額將大約有 75% 源自日本、中國內地及香港。

2.1　內地投資者對產品多樣化的需求殷切

自上海證券交易所（上交所）在 2005 年 2 月有首隻 ETF（追蹤上證 50 指數（上交所大型股指數）的表現）上市以來，內地 ETF 市場一直發展迅速。內地 ETF 的資產管理規模及數目於 2019 年第一季季末均達歷史新高，共有 205 隻 ETF，其資產管理規模達 5,831 億元人民幣（見圖 8）。當中 146 隻於上交所上市、59 隻於深圳證券交易所（深交所）上市。

37　資料來源：〈機構 ETF 買賣的全球趨勢〉（"Global trends in institutional ETF trading"），簡街資本（Jane Street）委託進行的調查報告，載於 Risk.net，2017 年 10 月 5 日。調查得到 210 家機構回應，美國、歐洲及亞洲分別佔 86 家、79 家及 45 家。

38　資料來源：〈亞太區 ETF：尋找臨界點〉（"APAC ETF: Finding the tipping point"），Broadridge 的簡報，載於新加坡投資管理協會午餐講座系列 —— 亞太區 ETF 機遇：新增長前沿（Investment Management Association of Singapore (IMAS) Lunchtime Talk Series — APAC ETF Opportunities: The Next Growth Frontier），2018 年 2 月 13 日。

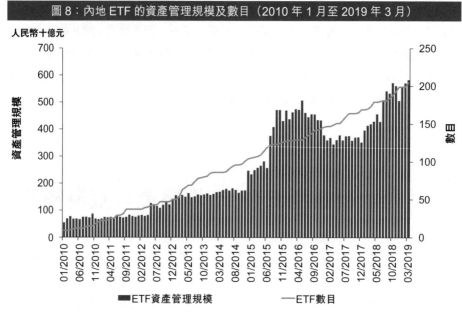

圖 8：內地 ETF 的資產管理規模及數目（2010 年 1 月至 2019 年 3 月）

註：資產管理規模的數字是中國證券登記結算有限公司紀錄中的 ETF 金額。
資料來源：Wind。

　　在 ETF 數目方面，截至 2019 年第一季季末，內地指數的股票 ETF 數目最多，共有 151 隻，佔總數的 74%，而海外股票指數的跨境 ETF 則只有 13 隻（佔總數的 6%）（見圖 9a）。在追蹤在岸股票的 ETF 當中，ETF 的相關指數涵蓋涉及內地股票的不同種類，包括智慧型投資策略（如規模、價值、增長或紅利）、行業、擁有權及市場（上海、深圳或跨市場）。而在 13 隻跨境 ETF 當中，9 隻追蹤香港股票的指數（包括 4 隻 H 股 ETF），3 隻追蹤美國股票的指數，1 隻追蹤德國股市。此外，投資於內地定息產品市場的 ETF 共有 37 隻——27 隻貨幣市場 ETF[39]（佔總數的 13%）及 10 隻債券 ETF（佔總數的 5%）。4 隻大宗商品 ETF 全部均為黃金 ETF（佔總數的 2%），追蹤於上海黃金交易所買賣的黃金合約。

39　貨幣市場 ETF 投資於短期定息投資產品，包括短期政府債券、短期商業票據、回購工具、銀行承兌及可轉讓定期存單。

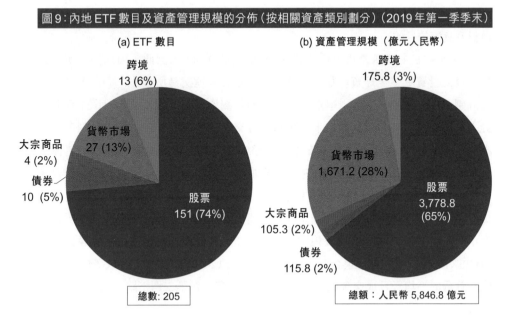

圖9：內地 ETF 數目及資產管理規模的分佈（按相關資產類別劃分）（2019年第一季季末）

(a) ETF 數目

- 跨境 13 (6%)
- 貨幣市場 27 (13%)
- 大宗商品 4 (2%)
- 債券 10 (5%)
- 股票 151 (74%)

總數：205

(b) 資產管理規模（億元人民幣）

- 跨境 175.8 (3%)
- 貨幣市場 1,671.2 (28%)
- 大宗商品 105.3 (2%)
- 債券 115.8 (2%)
- 股票 3,778.8 (65%)

總額：人民幣 5,846.8 億元

資料來源：Wind。

在資產管理規模方面，截至 2019 年第一季季末，貨幣市場 ETF 佔資產管理規模總額的 28%，僅次於股票 ETF 的 65%（見圖 9b）。貨幣市場 ETF 流動性極高，加上貨幣市場基金的孳息率通常較銀行一年期定期存款為高，因此深受投資者歡迎[40]。就內地的貨幣市場 ETF 而言，二級市場銷售所得的款項可於同一個交易日（T+0）內用於購買證券產品，或於下一個交易日（T+1）提取；至於在一級市場申購或贖回貨幣市場 ETF 單位，深交所及上交所的結算周期分別是 T+0 及 T+2[41]。因此，貨幣市場 ETF 成為提升孳息率及分散投資的另一種資產配置選擇。另外，上交所亦有僅可於一級市場作交易的場內貨幣市場基金（非 ETF），其申購/贖回的結算周期為 T+1[42]（相對於貨幣市場 ETF 的 T+0 或 T+2）。然而，現時內地所有

40　見〈貨幣 ETF 基金的優勢有哪些？〉，載於《中投在線》，2016 年 4 月 26 日；〈一文看懂場內貨幣基金〉，載於《新浪財經》網站，2018 年 9 月 5 日。

41　在深交所買賣的貨幣市場 ETF，新增的 ETF 單位可於同一個交易日（T+0）於一級市場贖回或於二級市場出售。在上交所買賣的貨幣市場 ETF，新增的 ETF 單位則可於第三個交易日（T+2）於一級市場贖回或於二級市場出售。

42　根據 Wind 的數據，2018 年年底上交所有 12 隻僅供於一級市場作交易的貨幣市場基金，其總資產管理規模為 219 億元人民幣。

上市基金均以人民幣作交易，換言之，內地投資者無法透過投資內地 ETF 達到分散投資貨幣的目的。

內地 ETF 主要涵蓋國內資產類別（佔總數的 94%）。因此，內地交易所正在探討容許投資更多全球資產類別的途徑，包括與其他交易所合作。「中日 ETF 互通計劃」的首批 ETF，分別追蹤對方市場的相關資產，於 2019 年 6 月 25 日分別於上交所及日本交易所集團旗下的東京證券交易所作跨境上市。在該計劃下，日本或中國的 ETF 發行商設立一種名為「連結 ETF」的全新基金單位類別，這種基金單位會以對方交易所的上市 ETF 為標的並投資至少 90% 於該標的 ETF，但要受合格境外機構投資者（QFII）或合格境內機構投資者（QDII）計劃的額度限制。該標的 ETF 須已上市一年以上，並須於原市場有足夠的流動性。首批於上交所跨境上市的四隻 ETF 當中，有三隻追蹤日經 225 指數，餘下一隻則追蹤東證股價指數（TOPIX），總資產管理規模合共 5.09 億元人民幣（2019 年 6 月 21 日數字）[43]。這些 ETF 首週的平均每日成交金額為 1.618 億元人民幣，在第二週下跌至 6,510 萬元人民幣[44]。

在內地的上市基金市場，還有上市開放式基金（LOF）與 ETF 一樣在交易所上市及買賣。LOF 與 ETF 類似，是一種流動性高而方便的工具，用以投資不同相關資產類別，包括股票、債券、混合策略（同時涉及股票和債券）、大宗商品及跨境投資（透過 QDII）。LOF 通常採納主動型投資策略，旨在跑贏基準，因此管理開支較 ETF 為高。

於 2019 年第一季季末，內地共有 298 隻 LOF（上交所 74 隻、深交所 224 隻），總資產管理規模約 3,583 億元人民幣[45]。採納股債混合策略的 LOF 在數目上及管理資產規模上的佔比均最多，分別佔 36% 及 68%（見圖 10），部分原因是相對內地的 ETF 而言，該等基金提供不同的風險回報。然而，如同內地的 ETF，LOF 仍以在岸的資產類別所佔比率最高，按數目及資產管理規模計分別為 89% 及 98%。

43　見〈首批 4 隻中日互通 ETF 今日上市〉，載於《新華網》，2019 年 4 月 22 日。

44　資料來源：Wind。

45　資料來源：Wind。

圖 10：內地 LOF 數目及資產管理規模按相關資產類別的分佈（2019 年第一季季末）

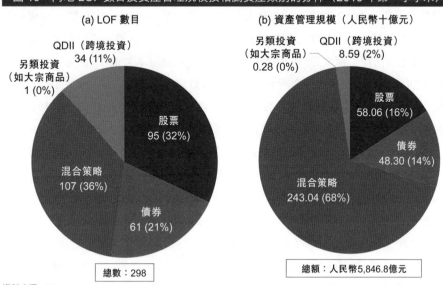

資料來源：Wind。

　　除 ETF 及 LOF 外，內地的上市基金亦包括分級基金[46]。母基金單位分為不同的子單位（分級基金），預期風險及收入各有不同，這些基金單位類別有部分或全部可供買賣。只有母基金單位能進行申購及贖回，而子基金單位卻不能。上交所的分級基金的母基金單位為 LOF[47]，深交所的母基金單位則未有上市及買賣。為保障散戶投資者，上交所及深交所自 2017 年 5 月設置了投資者買賣分級基金的門檻，包括過去 20 個交易日內的資產價值須達 300,000 元人民幣，並須與經紀簽訂投資風險揭示書。而投資者買賣 ETF 及 LOF 則沒有資產方面的最低要求。截至 2018 年年底，分級基金共有 122 隻（各有 A 類及 B 類子單位），其資產管理規模達 1,110 億元人民幣。分級基金的投資範圍與 ETF 及 LOF 類似，主要限於在岸的股票及債券。

46　還有另一類基金就是封閉式基金。然而，自內地基金行業改革後，近年封閉式基金已萎縮不少。於 2019 年 8 月 7 日，在交易所上市的封閉式基金只有兩隻（上交所及深交所各一隻）。資料來源：上交所及深交所網站的基金列表（於 2019 年 8 月 7 日瀏覽）。

47　母基金單位可由於同一個交易日購買的子基金單位所組成，並於同一個交易日贖回。參閱上交所及深交所於 2016 年 11 月 26 日分別發佈的《分級基金業務管理指引》，以及〈分級基金母基金能買嗎？分級基金母基金怎麼買？〉，載於《網貸》網站，2018 年 4 月 28 日。

在成交金額方面，ETF 稱冠內地的上市基金市場。ETF 在內地交易所的平均每日成交金額於 2018 年升至 184 億元人民幣（上交所：68 億元人民幣；深交所：115 億元人民幣），創下歷史新高，更遠高於 LOF 的 1.585 億元人民幣（上交所：1,570 萬元人民幣；深交所：1.428 億元人民幣）及分級基金的 11 億元人民幣（上交所：2,110 萬元人民幣；深交所：11 億元人民幣）[48]（見圖 11）。內地 ETF 於 2018 年的平均每日成交金額約佔在岸股票的平均每日成交金額的 5%[49]（見圖 12）。自 2011 年起，合資格的 ETF 可進行融資融券交易[50]。

深交所自 2019 年 4 月開始對其證券結算安排作出修改[51]，將深交所上市的 A 股、ETF 及 LOF 的交收周期與上交所上市的同類證券看齊 —— 證券結算會於同一個交易日（T+0）完成，款項結算則會於下一個交易日（T+1）完成。在此之前，深交所 ETF 的結算按貨銀對付交收方式於 T+1 進行（證券及款項結算均於 T+1 進行）。此外，在規則的相應修訂之前，追蹤上交所及深交所 A 股的跨市場 ETF 不可於同一個交易日（T+0）內購買 ETF 並贖回 A 股在市場出售，但在規則修訂實行後便會容許這樣做。投資者將可在 T+0 更頻繁地買賣 ETF，提升 ETF 二級市場的流動性。

48 資料來源：LOF 及分級基金的平均每日成交金額來自上交所及深交所的每月統計報表，載於上交所及深交所網站。

49 ETF 的平均每日成交金額按國際證券交易所聯會有關每家交易所的 ETF 總成交金額的統計數字來估算。在岸股市的成交金額來自 Wind。

50 可作融資融券（融券安排的使用者能沽空該證券）的 ETF 資格準則包括：上市後買賣至少 5 個交易日，過去 5 個交易日的平均資產管理規模達 5 億元人民幣及有至少 2,000 名基金持有人。參閱上交所及深交所於 2019 年 8 月 9 日分別發佈的有關融資融券的規則修訂（《融資融券交易實施細則（2019 年修訂）》）。

51 《關於做好深市 ETF 交易結算模式調整相關技術準備的通知》，深交所發佈，2019 年 4 月 8 日；《關於就《深圳證券交易所證券投資基金交易和申購贖回實施細則（2019 年修訂徵求意見稿）》公開徵求意見的通知》，深交所發佈，2019 年 8 月 23 日。

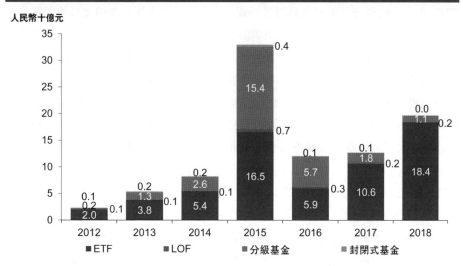

圖 11：內地上市基金按類別劃分的平均每日成交金額（2012 年至 2018 年）

資料來源：上交所及深交所。

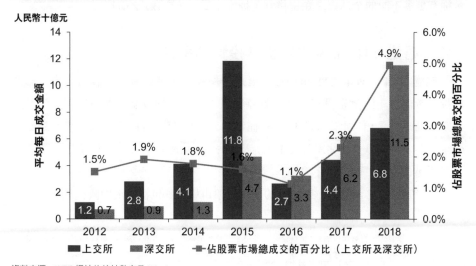

圖 12：內地交易所 ETF 的平均每日成交金額（2012 年至 2018 年）

資料來源：WFE 網站的統計數字及 Wind。

一級市場是另一個流動性的來源。就於內地上市的 A 股 ETF（包括跨市場 ETF）而言，在深交所的規則修訂實行後，一級與二級市場之間的套戥活動將可於同一個交易日（T+0）進行 —— 新申購的 ETF 單位將可於申購日期當天在二級市場出售，已取消 ETF 單位的贖回證券將可於贖回日期在二級市場出售。在 ETF 的沽空活動受限的情況下，這樣的安排有助這些 ETF 的價格發現。然而在一級市場，不同類別的 ETF 以及 ETF 與其他上市基金之間的結算周期仍不盡相同。就 LOF 而言，申購及贖回基金單位僅可以現金結算，因此可能會涉及買賣相關證券的額外成本。同樣，申購及贖回跨境 ETF 單位亦只能以現金結算。值得留意的是，跨境 ETF 及 QDII LOF 在現金結算時所收取現金的時間較長 —— 跨境 ETF 需時 4 至 7 天，QDII LOF 需時約 10 天[52]。

總括而言，內地投資者也許希望透過 ETF 達到投資多樣化。從內地 ETF 市場規模及交易活動的增長表明，內地投資者對 ETF 的興趣正不斷提升。然而，內地的上市基金當中以與國內資產（主要為股票）相關的基金佔大多數，而與全球資產相關的上市基金的供應則相當有限。與全球資產相關的上市基金在贖回時較遲收款亦可能限制這些基金的流動性。因此，若內地投資者可進入香港 ETF 市場，將會從有更多不同的全球資產類別供其選擇當中受益（見第 2.2 節）。

2.2　香港 ETF 市場作為全球投資的平台

經過 20 年，香港 ETF 市場已發展成一個吸引全球投資者的投資平台，讓全球投資者能以不同貨幣投資於多種資產類別，包括本地及全球股票、定息產品及大宗商品（見圖 13）。香港的 ETF 不僅以港元買賣（佔 ETF 以及 L&I 產品總數的 63%[53]），亦可以美元（佔總數的 14%）及人民幣（佔總數的 23%）進行交易。

香港的 ETF[54] 包括以下幾類：

- **股票 ETF**：覆蓋至少 80% 的全球股票市場[55]，包括追蹤中國內地 A 股、香

52　資料來源：李真、宋祖強與程炳光（華寶證券）〈跨境 ETF 交易機制及淨值計算詳解〉，載於《雪球》網站，2015 年 11 月 20 日；〈跨境 LOF 基金的交收規則〉，載於國泰君安網站。

53　資料來源：香港交易所網站上 ETP 證券名單的網頁，2019 年 6 月資料。

54　除非另有指明，本部分呈列的所有關於不同類別 ETF 的統計數據，均來自載於香港交易所網站的《ETF 市場概覽》，2019 年 6 月號。

55　Vanguard 所發行的香港上市 ETF 已覆蓋超過 80% 的全球股票市場（2018 年 3 月底數字）。資料來源：〈領航推出全港最全面的中國股票 ETF〉，Vanguard 新聞稿，2018 年 5 月 10 日。

港股票、亞太股票及其他海外股票的 ETF。截至 2019 年上半年底,股票
ETF 共有 90 隻,其資產管理規模達 2,797 億港元 (佔總額 86.0%)。

　　◦ 港股 ETF —— 追蹤恒生指數及其他香港指數,有些具有智慧型投資策
　　　略因子;

　　◦ A 股 ETF —— 不僅有追蹤內地旗艦指數的 ETF,亦有追蹤 MSCI 及富
　　　時等全球指數供應商編制的 A 股指數的 ETF;

　　◦ 亞太股票 ETF —— 追蹤印度、印尼、日本、韓國、馬來西亞、菲律賓、
　　　新加坡、臺灣及越南等國家和地區金融市場的股票指數;

　　◦ 其他海外股票指數 ETF —— 追蹤美國股票、歐洲股票及一些新興市場
　　　(如巴西及俄羅斯) 的股票,其中包括全球首隻追蹤納斯達克中國新經
　　　濟公司海外 50 指數 (涵蓋於香港地區及美國上市的內地新經濟公司)
　　　的 ETF (於 2018 年 9 月推出)。

- **L&I 產品:**提供相當於香港及海外股票指數最多 ±2 倍的表現。截至
  2019 上半年底,L&I 產品共有 22 隻,其資產管理規模達 73 億港元 (佔總
  額 2.2%)。

- **定息產品及貨幣 ETF:**追蹤內地及亞太市場的優質債券以及港元及美元貨
  幣市場的表現。截至 2019 年上半年底,定息產品及貨幣 ETF 共有 11 隻,
  其資產管理規模達 373 億港元 (佔總額 11.5%)。

- **大宗商品 ETF:**追蹤全球市場黃金及原油價格。截至 2019 上半年底,大
  宗商品 ETF 共有 6 隻,其資產管理規模為 11 億港元 (佔總額 0.3%)。

- **主動型 ETF:**香港監管當局於 2019 年 1 月起接受主動型 ETF 的申請,首
  隻主動型 ETF (追蹤美元貨幣市場) 於 2019 年 6 月 18 日上市 [56]。

56　資料來源:〈香港交易所歡迎首隻主動型 ETF 上市〉,香港交易所新聞稿,2019 年 6 月 18 日。

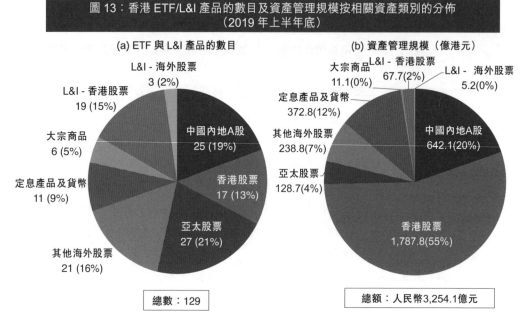

圖 13：香港 ETF/L&I 產品的數目及資產管理規模按相關資產類別的分佈
（2019 年上半年底）

(a) ETF 與 L&I 產品的數目

(b) 資產管理規模（億港元）

總數：129

總額：人民幣3,254.1億元

註：同一 ETF 的不同貨幣櫃枱合計當作一隻產品計算。

資料來源：《ETF 市場概覽》，2019 年 6 月號，載於香港交易所網站。

　　香港 ETF 的總資產管理規模於 2019 年上半年底約為 3,254 億港元（見圖 14），於 2019 年上半年的平均每日成交金額約為 53 億港元（約佔證券市場總成交金額的 5.4%）。

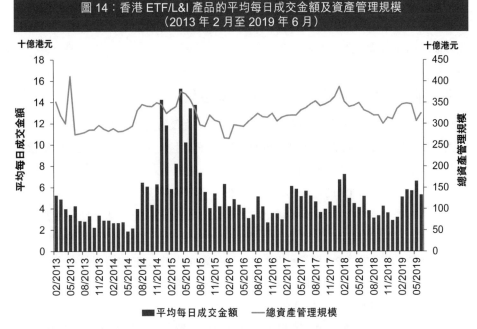

圖 14：香港 ETF/L&I 產品的平均每日成交金額及資產管理規模
（2013 年 2 月至 2019 年 6 月）

註：L&I 產品於 2016 年 6 月推出。

資料來源：《ETF 市場概覽》，2013 年 2 月號至 2019 年 6 月號，載於香港交易所網站。

　　機構投資者的成交額於 2016 年至 2017 年上半年期間佔香港 ETF 總成交額
的 81% 以上，而 ETF 的每筆交易平均金額由 2007 年的 224,000 港元倍增至 2017
年上半年的 459,000 港元[57]。機構投資者在香港 ETF 市場的高參與度可能是由於香
港有越來越多環球資產管理人 —— 香港獲發牌的資產管理公司由 2016 年年底的
1,300 家增加 14% 至 2017 年年底的 1,477 家，在 2018 年年底再增至 1,643 家[58]。此
外，這些資產管理公司對定息產品 ETF 感興趣，視之為另類的定息投資選擇 ——
2018 年的一項調查[59] 結果顯示，83% 的香港機構投資者已投資於定息產品 ETF，
比例高於股票 ETF 的 78% 及大宗商品 ETF 的 25%。

57　資料來源：〈證監會 2018 年 ETF 論文〉。

58　資料來源：證監會的 2017 年及 2018 年〈資產及財富管理活動調查〉，載於證監會網站。（2018 年的報告簡稱為
　　〈證監會 2018 年資產管理調查〉。）

59　資料來源：〈蓄勢待發：大中華 ETF 機遇〉（Ready to grow: ETF opportunities in Greater China），載於布朗兄弟
　　哈里曼（Brown Brothers Harriman）網站，2018 年 4 月。

香港 ETF 市場的特點，是提供多種資產類別的投資選擇和擁有廣闊的機構投資者基礎。香港市場在發行及買賣 ETF 方面的優勢，有利推進香港 ETF 市場的興旺發展。第 3 節將就此論述。

2.3 內地及香港 ETF 市場的互補性質

內地及香港 ETF 市場的產品構成、投資者基礎及市場准入制度均大不相同，但兩地市場同樣反映投資者對 ETF 的投資胃納正在上升。（有關內地 ETF 及 LOF 以及香港 ETF 的特徵概覽載於附錄。）

在內地，投資 ETF 的興趣不斷提升，但大部分 ETF 只追蹤國內資產，意味着內地的基金投資者也許對資產多樣化有所需求。然而，內地的基金投資者若要投資於全球資產，僅可透過 QDII 計劃、與香港的基金互認協議，以及數隻跨境上市的日本指數 ETF（見第 2.1 節）。相對之下，香港市場的 ETF 追蹤大量亞太及海外的股票以及其他全球資產類別，包括定息產品、貨幣及大宗商品。這些 ETF 產品都可以作為內地投資者的另類選擇。

在投資者基礎方面，內地市場大部分為散戶投資者，他們的交易在 2018 年佔 A 股市場成交額約 86%[60]。機構投資者的參與度相對不高，可能會限制 ETF 一級市場交易活動的增長（即申購/贖回 ETF 單位），而這些交易活動有助減少 ETF 的價格相對其資產淨值的溢價或折讓（即盡量達到公平定價）。香港市場的環球機構投資者基礎可補足內地市場的投資者組合，支持內地 ETF 市場的發展。

反觀香港 ETF 市場，雖然已有多樣投資於內地與全球資產類別的產品，但內地資產類別範圍仍然較窄。以 A 股 ETF 為例，約半數香港上市 A 股 ETF 均追蹤內地較受歡迎的 A 股指數——於 2019 年 6 月底的 25 隻 A 股 ETF 當中，有 8 隻追蹤滬深 300 指數，4 隻追蹤富時中國 A50 指數[61]。相比之下，內地 A 股不僅有追蹤大型指數的 ETF，亦有追蹤不同行業指數、股權結構及智慧型投資策略（如增長、價值、紅利等）的 ETF。若香港 ETF 市場能進一步提升產品組合的多元化（如透過准入內地 ETF 市場），將有助吸引來自全球各地的新投資者。

---

60 資料來源：Mee, K. 及 G. Ralston〈A 股市場的誘人特點〉（Compelling characteristics of the A-shares market），載於 Schroders 網站的 Insight 頁面，2019 年 3 月 27 日。

61 資料來源：香港交易所網站有關〈ETF 證券名單〉的網頁。

在交易活動方面，香港市場的 ETF 成交比較集中於為數不多的幾隻產品——在 2019 年上半年，五隻成交額最高的 ETF[62] 佔 ETF 總成交金額的 90%。內地投資者對不同資產類別的 ETF 投資取向可能有所不同。向內地投資者開放香港 ETF 市場或可促進香港 ETF 二級市場交易活動的增長。

由於內地與香港的產品及投資者具有互補性質，若內地與香港市場的互聯互通程度可擴展至 ETF 市場環節，或可滿足投資者不同的投資偏好、提升兩者 ETF 一級及二級市場的流動性，更可進一步加強香港作為亞太區 ETF 樞紐的優勢（見第 3 節）。

3 香港作為亞太區 ETF 發行及買賣樞紐的優勢

3.1 發行商在香港發行 ETF 的考慮事項

由於全球被動型投資的需求上升，香港 ETF 市場上發行商（即 ETF 管理人）的數目一直穩步增長。儘管激烈的競爭和不斷轉變的市況導致部分 ETF 管理人退出，但其數目仍由 2008 年的 7 家上升到 2012 年的 21 家，再增至 2018 年的 27 家[63]。香港 ETF 市場能吸引發行商的因素闡釋如下。

首先，香港擁有環球投資者基礎，其對 ETF 的需求與日俱增。香港資產及財富管理的資金於 2018 年有 62% 來自環球投資者[64]，包括那些將其亞太營運總部設於香港的跨國資產管理公司。這些環球投資者一直透過「滬港通」、「深港通」、「債券通」，以及 QFII 與 RQFII 計劃投資內地證券。市場對 A 股獲納入全球指數的樂觀情緒進一步加快資金的流入。除此之外，預料本地機構對 ETF 投資的需求也會

62　包括五隻分別追蹤富時中國 A50 指數、滬深 300 指數、恒指及恒生國企指數的 ETF。資料來源：香港交易所。

63　2008 年及 2012 年（5 月）的數字源自香港交易所於 2012 年 6 月 19 日一個關於 RQFII ETF 的講座中發表的簡報（見〈香港交易所主辦 RQFII ETF 講座〉，香港交易所 2012 年 6 月 19 日新聞稿）；2018 年（年底）數字來自香港交易所網站有關交易所買賣產品〈證券名單〉的網頁。

64　資料來源：〈證監會 2018 年資產管理調查〉。

上升。就如香港的養老金，當中強制性公積金（「強積金」）計劃的資產管理規模於 2018 年年底達 8,130 億港元，平均每年資金流入約 400 億港元[65]。強制性公積金計劃管理局（「積金局」）將 131 隻 ETF 以緊貼指數集體投資計劃的性質，納入香港強積金計劃的投資範疇內[66]。

其次，莊家數目逐步上升，有助支持市場的流動性。香港規定每隻 ETF 至少要有一名莊家（可以是自營交易商或投資銀行），而 ETF 莊家數目由 2008 年的 14 家增至 2012 年的 24 家，到 2018 年更進一步增至 33 家[67]。莊家透過一級與二級市場之間的套戥活動獲利，也由此間接為 ETF 二級市場上的買賣提供更多流動性。莊家有責任於出現闊差價時輸入莊家盤，維時至少 3 分鐘。莊家盤的差價由 4 至 32 個價位不等，視乎個別 ETF[68]。莊家必須於 90 秒內因應有關 ETF 輸入介乎 3 至 30 個價位的兩邊莊家盤，維時 3 分鐘。事實上，有些 ETF 通常按價位表上最窄的差價報價。市場參與者亦有建議將 ETF 價位表的價位收窄，這或可讓莊家提供更小差價的報價。莊家的這些責任有助理順於市場調整期時所產生的大量贖回指示。例如內地股市在 2015 年調整，雖然市場出現大量贖回，但 ETF 管理人都能夠有序回應，毋須動用任何處理贖回的手段（如暫停贖回或贖回門檻）[69]

第三，「在亞洲市場買賣亞洲產品」的交易理念突顯香港在亞洲資產 ETF 方面的地理優勢。除了所處時區的優勢，香港的資產管理公司在亞太資產投資方面經驗豐富，超過 65% 於香港管理的資產於 2014 年至 2018 年間均投資於亞太地區[70]。出於對沖需要，ETF 的流動性一般都會貼近可供相關資產買賣的市場。流動性愈高，ETF 的買賣差價（即交易成本）通常愈低。標的為國際市場資產相關的美國 ETF，由於相關市場於美國開市期間可能已收市，ETF 的買賣差價或會較大[71]。畢竟在相關市場開市期間，ETF 管理人及莊家可以在有關市場對沖風險，會較樂

65　資料來源：〈強積金計劃統計摘要〉，載於積金局網站，2018 年 12 月。

66　資料來源：積金局網站（2019 年 5 月 13 日資料）。

67　見註 63。

68　ETP 分為 5 組（A 至 E）以釐定相關的莊家責任。在一種 ETP 上市前，香港交易所會諮詢該 ETP 的發行商和莊家再劃定該 ETP 所屬組別。見香港交易所網站的資料文件〈交易所買賣產品莊家安排〉（只有英文版）（於 2019 年 6 月 26 日瀏覽）。

69　見〈證監會 2018 年 ETF 論文〉。

70　資料來源：〈證監會 2018 年資產管理調查〉。

71　見 Hammond, O. H. 與 M. Lieder〈破解 ETF 流動性的迷思〉（"Debunking myths about ETF liquidity"），載於摩根資產管理網站，2015 年 5 月。

意出手支持 ETF 的流動性；否則，相關市場收市後，他們就要自行估計 ETF 的公平值。因此，亞洲資產的 ETF 在香港市場的流動性一般會較在歐美時區的市場好。

　　第四，便利市場的監管環境支持各類 ETF 產品發行。香港市場的監管機構──證券及期貨事務監察委員會（證監會）致力兼顧市場發展與投資者保障[72]。為支持產品創新，證監會於 2016 年 2 月批准槓桿及反向產品的推出，以配合亞洲市場日益增長的需求。L&I 產品為經驗豐富及經常進行買賣的投資者而設，作為短期投資的工具。為了避免過度風險，證監會把槓桿因子的上限定於兩倍，同時要求在發行文件的前端披露環節中包含事先警告的信息，以及採用跟 ETF 不同的產品名稱[73]。自 2017 年 7 月起，證監會加快了投資基金的認可批核，將「簡單申請」及「複雜申請」的目標處理時間分別縮短至不超過 2 個月及不超過 6 個月[74]。2019年 1 月起，證監會更將認可範圍擴大至主動型 ETF[75]。證監會亦於 2018 年 12 月公佈了關於 UCITS 基金（包括在歐洲註冊成立的基金）的新條文，旨在為 2007 年起簡化了的跨境上市程序提供更為清晰的解釋[76]。監管機構一方面推出市場發展的措施，同時亦制定了若干保障措施以保護 ETF 投資者的權益。例如，證監會規定合成 ETF 產品需要在簡稱前加上記號 "X"，並註以星號（*）附加中英文警示，同時規定合成 ETF 產品的股票抵押品至少為對手方所承擔總風險的 120%[77]。

72　香港批核 ETF 為「集體投資計劃」受證監會《單位信託及互惠基金守則》（「單位信託守則」）規管，獲批核的 ETF 只要符合《香港聯合交易所（聯交所）上市規則》第二十章的規定即可上市。

73　見《有關槓桿及反向產品的通函》（*Circular on Leveraged and Inverse Products*）（只有英文版），證監會發佈，2016 年 2 月 5 日；《有關槓桿及反向產品的補充通函》（*Supplemental Circular on Leveraged and Inverse Products*）（只有英文版），證監會發佈，2019 年 3 月 14 日；〈槓桿及反向產品與 ETF 有何分別？〉，載於《錢家有道》網站，2019 年 5 月 8 日。

74　見證監會《致證監會認可單位信託及互惠基金的管理公司的通函── 推出試行優化基金獲得證監會認可後的申請程序》（*Circular to Management Companies of SFC-authorized unit trusts and mutual funds — Launch of pilot revamped process to enhance the processing of post authorization applications*）（只有英文版）（2017 年 6 月 30 日）。「複雜申請」指涉及新委任及／或改任主要營運商、證監會認可基金改變營運地點、證監會認可基金合併、證監會認可基金終止而涉及特定事宜或重大事宜及／或有關申請的政策影響者，其他的一概屬於「簡單申請」。

75　見〈Asia ETF Roundup (Industry) — December 2018 and January 2019〉，載於《晨星》網站，2019 年 2 月 14 日。

76　簡化了的跨境上市程序詳見〈證監會簡化專門性 UCITS III 基金的認可程序〉，載於證監會網站的新聞稿，2007 年 3 月 30 日。有關新條文的詳情見〈Hong Kong — The Revised Code On Unit Trusts And Mutual Funds〉，載於 Conventus Law 網站，2019 年 2 月 21 日。

77　見〈協助投資者識別合成 ETF 產品的新措施〉，證監會新聞稿，2010 年 11 月 18 日；〈為加強投資者保障而推行的本地合成 ETF 新措施〉，證監會新聞稿，2011 年 8 月 29 日。

3.2 支持香港 ETF 流動性的因素

首先，香港 ETF 一級和二級市場的流通性都非常高，而高流動性會吸引更多的交易。ETF 的平均每日成交金額由 2008 年的 18 億港元（佔證券市場總成交的 2.5%），上升一倍以上到 2018 年的 40 億元（佔證券市場總成交的 3.7%），若計及 L&I 產品，更達 44 億港元（佔證券市場總成交的 4.2%）（見圖 15）。相關資產的二級市場流動性一直保持充裕，可滿足 ETF 一級市場套利活動的需求。根據證監會一份研究報告 [78]，2012 年 6 月至 2017 年 6 月期間，恒指 ETF 於一級市場的交易並沒有抽走恒指成份股於二級市場在接近收市時（日終成交量）的流動性 —— 一級市場交易佔恒指日終成交量 [79] 的比例的中位數及第 90 百分位數分別僅為 1% 及 13%。同一研究也發現，高度流通的 ETF 市場有助減少 ETF 價格偏離資產淨值的幅度（溢價／折讓），約 80% 的香港上市 ETF 的溢價／折讓於 2013 年至 2017 年間都不超過 1%。

---

78 資料來源：〈證監會 2018 年 ETF 論文〉。
79 「恒指日終成交量」是指恒指成份股於下午 3 時 30 分至下午 4 時 10 分的成交量（以金額計）。

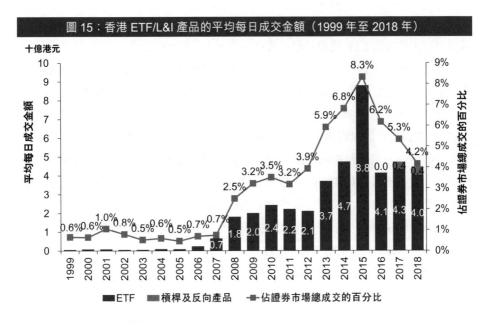

圖 15：香港 ETF/L&I 產品的平均每日成交金額（1999 年至 2018 年）

註：L&I 產品於 2016 年 6 月推出。

資料來源：香港交易所歷年的《市場資料》。

　　其次，香港 ETF 市場的投資總成本（TCO）具競爭力，支持經該市場進行「在亞洲市場買賣亞洲產品」。按照香港交易所的資料顯示 [80]，ETF 投資的投資總成本包括交易成本（如買賣差價及經紀佣金）及持有成本（如投資者層面的稅項及與追蹤差異 [81] 有關的成本，包括管理費、交易費用以及於投資及基金層面的稅項）。在交易成本方面，香港政府自 2015 年 2 月起將豁免印花稅的優惠擴大至香港所有的 ETF 產品。除此以外，香港莊家享有交易成本優惠，可免繳聯交所的交易費（0.005%）以及證監會的交易徵費（0.0027%），便利其莊家活動。至於持有成本，稅項是關鍵之一，因為投資者回報多寡受三層稅項影響——投資層面、基金層面及投資者層面。事實上，有研究 [82] 指出，香港逐步擴展的雙邊稅收協定網絡以及本

80　見《ETF 手冊》，載於香港交易所網站（於 2019 年 8 月 8 日瀏覽）。

81　「追蹤差異」是指於特定期間內 ETF 與相關指數之間的回報差異；「追蹤誤差」則通常是指於特定期間內 ETF 與相關指數之間回報差異的標準差數值。

82　資料來源：〈ETF 稅務報告 2019（香港）〉（"ETF taxation report for investors 2019 (Hong Kong)"）（只有英文版），香港交易所委託編制的報告，載於安永網站，2019 年。

地稅法，均有利於降低成本。同一研究報告總結，對於投資於日本股票及內地公司債方面，經香港 ETF 投資的稅務效益在主要市場中應為最高。

　　第三，香港 ETF 設多個貨幣櫃枱，便利資產管理人的配置。香港 ETF 能以多種貨幣交易，包括港元、人民幣及/或美元。於香港市場，以人民幣計價追蹤內地上市證券的 ETF 得以發行及買賣。香港交易所自 2012 年 8 月擴大其「人證港幣交易通」所支援的產品類別範圍後，以人民幣買賣的 ETF 的交投獲得了支持[83]。有些追蹤美國證券及環球指數的 ETF，除了以港元及人民幣交易外，還可使用美元進行交易，使投資者能採用不同貨幣交易，豐富了投資選擇。為進一步提高美元櫃枱的流動性，有市場參與者建議將美元櫃枱的價位表與港元櫃枱的價位表協調，又或允許市場偏離「一換一」規則（根據該規則，同一隻 ETF 的不同櫃枱每單位資產淨值必須相同），以期收窄同一 ETF 於不同櫃枱之間的價位差異[84]。

　　第四，香港的風險管理產品生態圈切合 ETF 投資者的需要。要對沖香港 ETF 的市場風險，市場上有按成交額計算的五大 ETF[85] 的股票期貨及期權可予買賣。投資者也可買賣恒指及恒生國企指數的指數期貨及期權，以及中華交易服務中國 120 指數及 MSCI 亞洲除日本淨總回報指數的指數期貨合約，用以對沖與香港、內地以及亞洲股票指數相關的 ETF 之市場風險。香港市場這些對沖工具的優點備受市場參與者認可，而他們建議進一步開發更多期貨及期權產品來對沖多元化的相關資產。至於黃金 ETF，香港市場提供以美元及人民幣（香港）計價的黃金期貨產品，可用以對沖金價波動的市場風險。若要對沖 ETF 的貨幣風險，香港市場則有可交收美元兌人民幣（香港）期貨及期權，以及歐元兌人民幣（香港）、日圓兌人民幣（香港）、澳元兌人民幣（香港）以及人民幣（香港）兌美元等現金結算期貨，可供投資 ETF 的港元、人民幣及美元櫃枱的投資者採用。

83　見香港交易所通告〈於 2012 年 8 月 6 日擴大「人證港幣交易通」支援的產品類別〉，2012 年 7 月 30 日。

84　在「一換一」規則下，美元櫃枱的價位相對於基金單位價格的百分比一般大於港元櫃枱，令為客戶提供跨櫃枱轉移服務的經紀擔心差價成本問題。

85　有關五大 ETF 資料見上文 2.3 節。

4　總結

ETF 投資日益盛行。ETF 比互惠基金在投資方面的優點包括方便、具成本效益，而且其流動性和透明度也更高。亞太地區是環球眾多市場中增長最快的 ETF 市場，背後動力來自投資者 (特別是內地投資者) 對投資組合多元化的需求不斷上升。

內地的 ETF 市場近年增長迅速，以滿足內地投資者對分散投資的需求，其中又以追蹤內地指數的股本證券 ETF 以及貨幣市場 ETF 為主導。相對於 ETF，LOF 則提供主動型回報，但當中以追蹤在岸證券的混合策略基金為主。內地投資者越益追求環球資產配置，並且越來越接受被動型投資的 ETF，而內地市場與環球資產相關的跨境 ETF 和 LOF 的供應卻不多，市場亦缺乏多種貨幣的投資選擇，這些都顯示內地投資者對參與香港 ETF 市場有潛在需求。

香港 ETF 市場可說是通向環球投資的門戶，特別是對內地投資者來說。香港 ETF 產品涉及環球市場多元化的相關資產類別，還有成熟的機構投資者基礎支持市場的流動性。這裏有來自世界各地的發行商與投資者，整個生態圈造就了香港 ETF 市場的發展。發行商可受惠於環球投資者的強勁需求、有效的莊家機制、「在亞洲市場買賣亞洲產品」的地利、監管環境相對便利等優勢。對投資者來說，他們可在流動性高的一級和二級市場交易 ETF，稅務優越之外，還有多種貨幣買賣單位的選擇，以及相關風險管理工具配套齊備等優點。

香港具備優勢擔當亞洲的 ETF 交易樞紐角色。香港 ETF 市場繼續脫穎而出，必須緊貼國際市場趨勢。市場優化措施可包括進一步涵蓋更多相關資產市場和支援產品創新，以切合投資者的不同需要。香港的 ETF 產品涵蓋領域甚廣，能滿足環球投資者的需要之外，還可切合內地投資者的潛在需求。與此同時，內地 ETF 涵蓋許多不同的內地資產，對環球投資者亦具有吸引力。若然內地與香港的 ETF 市場有機會互聯互通，將可令兩邊的 ETF 一級和二級市場投資者組合趨向平衡。若內地與香港的市場互聯互通可再進一步，擴大至包括 ETF 範疇，定將有助擴闊兩邊市場的投資者基礎並提升流動性，有利雙方相互增長發展。

附錄

內地 ETF 及 LOF 與香港 ETF 的比較

特徵	內地 ETF	內地 LOF	香港 ETF
基金單位數目	可變動		
二級市場交易	可以（交易時段內）		
莊家	通常有，但非強制要求	無	有（強制要求）
融資融券（融券容許沽空交易）	只限若干合資格基金		孖展（融資）交易無限制；僅合資格 ETF 可沽空
大手（大宗）交易門檻	≥ 200 萬個單位或 ≥ 200 萬元人民幣		無規模限制
二級市場結算	T 日證券交收；T+1 日現金結算		T+2 日
一級市場（申購及贖回）	• 實物及現金結算 • 僅限場內交易 • 面向所有投資者 • 高門檻（例如最低 50 萬至 100 萬個基金單位）	• 現金結算 • 場內或場外交易 • 面向所有投資者 • 低門檻（例如最低 1,000 個基金單位）	• 實物結算 • 僅限場外交易 • 僅面向參與經紀（及 CCASS* 參與者） • 高門檻（例如最低 100 萬個基金單位）
成本	• 管理費：資產淨值的 0.15%-0.8%	• 管理費：資產淨值的 0.1%-2.0%	• 管理費：資產淨值的 0.025%-0.99% • 豁免印花稅 • 沒有股息預扣稅
透明度	資產淨值每 15 秒更新	資產淨值每日收市時更新	資產淨值每 15 秒更新
策略	一般是被動型	大多是主動型	一般是被動型，但亦可以是主動型和槓桿及反向產品

* CCASS 指香港交易所證券市場的中央結算及交收系統。

資料來源：上交所、深交所、香港交易所及 Wind。

註：本研究報告參考了就 ETF 市場發展向香港 ETF 管理人諮詢所獲取的意見及反饋。

第15章

關於發展香港 ETF 市場和
跨境資產配置的政策建議

丁晨

南方東英資產管理有限公司總裁

1 國際 ETF 產品市場的發展趨勢

1.1 國際 ETF 產品市場的發展趨勢呈現的特點

1.1.1 美國市場規模最大，佔市場總規模 70% 以上，亞洲市場規模較小，但未來發展潛力巨大

　　交易所買賣基金（Exchange Traded Fund, 簡稱 ETF）是一種在交易所上市交易的、基金份額可變的一種開放式基金。1993 年，全球第一隻以標準普爾 500（S&P 500）指數為標的的交易所買賣基金發行[1]。之後的二十多年，全球 ETF 市場快速發展，截至 2018 年底，全世界 ETF 淨資產規模為 4.7 萬億美元，其中美國佔比 71%，歐洲佔比 15%，亞洲佔比 10%。美國市場蓬勃發展，規模最大，佔市場總規模 70% 以上（見圖 1）；亞洲市場規模較小，但未來發展潛力巨大。

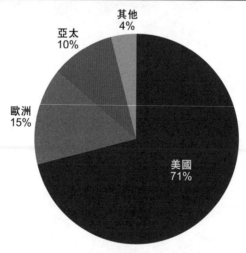

圖 1：全世界 ETF 淨資產各國/地區佔比（2018 地年年底）

其他 4%

亞太 10%

歐洲 15%

美國 71%

資料來源：2019 ICI Fact Book、南方東英。

1　資料來源：http://online.wsj.com/ad/focusonetfs/history.html。

　　據 Investment Company Institute（ICI）的數據統計，2018 年在美國註冊的投資公司的淨資產為 21.4 萬億美元，其中共同基金為 17.7 萬億美元，交易所買賣基金為 3.4 萬億美元，封閉式基金為 2,500 億美元，單位投資信託基金為 700 億美元。交易所買賣基金佔比 15.87%，在美國基金市場，ETF 體量排名第二。（見圖 2。）

圖 2：美國註冊的投資公司淨資產總額組成佔比（2018 年年底）

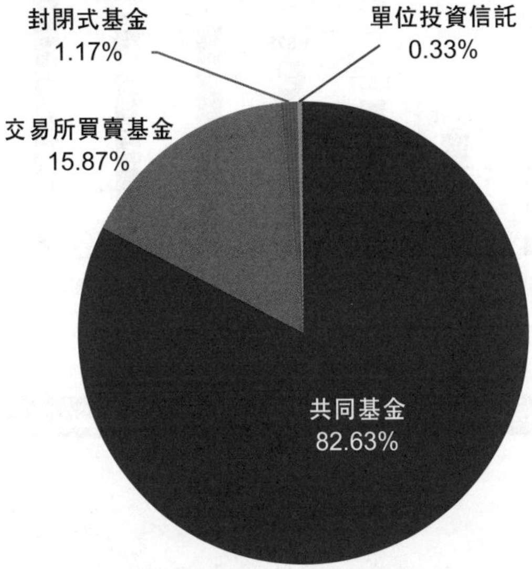

封閉式基金
1.17%

單位投資信託
0.33%

交易所買賣基金
15.87%

共同基金
82.63%

資料來源：2019 ICI Fact Book、南方東英。

　　截至 2018 年年底，美國市場共有 1,988 隻產品，淨資產總量達到 3.37 萬億美元，按資產規模計，佔全世界 ETF 體量的 71%。從 2009 年至今，美國 ETF 資產規模從 7,770 億美元增長至 3.37 萬億美元，年化增長率達 17.71%；美國 ETF 產品數量從 797 隻增長至 1,988 隻，年化增長率達 10.69%。（見圖 3 及圖 4。）

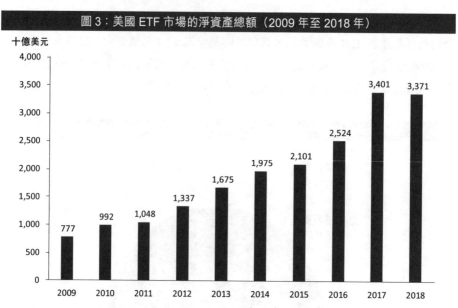

圖 3：美國 ETF 市場的淨資產總額（2009 年至 2018 年）

十億美元

資料來源：2019 ICI Fact Book、南方東英。

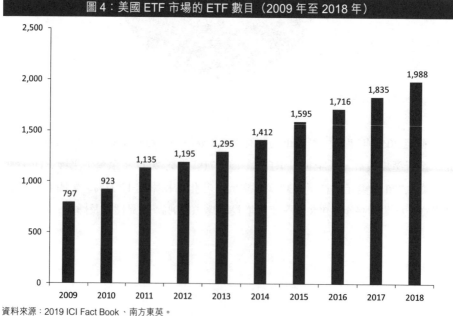

圖 4：美國 ETF 市場的 ETF 數目（2009 年至 2018 年）

資料來源：2019 ICI Fact Book、南方東英。

1.1.2　亞洲市場：日本韓國發展領先，中國內地、中國臺灣、 中國香港體量較小，其他亞洲國家起步更晚

亞洲各國的基金市場發展情況各不相同，亞洲 ETF 市場仍處於發展初期階段。據 ICI 數據，2018 年，以開放式基金淨資產計，日本和中國內地體量較大，分別佔比 40.28% 和 39.48%，韓國緊隨其後，佔比 10.34%。以開放式基金數目計，韓國和日本體量較大，分別為 40.18% 和 37.26%，其次是中國內地，佔比 15.02%。日本、韓國發展領先，體量較大，其他亞洲地區起步更晚且體量較小。

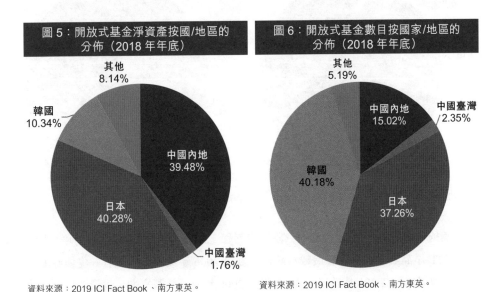

圖 5：開放式基金淨資產按國/地區的分佈（2018 年年底）

其他 8.14%
韓國 10.34%
中國內地 39.48%
日本 40.28%
中國臺灣 1.76%

資料來源：2019 ICI Fact Book、南方東英。

圖 6：開放式基金數目按國家/地區的分佈（2018 年年底）

其他 5.19%
中國內地 15.02%
中國臺灣 2.35%
韓國 40.18%
日本 37.26%

資料來源：2019 ICI Fact Book、南方東英。

1.1.3　美國債券 ETF 產品廣受投資者接受並應用，中國內地主要 機構投資者無法參與債券 ETF 市場

債券 ETF 以債券為標的，通過購買一籃子債券，實現對債券指數的追蹤或是債券組合的主動管理。2000 年，市場上第一隻債券 ETF 產品成立，標的為加拿大 5 年國債。2002 年，4 隻債券 ETF 在美國發行，分別為跟蹤雷曼 1-3 年期、3-7 年期、超 20 年期國債指數的 ETF，以及跟蹤高盛投資級企業債指數的 ETF。2007 年以前，全球一共有 6 隻債券 ETF 產品，總規模 200 億美元左右。2008 年以後，全球債券 ETF 的發展速度明顯加快。根據彭博的統計，截至 2018 年底，全球債券

ETF 共 1,176 隻，總規模達到 9,694 億美元。以發行地區統計，按規模排序，美國佔據絕對優勢，發行規模佔比達 71%；按數量排序，美國發行的債券 ETF 共 379 隻，佔比 32%，仍處於領先地位。

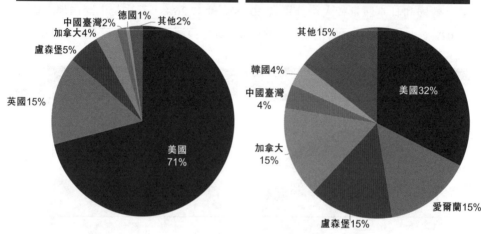

| 圖 7：全球債券 ETF 的規模按發行地區的分佈（2018 年年底） | 圖 8：全球債券 ETF 的數量按發行地區的分佈（2018 年年底） |

圖 7 標籤：德國 1%、中國臺灣 2%、其他 2%、加拿大 4%、盧森堡 5%、英國 15%、美國 71%

圖 8 標籤：其他 15%、韓國 4%、中國臺灣 4%、加拿大 15%、盧森堡 15%、美國 32%、愛爾蘭 15%

資料來源：彭博、南方東英。　　　　　　　　　資料來源：彭博、南方東英。

　　美國市場的債券 ETF 經歷了三段發展歷程。2000 年至 2006 年為市場培育期，債券 ETF 產品數量較少，發展較為緩慢，截至 2006 年美國市場僅有 6 隻債券 ETF 產品，總規模約為 200 億美元。2007 年至 2008 年為產品豐富期，投資者逐漸熟悉且接受債券 ETF，越來越多的公司開始發行債券 ETF，整個市場的體量大幅度提升，監管部門頒佈實施細則，促進了債券 ETF 的發展。2009 年至今為加速發展期，金融危機後債券市場流動性枯竭，債券 ETF 盤活了市場流動性。從 2009 年至今，債券 ETF 的交易量有了爆發性的增長，而交易量的增長又帶來了新資金的流入，使得債券 ETF 進入了快速發展的階段。

　　與美國相比，中國內地的債券 ETF 市場處於剛剛起步的階段。2013 年，中國內地市場推出第一隻債券 ETF 產品。截至 2019 年 5 月，內地共有 7 家基金公司推出 10 隻債券 ETF 產品，債券 ETF 產品總規模達到 115.75 億元。由於債券 ETF 產品選擇較少，同時因法規限制，銀行、保險等主要機構投資者參與債券 ETF 市場的積極性不高。

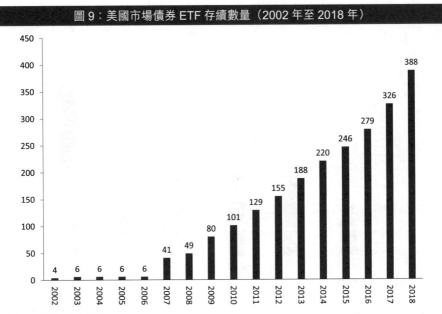

圖 9：美國市場債券 ETF 存續數量（2002 年至 2018 年）

資料來源：彭博、南方東英。

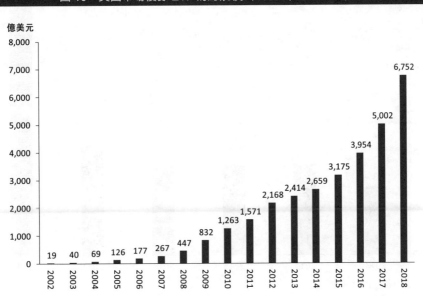

圖 10：美國市場債券 ETF 的總規模（2002 年至 2018 年）

資料來源：彭博、南方東英。

圖 11：中國內地市場債券 ETF 的規模變動情況（2013 年至 2018 年）

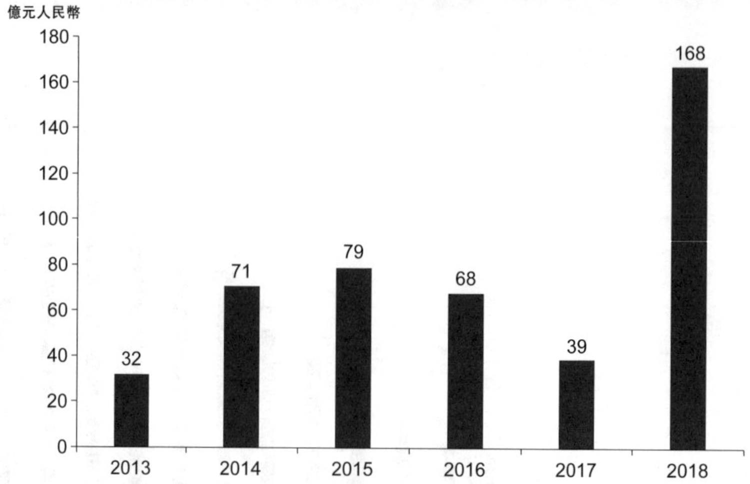

億元人民幣

- 2013：32
- 2014：71
- 2015：79
- 2016：68
- 2017：39
- 2018：168

資料來源：Wind、南方東英。

圖 12：中國內地市場債券 ETF 的數量變動情況（2013 年至 2018 年）

- 2013：3
- 2014：4
- 2015：4
- 2016：4
- 2017：5
- 2018：11

資料來源：Wind、南方東英。

　　然而，內地政府一直在努力推動債券 ETF 的發展。2019 年 5 月，中國證券監督管理委員會（中國證監會）與中國人民銀行聯合發佈《關於做好開放式債券指數證券投資基金創新試點工作的通知》，擬推出以跨市場債券品種為投資標的，可在交易所上市交易或在銀行間市場協議轉讓的債券指數公募基金。通知發佈後，中國證監會將推動試點債券指數基金的註冊審核工作。

　　2019 年 6 月，中共中央辦公廳、國務院辦公廳發佈《關於做好地方政府專項債券發行及項目配套融資工作的通知》，其中提到豐富地方政府債券投資羣體，推出地方政府債券交易型開放式指數基金，通過「債券通」等機制吸引更多境外投資者投資。政策的陸續出台表明了中國內地對債券 ETF 發展的支持，中國內地債券 ETF 的規模佔內地市場 ETF 總規模的比例僅為 4.4%，相比美國 14% 的比例[2]，未來存在着巨大的發展空間。

表 1：中國內地關於債券 ETF 的政策總結			
時間	文件名稱	發佈機構	要點
2013-02-08	上海證券交易所債券 ETF 業務指南[3]	上海證券交易所	針對採用現券申贖的單市場債券 ETF 的指導方案
2016-04-28	關於發佈《深圳證券交易所證券投資基金交易和申購贖回實施細則（2016 年修訂）》的通知[4]	深圳證券交易所	繼續暫免收取債券 ETF 和上市交易貨幣市場基金的交易經手費及交易單元流量費
2017-11-03	關於啟用現金申贖債券 ETF 業務相關代碼段的通知[5]	上海證券交易所	債券 ETF 交易經手費暫免收取
2019-01-11	關於修訂《深圳證券交易所證券投資基金交易和申購贖回實施細則》有關條款的通知[6]	深圳證券交易所	投資者當日競價買入的債券 ETF 份額當日可以賣出；投資者當日大宗買入的債券 ETF 份額當日可以大宗賣出，次一交易日可以競價賣出等
2019-05	關於做好開放式債券指數證券投資基金創新試點工作的通知[7]	中國證監會 中國人民銀行	擬推出以跨市場債券品種為投資標的，可在交易所上市交易或在銀行間市場協議轉讓的債券指數公募基金。通知發佈後，中國證監會將推動試點債券指數基金的註冊審核工作

2　資料來源：Wind。

3　資料來源：http://www.sse.com.cn/lawandrules/guide/zqznlc/c/c_20150912_3986154.shtml。

4　資料來源：http://www.szse.cn/aboutus/trends/news/t20160525_518721.html。

5　資料來源：http://www.sse.com.cn/lawandrules/guide/jyznlc/dealcode/c/c_20171103_4413981.shtml。

6　資料來源：http://www.szse.cn/lawrules/rule/tradetype/fund/t20190111_565175.html。

7　資料來源：http://www.csrc.gov.cn/pub/newsite/zjhxwfb/xwdd/201905/t20190520_356093.html。

（續）

表 1：中國內地關於債券 ETF 的政策總結			
時間	文件名稱	發佈機構	要點
2019-06-10	關於做好地方政府專項債券發行及項目配套融資工作的通知 [8]	中共中央辦公廳 國務院辦公廳	豐富地方政府債券投資羣體，推出地方政府債券交易型開放式指數基金，通過「債券通」等機制吸引更多境外投資者投資

1.1.4 美國槓桿、槓反 ETF 產品發展更加成熟

縱觀全球，美國是槓桿反向產品最發達的市場。在亞洲，日本、韓國和中國臺灣的槓桿反向產品發展較早，規模也較大。中國香港在 2017 年推出了第一批跟蹤恒生指數和恒生國企指數的槓桿反向產品。[8]

需要指出的是，槓桿反向產品包括 ETF 和交易所買賣票據（Exchange Traded Notes，簡稱 ETN），他們的對比列示於表 2。

表 2：ETN 與 ETF 的差別		
特徵	ETN	ETF
發行主體	證券公司或投資銀行	基金公司
本質	債權	基金
投資者持有物	發行人所發行的債權	基金份額
申購贖回	現金	現金或實物股票
發行人對標的資產的買進選擇	可選擇是否買進標的資產	必須買進標的資產
信用風險	有	無
到期期限	有	無
管理手續費	有	有

資料來源：南方東英。

美國是全球槓桿反向產品的主導者，總資產規模一馬當先。截至 2019 年 4 月 11 日，美國上市的槓桿反向產品（包括 ETF 和 ETN）共計有 286 檔，總規模達 518.7 億美元。其中 2 倍的槓桿反向產品達到 124 檔，3 倍的槓桿反向產品達到 103 檔，4 倍的槓桿反向產品達到 10 檔。（見表 3。）

8　資料來源：http://politics.people.com.cn/n1/2019/0610/c1001-31128317.html。

表 3：美國槓桿反向產品數據統計（2019 年 4 月 11 日）			
	ETF	ETN	總計
槓桿反向產品總資產規模（億美元）	369	150	519
槓桿反向產品總數	194	92	286
槓桿反向產品總規模佔比	71%	29%	100%
2 倍產品資產規模（億美元）	116	126	242
2 倍產品總數	77	47	124
2 倍產品規模佔比	48%	52%	100%
3 倍產品資產規模（億美元）	206	20	226
3 倍產品總數	84	19	103
3 倍產品規模佔比	91%	9%	100%
4 倍產品資產規模（億美元）	0	0.35	0.35
4 倍產品總數	0	10	10
4 倍產品規模佔比	0%	100%	100%

註：0.5 倍和 1.5 倍的產品因規模小而未有納入統計中。

資料來源：彭博。

　　日本與韓國是亞洲地區槓桿反向產品的先行者。截至 2019 年 4 月 11 日，日本共有 39 檔槓桿反向產品。相比於美國和韓國市場，日本監管機構相對保守，最高只允許發行 2 倍的槓桿反向產品。

　　韓國在亞洲地區槓桿反向產品領域起步比日本稍晚，但卻是當之無愧的領導者，不僅因為產品數量繁多及商品資產齊全，而且因為資產規模和日均交易量明顯高於區內其他市場。業界人士認為，韓國槓桿反向產品的成功受益於韓國投資者對高槓桿高波動產品的認可和偏愛。韓國與日本一樣，槓桿反向產品主要由相應期貨提供目標回報，管理方法相似。

　　中國臺灣在槓桿反向產品方面的發展比日本、韓國都晚，規模也相對較小。截至 2019 年 4 月 11 日中國臺灣共有 43 檔槓桿反向產品，規模接近韓國市場，主要由元大、富邦和國泰發行。中國臺灣的槓桿反向產品全部是 ETF，從 2019 年開始，中國臺灣監管機構允許合資格券商發行 ETN。

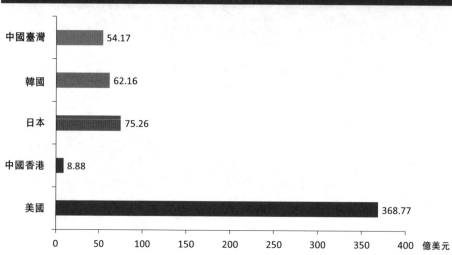

圖 13：美國和亞洲主要市場槓桿反向 ETF（不含 ETN）資產規模的比較
（2019 年 4 月 11 日）

資料來源：南方東英、彭博。

1.1.5 美國 ETF 產品標的資產更加豐富，包括股票、債券、大宗商品、外匯和其他類別資產，資產不僅包括美國資產，也包括全球資產

開放式基金包括共同基金、ETF 和機構基金，ETF 的標的資產構成和開放式基金類似。據 ICI 數據，2018 年，全球受監管的開放式基金共 118,978 隻，其中 34% 投資於股票，18% 投資於債券，2% 投資於貨幣市場，46% 投資於混合產品（平衡基金、房地產基金、保本基金等）及其他。在美國，ETF 覆蓋的資產類別不斷豐富，包括股票、債券、大宗商品、外匯和其他類別資產，資產不僅包括美國資產，也覆蓋全球資產。

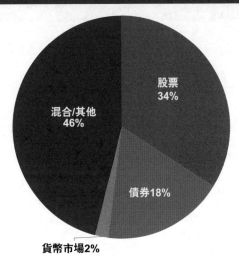

圖 14：全球受監管的開放式基金數量按標的資產類別的分佈（2018 年年底）

股票
34%

混合/其他
46%

債券18%

貨幣市場2%

資料來源：2019 ICI Fact Book、南方東英。

1.2　近年國際 ETF 市場快速發展的原因分析

1.2.1　美國投資者更加成熟，對 ETF 接受度高；
投資者包括機構投資者和零售投資者資金

　　美國金融市場更為發達與完善，投資者更為成熟，對不同種類的投資產品接受度更高。投資者除了機構投資者外，還包括 401k[9] 等零售投資者。據 ICI 統計，2018 年，790 萬（佔比約 6%）的美國家庭持有 ETF 產品；13% 持有共同基金的家庭也持有 ETF 產品；93% 持有 ETF 產品的家庭持有股票共同基金、股票、變額年金；66% 持有 ETF 產品的家庭持有債券共同基金、債券或固定年金；48% 持有 ETF 產品的家庭持有投資性房地產。

9　401k 計劃始於 20 世紀 80 年代初，是一種由僱員、僱主共同繳費建立起來的完全基金式養老保險制度。

表 4：投資 ETF 的家庭持有多種類的投資工具（2018 年年底）	
投資 ETF 的家庭同時持有的其他投資工具	家庭佔比
股票共同基金、股票、可變年金（總計）	93%
債券共同基金、債券、固定年金（總計）	66%
共同基金（總計）	90%
• 股票	87%
• 債券	53%
• 混合	44%
• 貨幣市場	57%
股票	74%
債券	26%
固定或可變年金	31%
投資性房地產	48%

資料來源：2019 ICI Fact Book、南方東英。

1.2.2　美國 ETF 發行服務商選擇更多，價格更具有競爭性

　　資產管理公司發行 ETF 產品，參與交易商（Participating Dealer）通過以大宗交易的形式向發行人申購 ETF 份額。在申購過程中，參與交易商以 ETF 對應的一籃子標的資產或現金換取 ETF 份額，之後，參與交易商可以把份額轉賣給投資者。除此之外，二級市場上還有做市商（可以是參與交易商本身），以保證市場的流動性。

　　在發行費用方面，中國香港因為可選擇的服務提供商數量有限，缺乏競爭，導致價格偏高，例如託管、法務、審計等費用。而美國服務提供商相對選擇更多，因此費用更具競爭性。

表 5：ETF 發行成本組成舉例（2018 年年底）		
費用類別	美國	中國香港
託管費	32%	60%
審計費	6%	20%
監管費	1%	7%
財務報告/文件服務費	4%	7%
分銷商費	7%	6%
董事費	14%	—
合規費	13%	—
財務費	13%	—
法務費	7%	—
保險費	3%	—

資料來源：南方東英。

1.2.3　成熟市場 ETF 退市非常普遍且廣為接受

隨着越來越多的創新型 ETF 產品的推出，在市場上流動性低、發展不好的 ETF 產品面臨着退市的風險。2013 年至 2017 年，美國 ETF 退市數量逐年增加。2017 年，退市數目達到 114 隻。2018 年，美國共有 84 隻 ETF 退市，退市數目略有回落。

在亞洲市場，尤其中國香港，ETF 發行商往往因為多種原因較少選擇退市。但隨着退市制度逐漸完善，市場開始接受 ETF 退市。近期也看到香港市場有 ETF 退市，相信這種趨勢會向成熟市場靠攏。

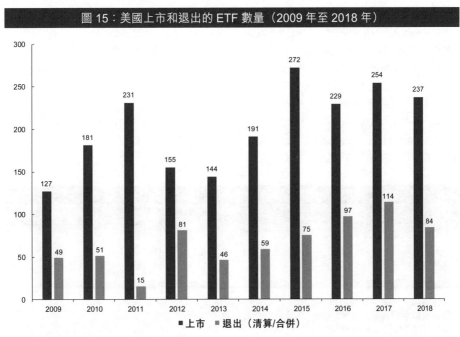

圖 15：美國上市和退出的 ETF 數量（2009 年至 2018 年）

資料來源：2019 ICI Fact Book、南方東英。

表 6：美國紐約證券交易所的 ETF 退市流程		
任務	負責方	預計時間線
董事會決議及清算計劃批准	投資經理	L-40
審核基金中預算和應計所需的費用	基金管理員	L-30
確定財務報告和審計要求	財務報告	L-15
可分配收入和稅收考慮的初步應納稅所得額計算	稅務	L-14
傳播新聞稿以獲得清算信息	投資經理	L-10
向紐約證券交易所提交退市計劃	投資經理	L-10
提交貼紙（公告）和郵寄貼紙	過戶代理機構	L-10
向存託清算公司（DTCC）通報活動	過戶代理機構	L-7
託管、基金管理、分銷協議的審查/修訂	投資經理	L-5
確認最後交易日和清算日期	投資經理	L-4
完成應稅收入的分配並向紐約證券交易所報告	過戶代理機構	L-1
確認存託清算公司（DTCC）的清算率	過戶代理機構	L-1
由紐約證券交易所提交證券交易委員會（SEC）表格 25，用於退市	紐約證券交易所	L+1
美國國家稅務局（IRS）多樣化測試	稅務	L+10
最終提交聯邦表格 1120-RIC	稅務	L+90

註："L"為清盤日（Liquidation Date）。
資料來源：紐約交易所、南方東英。

表 7：香港地區的 ETF 退市流程		
任務	負責方	預計時間線
內部批准擬議停止交易的計劃，建議終止、取消授權和退市	投資經理	T
審核基金中預算和應計所需的費用	項目管理和運營	T+3
確定財務報告和審計要求	運營和審計	T+3
將建議終止、取消授權、退市的申請提交給香港證券及期貨事務監察委員會（香港證監會）、香港交易及結算所（香港交易所）及受託人審閱及評論	香港證監會、香港交易所及受託人	T+7 ~ T+14
董事會決議終止上市	投資經理	T+25
發出受託人同意書	受託人	T+30
香港交易所確認 / 香港證監會同意	香港交易所/香港證監會	T+40
向投資者發佈關於建議停止交易、終止、自願取消授權和退市以及豁免嚴格遵守香港證監會有關守則的某些條文的公告及通知	投資經理	T+50（交易停止前至少提前 30 天通知）

(續)

表 7：香港地區的 ETF 退市流程		
任務	負責方	預計時間線
於公佈刊發後，在切實可行範圍內盡快向香港聯合交易所（香港聯交所）參與者及中央結算及交收系統參與者分別發出通告，告知他們建議停止交易、終止、自願取消授權及退市子基金（「終止子基金」）	香港交易所	T+53
在香港聯交所「終止子基金」單位交易的最後一天及「終止子基金」單位的創建及贖回的最後日期（「最後交易日」）	投資經理	T+80
在香港聯交所有關「終止子基金」的份額交易停止，且不再進一步創建及贖回「終止子基金」份額（「交易終止日期」），比如管理人將開始實現所有的「終止子基金」的投資且「終止子基金」將會停止跟蹤相關指數的那些日期	投資經理	T+81
投資者需要將投資者記錄為以香港結算代理人有限公司名義登記並在中央結算系統內持有以獲得最終分配及進一步分配（如有）的單位的實益擁有人的日期（「分配」記錄日期）	香港中央結算有限公司	T+85
最終分配公告	投資經理	T+100
在經理人向受託人及「終止子基金」的審計師諮詢後，最終分派將會支付給有關投資者（「最終分派日期」）	投資經理	T+105
當經理及受託人形成「終止子基金」不再擁有任何或有資產/負債或者實際資產/負債的意見時，「終止子基金」終止（「終止日期」）	投資經理/受託人	T+140（符合信託契約中事先通知條款的規定）
批准「終止子基金」的取消授權和退市 向投資者發出有關終止日期，「終止子基金」取消授權及退市日期的公告	香港證監會/香港交易所	於終止日期或緊接其後不久，即香港證監會及香港交易所分別批准取消授權及退市的日期
終止審計	審計師	終止後

資料來源：香港交易所、南方東英。

1.2.4　美國資本市場鼓勵創新，市場青睞首發優勢，不鼓勵抄襲跟風

　　美國市場鼓勵 ETF 創新，ETF 標的資產呈現多元化發展。美國的 ETF 標的資產共可分為六大類，分別是商品、混合、固定收益、股票、另類投資及貨幣。商品 ETF 下包含農業、能源、貴金屬、工業材料等不同種類，具體包括煤炭、棉花、黃金、鋼鐵、大豆、咖啡等。另類投資 ETF 下包含外匯、期貨、波幅、槓桿產品等。

　　除此之外，主題型 ETF 受到市場青睞，相關熱點包括電動車、人工智能、區塊鏈、線上零售等。例如，美國某 ETF 發行商旗下有以科技、人民、基礎建設為

主題的 ETF 產品。科技主題下，包括機器人和人工智能 ETF（投資研發機器人和運用人工智能的公司）、雲端計算 ETF（投資提供電腦雲端服務軟件的公司）、社交媒體 ETF（投資社交媒體公司）、鋰電池科技 ETF（投資整個鋰周期，從挖礦、精煉到電池生產）等；人民主題下，包括千禧一代 ETF（投資於受益於美國千禧一代人口增長帶來消費紅利的公司）、長命 ETF（投資於為長者提供醫療、藥物、生活設施等的公司）。

　　在美國市場，首發優勢明顯，第一個發行的基金往往會做大做強，而跟隨抄襲者往往沒有太多機會。

1.2.5　美國 ETF 基於全球資產配置

　　美國 ETF 產品基於全球資產配置，其中股票 ETF 覆蓋全球國家的重要指數。據彭博統計，截至 2019 年 6 月底，美國上市的股票 ETF 標的覆蓋全球各個市場，按照管理資產規模統計，北美市場佔 75.25%，國際市場（以美國投資者的角度，標的資產覆蓋全球，不包括美國市場資產）佔 15.46%，環球市場（以美國投資者的角度，標的資產覆蓋全球，包括美國市場資產）佔 5.57%，亞太地區佔 1.96%，歐洲市場佔 1.40%。

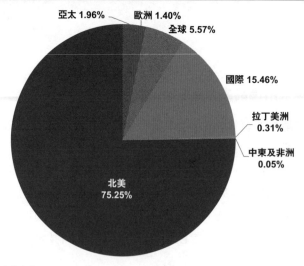

圖 16：在美國上市的股票 ETF 的資產管理總額按地區的分佈（2019 年 6 月底）

資料來源：彭博、南方東英。

　　據彭博數據，美國上市的股票 ETF 標的資產覆蓋全球多個國家與地區，包括美國、日本、中國內地、巴西、印度、加拿大、韓國、中國臺灣、德國、英國、中國香港、俄羅斯、澳洲、瑞士、法國、西班牙、墨西哥、沙特阿拉伯、新加坡等。對比來看，亞洲的 ETF 更多的是本國資產配置。

表 8：美國上市的股票 ETF 資產管理金額及佔比，前 20 名（2019 年 6 月底）		
序號	標的股票所屬國家 / 地區	資產管理金額（百萬美元）
1	美國	2,994,045.62
2	日本	21,872.10
3	中國內地	17,908.11
4	巴西	9,927.35
5	印度	8,706.78
6	加拿大	6,735.61
7	韓國	4,436.06
8	中國臺灣	2,992.92
9	德國	2,678.72
10	英國	2,423.93
11	中國香港	2,303.26
12	俄羅斯	2,094.68
13	澳洲	1,571.88
14	瑞士	1,389.60
15	法國	1,125.95
16	西班牙	1,081.16
17	墨西哥	870.15
18	沙特阿拉伯	854.54
19	新加坡	613.20
20	印度尼西亞	543.64

資料來源：彭博、南方東英。

1.3　國際 ETF 產品市場給予香港市場的經驗借鑒

1.3.1　增強投資者教育，鼓勵投資者使用 ETF 做資產配置

　　美國的投資者對共同基金（包括 ETF）的接受程度更高。據 ICI 統計，2018 年，美國共有 5,720 萬家庭持有美國註冊的開放式基金，1.016 億個人持有美國

註冊的開放式基金，退休市場總資產為 27.1 萬億，其中 8.2 萬億美元的美國個人退休賬戶（Individual Retirement Accounts，簡稱 IRA）和繳費確定型（Defined Contribution Plan，簡稱 DC Plan）資產投資於共同基金，佔比 30.26%。我們認為香港證券及期貨事務監察委員會（香港證監會）、ETF 發行商應進行全面完善的投資者教育，鼓勵投資者接受 ETF 產品，提高投資者使用 ETF 做資產配置的比例。

1.3.2 鼓勵更多服務提供商參與競爭，降低 ETF 成本

如 1.2.2 節中的分析，在發行費用方面，中國香港因為可選擇的服務提供商數量有限，缺乏競爭，導致價格偏高。然而 ETF 的優點之一是低成本，如果香港因為服務提供商價格高企導致在此掛牌的 ETF 成本居高不下，將影響它成為 ETF 中心的戰略目標。

所以，香港應該鼓勵更多的託管行、律師事務所、審計公司等 ETF 相關服務提供商參與競爭，共同努力降低 ETF 的運營成本，使香港 ETF 和成熟市場相比更有競爭力。

1.3.3 鼓勵市場創新，盤活市場流動性

香港市場目前存在創新型槓桿產品和槓桿反向產品，通過投資於股指期貨、互換合約等槓桿投資工具，實現每日追蹤目標指數收益的正向或負向的一定倍數。我們建議香港進一步加大創新力度，目前市場上的商品 ETF 標的資產僅包含黃金，建議未來引入更多種類的商品，如貴金屬、農業、能源等。

1.3.4 加大產品國際化配置，逐步納入其他市場基礎資產

與美國相比，香港 ETF 市場標的資產較為單一，股票佔比 85%，債券、貨幣市場和商品佔比 15%。香港市場的股票 ETF 標的資產集中在中國內地和中國香港資產，佔比一半。

表 9：香港上市 ETF 按標的資產所屬國家 / 地區的數目及佔比（2019 年 6 月底）		
國家 / 地區	ETF 數目	佔比
中國（包括香港、臺灣）	80	50.63%
環球	15	9.49%
歐洲	15	9.49%
亞洲	11	6.96%
美國	9	5.70%
亞太區（日本除外）	6	3.80%
韓國	5	3.16%
日本	5	3.16%
印度	4	2.53%
馬來西亞	1	0.63%
越南	1	0.63%
新加坡	1	0.63%
菲律賓	1	0.63%
俄羅斯	1	0.63%
巴西	1	0.63%
泰國	1	0.63%
印度尼西亞	1	0.63%

資料來源：彭博、南方東英。

　　香港如要發展成為亞洲甚至全球的 ETF 中心，需要令標的資產多樣化，並逐步增加其他市場的標的資產。

2 香港 ETF 市場的發展趨勢和新動力

2.1 香港 ETF 市場的概況

　　過去十年，能夠投資多種資產與不同策略的 ETF 成為市場寵兒。作為亞洲區內的國際金融中心，香港市場的 ETF 資產管理總額在過去十年內迎來了三倍的增長。在中國資本賬戶開放的背景下，跟蹤港股和中國內地 A 股市場的 ETF 產品、人民幣合格境外機構投資者 (RQFII)、股票通「滬股通」及「深港通」等成為資金流動的重要驅動力。截至 2019 年 7 月 8 日，香港市場共有 187 隻 ETF、資產管理金額為 4,575.11 億港元 [10]。市場上眾多參與者發行的 ETF 產品豐富了產品種類，帶來了香港 ETF 市場的發展。以市值計，香港為亞洲第二大的 ETF 市場，以 ETF 換手率計，香港為亞洲第四大的 ETF 市場 [11]。

2.2　目前香港 ETF 市場的結構

2.2.1　香港 ETF 市場產品的資產組成結構

香港 ETF 市場標的資產較為單一，主要是股票。根據香港交易所數據，截至 2019 年 7 月 11 日，以 ETF 數目計，標的資產中股票佔比 84.49%，定息產品及貨幣 (包括貨幣市場) 佔比 10.71%，商品佔比 2.67%，商品期貨佔比 2.67%，多元佔比 0.53%；以資產管理總額計，標的資產中股票佔比 89.76%，定息產品及貨幣 (包括貨幣市場) 佔比 9.63%，商品佔比 0.56%，商品期貨佔比 0.04%，多元佔比 0.01%。（見圖 17 及圖 18。）

10　資料來源：香港交易所網站。

11　資料來源：2019 ICI Fact Book。

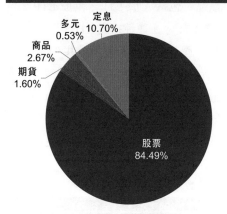

圖 17：ETF 數目按標的資產的分佈
（2019 年 6 月底）

定息 10.70%
多元 0.53%
商品 2.67%
期貨 1.60%
股票 84.49%

資料來源：彭博、南方東英、香港交易所網站。

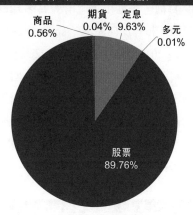

圖 18：ETF 的資產管理總額按標的資產的分佈（2019 年 6 月底）

商品 0.56%
期貨 0.04%
定息 9.63%
多元 0.01%
股票 89.76%

資料來源：彭博、南方東英、香港交易所網站。

若以實物資產、合成、期貨劃分，香港 ETF 市場實物資產佔據主要地位。根據香港交易所數據，截至 2019 年 7 月 11 日，以 ETF 數目計，標的資產中實物資產佔比 94.65%，合成佔比 3.74%，期貨佔比 1.60%；以資產管理總額計，標的資產中實物資產佔比 99.75%，合成佔比 0.21%，期貨佔比 0.04%。

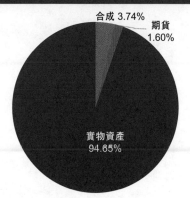

圖 19：ETF 數目按標的資產性質的分佈
（2019 年 6 月底）

合成 3.74%
期貨 1.60%
實物資產 94.65%

資料來源：彭博、南方東英。

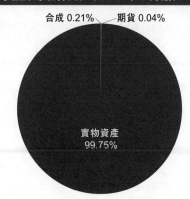

圖 20：ETF 的資產管理總額按標的資產性質的分佈（2019 年 6 月底）

合成 0.21%
期貨 0.04%
實物資產 99.75%

資料來源：彭博、南方東英。

2.2.2 香港市場目前存在的 ETF 產品

目前，香港市場上有豐富多樣的 ETF 產品，包括 A 股 ETF、港股 ETF、槓桿產品、槓桿反向產品、smart beta ETF、行業 ETF、主題性投資 ETF 等。

據 2019 年 6 月底的香港交易所數據，ETF 產品涉及的相關指數共 91 個，其中跟蹤滬深 300 指數的產品有 11 個，跟蹤富時中國 A50 指數的產品有 8 個，跟蹤彭博巴克萊中國國債加政策性銀行債券指數的產品有 5 個，跟蹤 NASDAQ 100 指數的產品有 5 個，跟蹤 MSCI 中國 A 股國際通指數的產品有 4 個。除此之外，還包括涉及恒生指數、MSCI 中國指數、歐元區 STOXX 50 指數、MSCI 印尼總匯報淨值指數、標普道瓊斯指數在內的多種 ETF 產品。

表 10：按交易金額排序前 20 名 ETF 產品（2019 年 1 月至 6 月）					
股份代號	ETF 名稱	ETF 經理	交易貨幣	資產管理金額（百萬港元）	成交金額（千港元）
2800	盈富基金	道富環球投資管理亞洲有限公司	港元	87,786	1,442,255,350
3188	華夏滬深 300 指數 ETF	華夏基金（香港）有限公司	港元	16,600	1,050,695,868
2822	南方富時中國 A50 ETF	南方東英資產管理有限公司	港元	17,269	980,422,993
2823	iShares 安碩富時 A50 中國指數 ETF	貝萊德資產管理北亞有限公司	港元	24,375	854,177,827
2828	恒生中國企業指數上市基金	恒生投資管理有限公司	港元	28,989	365,881,777
7500	南方東英恒生指數每日反向 (-2x) 產品	南方東英資產管理有限公司	港元	1,249	244,134,749
7300	南方東英恒生指數每日反向 (-1x) 產品	南方東英資產管理有限公司	港元	3,525	231,846,544
7200	南方東英恒生指數每日槓桿 (2x) 產品	南方東英資產管理有限公司	港元	244	62,069,909
82822	南方富時中國 A50 ETF	南方東英資產管理有限公司	人民幣	17,269	20,565,059
83188	華夏滬深 300 指數 ETF	華夏基金（香港）有限公司	人民幣	16,600	16,118,673
3053	南方東英港元貨幣市場 ETF	南方東英資產管理有限公司	港元	2,713	13,330,407

(續)

表 10：按交易金額排序前 20 名 ETF 產品（2019 年 1 月至 6 月）					
股份代號	ETF 名稱	ETF 經理	交易貨幣	資產管理金額（百萬港元）	成交金額（千港元）
7288	南方東英恒生中國企業指數每日槓桿（2x）產品	南方東英資產管理有限公司	港元	178	12,826,638
82833	恒生指數上市基金	恒生投資管理有限公司	人民幣	53,151	11,821,619
83010	iShares 安碩核心 MSCI 亞洲（日本除外）指數 ETF	貝萊德資產管理北亞有限公司	人民幣	462	11,304,521
9010	iShares 安碩核心 MSCI 亞洲（日本除外）指數 ETF	貝萊德資產管理北亞有限公司	美元	3,610	11,061,239
3147	南方東英中國創業板指數 ETF	南方東英資產管理有限公司	港元	525	10,916,298
2801	iShares 安碩核心 MSCI 中國指數 ETF	貝萊德資產管理北亞有限公司	港元	2,807	8,802,568
3167	工銀南方東英標普中國新經濟行業 ETF	南方東英資產管理有限公司	港元	914	8,398,372

資料來源：彭博、南方東英。

2.3　香港市場存在的差距

2.3.1　與美國市場相比，香港市場發行門檻更高

香港市場 ETF 發行門檻高於美國市場，美國 ETF 發行規模 200 萬美元起步，但在亞洲，ETF 的發行規模要上千萬美元起。美國在 2008 年金融危機後頒佈若干法案，限制金融機構的自營交易，旨在改變目前超級金融機構「大而不倒」的局面，有效防範系統性風險，同時保護金融市場中的弱勢羣體，避免金融消費者受到欺詐。所以，歐美投資銀行逐漸喪失給 ETF 提供種子基金的能力，導致香港 ETF 發行商因為找不到種子基金而發行困難。

2.3.2　香港與美國 ETF 市場發展階段不同

經過二十多年的發展，美國 ETF 市場處於成熟階段。在 ETF 發行、申贖、交易過程中，各個部門分工明確，這些服務均可通過外包形式外包給不同專業的

公司。與美國相比，香港 ETF 處於初步發展階段，角色分工不夠明確，ETF 發行商需要掌握多項技能以便應對發行、申贖、交易過程中的各種情況。未來，香港 ETF 市場會向美國的 ETF 市場的發展方向邁進。

2.3.3 與中國內地市場相比，香港市場費用上不佔優勢

深圳證券交易所對於債券 ETF、貨幣 ETF 暫免收取證券交易經手費，上海證券交易所也暫免收取貨幣 ETF、債券 ETF 的交易單元流量費。然而，香港交易所對貨幣 ETF 仍然按照普通股票標準進行收費，增加了普通投資者參與貨幣 ETF 的交易成本，或會阻礙香港貨幣 ETF 的發展。

香港市場的債券和貨幣市場 ETF 的交易量有限。2018 年全年的成交量分別僅為 16.5 億及 8 億港元。2019 年截至 5 月 31 日的成交量分別僅為 4.6 億及 8.2 億港元。

表 11：債券 ETF 及貨幣市場 ETF 的交易量總結		
	債券 ETF（港元）	貨幣市場 ETF（港元）
2018 全年總成交	1,652,393,536	803,993,300
2019 年 1 月至 5 月總成交	459,365,134	817,574,130

資料來源：彭博、南方東英。

投資者在二級市場交易 ETF，除了經紀佣金之外，還需要支付交易徵費、交易費、交易系統使用費，以及結算費。具體收費標準見表 12。

表 12：香港市場的 ETF 交易費用			
費用	徵費機構	收費基於	收費標準
交易徵費	香港證監會	交易金額	0.0027%
交易費	香港交易所	交易金額	0.005%
交易系統使用費	香港交易所	每筆交易	0.50 港元

資料來源：香港證券交易所。

由此可以計算得出，投資者交易債券及貨幣市場 ETF 所支付的費用（除經紀佣金之外）數額並不大。2018 年債券及貨幣市場 ETF 交易費用總計分別為 32 萬

及 15.6 萬港元。2019 年截至 5 月 31 日，交易費用分別僅為約 8.9 萬及 15.9 萬港元。相關交易費用一覽見表 13。

表 13：香港市場的 ETF 相關交易費用總額（2018 年及 2019 年 1 月至 5 月）		債券 ETF 交易費用（港元）		貨幣市場 ETF 交易費用（港元）	
費用	徵費機構	2018 年	2019 年 1 月至 5 月	2018 年	2019 年 1 月至 5 月
交易徵費	香港證監會	89,229	24,806	43,416	44,149
交易費	香港交易所	165,239	45,937	80,399	81,757
結算費	香港中央結算	66,096	18,375	32,160	32,703
總額		320,564	89,117	155,975	158,609

註：未計交易系統使用費。

資料來源：彭博、南方東英。

3　如何進一步發展香港 ETF 市場

3.1　關注投資者對 ETF 產品的訴求

3.1.1　提高槓桿、槓反產品的倍數

在成熟市場，槓桿反向產品在過去十年日漸受到投資者歡迎。從全球視角來看，美國是槓桿反向產品最發達的市場。在亞洲，日本、韓國和中國臺灣的槓桿反向產品發展較早，規模也較大。（見上文 1.1.4 節。）

在美國上市的槓桿反向產品倍數選擇較多，例如 3 倍的槓桿反向產品為 103 檔，4 倍的槓桿反向產品達到 10 檔。

香港市場在槓桿反向產品方面的發展起步晚於日本、韓國和中國臺灣市場，但是發展迅速，潛力巨大。現階段香港市場上的槓桿反向產品主要跟蹤恒生指數和恒生國企指數。產品的標的資產是股指期貨和現金管理工具（包括高質量的港幣貨幣基金、高流動性高評級的政府債券）。產品每日收盤時完成調倉交易，並系統

化管理現金部位，因此運作平穩，發行兩年錄得零意外。

目前香港市場最多只有 2 倍槓桿產品與 2 倍槓桿反向產品，而且標的資產只有恒生指數和國企指數。根據市場的反饋，香港可以提高槓桿反向的倍數並提供包括 A 股在內的其他標的資產，以滿足市場需要。

3.1.2　擴大 ETF 組成資產種類

根據 2.2.2 節的分析，香港應該大力發展債券 ETF 市場。包括在香港發行各主要久期的人民幣國債 ETF，以幫助建立境外人民幣收益率曲線，為香港進一步發展債券 ETF 奠定基礎。

在股票 ETF 方面，香港應鼓勵更多樣化的標的資產，讓香港 ETF 市場的參與者能夠將資產配置到更多的市場。

3.2　創新發展設計 ETF 產品

3.2.1　舉例目前香港市場上存在的創新 ETF 產品

目前香港市場上存在槓桿產品與槓桿反向產品，能夠追蹤 -1 倍、-2 倍及 +2 倍的市場指數表現。根據彭博，2019 年截至 6 月底，按槓桿和槓反產品交易金額拆分，南方東英的產品佔比 98%，三星資產、未來資產、易方達基金、華夏基金的產品合計佔比近 2%；按槓桿和槓反產品資產管理總額拆分，南方東英佔比 82%，三星資產佔比 8%，華夏基金佔比 5%，未來資產佔比 3%，易方達基金佔比 2%。

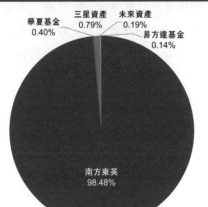

圖 21：槓桿和槓反產品交易額按發行人的分佈（2019 年 1 月至 6 月）

華夏基金 0.40%
三星資產 0.79%
未來資產 0.19%
易方達基金 0.14%
南方東英 98.48%

資料來源：彭博、南方東英。

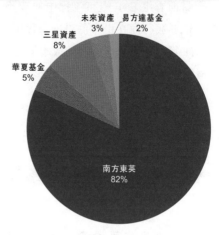

圖 22：槓桿和槓反產品資產管理總額按發行人的分佈（2019 年 6 月底）

未來資產 3%
易方達基金 2%
三星資產 8%
華夏基金 5%
南方東英 82%

資料來源：彭博、南方東英。

我們認為可以引入跨市場的標的資產，與國內 ETF 產品相連，或與國際 ETF 產品相連。跨市場 ETF 相連可以參考「中日 ETF 互聯互通」機制的發展經驗。

2018 年 10 月，中國證監會與日本金融廳簽署合作諒解備忘錄。2019 年 4 月 22 日，中日雙方在中日資本市場論壇上簽署了中日 ETF 互通協議，雙方約定合作建立兩地互通機制，計劃分別上市以對方市場 ETF 為投資標的的基金。2019 年 6 月 25 日，上海證券交易所舉行中日 ETF 互聯互通開通儀式，4 隻中日 ETF 互通產品在上交所成功上市。同日，日方有 4 隻 ETF 互通產品登錄東京證券交易所。

在中日 ETF 互通機制下，雙方分別上市以對方市場 ETF 為投資標的的基金，具體由雙方基金公司分別通過合格境內機構投資者（QDII）和合格境外機構投資者（QFII）機制，設立跨境基金，並將全部或絕大部分基金資產投資於對方市場具有代表性的 ETF 產品。與「滬港通」和「深港通」不同的是，「滬港通」和「深港通」是內地與香港雙方投資者直接投資對方市場的股票，而中日 ETF 互通使得投資者可通過投資本地基金產品來投入對方基金產品，從而實現跨境投資。中日 ETF 互通以聯接基金的形式，連接了中國內地與日本的金融市場。香港可以學習「中日 ETF 互聯互通」的模式，從而與中國內地的 ETF 市場對接。

3.3 改善 ETF 退市機制以完善 ETF 市場架構

目前香港 ETF 退市流程中如涉及 A 股停牌情況，會導致流程複雜，耗時冗長。ETF 發行商在退市的長期過程中，仍要支付高額的指數費、託管費等費用，給 ETF 發行商帶來不必要的成本。建議香港簡化 A 股 ETF 在股票停牌情況下的退市機制，為 ETF 發行商減輕負擔。

4 相關政策建議總結

(1) 對貨幣基金免徵中央結算系統（CCASS）等雜費

如 2.3.3 節中所分析，與中國內地相比，香港市場的交易及結算費用上不佔優勢，香港市場存在着交易徵費等費用，雖然數額不大，但對於債券和貨幣市場而言，費用的減免也會起到提升吸引力的作用。因此我們建議香港交易所等機構學習上海證券交易所和深圳證券交易所的做法，免除債券及貨幣市場 ETF 的交易結算等費用，普惠香港零售投資者。

(2) 擴大 MPF 投資產品種類

強積金即強制性公積金（Mandatory Provident Fund Schemes，簡稱 MPF），是香港政府在 2000 年 12 月 1 日正式實行的一項政策，強制香港所有僱員成立投資基金以作退休之用。如今，MPF 還無法讓投資者參與 A 股 ETF，等同於將香港僱員排除在 A 股市場之外。我們建議擴大 MPF 投資產品種類，比如允許投資 A 股 ETF，以分享中國經濟發展帶來的紅利增長。

(3) 擴大 ETF 標的資產種類

目前香港 ETF 市場的標的資產集中於股票，較為單一，我們建議未來能夠擴大 ETF 標的資產種類，納入更多的商品、期貨、定息產品、債券，以及引入外匯等，使 ETF 種類多樣化，讓香港投資者做到全球資產配置。

(4)　提高槓桿、槓反產品倍數

目前香港市場最多只有 2 倍槓桿產品與 2 倍槓桿反向產品，而且標的資產只有恒生指數和國企指數。根據市場的反饋，香港可以提高槓桿反向的倍數並提供包括 A 股在內的其他標的資產，以滿足市場需要。

註：南方東英徐小雅、柳晨寧對本文也有貢獻。

後 記

把握全球金融格局變化的大趨勢，
促進互聯互通和新經濟融資創新

在當前動盪的國際金融環境下，以「滬港通」、「深港通」和「債券通」為代表的互聯互通機制，奠定了香港連接東西的獨特金融樞紐地位，也為中國資本市場的對外開放探索出一條新路徑。

隨着 A 股加入 MSCI 的國際指數、中國債券納入彭博等主要國際債券指數帶來的重大趨勢性轉變，越來越多的海內外市場參與者正不斷參與到現時由「滬港通」、「深港通」和「債券通」組成的「互聯互通」體系當中，滿足其日益增長的中國或全球資產配置的需求，連同中國內地國民財富持續累積帶來的全球化、多元化配置趨勢，「互聯互通」正逐步成為跨境資本全球配置的重要新平台。與此同時，在全球面臨經濟結構轉型的新趨勢下，香港在新經濟融資創新方面不斷積極探索，積累了亞洲時區可供參考的探索經驗，促進香港市場逐步形成新的、圍繞新經濟企業的生態圈，為參與「互聯互通」的海內外投資者提供了新的標的。

為了回應市場對互聯互通和新經濟融資的廣泛關注，我們特意組織在相關領域具有豐富業務經驗的專業人士和研究力量，編撰了這本名為《互聯互通與香港新經濟融資創新》的圖書，圍繞近年來促進香港和內地金融業開放創新的最為關鍵的動能：「互聯互通」、「新經濟融資」、「全球資產配置」展開深入討論，系統梳理互聯互通的獨特設計理念，總結香港交易所新經濟融資的改革探索經驗，為市場參與者更好地把握「互聯互通」和新上市制度改革帶來的制度紅利，提供系統的專業研究支持。

作為市場上第一本深入闡述香港與內地的互聯互通和新經濟融資創新的中、英雙語著作，本書在以下幾個方面具備獨特性和創新性：

第一，本書邀請具體的參與者和設計者，以自身的經驗對「互聯互通」框架進行系統性、具深度的梳理和總結，讓市場參與者更加深入理解「互聯互通」在中國資本市場開放方面所帶來的積極影響。

作為中國資本市場開放的創新制度安排，「互聯互通」以最小的制度成本推動內地資本市場的雙向開放，在不增加資本流出風險的同時，實現了中國內地與世界金融體系的有效對接。近幾年來，「互聯互通」平穩運行，體系不斷延伸，從「滬港通」、「深港通」到「債券通」，亦在主要國際指數覆蓋中國內地資產的過程中起到重要的推動作用。

在這樣的背景下，海內外市場急需系統了解「互聯互通」的總體框架和運作邏輯，為數萬億元計的國際和中國資本尋求更便利的配置渠道，為金融資源對接新經濟企業探索更好的機制安排。本書第一章由「互聯互通」機制的設計者與推動者、香港交易所集團行政總裁李小加先生主筆，系統總結了「互聯互通」的運作機制，分享了他近年來推動香港上市制度改革與新經濟行業融資的設計理念和新思路，彰顯他作為一位國際著名金融家的宏觀視野。同時，第一篇也邀請了海外市場資深專家野村國際（香港）有限公司中國首席經濟學家陸挺先生和交銀國際控股有限公司研究部主管洪灝先生，深入探討了香港在新經濟公司融資創新方面的獨特定位、「互聯互通」框架對境內外市場的影響、對中國資本市場改革開放的獨特意義等諸多市場關注的重要議題，以基於扎實的市場數據深度分析，讓市場參與者多角度了解「互聯互通」在中國資本市場開放中發揮的重要作用。

第二，本書系統介紹了香港交易所的上市制度和流程、特別是針對 2018 年 4 月新經濟公司上市的制度改革，為市場全面了解香港交易所新上市制度和實務操作提供了權威的參考資料。

新經濟企業崛起已成為引領全球資本市場結構轉型的一大趨勢。2018 年 4 月香港交易所推出上市制度改革，對新經濟企業融資起到了重要的推動作用，也使香港 IPO（首次公開招股）集資金額再次榮登全球榜首。越來越多的新經濟公司正在着手進行香港的上市安排，迫切需要對香港交易所新的上市制度有全面、清晰的了解。

為了給海內外市場提供一份全面、易於查閱的上市參考資料，本書的第二篇全面介紹了香港主板上市的一般要求、上市審核流程特點、主板上市定價與發行等市場關注的要點，又重點介紹了與香港交易所 2018 年上市規則改革以及與生物科技行業相關的上市實務操作。另外，本篇有兩個章節從不同角度探討了同股不同權的法律意義和影響，讓市場深入了解同股不同權架構在新經濟公司成長中所起到的獨特作用。

為了讓市場更全面地把握香港在新經濟融資方面的創新趨勢，本篇還專門邀請了花旗環球金融亞洲有限公司和德勤中國的研究力量，綜合介紹了中國企業在海外發債、海外併購的具體案例，分享香港多層次融資渠道對新經濟企業發展、海外併購帶來的支持作用。

第三，本書專門在第三篇多角度說明了「互聯互通」作為海內外資金進行國際化配置的重要平台功能，為市場參與者深入了解「互聯互通」機制在國際資金配置中的實際操作意義提供不同思路。

「互聯互通」機制創新性地促進了國際資本流入中國內地市場，也為內地投資者進行國際資產配置提供了低成本的平台，而新經濟融資創新，則為資產跨境配置提供了重要的標的。如何在「互聯互通」和「新經濟融資」創新框架下為跨境資產配置和相關產品創新找到新思路一直是海外內市場熱烈討論的議題。

本書第三篇邀請了指數公司 MSCI 的中國研究主管魏震和南方東英資產管理有限公司總裁丁晨各抒己見，分享了他們對「互聯互通」機制下投資 A 股市場與交易所買賣基金 (ETF)、進行資產跨境配置的專業觀點，同時也向讀者介紹了基於「一帶一路」板塊和生物科技行業開發的相關指數，便於投資者充分運用好相關的工具與平台。股票互聯互通優化中的大宗交易機制與交易所買賣基金互聯互通帶來的發展機遇也在第三篇進行了探討。

本書圍繞「互聯互通」、「新經濟融資創新」、「全球資產配置」三個主題並行展開，互相呼應，以香港交易所首席中國經濟學家辦公室的研究力量為主要組織者和起草者，同時邀請香港交易所的相關業務部門和海內外不同業務領域富有經驗的知名金融機構相關負責人和資深專家共同參與寫作，得以形成這樣一本具有專業性和權威性、系統梳理「互聯互通」、「新經濟融資創新」、「全球資產配置」的專業著作。

在此，我要特別感謝香港交易所集團行政總裁李小加先生的鼎力支持，他本人專門為本書撰寫了序言和第一章，分享了他對其一直不遺餘力推動的「互聯互通」體系和新經濟融資的獨到而又深刻的見解。同時,香港交易所的合規團隊、法務團隊、企業傳訊團隊、翻譯團隊、相關業務部門也提供了大量幫助,正是他們的通力合作和建議,本書才得以成功出版。另外,本書的出版發行得到了商務印書館 (香港) 有限公司高效率的專業配合,將本書及時推向市場,在此一併表示誠摯的謝意!

今天的香港與祖國內地緊密相聯,唇齒相依,香港打造出的「互聯互通」為中國連接世界市場提供了獨一無二、風險可控的開放平台,既利於香港的繁榮,更利於國家的開放發展。「互聯互通」下香港不僅是內地連通世界的門戶,還可以與內地金融體系進一步融合,成為內地引入新的資金流量和持續發展的新動力,為再次助力中國資本市場開放揭開新篇章。

由於「互聯互通」這一創新發展涉及到市場的方方面面,而且還在繼續發展演變之中,書中缺點錯漏在所難免,敬請廣大讀者批評指正。

巴曙松 教授

香港交易及結算所有限公司　首席中國經濟學家

中國銀行業協會　首席經濟學家

2019 年 10 月

Part 1

The new trend of Connectivity and new-economy financing

Chapter 1

Mutual Market Access: Hong Kong as a financial hub connecting China and the world

Charles LI

Chief Executive

Hong Kong Exchanges and Clearing Limited

Introduction

Hong Kong's economic and financial development over the past decades is a legendary blend of oriental and western cultures through the continuous exploration of a unique identity. So far, Hong Kong has made tremendous contributions to China's economic boom by capitalising on its unique geographical location and its free and international financial framework. It has seized the opportunity in the changing world economy, leveraged trends and undergone a radical transformation that has turned itself into a globally-recognised international financial centre and a regional hub connecting China and the world. Hong Kong has capitalised on its unique position under the "One Country, Two Systems" framework to promote a mutual market under the Connectivity model, thereby achieving the maximum benefits of capital market opening for the Mainland with minimum institutional costs. This chapter will illustrate the main reform concepts, measures and effects of the Connectivity regime and new-economy financing reform promoted by HKEX in the past ten years that provide new models and impetus for the openning-up of China's capital market and economic transformation.

1   Mutual Market Access: The times backdrop

In the early days of China's reform and opening up, Hong Kong provided valuable capital for China's economic development. By facilitating China's entrepôt trade and foreign direct investment (FDI) in China, the city became a fund-raising and investment hub for the Mainland and the world. The next 30 years saw increasing interactions between the global financial and economic system on one side and China's market on the other, with China becoming an important pillar for driving global economic growth. In 2016, the Renminbi (RMB) was included in the International Monetary Fund's Special Drawing Rights (SDR) basket of currencies for the first time, a milestone signifying China's deep integration with the global financial community. At the same time, capital flows to and from China gradually brought about important changes in many aspects with worldwide implications:

- First, China became affluent with abundant capital supply compared to capital shortfalls in the past. Hopefully it will become a new international currency liquidity provider in Asia and in countries along the Belt and Road;
- Second, the form of foreign trade in China has changed. China has matured from a dominant exporter of goods to an exporter of both goods and capital. China's global allocation of assets will trigger a new wave of global capital flows;
- Third, China will grow from a rule-taker into a rule-maker as well, with increasing influence in global finance. It will need effective platforms and tools to match its increasing weight in international matters.

The internationalisation of the RMB and the globalisation of China's capital will inevitably promote the two-way flows of cross-border funds and the two-way allocation of resources and assets. International products, prices or investment flows will "go in" to China and China's products, prices or investment flows will "go out" to the world. This will make China's wealth and investment allocation more diversified and international, ensuring China is able to participate in and influence international pricing and standard-setting more effectively.

Against this backdrop, Hong Kong's positioning is bound to change drastically. Hong Kong can fully capitalise on its position under the "One Country, Two Systems" framework and its unique advantages as a free port for international capital to become a crucial gateway for the application of the RMB in cross-border investment and financing, risk management and financial innovation. While there remain huge gaps between the Mainland and international financial markets in terms of institutional systems and operations, Hong Kong can help accelerate the internationalisation of the RMB through the two-way opening-up of the Mainland and Hong Kong capital markets through innovative connectivity channels like Stock Connect (Shanghai Connect and Shenzhen Connect) and Bond Connect. Hong Kong will play a key role in China's further internationalisation and economic transformation, reinforcing its role as a "converter" connecting the Mainland and international markets.

2 Mutual Market Access: Leading to an innovative two-way opening-up of the "Mutual Market"

To accelerate the integration of the Mainland's economy and currency with the world, the gradual opening-up of the domestic market and capital account is necessary. However, there are currently vast differences between the Mainland and international financial markets in their institutional systems, investment practices and financial infrastructure. This is partly due to long-established market practices as well as innovation and exploration happening in each market. It is not a matter of which market is better; they both developed differently with varied room for experimentation. With that in mind, it is unlikely for the Mainland to fully turn away from its systems and practices to follow international rules, or for foreign capital to eschew established international practices and follow the Mainland model. This underscores the importance of establishing a new innovative connection through an orderly and controllable cross-border trading platform that complies with international rules and fulfills China's requirements to open up its financial market, promote the healthy and steady interaction of Mainland capital, assets and resources with the international financial system, and facilitate the gradual integration of the Mainland and international capital markets.

The unique design of the "One Country, Two Systems" framework provides a unique, non-replicable model of connecting China with the world, allowing both sides to interact with great depth and breadth without drastically changing their systems. Under "One Country", Hong Kong can collaborate with the Mainland more intensively than any other country or region, giving the Mainland effective control over its comprehensive market opening reforms. "Two Systems" preserves Hong Kong's market mechanisms and institutional frameworks that are trusted and respected globally and in line with international financial market practices. This allows Hong Kong a high degree of flexibility to provide connectivity between Mainland China and the international market and help the Mainland achieve its "going out" and "coming in" goals for financial development that would not yet be fully achievable in its existing system.

This is precisely the focus of the Mutual Market Access programme, or financial market connectivity, which HKEX has been rapidly developing in recent years. The significance of the Mutual Market Access model is that it pioneered an innovative two-way opening-up mechanism and created a mutual market that meets the requirements of regulators on both sides of the boundary. Through connections between the financial infrastructure of the exchanges and clearing houses in Mainland China and Hong Kong, an open and risk-controllable platform is formed that brings together international and Mainland investors and companies in a closed-loop system. Participants in this system include international and Mainland institutional investors, the enormous retail investor base in Mainland China and quantitative traders and traditional asset managers. This design can accommodate multiple asset classes including stocks, exchange traded funds (ETFs), fixed-income products, currency products, derivatives and commodities that meet the needs of both international investors interested in Mainland exposure and Mainland investors looking for international products. Through this transparent, safe and controllable channel, Mainland enterprises and residents can invest and trade in multiple markets like global equities, financial risk management and world commodity pricing with minimal adaptation costs. This effectively allows flows and allocation of international and Mainland capital and assets at multiple levels without altering the market systems and rules and trading practices of the participants on either side.

3 Mutual Market Access: A proven way to connect China and the world while controlling risk

Since the launch of Shanghai Connect in 2014, the Mutual Market Access programme has been steady, reliable and a key catalyst in promoting the two-way opening-up of the Mainland capital market in a risk-controlled framework. It is unique in the following aspects:

(1) Maximum benefits to the market with minimal institutional costs under the principle of "home rules apply"

The Mutual Market Access programme allows investors in Hong Kong and Mainland China to trade and settle shares listed on the other market via the exchange and clearing house in their local market. This programme requires no change to the existing local systems, rules and trading practices, thereby incurring minimal costs to change the two respective systems. Through the Southbound Trading Link, Mainland investors can diversify their investments abroad through local brokers on their own with ease. The Northbound Trading Link under Shanghai Connect and Shenzhen Connect (collectively called "Stock Connect") allows international investors (including a large number of international passive funds tracking MSCI indices) to invest in the Mainland market via Hong Kong, following international practices and through international brokerages, thereby introducing more international capital flows to the Mainland market.

(2) Clearing and settlement in a closed loop enables open and risk-controllable repatriation of funds

Funds circulated in the connectivity programme can only be used to buy and sell stocks within the prescribed scope. Funds received from selling the stocks must be returned along the original route through the local settlement system and will not remain in the other market in the form of other assets, thus preventing capital outflow. All RMB conversions take place in the Hong Kong offshore market, minimising the impact on the onshore RMB exchange rate and China's foreign exchange reserves, which helps facilitate the internationalisation of the RMB while eliminating the potential risk to the onshore market.

(3)   Clearing and settlement in net amount for minimising cross-border fund flows

Under the Mutual Market Access, funds for clearing and settlement of shares transactions will take place in the investors' local market and only the net amount will be settled between the clearing houses in the two markets. According to this mechanism, over the past four years or so, the total trade value under the scheme was nearly RMB 15 trillion while the actual sum of funds remitted across the border was only slightly over RMB 100 billion. This substantially reduces fund flows between markets and the incidental risk of cross-border capital flows, preventing fluctuations in the currency market.

(4)   The Mutual Market Access programme covers a wide range of unique investment products; in particular, the Southbound link provides a wealth of asset allocation options for Mainland investors

As of the end of April 2019, there were a total of 480 stocks eligible for Southbound trading under Stock Connect, accounting for 88% of the total market capitalisation of Hong Kong's Main Board. Among these are large Chinese financial, energy and telecommunications companies, such as the top four Mainland banks, China Petroleum and China Mobile, which play a vital role in China's economic boom, as well as leading technology companies such as Tencent. In April 2018, Hong Kong further introduced new listing rules to facilitate the listing of new-economy companies. More innovative companies are attracted to list in Hong Kong and be possibly included in the Mutual Market Access programme for Mainland investors to invest in. This facilitates diversified asset allocation by investors and the sharing of economic benefits.

Table 1. Top 10 stocks in Stock Connect Southbound trading in 2018 by turnover value (buy and sell)			
Shanghai Connect		Shenzhen Connect	
Stock name	Share of Shanghai Connect total turnover	Stock name	Share of Shenzhen Connect total turnover
TENCENT	10.3%	TENCENT	8.0%
ICBC	3.6%	SUNAC	3.0%
CCB	3.4%	GENSCRIPT BIO	2.8%
SUNAC	3.1%	ZTE	2.6%
PING AN	2.7%	KINGDEE INT'L	1.8%
GEELY AUTO	2.4%	CCB	1.5%
HSBC HOLDINGS	2.0%	GEELY AUTO	1.4%
ABC	1.9%	SMIC	1.4%
CM BANK	1.8%	SUNNY OPTICAL	1.4%
CPIC	1.7%	PING AN	1.4%
Top 10 stocks	32.8%	Top 10 stocks	25.3%
Total turnover: HK$1.82 trillion		Total turnover: HK$1.01 trillion	

Note: Individual percentages may not add up to the total of the top 10 stocks due to rounding.
Source: HKEX.

The Mutual Market Access scheme has established a risk-controllable and transparent channel for two-way fund flows and capital allocations. It is a milestone in the opening-up of China's capital account. Since its launch more than four years ago, the programme has been highly effective in helping Chinese nationals allocate their wealth globally without leading to capital outflows. From launch to June 2019, net cross-border fund flows of RMB 36.2 billion (Southbound) was recorded, supporting a cumulative turnover (Southbound and Northbound, buy and sell) of RMB 22 trillion[1] (more than 600 times the net figure). In June 2019, the average daily turnover (one-sided) value (ADT) for Southbound trading accounted for 5.0% of that of the Hong Kong stock market, and Mainland investors' holdings of Hong Kong stocks in Southbound trading amounted to HK$914 billion. During the same period, Northbound trading (Shanghai and Shenzhen) saw a cumulative turnover of RMB 14.3 trillion, and through this, overseas investors were holding RMB 1.04 trillion worth of Mainland stocks. Thanks to the Mutual Market Access programme, Hong Kong has gradually become an important platform for the global allocation of funds from the Mainland and for the entry of international capital into the Mainland capital market (see Figures 1 to 4).

---

1   Source: HKEX.

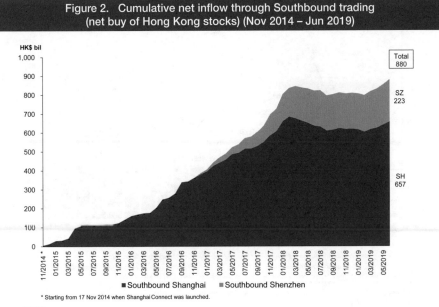

Figure 1.   Cumulative net inflow of funds through Northbound trading (net buy of Shanghai/Shenzhen stocks) (Nov 2014 – Jun 2019)

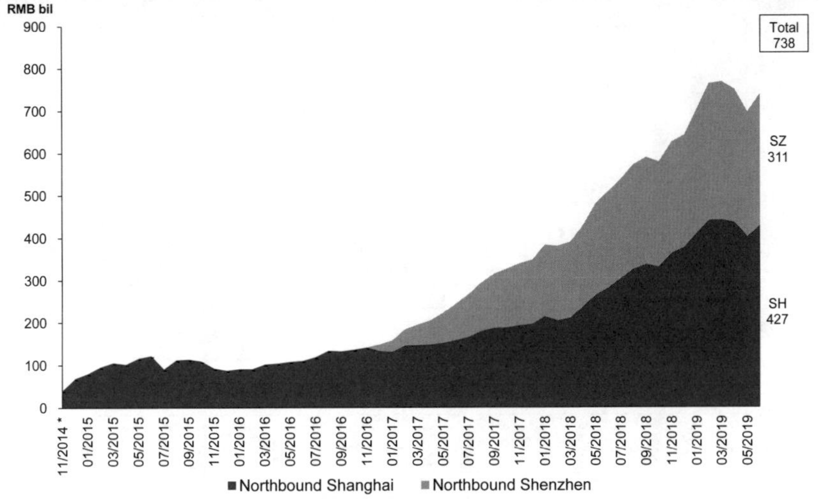

* Starting from 17 Nov 2014 when Shanghai Connect was launched.

Source: HKEX.

Figure 2.   Cumulative net inflow through Southbound trading (net buy of Hong Kong stocks) (Nov 2014 – Jun 2019)

* Starting from 17 Nov 2014 when Shanghai Connect was launched.

Source: HKEX.

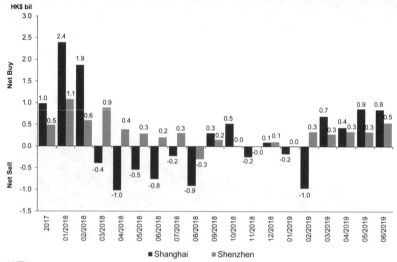

Figure 3. Monthly net inflow/outflow through Southbound trading (Shanghai and Shenzhen) (2017 – Jun 2019)

Source: HKEX.

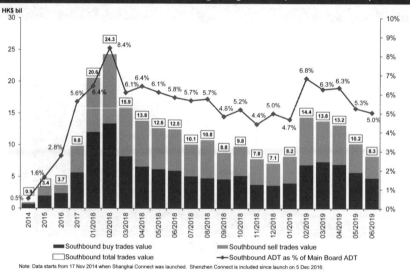

Figure 4. Average daily turnover value of Southbound trading and percentage share in the total turnover value of the Hong Kong market (2014 – Jun 2019)

Note: Data starts from 17 Nov 2014 when Shanghai Connect was launched. Shenzhen Connect is included since its launch on 5 Dec 2016.
Source: HKEX.

The Mutual Market Access programme has proven to be a reliable channel for investing in the A-share market. When considering the inclusion of China A-shares in its indices, index compiler MSCI believed that the Mutual Market Access programme had eliminated the obstacles for stock trading and capital inflows and outflows[2]. In recent years, Mainland funds have been making use of the Mutual Market Access programme to reach out to overseas markets. Social security funds and insurance funds in the Mainland are also making use of this risk-controllable channel under Southbound trading to conduct cross-market investments. This demonstrates that the Mutual Market Access programme has gained the trust of Mainland and overseas investors and has won international recognition for its convenience and accessibility.

As far as Mainland investors are concerned, even if China's capital account is fully opened, Mutual Market Access remains an important investment channel for Mainland investors with diverse risk profiles and investment needs. The market is always made up of investors with varied investment experience and objectives, which I like to assign to three groups: postgraduates, undergraduates and high school students. Postgraduates are those with the ability, resources, channels, willingness and experience of investing overseas; high school students are not interested in overseas markets whether they are accessible to them or not; while undergraduates are those with relatively more funds and investment experience along with the need to allocate some of their assets abroad, but do not dare to invest *directly* in overseas markets. Despite the gradual opening-up of China's capital market, the huge differences in the trading systems and cultures in different markets will remain for a long time. As long as Mutual Market Access can continue expanding its product range, providing complex options like derivatives as well as easy-to-trade and passive investments like ETFs to accommodate different risk profiles and investment needs, Stock Connect will thrive as a dynamic two-way channel connecting China and the world.

For listed companies, Mutual Market Access further strengthens the financing function of the capital market, allowing the interaction of investors and capital from Mainland China and international markets, thereby helping create a mutual market of enhanced scale, depth and liquidity, as well as a larger and more diverse investor base. Through this, market liquidity will be increased, facilitating refinancing activities and the formation of a diverse and multi-tiered market structure. In April 2018, new listing rules were introduced to allow the listing of new economy companies, attracting more innovative companies to float and be included in the Mutual Market Access system. This is substantially expanding the investor base of these listed new-economy companies, an important strategic significance to

---

2   See "Consultation on further weight increase of China A shares in the MSCI indexes", MSCI, December 2018.

their business expansion and financing in the international market.

And for regulators, the Mutual Market Access programme steps up interactions between the regulatory bodies on both sides, accelerating the integration of the Mainland capital market with its international counterparts. Exchanges are playing to their strengths to form a complementary market structure which not only achieves connectivity but also maintains the special rules and trading features of the local markets. Market participants and regulators are able to gain more in-depth knowledge and understanding of the operation and culture of each market. Such mutual understanding and learning will help further promote mutual reforms and development of both markets.

4   Mutual Market Access: A blueprint for connecting multiple capital market segments

As a channel that helps facilitate the opening-up of China's capital account in a controlled and manageable way, Mutual Market Access is scalable to accommodate more asset classes. Mainland and foreign investors will be able to purchase a wide range of financial products from China and all around the world in this all-encompassing "department store" of financial products. They can not only find stocks, but also fixed-income, currency, and commodity products; these products aren't just Mainland products for international investors, but also international products for Chinese investors; they can also find international companies listed in Hong Kong as well as products listed in other international markets like ETFs or via Southbound trading in Bond Connect, for example. The abundant supply of Chinese, Asia-Pacific and international products on Hong Kong's shelves will attract increasing amounts of investment capital from China and countries around the world to the Hong Kong market, making Hong Kong an integrated financial centre with multiple asset classes.

Northbound trading under Bond Connect launched as an extension of the Mutual Market Access model to the bond and fixed-income market in 2017. It integrates the onshore and offshore bond markets by linking up the infrastructure in the Hong Kong and

Mainland bond markets, allowing overseas institutions to hold Mainland bonds directly through offshore infrastructure in accordance with international rules and familiar trading practices. The connectivity of Bond Connect is manifested by its seamless and innovative connection of China's "direct holding" system with the multi-level "nominee holding" arrangement in international bond markets, blending in the "see-through" model under China's depository arrangement. This effectively reduces the conversion cost at the operational level to connect different market systems.

The overall framework of Bond Connect basically follows the relatively closed-loop design of the Mutual Market Access scheme, aiming to facilitate and accelerate the opening-up of the Chinese bond market within a risk-controllable framework. Since launch in July 2017, 845 international institutions or funds covering 27 jurisdictions across the world have taken part in the scheme as of the end of April 2019, a significant increase from 503 at the end of 2018. Average daily turnover amounted to RMB 5.31 billion[3]. As shown in Figure 5, bond holdings of foreign investors in the China Interbank Bond Market (CIBM) has risen steadily since the launch of Bond Connect.

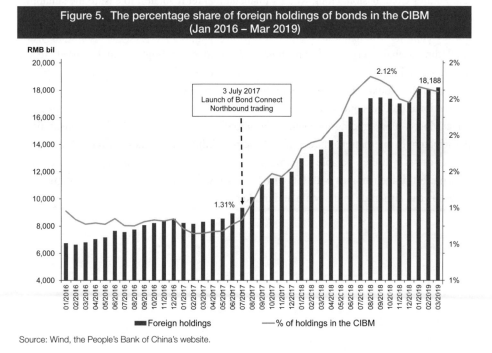

Figure 5. The percentage share of foreign holdings of bonds in the CIBM (Jan 2016 – Mar 2019)

Source: Wind, the People's Bank of China's website.

---

3   Source: Bond Connect Company Limited.

At present, the business scope of Bond Connect covers both the primary and secondary markets. Policy banks in the Mainland, like the Agricultural Development Bank of China, have established overseas bond issuance mechanisms under Bond Connect and publish bond issuance information on the website of the Bond Connect Company Limited (BCCL), making it more convenient for overseas institutions to subscribe and invest in onshore RMB bonds. Bond Connect will launch a Southbound link when conditions become mature, which will promote the internationalisation of the RMB and enhance domestic residents' return on assets and their global asset allocation while maintaining information transparency, safety and controllability on the regulatory front.

Regulators on both sides have also indicated that they will expand the Mutual Market Access programme to cover ETFs in the future, bringing more passive institutional investors and capital flows to the Mainland and providing Mainland investors with a wider range of choices for their global asset allocation. These ETFs will not be limited to ETFs tracking Hong Kong-listed securities, but also those that track international (especially Asia-Pacific) assets. This way, as long as the ETF issuers are regulated by the Securities and Futures Commission, Hong Kong's financial market regulator, to ensure the suitability of the products, there is no need to have geographical restrictions on the stocks the ETFs are investing in. Mainland investors can then buy unique ETF products around the world (including global energy, emerging markets, and more) through Mutual Market Access according to their needs, resulting in a more effective global asset allocation.

Under the Mutual Market Access framework, Hong Kong can continue adding attractive international products and features into the market. This could include ETFs with overseas underlying, a block trading mechanism, or allowing and facilitating Mainland institutions' participation in international new issues. The possibilities are endless. Chinese companies and investors will then be able to participate in global pricing and trading conveniently in the comfort of their homes with lots of overseas investment choices whereas international investors can also trade in the Mainland market conveniently.

5   Creating an offshore product ecosystem based on Mutual Market Access and developing Hong Kong into an offshore RMB product trading and risk management centre

The full expansion of "money" and "products" also implies the need for a full-scale upgrade of the functions of the "market".

Mutual Market Access is not only a platform for two-way closed-loop cross-border investments, but it's also an incubator for multi-level risk management products and financial instruments. It is our vision to develop an offshore RMB financial product ecosystem, substantially strengthen risk management and post-trade custody and clearing services, and fully upgrade the functions of the "market".

We will inevitably see huge fluctuations in asset prices in China when the RMB is trying to integrate with the rest of the world. In our contact with market participants, we found that a considerable number of international investors interested in holding assets in China (including stocks, bonds and currencies) are not active in Chinese investment due to the lack of hedging tools. On the other hand, Chinese investors who have started holding more and more overseas assets also need corresponding tools to manage exchange rate risk. According to market surveys, international participants are of the view that, compared to other factors, exchange rate volatility is the major policy factor that will impact the RMB's internationalisation. This shows that for Mutual Market Access to further develop vertically, the only way out is to upgrade the functions of the market under Mutual Market Access and build a financial ecosystem that includes trading, clearing and settlement and risk management.

A deep, diversified and developed financial product market is instrumental in giving significant backing for the internationalisation of the RMB. In recent years, HKEX has been proactive in product innovation, taking into account market feedback. A number of

RMB-denominated derivatives and RMB exchange rate derivatives have been introduced
since 2012, to serve Mainland and overseas institutions' needs to manage interest and
exchange rate risks associated with RMB assets (see Table 2). In 2018, the turnover of
RMB currency products reached a record high of 1.809 million contracts, up 137% from
2017. The average daily trading volume of RMB currency futures in the first six months of
2019 was 10,833 contracts, making the contract an important risk management option for
international investors (see Figure 6).

Table 2. Major types of exchange traded derivatives in Hong Kong		
Launch date	Product	Feature
September 2012	USD/offshore Renminbi futures (i.e. USD/CNH Futures)	Providing the market with a currency risk management tool and investment tool in the course of RMB internationalisation
December 2014	RMB-denominated commodity futures contracts (including aluminum, copper, zinc, lead, nickel and tin)	The first metal products outside Mainland China which provide for RMB exposure in the underlying assets, supporting RMB benchmarking for the metals in the Asian time zone
May 2016	RMB-denominated currency futures of off-shore RMB against euro, Japanese yen and Australian dollar — EUR/CNH, JPY/CNH and AUD/CNH Futures, and the USD-denominated CNH/USD Futures	Further enriching the RMB exchange rate product line
March 2017	USD/CNH Options	Help market participants to more effectively hedge currency risk
April 2017	5-Year China Ministry of Finance Treasury Bond Futures (MoF T-Bond Futures) (This pilot scheme was suspended on the expiry of the contract in December 2017. New RMB interest rate products will be re-launched in due course.)	The first offshore futures on domestic Chinese government bonds by way of a pilot scheme

Source: HKEX.

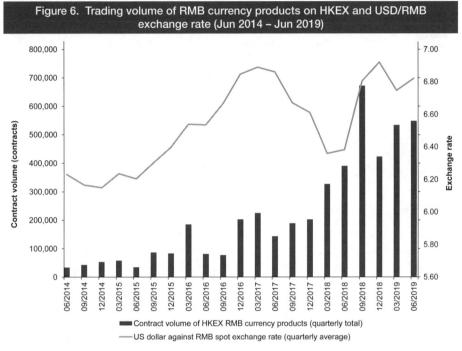

Figure 6. Trading volume of RMB currency products on HKEX and USD/RMB exchange rate (Jun 2014 – Jun 2019)

■ Contract volume of HKEX RMB currency products (quarterly total)
── US dollar against RMB spot exchange rate (quarterly average)

Source: Wind.

Following the inclusion of A shares into the MSCI Emerging Markets Index, Chinese bonds were admitted into the Bloomberg Barclays and other bond indices that are widely tracked by global funds in 2019. More and more medium and long-term institutional investors around the world, including central banks, sovereign wealth funds and international pension funds, will track these global indices and passively increase their allocations into Mainland assets of stocks and bonds based on changes in these indices. This will lead to a shift in asset allocation of the nearly US$4 trillion worth of assets under management worldwide, generating demand for adequate and liquid tools for hedging related risks. Data shows that after the pilot launch of the MOF T-Bond Futures in Hong Kong in April 2017, the daily trading volume of exchange traded RMB bonds increased significantly. Overall liquidity and pricing efficiency of the market were boosted (see Figure 7). In addition, the turnover of RMB currency futures in Hong Kong kept rising with the increasing two-way movements of the RMB exchange rate (see above), indicating investors' increasing hedging needs against a widening trading band of the RMB.

Figure 7. Daily turnover of exchange traded RMB bonds (Jan – Jul 2017)

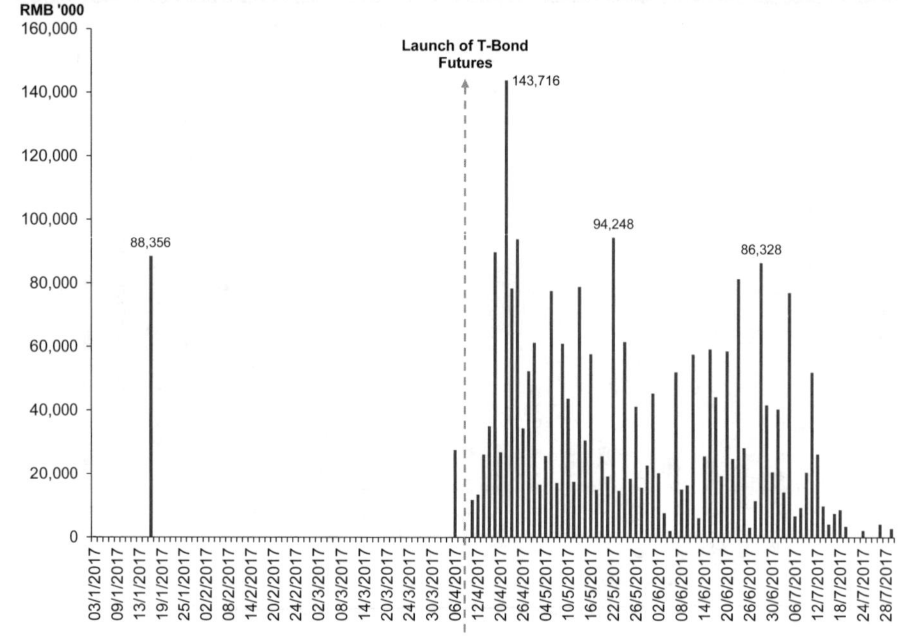

Source: HKEX.

Hong Kong is well-positioned as an offshore RMB market for its professional services, sound financial infrastructure and the provision of a wide range of useful tools for risk management. The offshore RMB derivatives product suite developed by HKEX is very actively traded, providing high liquidity and market depth for investors. The price formation mechanism and clearing and settlement system for the on-exchange offshore RMB products in Hong Kong also effectively ensure that the offshore exchange rates track the onshore exchange rates closely at final settlement. The trading of offshore RMB products thus could contribute to a larger impact of onshore RMB prices in the global market and keep the price-setting process onshore. Mutual Market Access has induced higher demand for risk management and promoted the further development and enrichment of offshore RMB derivatives to meet the needs of the real economy. This will be conducive to the gradual development of Hong Kong to become an offshore RMB products trading and risk management centre, and the facilitation of the RMB's greater international use and internationalisation.

6 Let Mutual Market Access be the catalyst for China's economic transformation and enterprise innovation

The development of the multi-tiered capital market in the Mainland has been very successful, now the second largest capital market in the world[4] in terms of market capitalisation. However, the reforms of the capital market in the Mainland are somewhat constrained by the dominance of retail investors. The current share issuance review and approval system, trading system and investor composition still fall short of international standards and fail to fully meet the financing needs of many enterprises, especially new-economy and innovative companies. The Science-Technology Innovation Board of the Shanghai Stock Exchange (now called the SSE STAR Market) has been rolled out in the Mainland, another indication that reforms will continue and China will be able to provide new-economy companies with more funding sources and greater flexibility.

The revised Listing Rules in Hong Kong in 2018 kept the city on pace with the practices of other major financial markets[5] and includes provisions for the listing of biotech companies and companies with weighted voting rights structures. These companies are also being included in the Mutual Market Access programme to draw a more diverse investor base, enabling Mainland companies listed in Hong Kong to reach a wider group of investors from around the world.

After the new Listing Rules took effect, 50 new-economy companies listed in Hong Kong by the end of June 2019, raising a total of HK$167.5 billion, accounting for 52% of the total funds raised from initial public offerings (IPOs) on our market during the period. Out of these 50 companies, 14 are biotechnology companies (including eight pre-revenue

---

4   According to the statistics of the World Federation of Exchanges (WFE) as of May 2019, the total market capitalisation of the listed companies on the Shanghai and Shenzhen stock exchanges amounted to almost US$7.6 trillion, second only to the combined market capitalisation of the New York Stock Exchange and NASDAQ in the United States.

5   Capital markets around the world are competing to attract the listing of innovative companies. Data shows that 30 out of the world's 46 largest national stock markets have listed companies that are adopting or had adopted a dual-class share structure. (See HKEX research report, "Listing regime reforms for dual-class share structure and biotech industry", published on the HKEX website, 15 November 2018.)

companies that raised a total of HK$23.5 billion through the new listing chapter) that raised a total of HK$47.5 billion. In the new era where new technologies are driving the development of the world economy, HKEX, with listing criteria in line with international standards and the innovative Mutual Market Access programme, can help boost the growth and development of the many innovative enterprises in China. The power and influence of Chinese companies in the global market will thereby be enhanced, giving new impetus to China's economic growth.

7   Conclusion

With the mutual market created under the Mutual Market Access programme, Hong Kong can capitalise on its unique market system and financial environment to introduce high-quality assets from the international market and make them available to Mainland investors seeking a broader allocation of assets. Together with a deep understanding of the financing needs of the Mainland, Hong Kong will also be able to grow the international investor base for high-quality Chinese companies so that these companies can go international in investment and fund-raising under the connectivity scheme, facilitate the global allocation of Chinese resources, and enable quality enterprises to raise funds and share the wealth they create. In the current global financial environment with such a high degree of openness and integration, the mutual market developed in Hong Kong through Mutual Market Access is a risk-controlled, two-way opening with genuine market breadth and depth. This will help attract new capital to Hong Kong and new drivers for sustainable growth, boosting the international competitiveness of Chinese enterprises and providing new impetus for China's economic transformation.

Chapter 2

Financing innovation for the new economy and Hong Kong's explorations

Ting LU

Chief China Economist, Global Markets
Nomura International (Hong Kong) Limited

Lisheng WANG

China Economist, Global Markets
Nomura International (Hong Kong) Limited

Jing WANG

China Economist, Global Markets
Nomura International (Hong Kong) Limited

Summary

Technology has been reshaping the global economy rapidly. By improving productivity, industrial structure, and people's living standards, technological development leads to a rise of new-economy sectors and shifts part of the focus of global capital markets to new-economy investment and financing. As one of the world's top three financial hubs, Hong Kong has played a key role in facilitating the offshore financing of companies from Mainland China and other Asia-Pacific regions, and meanwhile, it keeps exploring innovative financing approaches for the new economy. We expect Hong Kong to make more significant contributions to financing China's blossoming new-economy sector in the coming decade. This chapter discusses the recent trends of new-economy sectors, funding difficulty of new-economy companies, and Hong Kong's efforts in searching for financing solutions for new-economy companies. We specially focus on Hong Kong's unique role in helping Mainland Chinese new-economy companies to fund their domestic investments, to conduct their offshore mergers and acquisitions, and to hedge exchange rate risks.

1   The world's rising new-economy sectors

"New-economy" sectors, as defined by the Organisation for Economic Co-operation and Development (OECD)[1], refer to "sectors of an economy that are producing or intensively using innovative or new technologies". This relatively new concept applies particularly to industries where people increasingly depend on computers, telecommunications and the Internet to produce, sell and distribute goods and services. The scope of new-economy sectors is not limited to one or two specific sectors, it could cover an Internet company in the information and communication technology industry that adopts new digital computing technologies, a biotech company that manufactures new drugs and medicines, or a company in the financial service industry that develops new financial products and services.

The new-economy has risen to be a crucial driver of global economic development. In particular, the digital economy, as a major form of the new economy, has transformed the pre-existing economic landscape and its impact is increasingly felt in many aspects of the global economy. According to a joint research led by Huawei and Oxford Economics in 2017[2], the digital economy across the world has grown 2.5 times faster than the global gross domestic product (GDP) over the past 15 years, taking its overall size to US$11.5 trillion in 2016 which accounted for 15.5% of the global GDP. Based on their projections, the size of digital economy is expected to amount to US$23 trillion in 2025, accounting for an estimated 24.3% of the global GDP. By countries, in 2016 the US and the European Union contributed respectively 35% and 25% to the global digital economy, while China's share rose to 13% in 2016 from just 4% in 2000. They also forecasted that, on a 3-year investment horizon, the long-term return on digital-related investments is 6.7 times higher than that on non-digital investments, owing to more significant spill-over effects.

---

1   See OECD's Glossary of Statistical terms (https://stats.oecd.org/glossary/detail.asp?ID=6267).
2   See Huawei & Oxford Economics. (2017) *Digital Spillover: Measuring the true impact of the digital economy* (https://www.huawei.com/minisite/gci/en/digital-spillover/files/gci_digital_spillover.pdf).

2   The rapid catch-up of new-economy sectors in China

Thanks to the accession to the World Trade Organisation (WTO), the flagship "Going Global" strategy, and the opening-up of the domestic market, the Chinese economy has been increasingly integrated into the global supply chains and interconnected with the rest of the world. According to WTO, China ranked the world's largest exporter in terms of merchandise trade in 2018 (up from 11th in 1995). And according to the World Bank, China has been the world's second largest economy in terms of GDP since 2010, just behind the US, up significantly from 8th in 1995. Meanwhile, China's industrial structure has been rebalancing towards new-economy sectors, with companies specialised in new technologies, new energy and new business models rapidly emerging and growing.

Similar to their impact on the global economy, new-economy sectors have become a key driving force for China's economic growth. According to China's National Bureau of Statistics (NBS), GDP in China's new-economy sectors, which include new industries, new forms and models of businesses, grew by 12.2% in 2018, 2.5 percentage points higher than nominal GDP growth in the same period, with their share in national GDP rising to 16.1% in 2018 from 15.7% in 2017. The New Economic Development Index (NEDI) of China, released annually by the NBS since 2015, tracks the development trends of new-economy sectors based on a wide range of industry-level indicators such as the number of broadband Internet users, the expenditure on research and development (R&D), and the express delivery volume. The NEDI jumped to 270.3 in 2018 from 210.1 in 2017, almost three-fold of the level in 2014 (100.0), indicating a rapid growth of China's new-economy sectors. Breaking down by sector, the big jump of NEDI was led by the Internet Economy sub-index, which surged to 605.4 in 2018 from 362.1 in 2017, contributing 80.8% of the rise in the NEDI. The contributions from Innovation, Transformation and Upgrade, Economic Vitality, and Intellectual Capacity sub-indices were 10.3%, 3.9%, 2.6% and 2.4% respectively.

In addition to a sizable domestic market and an advanced infrastructure network, the Chinese government's policies, including fiscal subsidies, have been offering strong support to the robust development of the new-economy sectors. However, with a financial system dominated by large commercial banks, financing conditions are much more favourable for state-owned enterprises than for private enterprises. As a result, the growth of new-economy companies in Mainland China still largely rely on self-funding and offshore financing, which have been insufficient to meet their special financing demand. A more market-based financing channel is a prerequisite for sustaining their rapid development.

3   Financing remains a headache for China's new-economy sectors

Although many new-economy companies have decent returns on capital and great development potentials, these companies are less favoured by the traditional banking system and capital markets. The reasons include the lack of collaterals, limited credit records, high uncertainty in their future revenues, and unavailability of implicit government guarantees compared to their large, especially state-owned, counterparts.

Some new-economy companies have been overly dependent on financing through shadow banking channels, such as trust loans, entrusted loans and peer-to-peer loans, since 2009 when the Chinese government allowed shadow banking to rapidly expand to stabilise economic growth. However, financing through these channels started falling in the past two years due mainly to tighter financial regulations and the forceful deleveraging campaign since mid-2017. Moreover, in a risk-off regime given slowing domestic growth and elevated uncertainty around US/China trade tensions, Chinese new-economy companies, especially those with high exposure to cross-border businesses, are facing increasing difficulty in external financing.

One example is equity market financing in China's A-share market. Founders of new-economy companies have to exchange some of their controlling stakes and voting rights to raise funds from external investors. In practice, the loss in their controlling stakes and voting rights is usually proportional to the amount of funds they raised. This institutional setup in Mainland China's equity market in many cases dampens these founders' desire to adopt equity financing, especially if they have strong expectations on their companies' future growth potential. The limited supply of high-quality new-economy shares results in high prices of these shares, which in turn discourage external investors.

The financing difficulty for Mainland China's new-economy companies tend to be more challenging when domestic financial conditions are tightened, especially during the deleveraging campaign in the period of 2017 to 2018. Nevertheless, some of these new-economy companies are quite attractive to international investors, due to their high technology, new business models, and prospects of high growth.

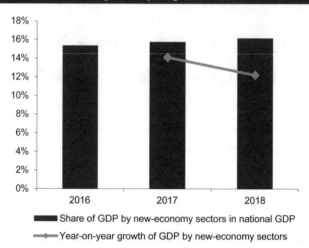

Figure 1. China's new-economy sectors: Share in the national GDP and year-on-year growth

Source: Wind and Nomura Global Economics.

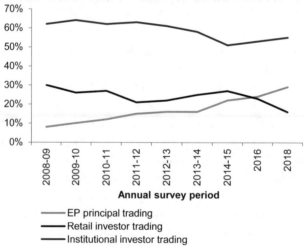

Figure 2. Breakdown of trading value in Hong Kong's equity markets by investor type

Note: EP principal trading refers to trading in the Stock Exchange Participant firm's own account.
Source: HKEX's annual Cash Market Transaction Survey throughout the period.

4   Hong Kong's financial innovations for new-economy companies

Benefitting from a developed financial system, a rich pool of financial and legal professionals, and the high connectivity with global markets, Hong Kong has taken a leading role in financial innovations to facilitate funding for the new economy. Over the past decade, we have seen the listing regime reform launched by Hong Kong Exchanges and Clearing Limited (HKEX) in 2018, the Stock Connect and Bond Connect schemes that link markets between Mainland China and Hong Kong, and the offshore dollar bond markets for serving companies throughout the world, especially from Mainland China.

4.1   HKEX's reform to facilitate the equity financing of new-economy companies

To attract and accommodate the listing of companies in new-economy sectors, HKEX rolled out a big reform on its listing regime on 30 April 2018. As a major part of the reform, HKEX added three new chapters in the Main Board Listing Rules of its wholly-owned subsidiary, the Stock Exchange of Hong Kong (SEHK), to effect the followings: (1) permitting the listing of biotechnology (biotech) companies that do not meet the Main Board's financial eligibility tests, or the listing of pre-revenue biotech companies; (2) permitting the listing of companies with weighted voted rights (WVR) structures; and (3) establishing a new secondary listing route for companies in Greater China and international companies that seek a secondary listing in Hong Kong.

According to Charles Li[3], the HKEX Chief Executive, the reform was of great significance and culminated in the single most transformative change to the Hong Kong market in a quarter century, dating back to the introduction of the H-share regime in 1993. He also mentioned in his subsequent blog that new-economy companies raised HK$136 billion through initial public offerings (IPO) in 2018, accounting for nearly half of the total IPO funds raised, partly owing to the reform[4].

---

3   See "Dawn of a new era", *Charles Li Direct* on the HKEX website, 29 April 2018 (https://www.hkexgroup.com/Media-Centre/Charles-Li-Direct/2018/Dawn-of-a-New-Era?sc_lang=en).

4   See "Reflections on Hong Kong's IPO market to close out 2018", *Charles Li Direct* on the HKEX website, 20 December 2018 (https://www.hkexgroup.com/Media-Centre/Charles-Li-Direct/2018/Reflections-on-Hong-Kongs-IPO-market-to-close-out-2018?sc_lang=en).

Take biotech companies as an example. Biotech companies usually require massive investment in R&D before any revenue is generated, as they have to go through a long process of clinical trials and regulatory approvals. The new listing rules allow innovative biotech companies to bypass the financial eligibility tests of the Main Board to acquire the equity funding they need. The SEHK has formed a Biotech Advisory Panel consisting of industry experts to assist in its review of listing applications from biotech companies applying under the new regime.

Hong Kong is known as a market that treats investors fairly and protects their interests to the greatest extent possible from potential malfeasance. In line with the current regime in respect of investor protection, the new listing rules for innovative companies with WVR structures have significantly changed the ways available for shareholders to acquire their controlling stakes. In addition to the traditional ways based on simple financial capital injections, the new listing rules allow shareholders to use other ways, such as human capital (like intellectual property and new business models) to establish their control, without any change to how minority shareholders' interests are being protected.

To help innovative companies, including pre-revenue biotech companies and innovative companies with non-standard share structures, that already have primary listings on the New York Stock Exchange, NASDAQ or London Stock Exchange, to benefit from more flexible listing requirements, the new listing rules established a new secondary listing route for companies in Greater China and international companies that seek a secondary listing in Hong Kong. The change creates a new financing channel for public companies from the new-economy sectors, especially those from Mainland China, and provides Mainland investors with more access to these companies through the Connectivity framework between the Mainland and Hong Kong.

According to HKEX's statistics, a total of 40 new-economy companies conducted new share issues and listed in Hong Kong in just one year after the reform (for the period of 30 April 2018 to 30 April 2019), most of which were from Mainland China. These newly listed companies raised HK$150.4 billion in total, accounting for 53% of the total IPO funds raised during the same period. In particular, in July 2018 smartphone maker Xiaomi became the first company with dual-class share structure listed in Hong Kong. It was followed in September 2018 by the online food delivery service Meituan Dianping, another company with a WVR structure. Meanwhile, nine biotech companies, including Shanghai medical technology platform WuXi AppTec and Beijing cancer drug manufacturer Beigene, have gone public in Hong Kong, raising HK$32 billion in total. Seven out of the nine are pre-revenue biotech companies, raising a total of HK$22.3 billion through the new listing chapter.

4.2　Connectivity framework sets to amplify the benefits of the new listing regime

Over a long history in time, China's economic growth has been offering Hong Kong tremendous opportunities to develop and advance its financial market. The Stock Connect scheme, comprising the Shanghai-Hong Kong Stock Connect and the Shenzhen-Hong Kong Stock Connect, represents a historic development for Mainland China and Hong Kong, as it provides a unique platform for investors in Mainland China and Hong Kong to mutually access each other's stock market via their own brokers. The Stock Connect scheme is set to enhance the benefits of the new listing regime and attract more new-economy companies from Mainland China, as it not only allows new-economy companies in Mainland China to access the equity financing in Hong Kong, but also provides a channel for these companies to access financing from Mainland investors as well.

Hong Kong is a leading international financial centre on the doorstep of Mainland China. When compared to Mainland exchanges, the Hong Kong capital market is much more international and market-oriented. When compared to other major international listing venues, in particular the US bourses, Hong Kong is close to the Mainland and more connected to it in terms of culture, language and trading habits. As such, Hong Kong has been a compelling listing and fund-raising venue for Mainland companies seeking to go public in offshore markets. As of December 2018, 1,146 Mainland companies were listed in Hong Kong, comprising H-share, red-chip and private companies, with a total market capitalisation of around US$2.6 trillion, or 68% of the market total[5].

Thanks to the new listing regime rolled out in 2018, Hong Kong has become a more attractive listing venue for new-economy companies from Mainland China. Meanwhile, the Stock Connect scheme between Mainland China and Hong Kong could amplify the merits of the new listing regime by allowing new-economy companies to access a diverse global investor base, including onshore investors who have more understanding of the growing trend and potential of Mainland new-economy companies, and offshore investors who are mainly institutional investors. According to HKEX's Cash Market Transaction Survey 2018, institutional investors (local and overseas) contributed 55% to total market turnover in the HKEX securities market in 2018. In contrast, individual investors contributed 82% of the Shanghai Stock Exchange's total market turnover in 2017, according to the Shanghai Stock Exchange 2018 Statistical Yearbook. As institutional investors usually provide

---

5　Based on information from The Hong Kong Trade Development Council (http://hong-kong-economy-research. hktdc.com/business-news/article/Market-Environment/Economic-and-Trade-Information-on-Hong-Kong/etihk/ en/1/1X000000/1X09OVUL.htm).

more sticky investment than individual investors, Hong Kong could provide more stable financing for new-economy companies along their growth trajectory.

4.3   Bond Connect facilitates foreign access to Mainland new-economy companies

Bond Connect is a new Mainland-Hong Kong mutual market access scheme that allows investors from Mainland China and overseas to trade in each other's bond markets through connection of financial infrastructure institutions between Mainland China and Hong Kong. Northbound Trading commenced on 3 July 2017, allowing overseas investors from Hong Kong and other regions to invest in the China Interbank Bond Market (CIBM) through mutual access arrangements in respect of trading, custody and settlement. According to the People's Bank of China (PBoC) and the Hong Kong Monetary Authority (HKMA), Southbound Trading will be explored at a later stage in coming years, to allow Mainland Chinese investors to invest in bond markets in Hong Kong and the rest of the world.

Prior to the launch of Bond Connect, foreign investors needed to go through a lengthy and complicated process of opening trading accounts, applying for Renminbi (RMB) quotas and finding a clearing agent to deal with related international settlement. Through Bond Connect, foreign investors are able to trade bonds on the CIBM directly and hence have an easier access to bond-related investment in China's new-economy sectors. Eligible investors for the scheme are mainly financial institutions, such as banks, insurance companies, brokerages and asset management companies. Mainland China's bond market is now valued at RMB 62.1 trillion as of July 2019, of which around RMB 1.7 trillion was held by foreign investors, accounting for 2.7% of the total; most of the foreign holdings are in government bonds and policy bank bonds[6]. The proportion is much lower than in major advanced countries, such as the US (25%) and Japan (12%), and also well below that in some emerging market economies, such as Korea (6%) and Brazil (5%)[7], signalling a huge potential for Mainland China to further open up its bond market and attract more foreign investors.

---

6   Source: Wind database, data is regularly released by ChinaBond.

7   See China's State Administration of Foreign Exchange (2019), "2018 annual report on China's balance of payment" (http://m.safe.gov.cn/safe/file/file/20190329/bdcf1afe0d5d41838f9f9554721494c6.pdf).

4.4   The rising offshore corporate dollar bond market for Chinese companies

In recent years, a rising number of Chinese companies flocked to offshore bond markets, especially Hong Kong's dollar bond markets, to raise hard-currency funds. These markets provide companies from Mainland China, and those from around the world, an alternative debt financing solution to fund their domestic investment projects, to facilitate their offshore mergers and acquisitions, and to hedge the risk of exchange rate fluctuations for their cross-border businesses. The channel may be especially helpful for Chinese new-economy enterprises which usually have limited access to onshore bank lending and direct financing markets. From a macro perspective, it may also facilitate, to a certain extent, the transformation of China's industrial structure towards high-tech manufacturing and service industries.

We define Chinese "offshore corporate dollar bonds" (OCDBs) as all bonds issued in currencies other than RMB in the offshore markets by Chinese companies (non-government and non-policy banks)[8]. Chinese companies started issuing OCDBs in the 1990s but OCDB issuance was quite limited before 2008 (e.g. at just around US$8.4 billion in 2008). Over the past decade, Chinese companies have experienced three periods of explosive growth in OCDB issuance. These are reviewed in the following sub-sections.

4.4.1   The first wave: Q3 2009 to Q2 2011, led by property developers

The Chinese government launched the "Four-Trillion Yuan Stimulus" amid the 2008 global financial crisis to stem the growth slowdown, and gradually tapered it between late 2009 and 2010. Facing increasing barriers in raising funds onshore, Chinese corporates began to take advantage of both the extremely low funding costs in the offshore dollar bond markets and the stable RMB exchange rate against the US dollar (USD). The pace of Chinese OCDB issuance significantly picked up in late 2009, with the quarterly average issuance surging to US$8.6 billion between Q3 2009 and Q2 2011 from US$0.7 billion in the preceding four quarters. This was mainly led by property developers due to a booming property market but tighter credit supply for them. The wave receded in late 2011 as the Chinese government took increasingly stringent policy measures to cool the red hot property market during 2010 to 2011.

---

8   For data on OCDB given in this section, see Ting Lu, Wendy Chen and Lisheng Wang. (2018) "China Monthly: Exploring more obscure financing channels", *Nomura Asia Insights*; Ting Lu, Lisheng Wang, Wendy Chen and Jing Wang. (2018) "A bumpier ride ahead for China's offshore corporate dollar bonds", *Nomura Asia Insights*; and Ting Lu, Lisheng Wang and Jing Wang. (2019) "China: An update on corporate dollar bond financing", *Nomura Asia Insights*.

4.4.2 The second wave: Q1 2012 to Q2 2014, driven by RMB appreciation and arbitrage of interest rate differentials

China pegged its RMB to the USD in late 2008 during the global financial crisis and ended the peg in mid-2010. In the after-crisis era, while borrowing costs declined to historical lows in developed market economies, market interest rates in China surged due to the rapid rise of local government financing vehicles (LGFVs), which are insensitive to borrowing costs, and the rapidly rising zeal of China nationals in buying properties. For a time, the arbitrage trade (borrowing in USD offshore, converting to RMB onshore) boosted hot money flowing into Mainland China, pushing the RMB exchange rate against the USD to its 20-year peak in early 2014 (USD/RMB at around 6.04 in mid-January 2014). According to the PBoC, China's foreign exchange reserves surged all the way to around US$4 trillion in mid-2014.

Against this unique background, the second wave of Chinese OCDB issuance began in Q1 2012, contributing directly to the hot money inflow into Mainland China. Based on our estimates, the quarterly average OCDB issuance jumped to US$21 billion between Q1 2012 and Q2 2014 from US$8.5 billion in the preceding four quarters. The surge of OCDB issuance was mainly driven by general sectors, which we defined as all sectors excluding governments, policy banks, property developers, financial institutions and LGFVs, broadly including consumer, technology, industrial and energy sectors, as well as certain new-economy sectors. Meanwhile, OCDB issuance by property developers and financial institutions also rose significantly.

Since mid-2014, the RMB started to depreciate against the USD and onshore borrowing costs declined due to a growth slowdown and policy easing. As a result, Chinese corporates, especially property developers, slowed down their OCDB issuance and even repaid part of their high-yield offshore bonds to cut costs. The boom of OCDB issuance ended soon.

4.4.3 The third wave: Q2 2016 to present, due partly to the government's lifting of restrictions

Chinese OCDB issuance rebounded in Q2 2016, rising from US$129.8 billion in 2016 to US$249.4 billion in 2017, and remained elevated in 2018 (US$209.6 billion) and H1 2019 (US$130.8 billion). The quarterly average issuance reached US$54.3 billion between Q2 2016 and Q2 2019, up from US$30 billion in the preceding four quarters.

We believe that the following factors were the causes of the surge of Chinese OCDB issuance since Q2 2016. Owing to the significant capital flight after the ill-timed devaluation in August 2015, the Chinese government greatly eased restrictions on corporates' foreign borrowing to ease the capital outflow pressures. The lower offshore funding costs and the rising onshore yields since mid-2016 provided OCDB issuance with another boost. After September 2016, owing to concerns regarding runaway property bubbles and ever-rising local government debt, the Chinese government once again tried to rein in credit supply to property developers and LGFVs. With tighter credit conditions and rising borrowing costs onshore, Chinese corporates found the offshore dollar bond market increasingly attractive.

Compared with the previous booms, OCDB issuance by "general sectors" remained solid in this round, while the biggest driver came from Chinese financial institutions. Chinese financial institutions' outstanding OCDBs leaped from US$54.8 billion at end-2014 to US$280.5 billion at the end of Q2 2019. Chinese banks, brokers and financial conglomerates tapped the offshore dollar bond markets out of various motivations: banks had a need to replenish their capital base, partly for facilitating either their own or their clients' cross-border mergers and acquisitions; brokers eagerly got hold of cash for their collateralised loan business onshore and offshore; and financial conglomerates had an ambition for global business expansions.

Based on our estimates, the size of outstanding Chinese OCDB reached US$841.6 billion as of the end of Q2 2019, which is more than triple its end-2014 size (US$268.7 billion), about half of which were issued via the Hong Kong market. By issuer type, the size of outstanding Chinese OCDBs by "general sectors" was US$322.5 billion as of the end of Q2 2019, accounting for 33% of total Chinese OCDBs, while those by financial companies, property developers and LGFVs were US$280.5 billion (33%), US$166.8 billion (20%) and US$61.9 billion (7%) respectively.

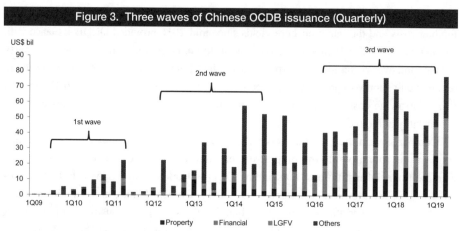

Figure 3. Three waves of Chinese OCDB issuance (Quarterly)

Source: Bloomberg and Nomura Global Economics.

4.4.4 New-economy sectors also benefited from the rise of OCDB markets

A proportion of issuers in the new-economy sectors, such as high-tech manufacturing, e-commerce business, clean and renewable energy, have already joined Hong Kong's dollar bond market. Specifically, issuing offshore dollar bonds could have some branding effects, especially for small- and medium-sized new-economy companies, which may be helpful for increasing their international investor base and thus lowering their financing costs.

Take green bonds, i.e. bonds specifically earmarked to be used for climate and environmental projects, as an example. In recent years, global green finance has grown exponentially, driven by the strong demand for green funding coming from Mainland China, and the strong interest for green investment opportunities from international investors.

Hong Kong is the gateway between the Mainland China and the rest of the world, and hence is uniquely positioned to connect these two markets to capture the opportunities from green finance. Benefitting from these advantages, Hong Kong has grown to be one of the largest green bond markets in the world, with major issuers changing from local institutions to Mainland Chinese and overseas institutions. According to the Hong Kong Green Bond Market Report released by the HKMA in February 2019, green bonds arranged and issued in Hong Kong surged to US$11 billion in 2018 from US$3 billion in 2017, recording a 237% growth. The outstanding green bonds issued by Mainland Chinese and overseas institutions reached US$9 billion in 2018, accounting for 83% of the total amount of green bond issuance in Hong Kong in the year.

We believe that new-economy companies may benefit more from Hong Kong's offshore dollar bond market, amid further integration of global financial markets, continued global growth rebalancing towards new-economy sectors, and more regional cooperation, especially between Mainland China and Hong Kong (e.g. through the development of the Greater Bay Area), and between China and the rest of world (e.g. through China's market opening and the Belt and Road Initiative).

Chapter 3

Stock Connect ushers in a new era for China's capital market

Hao HONG

Head of Research and Managing Director
BOCOM International Holdings Company Limited

Summary

The increasing quotas of Stock Connect, the expansion of the eligible stock range and the establishment of the Northbound see-through regulatory mechanism provide investors with more diversified choices while protecting their interests. With further development and optimisation, Mainland-Hong Kong Stock Connect plays an increasingly significant role in the opening-up of Mainland China's capital market.

Impact of Mainland-Hong Kong Stock Connect on A shares

- Northbound trading under Stock Connect curbs speculative activities in the A-share market. Overseas institutional investors trading through Shanghai and Shenzhen Connect adhere to value investment principles and eligible Northbound securities are less volatile than the overall Mainland market. However, speculative "pump and dump" associated with the Mainland's retail-driven market is not spreading to Hong Kong. Mainland investors accessing the Hong Kong market through Southbound trading seem assimilated, and sometimes behave like overseas investors. Overseas investors' capability to process information and select stocks will improve the efficiency of price discovery in the Mainland market. The growing participation in the Mainland market by international investors will prevent significant deviation in the prices of A shares from underlying fundamentals.

- Stock Connect fails to eliminate A-H share price gap. Limited channels for price arbitrage, restrictions on cross-border capital flows, and different investor structures may have led to the A-H premium. In addition, only small fractions of both the Southbound and Northbound daily quotas are being used by investors, which is much lower than expected. Inactive fund flow limits its impact on A and H shares' valuation convergence. However, the A-H share price gap is narrower for a dual-listed security of higher dividend payment, higher institutional ownership, lower volatility, and lower turnover rate. Theoretically, value investment strategies based on fundamental analysis which are used by overseas investors should be embraced by more and more Mainland investors, given the increasing participation of international investors in Mainland markets. As a result, price differentials between A and H shares should be eliminated if China further opens up its capital market.

- Stock Connect has enhanced the correlation between Mainland and Hong Kong markets. The Shanghai Composite Index and the Hang Seng Index have become more correlated after the launch of the Shanghai-Hong Kong Stock Connect, while the Shenzhen Composite Index and the Hang Seng Index have become less correlated after the launch of Shenzhen-Hong Kong Stock Connect. Given the same macro-economic conditions,

Stock Connect is one of the factors affecting the correlation between the Mainland and Hong Kong markets.

Stock Connect marks a sound step towards the internationalisation of the Mainland capital market

- Stock Connect is an important exploration in the internationalisation of the Mainland stock market. In recent years, channels for overseas investment have been significantly expanded. The expansion of Stock Connect and the optimisation of the scheme pave the way for the internationalisation of A shares. Since the formal inclusion of A shares into the MSCI Emerging Markets Index in 2018, the funds flowing through Northbound trading under Stock Connect have surged in volume. In the long term, overseas funds are expected to bring trillions of incremental funds in Renminbi (RMB) to the A-share market, and these funds will gradually become an important part of the Mainland capital market. As a result, the internationalisation level of the A-share market will be further enhanced.
- Stock Connect is an important part of RMB internationalisation. Stock Connect, in which funds flow and net settlement are conducted in RMB. This not only helps to accelerate the internationalisation of the currency, but also effectively controls the risk associated with fluctuations in cross-border capital flows. As a fund-flow bridge between Hong Kong and the Mainland, it provides investors with greater convenience to comprehensively invest in the Mainland's capital market. It not only expands the investment channels of offshore RMB, but also increases the liquidity of offshore RMB in Hong Kong.

1 Development of Shanghai- and Shenzhen-Hong Kong Stock Connect

1.1 Orderly opening-up of Stock Connect with improvement of the mechanism

In terms of quota control, Shanghai-Hong Kong Stock Connect (Shanghai Connect) initially set two thresholds, including (1) aggregate quota — RMB 250 billion (Southbound trading) and RMB 300 billion (Northbound trading), and (2) daily quota — RMB 10.5 billion (Southbound trading) and RMB 13 billion (Northbound trading). With the launch of Shenzhen-Hong Kong Stock Connect (Shenzhen Connect) by the end of 2016, the aggregate quota limit was abolished, but the daily quota was maintained. In May 2018, the daily quota limit of Stock Connect was quadrupled to RMB 42 billion (Southbound trading) and RMB 52 billion (Northbound trading) as A shares entered the MSCI Emerging Markets Index. Orderly relaxation of the quota limit reflects the further coordinated development of the Mainland and Hong Kong markets under the connectivity mechanism, which plays a pivotal role in the opening-up of the Mainland capital market.

In terms of eligible stocks and investors, Shanghai Connect initially covered only around 500 large- and mid-cap stocks, including constituent stocks in CSI 180 and SSE 380 indices, and A+H shares listed on both the Shanghai Stock Exchange (SSE) and the Stock Exchange of Hong Kong (SEHK). Individual investors with an asset balance of not less than RMB 500,000 in their securities account will be allowed for Southbound trading. However, the stock coverage as a result of Shenzhen Connect has been expanded to include small- and mid-cap stocks. Around 900 stocks including stocks listed on the Main Board, SME Board and ChiNext of the Shenzhen Stock Exchange (SZSE), with a market capitalisation of more than RMB 6 billion, are eligible for Northbound trading. Investor criteria have also been raised: only professional institutional investors can invest in the ChiNext market. The expansion of the eligible stock range will provide investors with more diversified choices while protecting their interests.

In September 2018, the Northbound see-through mechanism for Stock Connect was officially launched, followed by the introduction of an investor identification regime for Northbound trading in October, in order to fence off transactions that do not meet the requirements. Regulation has attracted attention at the same time as the expansion of the Stock Connect quota to attract more foreign investment to the A-share market. The establishment of the see-through mechanism will improve the regulatory efficiency of cross-border markets, maintain the stable operation of the market, and create a favourable trading environment for investors.

2　Impact of Mainland-Hong Kong Stock Connect on A shares

2.1　Stock Connect curbs speculative activities in A shares

Stock Connect enhances the reciprocal opening of the Mainland and Hong Kong markets, attracts overseas investors to invest in Mainland A shares, and facilitates the entry of Mainland investors into overseas markets. It was also expected that Stock Connect would induce the entry of new market participants with different investment ideologies and preferences, as a result of which investment cultures of the local markets may be reshaped. In reality, overseas investors trading through Shanghai and Shenzhen Connect still adhere to value-investing principles and eligible Northbound securities tend to be less volatile than the overall Mainland market. The A-share market seems to be more like Hong Kong under Stock Connect. However, the kind of speculative "pump and dump" associated with the Mainland's retail-driven market has not spread to Hong Kong. Mainland investors accessing the Hong Kong market through Southbound trading seem assimilated, and sometimes behave like overseas investors.

Overseas investors' capability to process information and select stocks will improve the efficiency of price discovery in the Mainland market. The growing participation in the Mainland market by international investors will prevent prices of A shares from deviating significantly from underlying fundamentals caused by "pump and dump" activities. Compared with Mainland investors, overseas investors take a medium- to long-term view and focus more on sustainability. The increased presence of these international institutions has likely helped to tip the scales away from sentiment-driven local retail investors, and the A-share market will become less volatile. Meanwhile, overseas investors favour listed companies with greater transparency, better financial conditions and lower volatility.

Therefore, overseas investors and less volatile markets are related by mutual causality.

To assess the impact of Shanghai Connect on the Mainland market, we divide the Mainland securities into two groups. The first group includes eligible securities for Northbound trading under Shanghai/Shenzhen Connect; the rest of SSE/SZSE securities are grouped into the second one. The sample period of January 2005 to July 2019 is divided into two sub-periods in order to investigate how the introduction of Shanghai Connect impacts the Shanghai market's behaviour. The first sub-period, which is referred to as the pre-Shanghai Connect period, is from 2005 to 2014. The second sub-period, which is referred to as the post-Shanghai Connect period, is from 2015 to July 2019. Then, we compare the average annualised volatility and turnover rates[1] of the two groups of securities in each sub-period to find out if Shanghai Connect efficiently reduces volatility and excessive trading in the Shanghai market. The same method is applied to examine the impact of Shenzhen Connect as well.

The growing participation in the A-share market by overseas investors has indeed reduced volatility and the risk of excessive trading in the Mainland market. After the launch of Shanghai Connect in November 2014, the volatility of Northbound securities under Shanghai Connect has increased, but to a lesser degree than other SSE securities. After the launch of Shenzhen Connect in December 2016, the volatility of securities under Shenzhen Connect has declined, and to a larger extent than other SZSE securities. It means that Stock Connect prevents market instability from excessive speculation such as chasing winners and cutting losers. On the other hand, Stock Connect also helps to avoid excessive trading in the A-share market since the turnover rate of securities under Shanghai Connect rose slightly, and to a lesser extent than other SSE securities, and the turnover rate of securities under Shenzhen Connect declined, and to a larger extent than other SZSE securities. Overseas investors are more likely to invest for long-term value creation with "buy and hold" mentality, and their preference to trade less frequently can help suppress the turnover rate of the A-share market.

---

1   For SSE/SZSE securities, the turnover rate refers to the average daily turnover rate in each year.

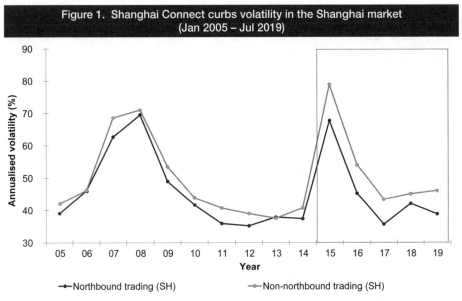

Figure 1. Shanghai Connect curbs volatility in the Shanghai market (Jan 2005 – Jul 2019)

Source: Wind, BOCOM International.

Figure 2. Shenzhen Connect lowers volatility in the Shenzhen market (Jan 2005 – Jul 2019)

Source: Wind, BOCOM International.

387

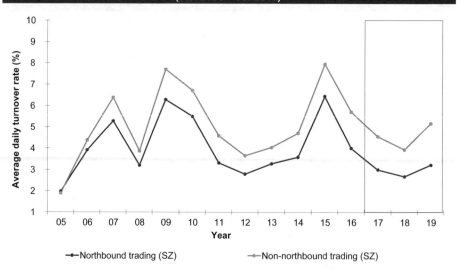

Figure 3. Shanghai Connect curbs turnover rate in the Shanghai market (Jan 2005 – Jul 2019)

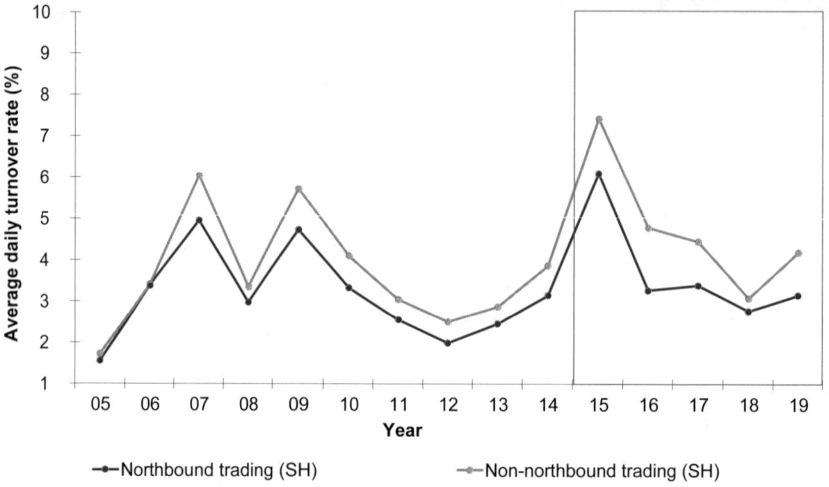

Source: Wind, BOCOM International.

Figure 4. Shenzhen Connect decreases turnover rate in the Shenzhen market (Jan 2005 – Jul 2019)

Source: Wind, BOCOM International.

Mainland investors accessing the Hong Kong market via Stock Connect seems to adopt a "trading" rather than "holding" approach, but they do not add to the volatility of Southbound securities under Stock Connect. After the launch of Shanghai Connect in November 2014, the average volatility of eligible securities for Southbound trading has decreased, and to a larger extent than other SEHK securities. After the launch of Shenzhen Connect in December 2016, the scope of eligible Southbound securities has expanded and the average volatility has decreased further along with other SEHK securities, and to a similar extent. It shows that the speculative Mainland investment style of chasing winners and cutting losers does not impact the Hong Kong market. However, funds from the Mainland market trade frequently. After the launch of Stock Connect, the average turnover rate[2] of Hong Kong Connect securities has decreased, but to a lesser extent than other SEHK securities.

Figure 5. Shanghai Connect decreases volatility in the Hong Kong market (Jan 2005 – Jul 2019)

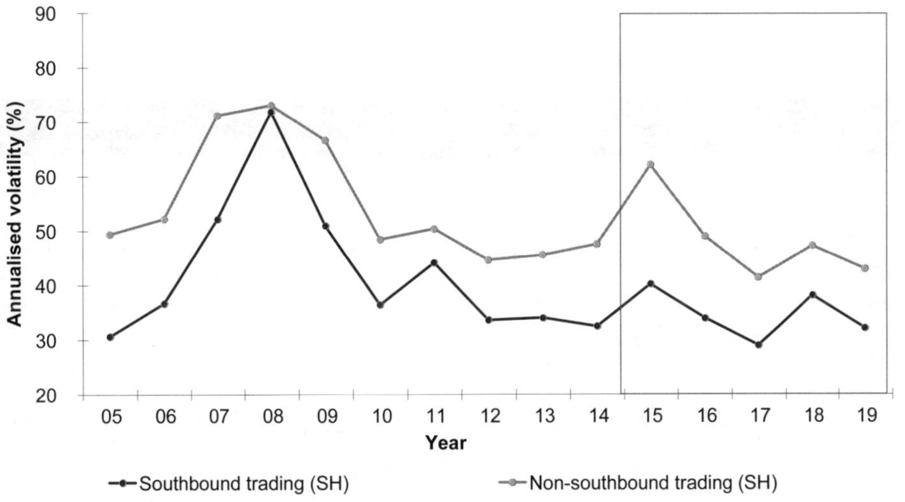

Source: Wind, BOCOM International.

---

2   For SEHK securities, the turnover rate refers to the average annualised turnover rates.

Figure 6. Shenzhen Connect does not add to volatility in the Hong Kong market
(Jan 2005 – Jul 2019)

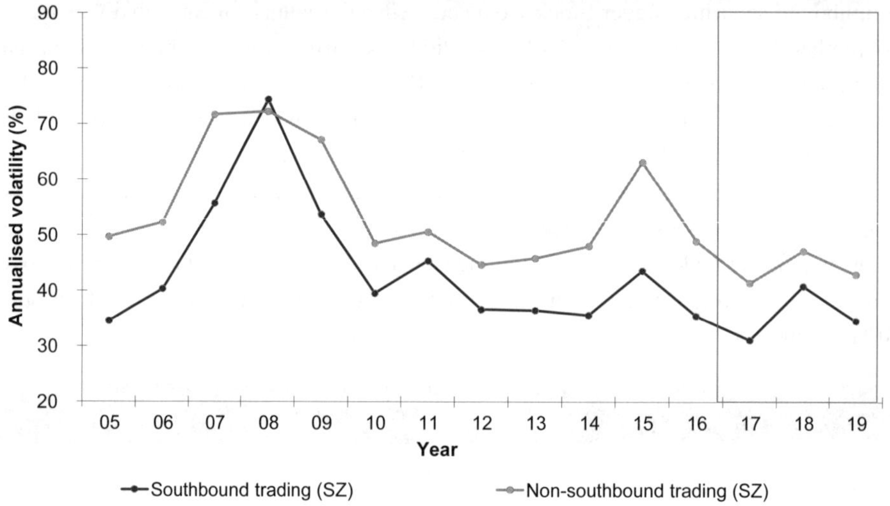

Source: Wind, BOCOM International.

Figure 7. Shanghai Connect increases turnover rate in the Hong Kong market
(Jan 2005 – Jul 2019)

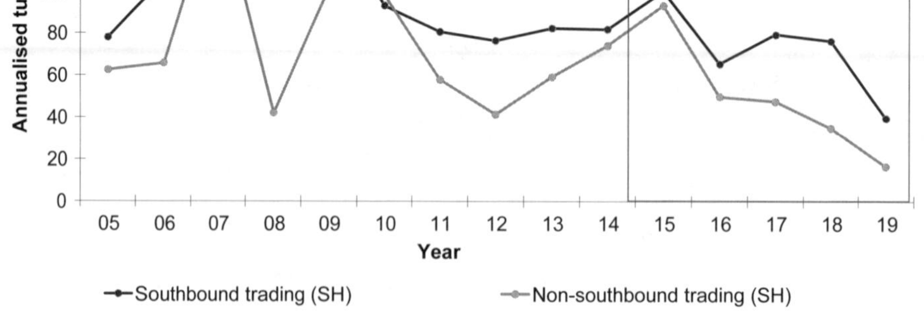

Source: Wind, BOCOM International.

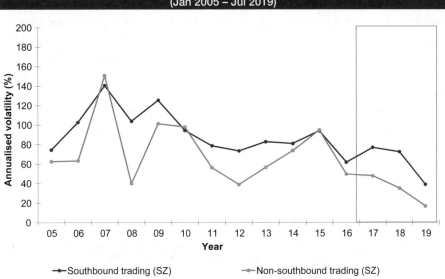

Figure 8. Shenzhen Connect increases turnover rate in Hong Kong market
(Jan 2005 – Jul 2019)

Source: Wind, BOCOM International.

2.2   Stock Connect fails to eliminate A-H share price gap

Although A and H shares are identical with respect to shareholder rights, such as voting and profit-sharing, large and persistent price differentials between the dual-listed A-H stocks have existed for a long time. Some expected that the prices of A and H shares would converge with the development of Stock Connect, since the expectation of price convergence would attract investors to buy the undervalued and sell the overvalued. Although the stock valuations of the Mainland and Hong Kong markets used to be different, due to varied investment style and investor structures, it was believed that such differences will be eliminated over time with the increasing cross-border capital flows under Stock Connect, so that price gaps between A and H shares would diminish as well. However, in reality, A shares have always been trading at a premium to H shares since the launch of Stock Connect and the price gaps have not been removed as expected.

Limited channels for price arbitrage, restrictions on cross-border capital flows, and different investor structures may have led to A-H price premium. Since the launch of Shanghai Connect, the prices of A and H shares remain divergent instead of converging, and the Hang Seng China AH Premium Index has been oscillating within a range of 110-150 points. While Stock Connect is the reciprocal opening of both markets with Mainland investors accessing the Hong Kong market via Southbound trading and overseas investors accessing the A-share market via Northbound trading, there is no formal channel for price arbitrage by cross-market conversion between the A and H shares of a dual-listed company. In addition, cross-border capital flows are regulated strictly and only small fractions of both the Southbound and Northbound daily quotas are being used by investors, which is much lower than expected. Besides, the inactive fund flows have limited impact on A and H shares' valuation convergence. We also noted that Mainland investors favour medium and small caps as well as growth stocks, while such stocks are not popular in Hong Kong. This may have resulted from the different investor structures and investment styles of these two markets. The Mainland market is driven by speculative retail investors who account for 80% of trading value in the A-share market, while the Hong Kong market is dominated by institutional investors adopting value investment and fundamental analysis — over 80% of trading value in the Hong Kong cash market is attributed to institutional investors (principal trading of Stock Exchange Participants (EPs) is included)[3].

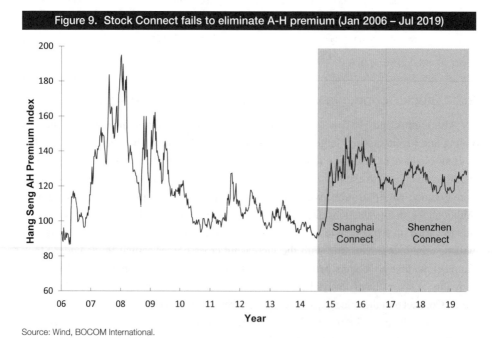

Figure 9. Stock Connect fails to eliminate A-H premium (Jan 2006 – Jul 2019)

Source: Wind, BOCOM International.

---

3   Source: HKEX's *Cash Market Transaction Survey 2018*, published on HKEX's website, July 2019.

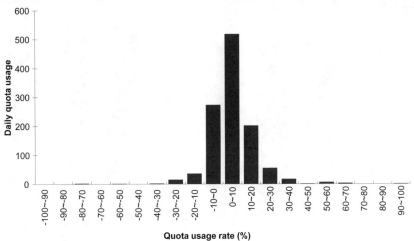

Figure 10. Daily quota usage of Shanghai and Shenzhen Connect is lower than expected (17 Nov 2014 – 31 Jul 2019)

Note: A negative quota usage rate means that sell trades are more than buy orders.

Source: Wind, BOCOM International.

Figure 11. Distribution of Hong Kong cash market trading value by investor type (2010 – 2018)

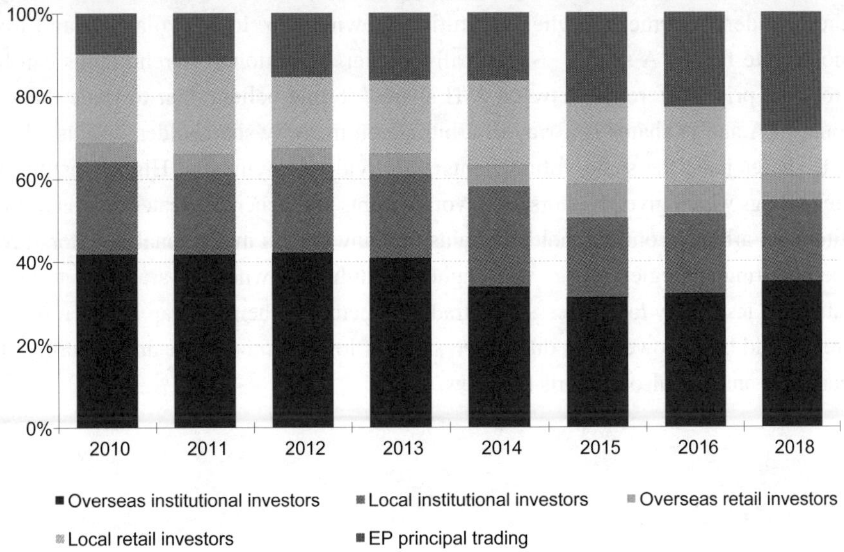

- Overseas institutional investors
- Local institutional investors
- Overseas retail investors
- Local retail investors
- EP principal trading

Source: HKEX Cash Market Transaction Surveys.

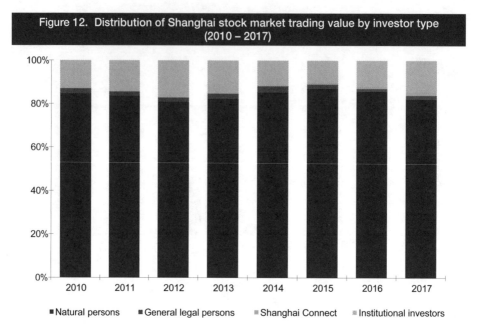

Figure 12. Distribution of Shanghai stock market trading value by investor type (2010 – 2017)

■ Natural persons　　■ General legal persons　　■ Shanghai Connect　　■ Institutional investors

Source: SSE Fact Books, Wind.

However, the A- and H-share price gap has narrowed for a dual-listed company of higher dividend payment, higher institutional ownership, lower volatility and lower turnover rate for the A shares. Specifically, higher institutional shareholdings can help narrow the price difference between A/H shares. Some believe that to trade the same amount of A and H shares is a way of arbitrage on majority shareholders' rights, since A and H shares have the same voting rights and dividend payments. Higher institutional shareholdings which give investors more voting rights will generate greater returns from the institutional arbitrage on shareholders' rights. Meanwhile, as institutional investors favour value-investing strategies, stocks with higher institutional ownership are likely to be less volatile and less likely to be excessively traded. Therefore, the price gap between A and H shares should be narrower for a dual-listed stock of lower turnover rate and volatility. Our quantitative analysis also supports this view.

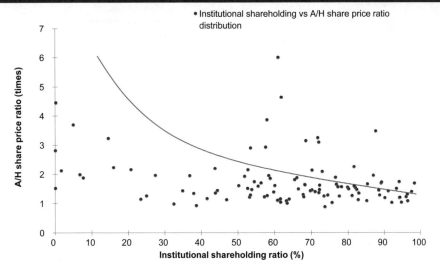

Figure 13. Institutional shareholding is negatively correlated with A-H premium (Jan 2018 – Dec 2018)

Source: Wind, BOCOM International.

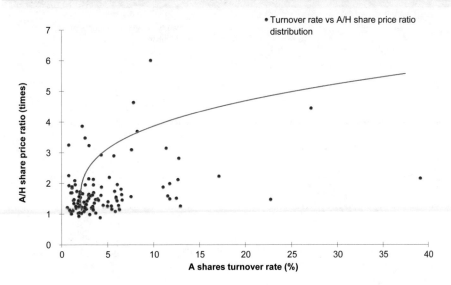

Figure 14. Turnover rate is positively correlated with A-H premium (Jan 2018 – Dec 2018)

Source: Wind, BOCOM International.

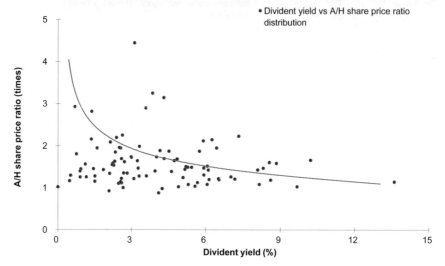

Figure 15. Dividend payout is negatively correlated with A-H premium
(Jan 2018 – Dec 2018)

Source: Wind, BOCOM International.

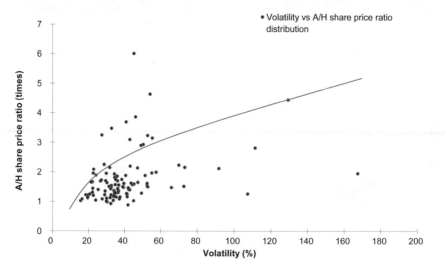

Figure 16. Volatility is positively correlated with A-H premium
(Jan 2018 – Dec 2018)

Source: Wind, BOCOM International.

The A-share market is to be subject to greater influence from overseas investors through reshaping the domestic investment approach and trading strategies. Theoretically, value-investing strategies that are based on fundamental analysis used by overseas investors should be embraced by more and more Mainland investors, given the increasing participation of international investors in the Mainland market. As a result, price differentials between A and H shares should be eliminated if Mainland China further opens up its capital market. Stock Connect, as a major part of China's opening-up, will accelerate capital flows between the Mainland and Hong Kong markets with its further expansion. In the short term, it is still difficult to achieve complete convergence of AH stock valuations due to the restrictions of investment quota and market openness. However, as the country pushes for greater opening-up of its capital market, A and H shares will converge to their equilibrium valuation. The valuation convergence between A and H shares of dual-listed companies will depend on the degree of market openness and investor structure improvement.

2.3 Stock Connect strengthens the correlation between the Mainland and Hong Kong markets

Stock Connect enhanced the correlation between the Mainland and Hong Kong markets. Theoretically, stock market synchronisation can be explained by macro-economic conditions, financial market contagion and investor behaviour. Financial markets are pro-cyclical and contagious. Given an increase in the strength of economic linkages across world economies, it is very likely that region-specific shocks will be transmitted internationally with greater scope and strength through capital flows, exchange rates and market sentiment. Meanwhile, a rising correlation of countries' business cycles also strengthens stock market co-movements. As a result, stock markets co-move more frequently than before. Besides, correlation between stock markets can also be strengthened by herding effects as market participants imitate each other and follow a collective action due to information asymmetry.

Stock Connect plays a more important role in the synchronisation between the Shanghai and Hong Kong markets than between the Shenzhen and Hong Kong markets. Theoretically, a greater degree of financial integration may induce more correlated stock markets. In reality, the SSE Composite Index and the Hang Seng Index have higher co-movement after the launch of Shanghai Connect. If we focus on eligible Northbound stocks under Shanghai Connect, their correlation with Hong Kong has increased even more. However, the SZSE Composite Index and the Hang Seng Index have become less correlated after the launch of Shenzhen Connect. The correlation between Hang Seng

Index and eligible Northbound stocks under Shenzhen Connect even changed from positive to negative at times. This suggests that Stock Connect is one of the factors affecting the correlation between the Mainland and Hong Kong markets, given the same macro-economic conditions.

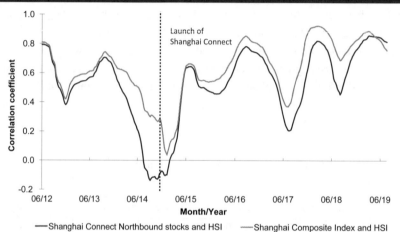

Figure 17. Shanghai Connect strengthens the correlation between Shanghai and Hong Kong markets (Jun 2012 – Jul 2019)

Source: Wind, BOCOM International.

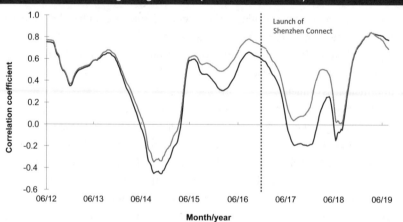

Figure 18. Shenzhen Connect weakens the correlation between Shenzhen and Hong Kong markets (Jun 2012 – Jul 2019)

Source: Wind, BOCOM International.

From the perspective of tail effect, the probability of synchronous rises of the Mainland and Hong Kong markets is higher than that of synchronous drops. However, the probability of synchronous slump is higher than that of synchronous boom. Since the launch of Shanghai Connect (based on data for the period from 17 November 2014 to 31 July 2019), the probability of the Shanghai stock market rising in step with Hong Kong is 37%, which is higher than that of falling in step (30%). The probability that both markets rise more than 2% on the same day is 1.6%, less than the probability that they fall more than 2% on the same day, which is 2.0%. Since the launch of Shenzhen Connect (based on data for the period from 5 December 2016 to 31 July 2019), the probability of the Shenzhen stock market rising in step with Hong Kong is 37% and that of falling together is 31%. The probability that the two markets rise by more than 2% on the same day is 0.8%, less than the probability that they fall by more than 2% on the same day, which is 1.4%.

Figure 19. With Shanghai Connect, the probability of synchronous slump is higher than synchronous boom in Shanghai and Hong Kong markets (Nov 2014 – Jul 2019)

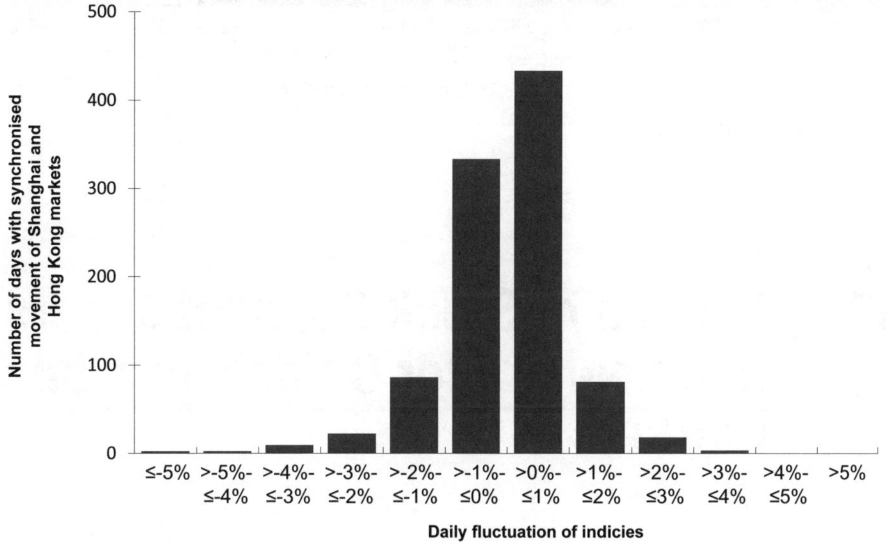

Source: Wind, BOCOM International.

Figure 20. With Shenzhen Connect, the probability of synchronous slump is higher than that of synchronous boom in Shenzhen and Hong Kong markets (Dec 2016 – Jul 2019)

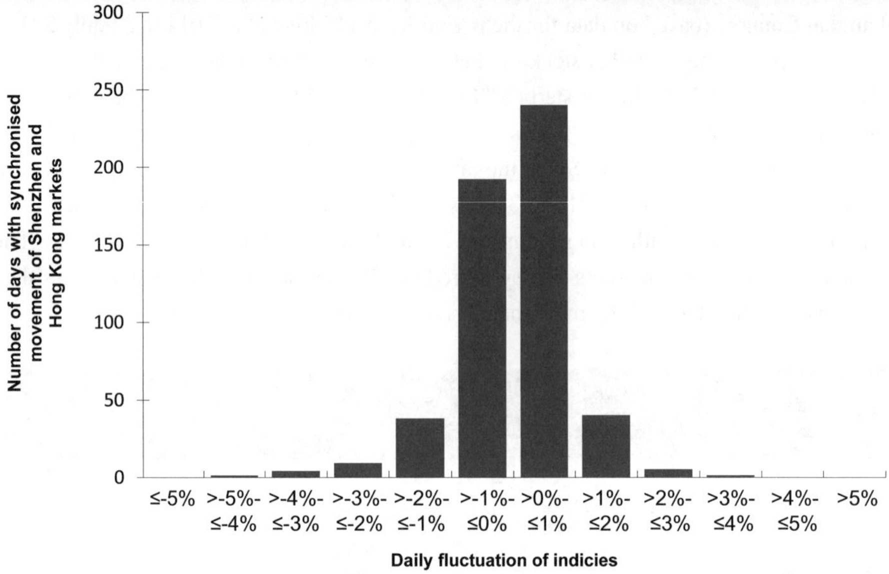

Source: Wind, BOCOM International.

3 Launch of Stock Connect ushers in a new era for China's capital markets

3.1 Stock Connect marks a sound step towards the internationalisation of the Mainland stock market

Shanghai and Shenzhen Connect, Renminbi Qualified Foreign Institutional Investor (RQFII) and Qualified Foreign Institutional Investor (QFII) schemes, and the conversion of B shares into H shares are the footprints of the two-way opening-up of the Mainland capital market. In recent years, channels for the entry of foreign investments have been drastically expanded. The investment quota of RQFII increased from RMB 10 billion to nearly RMB 300 billion in less than three years, and even doubled in less than 12 months in 2014. Recently, the State Administration of Foreign Exchange (SAFE) has also delivered the message that it will "reform the QFII and RQFII schemes, expand the investment scope, and explore appropriate QFII quota relaxation and even the removal of QFII quota management". The continuous enhancement of the Stock Connect scheme has made it an important milestone of the opening-up of the Mainland capital market.

The Shanghai Connect was officially launched in November 2014 with an initial aggregate quota of RMB 550 billion and a daily Northbound and Southbound combined quota of RMB 23.5 billion. The restriction of aggregate quota was removed upon the launch of Shenzhen Connect in December 2016, and the daily quota was further quadrupled in April 2018 as A shares successfully made their way into the MSCI Emerging Markets Index. The gradual expansion of the quota greatly improves market liquidity, and also reflects overseas investors' recognition of, and confidence in, the achievements of Stock Connect. In addition, the introduction of the investor identification regime for Northbound trading and the inclusion of companies with weighted voting rights into Southbound trading show China's determination to further open up its capital market.

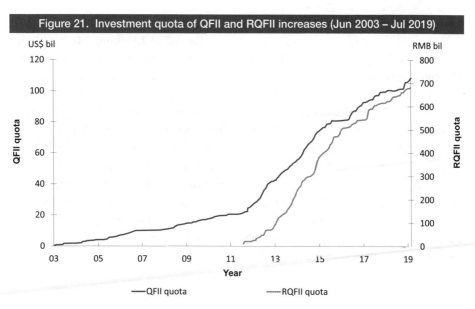

Figure 21. Investment quota of QFII and RQFII increases (Jun 2003 – Jul 2019)

Source: Wind, BOCOM International.

Figure 22. Foreign funds continue to flow into A-share market through Stock Connect (Nov 2014 – Jul 2019)

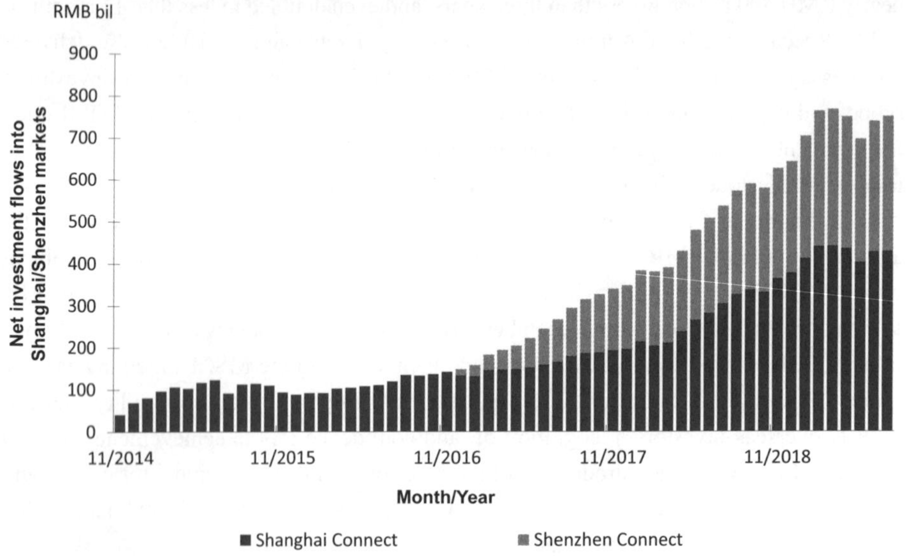

Source: Wind, BOCOM International.

Since the formal inclusion of A shares into the MSCI Emerging Markets Index in 2018, Northbound fund flows through Stock Connect have increased significantly. The cumulative turnover in 2019 up to 31 July had reached RMB 5.6 trillion, with an average daily turnover of RMB 41 billion, which doubled that in 2018. The successful launch of Stock Connect has facilitated the inclusion of A shares into the global index system, and also marked significant headway in the opening-up of the Mainland capital market. Following the inclusion of A shares into MSCI and the subsequent increase in inclusion factor, A shares have also been added to the FTSE Russell Index in June 2019. Furthermore, A shares are expected to be included in S&P Dow Jones Indices in September the same year. In the long term, overseas funds are expected to bring trillions of incremental funds in RMB to the A-share market, and these funds will gradually become integral to the Mainland capital market. As a result, the internationalisation level of the A-share market will be further enhanced.

3.2   Stock Connect is an important part of RMB internationalisation

After World War II, the US dollar became the leading international currency in the new international monetary system established through the Bretton Woods agreement. Afterwards, the US significantly increased the supply of offshore dollars through the Marshall plan, creating a cycle of onshore/offshore dollar flows. Although the collapse of the Bretton Woods system in the 1970s dented market confidence in the US dollar, the petrodollar system finally elevated the US dollar to the pricing and clearing currency for international trade. On the other side, the reason behind the failure of the internationalisation of the Japanese yen is that Japan failed to maintain its dominant position in the development of the industry chain in Asia. Of course, that Japan was unable to increase the supply of offshore yen only through the expansion of yen loans is another major factor. China could draw lessons from the success and failure of the internationalisation of these two major currencies.

The launch of cross-border trade settlement in RMB to facilitate international business activities and increase countries' willingness to hold RMB is an essential step in the internationalisation of the RMB. The introduction of Shanghai Connect, in which Northbound funds flow and net settlement are conducted in RMB, has minimised the impact on the exchange rate of onshore RMB. As a result, Stock Connect not only helps to accelerate RMB internationalisation, but also effectively contains the risk associated with fluctuations in cross-border capital flows. Hong Kong, as the largest offshore RMB centre, helps Stock Connect to realise the investment cycle of the RMB. The subsequently launched Shenzhen Connect also replicated the cross-border trade settlement mechanism of its precursor. Besides becoming a globally-recognised pricing unit, settlement currency and storage currency, popularising the use of the RMB in investment activities in the financial sector is also an important part of RMB internationalisation. Stock Connect, as a fund-flow bridge between Hong Kong and the Mainland, provides investors with greater convenience to comprehensively invest in the Mainland capital market. It not only expands the investment channels of offshore RMB, but also increases the liquidity of offshore RMB in Hong Kong.

Part 2

The listing reform and financing innovation for new-economy companies

Chapter 4

Initial public offering (IPO) and listing process on the SEHK with highlights

Chief China Economist's Office

and

Global Issuer Services

Hong Kong Exchanges and Clearing Limited

Summary

This chapter presents in a succinct way the initial public offering (IPO) and listing process of enterprises on the Stock Exchange of Hong Kong, i.e. on the securities market operated by the Hong Kong Exchanges and Clearing Limited. Enterprises, in particular Mainland enterprises, seeking a listing in Hong Kong may decide on a choice among various shareholding structures.

Apart from the listing requirements of the Main Board and GEM, the chapter also introduces the professional services that an enterprise requires in the IPO and listing process, and the ways for equity re-financing after listing.

1 General requirements for listing in Hong Kong

In general, companies in any business sector seeking a listing on the Main Board of the Stock Exchange of Hong Kong Limited (the "SEHK", or the "Exchange")[1] shall comply with the set of conditions set out in Chapter 8 of the *Rules Governing the Listing of Securities on the Exchange* (the "Listing Rules", or the "Rules"). It should be noted that the Exchange retains an absolute discretion to accept or reject an application for listing and that compliance with the relevant conditions may not of itself ensure an applicant's suitability for listing.

1.1   Main Board listing conditions

The Main Board market in Hong Kong is designed for more established companies with a certain level of market capitalisation/revenue/profit scale. Listed companies range from conglomerates, banks, property developers to Internet companies and healthcare service providers. Companies applying for listing on Main Board shall meet the following basic conditions:

(1) Financial requirements (to meet at least one of the following three sets of test)[2]

Profit test
- the profit attributable to shareholders must, in respect of the most recent financial year before listing, be not less than HK$20 million;
- the profit attributable to shareholders must, in respect of the two preceding years, be in aggregate not less than HK$30 million; and
- a market capitalisation of at least HK$500 million at the time of listing[3].

Market capitalisation/revenue test
- revenue of at least HK$500 million for the most recent financial year before listing; and

---

1   Unless otherwise stated, references in this chapter to "listing" means an initial public offering (IPO) and listing on the Main Board of the SEHK.
2   See Listing Rule 8.05.
3   See Listing Rule 8.09(2).

- a market capitalisation of at least HK$4 billion at the time of listing.

Market capitalisation/revenue/cash flow test

- revenue of at least HK$500 million for the most recent financial year before listing;
- positive cash flow from operating activities that are to be listed of at least HK$100 million in aggregate for the three preceding financial years; and
- a market capitalisation of at least HK$2 billion at the time of listing.

(2) Other requirements

Apart from the above financial requirements, companies applying for listing in Hong Kong shall also meet a set of basic requirements in relation to the business history, public float, management and control, including the followings:

- trading record period of at least three financial years (a shorter trading record period may be considered in exceptional cases[4]);
- minimum public float (usually 25%, but may be reduced to 15% at discretion if the market capitalisation at the time of listing is more than HK$10 billion);
- there must be at least 300 shareholders at the time of listing[5];
- no change to the management in the three preceding financial years;
- no change to the ownership and control of the company in the most recent financial year;
- there must be at least three independent directors representing at least one-third of the board[6];
- at least one independent non-executive director (INED) must be ordinarily resident in Hong Kong in the case of H-share applicant;
- at least two of the executive directors must be ordinarily resident in Hong Kong[7];
- a relevant professional be appointed to discharge the functions of a company secretary[8];
- the latest financial period reported on by the reporting accountants must not have ended more than six months before the date of the listing document.

4　See Listing Rule 8.05A.
5　See Listing Rule 8.08.
6　See Listing Rules 3.10 and 3.10A.
7　See Listing Rule 8.12.
8　See Listing Rule 3.28.

1.2   Shareholding structures for listing in Hong Kong

1.2.1   Jurisdictions acceptable as place of incorporation

Applicants seeking to list in Hong Kong must first be incorporated in an "acceptable jurisdiction"[9] admitted by the Exchange. According to the Listing Rules, companies incorporated in Hong Kong, Mainland China, Cayman Islands and Bermuda are eligible to apply for listing. And as of 30 June 2019, there are 27 other "acceptable jurisdictions" for overseas companies, including England & Wales, the British Virgin Islands, France, Italy, Russia, Japan, Singapore, Australia, etc. Overseas applicants shall demonstrate that the standards of shareholder protection of its place of incorporation are at least equivalent to those provided in Hong Kong, or they may have to vary its constitutive documents to provide an equivalent protection.

1.2.2   Shareholding structures of Mainland companies seeking to list in Hong Kong

For applicants based in Mainland China and with principal businesses there, the SEHK accepts a number of various shareholding structures, including:

- **Red-chip structure** — the listed entity is an offshore holding company injected with domestic equity or assets to achieve an indirect overseas listing. The red chip structure includes two modes:
  - (1) **State-owned red-chip** — the actual controller of the Mainland enterprise, assets or businesses is usually a government entity.
  - (2) **Private red-chip** — the actual controller of the Mainland enterprise, assets or businesses is usually an individual.
- **H-share structure** — the listed entity is a joint-stock company incorporated in Mainland China which has its domestic equity directly listed on the SEHK. H-share listings can be done in four ways:
  - (1) **First A and then H** — an A-share company already listed in the Mainland to dually list in Hong Kong;
  - (2) **First H and then A** — a company already listed on the SEHK seeking an A-share dual listing in the Mainland;
  - (3) **A+H at the same time** — a company seeking to list its A-shares in the Mainland and H-shares on the SEHK simultaneously;

---

9   See "List of Acceptable Overseas Jurisdictions" on HKEX's official website for details.

(4) **NEEQ + H** — a company already listed on the National Equities Exchange and Quotations (NEEQ) market in the Mainland to list on the SEHK as well without prior delisting from the NEEQ.

- **Spin-off** — companies already listed in Hong Kong, Mainland China or overseas may spin off a subsidiary or certain businesses for a listing in Hong Kong.

1.2.3 Red-chip structure

Under the red-chip model, an offshore holding company, i.e. a special-purpose vehicle (SPV), is formed and injected with assets and interests of a Mainland enterprise and then listed overseas in its name for fund-raising purpose. Red-chips are commonly classified into "state-owned red-chips" and "private red-chips". Statistics show that most of the overseas-listed Mainland enterprises adopt the red-chip model, especially the private ones.

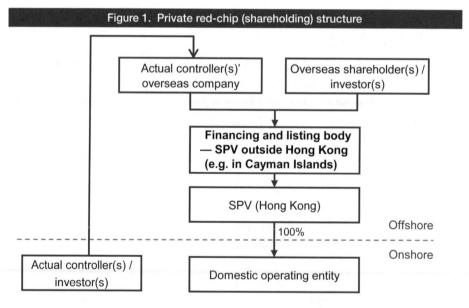

Figure 1. Private red-chip (shareholding) structure

Note: For illustration only. There may be differences in practice.

1.2.4   Variable Interest Entity (VIE) structure

The SEHK adopts a disclosure-based approach in reviewing listing applications submitted under the VIE structure, considering major factors including the followings:

- The VIE structure shall be "narrowly tailored" and only be used to address limits on foreign ownership, such that the contractual arrangements shall be terminated when those limits no longer exist[10];
- Minimise the potential of conflict with relevant laws and regulations of the People's Republic of China (PRC);
- Where practicable, appropriate regulatory assurance should be obtained from the relevant regulators by the listing applicant;
- Terminate the contractual arrangements as soon as the law allows the business to be operated without them;
- Ensure that the contractual arrangements: (1) include a power of attorney by which the domestic operating entity's shareholders grant to the listing applicant's directors and their successors the power to exercise all rights of the domestic operating entity's shareholders, and ensure that the power of attorney does not give rise to any potential conflicts of interest; (2) contain dispute resolution clauses; (3) encompass the power to deal with the domestic operating entity's assets;
- Consider the latest developments in the Foreign Investment Law in China.

When reviewing the application documents, the SEHK requires:

- A confirmation from the applicant's sponsor that the applicant is eligible for listing, and shall disclose details of the contractual arrangements and related risks in the prospectus as required;
- An affirmative opinion from the applicant's legal advisor;
- A confirmation from the applicant's reporting accountant that the consolidation of the domestic operating entity's financial results is in line with current accounting standards.

Companies seeking to list with a VIE structure with novel issues are recommended to conduct pre-listing consultation and communications (Pre-A1 Submission) as early as possible and adjust their listing structure accordingly.

---

10 See HKEX Listing Decision HKEX-LD43-3.

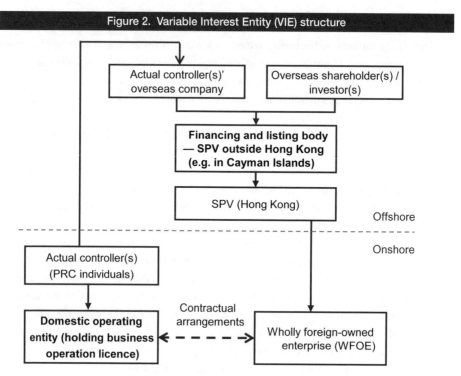

Figure 2. Variable Interest Entity (VIE) structure

Note: For illustration only. There may be differences in practice.

1.2.5 H-share structure

H-shares are shares that are issued in Hong Kong to foreign investors and listed on the
SEHK by joint-stock companies legally incorporated in Mainland China. Mainland laws
such as the *PRC Securities Law* and the *Special Provisions of the State Council Concerning
the Floatation and Listing Abroad of Stocks by Limited Stock Companies* also have
provisions regarding the issuance of H-shares.

At present, companies applying for listing in Hong Kong under the H-share structure
need to go through dual approval procedures, both domestic and overseas, with the China
Securities Regulatory Commission ("CSRC") and the SEHK.

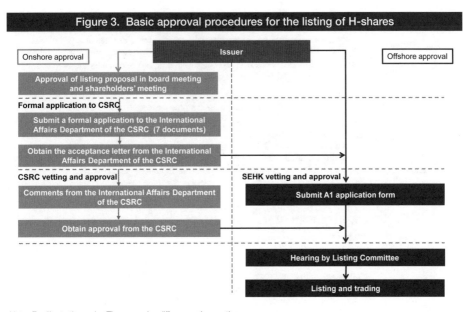

Figure 3. Basic approval procedures for the listing of H-shares

Note: For illustration only. There may be differences in practice.

During its preliminary review, the International Affairs Department of the CSRC will raise matters of special concerns in the feedback meeting for discussion. After receiving the feedback, the applicant co-ordinates with the relevant intermediaries to prepare a reply as required, and submits the relevant documents to the processing department within the specified time frame. CSRC reviewers shall write a review report, perform the approval procedures involved in the administrative licensing for approving or rejecting an offshore IPO and listing or a follow-on offshore re-financing. The processing department will then issue approval documents to the applicant.

According to the *Highlights on the Review and Approval Process of Overseas Public Offerings and Listing (Follow-on Refinancing Included) by Joint-Stock Companies* issued by the CSRC on 23 July 2019, the key points to note for an H-share listing are set out in Table 1 below:

Table 1. Key points on the approval of H-share listings (as issued by the CSRC)	
	1. Whether the business scope of the issuer and its subsidiaries involves areas where foreign investments are prohibited or restricted by China, and whether the relevant foreign investment access policy has been complied with both before and after the overseas shares issuance and listing.
Foreign entry and macro-economic control and industrial policy	2. Whether the issuer's business scope falls within any of the following conditions: (1) the principal business is real estate business, or (2) the real estate business (consolidated) accounts for 50% or more of the operating income, or (3) the revenue and profit of the real estate business are the highest among all businesses, in average accounting for 30% or more of the company's total revenue and profit. And if positive for any of the above, whether there is any violation of the *Notice of the General Office of the State Council on Further Improving the Regulation of the Real Estate Market* ([2013] No. 17 of the General Office of the State Council) or *Notice of the State Council on Resolutely Curbing the Soaring of Housing Prices in Some Cities* ([2010] No. 10 of the State Council), and whether it had been named as a "real estate development enterprise and intermediary agent in breach of the laws or regulations" by the Ministry of Housing and Urban-Rural Development.
	3. Whether the issuer and its subsidiaries have violated the Market Access Negative List; whether they are industries with overcapacity; whether there are violations of the rules, incompliance with legal procedures or key industry adjustment and revitalisation plans and related industrial policies, unauthorised approvals or approvals which are against the regulations or other cases of violation of [2013] No. 41 of the State Council, *Guiding Catalog for Adjustment in the Structure of Industries*, [2013] No. 67 of the General Office of the State Council, [2009] No. 386 of the People's Bank of China, [2016] No. 34 of the General Office of the State Council and the relevant documents issued by the State Council and the relevant departmental regulations.
	4. Whether the issuer and its subsidiaries have in the past year violated the *Notice of the State Council on Further Strengthening the Work Safety of Enterprises* ([2010] No. 23 of the State Council); whether they have violated national and local environmental protection requirements.
Operational compliance	5. Whether the issuer and its subsidiaries have in the past two years violated the *Securities Law*, the *Securities Investment Fund Law, Regulation on the Administration of Futures Trading, Special Provisions of the State Council Concerning the Floatation and Listing Abroad of Stocks by Limited Stock Companies* (Order of the State Council (No. 160)) and *Circular of the State Council Concerning Further Strengthening the Administration of Shares Issuance and Listing Overseas* ([1997] No. 21 of the State Council) and other securities and futures laws and regulations.
	6. Whether the issuer and its subsidiaries and the relevant securities service providers engaged by the issuer are subject to ongoing investigations by the administrative bodies for suspected violation of laws or regulations, or subject to ongoing investigations by the judicial authorities; whether they are currently subject to the CSRC's actions of imposing a restriction on their business activities or imposing an order of suspension for reorganisation, or of designated custody or takeover or other regulatory measures.
	7. Whether the shares subscribed by the issuer's promoters are fully paid, and whether the procedure of transferring the property rights has been completed regarding the capital contributions by the promoter. Whether the issuer is a "seriously dishonest entity" under the Guiding Opinions ([2016] No. 33 of the State Council).

(continued)

Table 1. Key points on the approval of H-share listings (as issued by the CSRC)	
Shareholding structure and corporate governance	8. Whether the issuer has stated in its articles of association what is required under the *Mandatory Provisions for the Articles of Association of the Companies to be Listed Overseas*. Whether there is any association or concerted action between the issuer's existing shareholders and the current issuance targets/target subscribers of the issue (if applicable). Please disclose the shareholding control relationship between the major shareholder(s) who, alone or together, hold(s) 5% or more of the shares of the issuer and the target of the current IPO/offer (if applicable) (including but not limited to the controlling shareholders and actual controllers).
	9. Whether the issuer and its subsidiaries have established and perfected complete and standardised rules and regulations for confidentiality and file management and whether they are in compliance with the *Provisions on Strengthening the Confidentiality and File Management related to Offshore Securities Issuance and Listing*.
The current issuance	10. In the current issuance, whether complete internal decision-making procedure has been implemented and whether the necessary internal approval and authorisation have been obtained; whether the necessary external approval such as the regulatory opinion (if applicable) issued by the industry regulatory authorities has been obtained; whether the issuance targets (if applicable) and issuance currencies comply with the relevant requirements of the *Special Regulations of the State Council on the Overseas Offering and Listing of Shares by Joint-Stock Limited Companies* (Order of the State *Council* (No. 160)).
	11. Whether the issuance and fund-raising project has obtained the necessary approval, permission or filing documents (if applicable) and whether it meets the relevant regulatory provisions of fixed asset investment administration; whether it meets the national and local environmental protection requirements. Whether the funds raised from this offshore issuance will be directed to the illegal land use projects stipulated in the *Notice on Promoting Land Saving and Intensive Use* ([2008] No. 3 of the State Council), or to the eliminated or restricted categories of industries stated in the *Guiding Catalog for Adjustment in the Structure of Industries*, or to the forbidden items provided in the *Market Access Negative List*; if it involves overseas investment, whether it is in line with the overseas investment direction stated in [2017] No. 74 of the General Office of the State Council and whether the overseas investment approval or filing procedures have been followed in accordance with the law.
Matters applicable to designated targets/ applicants	1. Whether the current issuance complies with the *Guideline on the Review, Approval and Supervision of Domestic Enterprises in China Applying to List on the Hong Kong Growth Enterprise Market* issued by the CSRC ([1999] No. 126). (Applicable to domestic companies seeking to list on Hong Kong GEM board)
	2. Whether the current issuance is in breach of the *Circular on Issues in Relation to Regulating Overseas Listing of Subsidiaries of Domestic Listed Companies* issued by the CSRC ([2004] No. 67). (Applicable to subsidiaries of domestic listed companies seeking to list overseas)
	3. Whether the current issuance is in breach of the *Administrative Measures on the Pilot Scheme for Preference Shares* issued by the CSRC (Order of CSRC No. 97), the *Guidance Opinions on the Issuance of Preference Shares by Commercial Banks for Replenishing Tier 1 Capital* ([2014] No. 12 of the China Banking Regulatory Commission). (Applicable to domestic companies seeking to issue preference shares overseas)
	4. Shareholders already holding the issuer's shares before the current issuance and directors, supervisors and senior executives holding the issuer's shares (directly or indirectly) shall specifically undertake that they will not transfer their shareholdings within 1 year from the date the issuer's shares are listed and traded on the overseas exchange. Please supplement with the provision of such specific undertakings. (Applicable to domestic applicants not listed in Mainland China)

417

(continued)

Table 1. Key points on the approval of H-share listings (as issued by the CSRC)	
Matters applicable to designated targets/ applicants	5. Whether the number of shareholders of the issuer and its controlling shareholders, actual controllers, and significant majority-owned subsidiaries exceeds 200 according to *Guideline No. 4 for the Supervision of Unlisted Public Companies — Review Guidelines on Issues Concerning Administrative Licensing Applications of an Unlisted Joint-Stock Limited Company with Over 200 Shareholders.* If yes, please provide the relevant application documents and fulfill the relevant procedures. A "significant majority-owned subsidiary" refers to a majority-owned subsidiary whose operating income, net profit or total assets account for over 50% of the totals on the consolidated statement during the reporting period, and also one which has significant influence on the current and future operation and financial status of the group as a whole. (Applicable to domestic applicants not listed in Mainland China)
	6. Whether there is any direct or indirect shareholding or warehousing arrangement by the employees of the issuer. If yes, whether it is in breach of the *Notice on the Regulation of Internal Staff Shareholdings in Financial Enterprises* ([2010] No. 97 of the Ministry of Finance) and other relevant regulations. The issuer's relevant senior executives and other individuals holding more than 50,000 employee shares are required to give specific undertakings on the lock-up period and the sale limit of share transfers in accordance with relevant regulations. (Applicable to financial enterprises)
	7. Whether the non-overseas listed shares have been registered and in the custody of the China Securities Depository and Clearing Co., Limited pursuant to the *Notice of the CSRC on the Centralised Registration and Custody of Non-overseas Listed Shares of Overseas Listed Companies* ([2007] No. 10 of the CSRC). (Applicable to H-share listed companies seeking follow-on refinancing)

Source: Official website of the CSRC.

1.3   Listing of H-shares: An update

1.3.1   H-share companies have become an important part of the Hong Kong stock market

Starting from June 1993, companies incorporated in Mainland China are able to list in Hong Kong under the Exchange's Listing Rules. Tsingtao Brewery (stock code: 00168) became the first Mainland company to issue H-shares and list in Hong Kong in July 1993. After 2000, large Mainland enterprises, including the four major state-owned banks, China Unicom and CNPC (China National Petroleum Corporation), launched their H-share IPOs, unfolding a boom in Mainland enterprises' listing in Hong Kong.

As of 30 June 2019, a total of 1,197 Mainland enterprises were listed in Hong Kong (including 274 H-share companies), accounting for about 68% of Hong Kong's market total in terms of market capitalisation, and about 80% in terms of average daily turnover value (ADT) in 2019 up to June. The number of listed A+H share issuers reached 112, of which 90 and 22 were also listed on the Shanghai Stock Exchange (SSE) and the Shenzhen Stock

Exchange (SZSE) respectively.

Funds raised through all 274 H-share IPOs amounted to HK$1.81 trillion, almost 80% of which came from IPOs in the major sectors of industrials (69 in number), financial (66) and consumer discretionary (25). Hong Kong's capital market is increasingly capable in serving Mainland enterprises.

Table 2. Companies listed on the SEHK							
Type of listed company	Number			Total market capitalisation (End-Jun 2019)		ADT (Jan-Jun 2019[11])	
	Total	Main Board	GEM	HK$ million	% of total	HK$ million	% of total
Mainland China	1,197	1,093	104	22,247,751.84	67.98	55,134.84	79.89
H-shares	274	250	24	6,376,083.15	19.48	24,257.63	35.15
Red-chips[12]	176	171	5	5,595,665.82	17.10	7,908.10	11.46
MPEs[13]	747	672	75	10,276,002.88	31.40	22,969.11	33.28
Hong Kong	1,006	765	241	7,921,907.18	24.21	9,484.60	13.74
Overseas	179	138	41	2,556,988.79	7.81	4,393.95	6.37
Total	2,382	1,996	386	32,726,647.81	100.00	69,013.39	100.00

Source: HKEX.

In addition, since the launch of the Shanghai-Hong Kong Stock Connect (Shanghai Connect) in 2014 and the Shenzhen-Hong Kong Stock Connect (Shenzhen Connect) in 2016, the mutual market access between the stock markets in Hong Kong, Shanghai and Shenzhen has been promoting the two-way healthy development of the capital markets across the border. As of the end of June 2019, 118 of the 326 Hong Kong stocks included in the Shanghai Connect, and 144 of the 481 Hong Kong stocks in Shenzhen Connect were H-shares.

1.3.2 Successful implementation of the H-share full circulation pilot programme

The share capital structure of H-share issuers includes "foreign shares" which are tradable in the Hong Kong stock market, as well as "domestic shares" such as the original legal person shares and state-owned shares before listing. "Domestic shares" are non-tradable before they are listed and traded in the domestic market. Full circulation of H-shares

---

11 Turnover value included listed ordinary shares and preference shares.
12 Red-chips here refer only to those listed in Hong Kong with a "state-owned red-chip" structure.
13 MPEs refer to Mainland private enterprises that have no H-shares.

will turn these non-tradable "domestic shares" into "foreign shares" available for public trading on the Hong Kong stock market.

According to the *CSRC Pilot Programme for the Deepening Reforms on the Overseas Listing Regime with the "Full Circulation" of H Shares* issued on 29 December 2017, the pilot programme for the full circulation of H-shares will be implemented in an orderly manner, "enterprise by enterprise", in accordance with the principle of active, steady and gradual progress.

Given that Legend Holdings (03396), AVIC (02357) and Weigao (01066) have successfully completed their pilot programmes of H-share full circulation in 2018, CSRC Vice Chairman FANG Xinghai publicly stated that the programme shall be launched in full swing as soon as possible so as to allow qualified H-share companies to apply for full circulation, and facilitate overseas listing of domestic innovative enterprises.

In June 2019, the CSRC Chairman, YI Huiman, said at the Lujiazui Forum in Shanghai that "the capital market will be further opened up, with impetus given to the comprehensive promotion of H-share circulation reform". With good progress of the pilot programme, H-share companies listed on the SEHK can expect a wider and stronger capital market support in the future.

1.3.3 Pilot programme for NEEQ companies to list in Hong Kong under the "NEEQ + H shares" model

In April 2018, HKEX signed a Memorandum of Cooperation with NEEQ Co., Ltd. which operates the NEEQ system ("New Third Board") in the Mainland to allow companies qualified for listing in Hong Kong to list in both places simultaneously in the form of "NEEQ + H shares". In other words, NEEQ companies can apply for listing on the SEHK without prior delisting from NEEQ.

According to statistics, as of 30 June 2019, more than 25 NEEQ companies (or their spin-off subsidiaries) have submitted listing applications to the SEHK. Seven have completed listing in Hong Kong, among which Junshi Pharma-B (01877) has successfully landed on Hong Kong's capital market in the form of "NEEQ + H shares". The success of this "NEEQ + H shares" programme provides insights and valuable experience in mobilising both domestic and overseas capital markets to serve H-share issuers.

1.3.4 Closer connection between B and H shares

B-shares generally refer to the RMB-denominated foreign-funded shares listed and traded on the SSE or the SZSE that are subscribed and traded in foreign currencies. As of 30 June 2019, there were 50 B shares listed on the SSE and 47 on the SZSE.

In recent years, some B-share listed companies have chosen to delist from the B-share market and convert their shares into H shares to be listed on the SEHK. CIMC (02039) was the first to have successfully entered the Hong Kong capital market in this way in August 2012 by converting its B shares to H shares. After that, Vanke (02202) and Livzon Pharma (01513) also delisted from the B-share market and changed respectively from an "A+B" share company to an "A+H" share company in 2014. Apart from these, a number of B-share companies have chosen to be listed on the SEHK in other ways, such as Chenming Paper (01812) listed in the form of a "A+B+H" share company, and Yitai Coal (03948) listed in the form of a "B+H" share company.

As these "B to H" and "B+H" models go popular, the connection between Hong Kong and the Mainland capital markets will be further deepened, providing more channels for the multi-currency and international equity financing of the Mainland enterprises.

1.4   Listing in Hong Kong: Professional parties involved

1.4.1   Sponsors and Compliance Advisors

An applicant who seeks to list on the Main Board of the SEHK must appoint at least one independent sponsor licensed by the Securities and Futures Commission (SFC) to assist with its listing application[14]. A sponsor, once appointed or dismissed, the listing applicant must notify the SEHK in writing as soon as possible, regardless of whether a listing application has been submitted. A listing application must not be submitted to the SEHK by a new applicant in less than two months upon the formal appointment of a sponsor. Duties of a sponsor usually include:

- Conduct overall planning and coordination of listing;
- Assist in preparation of a financing plan for the applicant;
- Conduct reasonable due diligence;
- Review the profit forecast and cash flow forecast prepared by the applicant;
- Be deeply and closely involved in the preparation of application materials for listing for the applicant;
- Liaise with regulators on behalf of the applicant and accompany it to meetings.

In addition, an applicant must appoint a Compliance Advisor for a term commencing on the date of listing and ending on the date of announcement of its financial results for the first full financial year after listing.

---

14 See Listing Rule 3A.02 for details.

1.4.2 Underwriters

An underwriter of an IPO mainly involves in the issue and sale of shares. The appointment of underwriters of different types may help absorb, to a certain extent, the sales pressure and lower offer risks, especially for large offers or in a downbeat market. Duties of an underwriter usually include:

- Assist in the preparation of roadshow materials for the applicant;
- Arrange analysts to compile research reports;
- Identify potential investors for the applicant;
- Organise roadshows and promotion activities;
- Conduct public offering and international placing;
- Fully underwrite the IPO subscription of the securities;
- Assist in determining IPO price;
- Provide post-listing share price support.

1.4.3 Solicitors

Due to the different legal systems of the Mainland and Hong Kong, application for listing in Hong Kong normally involves four teams of solicitors, which are solicitors representing the issuer (both local[15] and overseas solicitors[16]) and solicitors representing sponsors and underwriters (both local and overseas solicitors). Duties of a solicitor team usually include:

- Draft prospectus;
- Assist in restructuring the company;
- Assist in due diligence work;
- Advise on legal matters of the applicant in the course of due diligence work, in accordance with listing requirements;
- Provide legal advice to sponsors and underwriters on compliance matters of the applicant;
- Assist the applicant and its sponsors in responding to regulators' enquiries.

1.4.4 Accountants

An applicant shall engage professional accountants to audit its financial reports and

---

15 Local solicitors refer in general to those who have background in Mainland laws.
16 Overseas solicitors refer in general to those practicing in Hong Kong, Cayman Islands and the United States (if securities laws in the US is involved) and other jurisdictions popular for overseas operations (if applicable).

assist with its IPO. Duties of an accountant usually include:

- Prepare audit reports;
- Assist the applicant in taxation matters (if the applicant do not have an independent tax consultant);
- Review the profit forecast and cash flow forecast prepared by the applicant;
- Assist the sponsors in responding to regulators' enquiries.

In addition, a Hong Kong IPO may involve professional services provided by other external parties as follows:

- Industry consultants: mainly responsible for industry research and industry analysis reports etc.;
- Property valuers: responsible for issuing valuation reports on the applicant's property interests or other assets, etc.;
- Special asset valuation: valuation of special assets, such as mine valuation;
- Printers: mainly responsible for the proofreading, translation and printing of prospectuses, etc.;
- Public relations firms: arrange roadshows, media activities and listing ceremonies, etc.

1.5   Conditions for a GEM listing

GEM has been positioned as a market for small and medium-sized companies. It is a market with lower listing eligibility criteria but similar continuing obligations compared to the Main Board. On 15 December 2017, the Exchange published its GEM consultation conclusions with amendments to the GEM Listing Rules that removed the streamlined process where GEM issuers could be transferred to the Main Board, and repositioned GEM as a stand-alone market for small and medium-sized companies[17]. Upon such amendments, applicants seeking a GEM listing shall meet the following basic requirements:

- Aggregate operating cash flow over the two preceding years ≥ HK$30 million[18];
- Market capitalisation ≥ HK$150 million;
- A minimum public float (normally 25%; may reduce to 15% at discretion if market capitalisation at the time of listing > HK$10 billion);
- At least 100 shareholders after listing[19];
- Largely the same management in the latest two financial years;

---

17 The Consultation Conclusion took effect on 15 February 2018.
18 See GEM Listing Rule 11.12A for details.
19 See GEM Listing Rule 11.23(2)(b) for details.

- Same ownership and control in the latest financial year;
- At least three independent directors, which shall account for at least one-third of the Board;
- At least one independent non-executive director who shall be ordinarily resided in Hong Kong in case of H-share applicant;
- Publication of quarterly financial reports.

Eligible GEM issuers may apply for a transfer of listing to the Main Board[20]. There were ten successful transfers of listing of issuers from GEM to the Main Board in 2018, and eight in the first half of 2019.

1.6   Backdoor listing in Hong Kong

In recent years, the Exchange has noted an increase in market activities related to the trading, and creation, of shell companies on the back of demands for backdoor listing through listed shells. To maintain market quality, the Exchange has adopted a more stringent approach in applying the Listing Rules to crack down on ever changing shell activities, including backdoor listing, corporate actions involving disposals of businesses of listed issuers, and not having a sufficient level of operations. On 29 June 2018, the Exchange published the *Consultation Paper on Backdoor Listing, Continuing Listing Criteria and Other Rule Amendments*, and consulted the market on proposed Listing Rules amendments to address market concerns on backdoor listing and shell activities.

On 26 July 2019, the Exchange published its consultation conclusions on backdoor listing and other shell activities. Consultation proposals are to be implemented with modifications reflecting comments received, and with related Listing Rules amendments to take effect on 1 October 2019[21]. In this connection, the Exchange also issued three new guidance letters on the application of the amended Listing Rules[22].

Major amendments of the Rules relating to backdoor listings include those in sub-sections below.

1.6.1   Definition of reverse takeover (RTO) transactions

Principle-based test

The six assessment factors under the principle-based test in the relevant Guidance

---

20 See Chapter 9A for details.

21 The Consultation Conclusions also applied to GEM Listing Rules. See Amendments to the GEM Listing Rules (Update No. 62) for details.

22 See HKEX Guidance Letter HKEX-GL104/105/106-19 for details.

Letter[23] are codified into the Listing Rules with modifications made to the last two factors:

- Size of transaction;
- Quality of business/assets to be acquired;
- Nature and scale of the issuer's business;
- Fundamental change in principal business;
- Change in control or *de facto* control;
- Series of transactions and/or arrangements, including acquisitions, disposals and/or change in control or *de facto* control that take place in reasonable proximity of time (normally within 36 months) or are otherwise related.

Bright-line tests

The bright-line tests are modified to cover very substantial acquisitions from an issuer's controlling shareholder within 36 months after a change in control of the issuer; and to restrict disposals (or distributions *in specie*) of all or a material part of the issuer's business proposed at the time of or within 36 months after a change in control of the issuer. The Exchange may also apply the restriction to disposals (or distributions *in specie*) at the time of or within 36 months after a change in *de facto* control (as set out in the principle-based test) of the issuer.

Backdoor listing through large-scale issue of securities

The relevant Guidance Letter[24] is codified into the Listing Rules to disallow backdoor listing through large-scale issuance of securities for cash, where there is, or will result in, a change in control or *de facto* control of the issuer, and the proceeds will be applied to acquire and/or develop new business that is expected to be substantially larger than the issuer's existing principal business.

1.6.2 Tightening the compliance requirements for RTOs and extreme transactions

Extreme transactions

The "extreme very substantial acquisitions" requirements in the relevant Guidance Letter[25] are codified into the Listing Rules and this category of transactions is renamed "extreme transactions". Additional eligibility criteria are also imposed on issuers who may use this transaction category: (1) the issuer must operate a principal business of a substantial size; (2) the issuer must have been under the control or *de facto* control of the same

---

23 See HKEX Guidance Letter HKEX-GL78-14 for details.
24 See HKEX Guidance Letter HKEX-GL84-15 for details.
25 See HKEX Guidance Letter HKEX-GL78-14 for details.

person(s) for a long period of time (normally not less than 36 months) and the transaction
will not result in a change in control or *de facto* control of the issuer.

Requirements for RTOs and extreme transactions

The Listing Rules are modified to require the acquisition targets in an RTO or extreme
transaction to meet the requirements of Rule 8.04 and Rule 8.05 (or Rule 8.05A or 8.05B),
and the enlarged group to meet all the new listing requirements in Chapter 8 of the Listing
Rules other than Rule 8.05. Where the RTO is proposed by a Rule 13.24 issuer, the
acquisition targets must also meet the requirement of Rule 8.07.

During the consultation period, the Exchange received a total of 121 responses from a
broad range of respondents, which were supportive of the initiatives to address backdoor
listing and shell activities. The consultation and related amendments to the Listing Rules
will also help the Exchange enhance the quality and sustainability of the Hong Kong
market.

2 Application and vetting process for enterprise listing in Hong Kong

2.1 Roles and duties of the SEHK and the SFC

The SEHK adheres to the disclosure-based approach with transparent processes and
predictable timetables on listing approvals. It has a statutory obligation under the Securities
and Futures Ordinance to ensure, so far as reasonably practicable, that Hong Kong markets
are fair, orderly and informed.

The SFC plays a leading role in market regulation and certain areas of listing regulation
and a complementary role in cases involving corporate misconduct through the exercise of
its statutory powers of investigation and enforcement.

The SFC has a statutory duty to supervise and monitor the SEHK's performance of
its listing-related functions and responsibilities. The staff of the SEHK and the SFC meet
regularly to discuss listing-related matters. Under the dual filing regime, the SEHK passes
copies of materials submitted by listing applicants to the SFC. The SFC may object to
a listing if the disclosure in the listing materials appears to contain false or misleading
information. The SFC also conducts a periodic audit of the performance of the SEHK in its

regulation of listing-related matters.

The SEHK's listing-related functions are discharged by the Listing Department and the Listing Committee, including the formulation, announcement and enforcement of the Listing Rules. All changes to the Listing Rules and policy decisions that have mandatory effect or general application are to be approved by the SFC.

2.2   Overview of the IPO vetting process

When a company comes to a certain stage of development and intends to be listed in Hong Kong, it shall go through procedures at different stages from pre-listing preparation to listing application (A-1 filing), vetting and hearing, and finally listing and trading of shares in the market, as indicated in Figure 4 below.

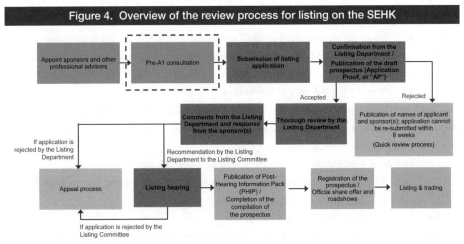

Figure 4. Overview of the review process for listing on the SEHK

Note: For illustration only. There may be differences in practice.

(1) Step 1: Appointment of sponsors and other professional advisors

Appointing an experienced team of professional advisors is crucial to the success of an IPO. Professional advisors normally include sponsor(s) and underwriter(s), domestic and overseas lawyers, accountants, etc. It is required to appoint a sponsor at least two months before submission of an IPO application and notify the SEHK in writing within five business days of its appointment. The applicant should discuss with professional advisors whether a listing is right for it, the listing structure, the time and cost, division of labour in the listing team, and the potential issues, challenges and obligations the applicant may face once listed.

(2) Step 2: Listing preparation process

Sponsors conduct due diligence on the applicant and assist in the drafting of the prospectus (aka Application Proof, or "AP", in this stage). The prospectus must contain all the significant information that an investor reasonably requires to make an informed investment decision. During the preparation process, sponsors would assist the applicant in formulating or enhancing medium- to long-term strategies and making business and financial forecasts for underwriter(s) to create valuation models, determine the use of proceeds and sort out highlights to pitch investors.

(3) Step 3: Submission of listing application to Listing Department

The listing application (A1), which includes the AP, is submitted to the Listing Department. Assuming the A1 is substantially complete, the Listing Department will confirm receipt and publish both Chinese and English versions of the AP online.

Applicants are encouraged to seek informal and confidential guidance from the Exchange regarding novel and specific issues prior to the formal submission of a listing application. This is usually referred to as "Pre-A1 Submission". It can help issuers, based on their actual circumstances, understand whether the proposed listing is in line with the requirements in a timely manner and also their key considerations for listing in Hong Kong.

(4) Step 4: Vetting by the Listing Department

The Listing Department will subsequently conduct detailed vetting of the A1 based on eligibility, suitability for listing, business sustainability, compliance of rules and regulations and sufficiency of disclosures.

First round of comments will be provided as soon as possible, normally within 15 business days, from receipt of application. Time needed for subsequent review will depend on the applicant's response time and quality of response.

(5) Step 5: Hearing by the Listing Committee

The Listing Committee will review the application and determine if it is suitable for the applicant to proceed with its IPO. As required under the Listing Rules, the Listing Committee shall comprise 28 members, among which at least 8 members shall represent investors' interests; 19 members shall appropriately represent the listed issuers and market participants including lawyers, accountants, corporate finance consultants and other senior personnel, nominated in appropriate proportions; and the remaining member shall be the Chief Executive of HKEX. Major duties of the Listing Committee include[26]:

- Listing review;
- Vetting of listing applications and oversight of the Listing Department;

---

26 See the official website of the SEHK for details.

- Approval of IPO listing and rule waiver applications;
- Passing, amendment or modification of decisions made by the Listing Department and the Listing Committee;
- Provision of policy advice and comments for important policies and amendments to the Listing Rules.

(6) Step 6: Marketing and sales of new shares

At this stage, the applicant would be required to meet the minimum standard for public shareholding and the number of shareholders, but the SEHK will not interfere in the pricing process of the applicant.

The underwriter(s) or underwriting group are generally responsible for assisting the applicant in listing preparation and marketing, including investor education, communication with potential cornerstone investors, analyst roadshows and IPO roadshows by management. Allocation of IPO shares in Hong Kong falls into two tranches "international placing" and "Hong Kong public offering", subject to adjustment as appropriate according to the clawback mechanism and share reallocation mechanism. International placing and public offering usually account for 90% and 10% respectively of the total number of newly issued shares.

Many companies have chosen to introduce cornerstone investors when they have IPOs in Hong Kong. IPO shares under the placing tranche are preferentially placed to certain investors, usually referred to as "cornerstone investors". Principles for placing to cornerstone investors generally include[27]:

- Placing must be carried out at the IPO price;
- Shares placed during the IPO are subject to a lock-up period for at least 6 months in general from the listing date;
- Each investor has no representative in the board of directors of the listing applicant and is independent of the listing applicant, and its connected persons as well as their respective associates;
- Details of the placing (including identities and background of the investors) must be disclosed in the listing documents;
- If an investor will be regarded as a member of the public according to Rule 8.24 of the Listing Rules (Notes 2 and 3 to Rule 11.23 of the GEM Listing Rules), the shares placed to such investor will still be part of the public float as referred to in Rule 8.08 of the Listing Rules (Rule 11.23 of the GEM Listing Rules) regardless of the lock-up for a period of at least 6 months as mentioned above;

---

27 See HKEX Guidance Letter HKEX-GL51-13 for details.

- Save for the guaranteed preferential share placement, cornerstone investors should not be given any direct or indirect interests.

(7) Step 7: Listing and post-listing trading

Upon completion of pricing and share allocation to institutional and retail investors, shares of the applicant would be listed for trading on the SEHK.

Upon listing, the share price of an issuer would be subject to fluctuation due to different reasons. The underwriting group would usually designate one or more underwriter(s) as the Stabilisation Agent responsible for buying shares of the newly listed company when the share price becomes lower than the IPO price in order to stabilise the post-listing share price performance for a defined period of time.

For Hong Kong IPOs, the listing applicant may grant an "over-allotment option" to its underwriter(s) which usually equals to 15% of the total number of shares originally issued in the IPO and is to be exercised within 30 days after the Hong Kong public offering is closed; such process is called "Green Shoe" mechanism. The Stabilisation Agent may assess secondary market investors' demand for shares of the issuer based on the post-listing share price movement, and decide at its own discretion whether to exercise the over-allotment option in order to mitigate volatility in the market, bringing mutual benefits to all parties concerned including the listed company, investors and underwriters.

3 Hong Kong listed company regulation and equity refinancing

3.1 Listed company regulation

The SEHK adheres to the basic principle for listed issuer regulation to ensure issuers' ongoing compliance with the Listing Rules in order to maintain a fair, orderly and efficient market. The regulatory functions of the SEHK are mainly reflected in three aspects, namely supervision, vetting and guidance. For example, the SEHK is responsible for monitoring media reports and share prices of listed issuers, handling complaints, vetting waiver applications, suspending and resuming the trading of shares, handling enquiries from listed companies regarding the Listing Rules, providing guidance to listed companies, and market education, etc.

The regulation of listed companies in Hong Kong is effected through enforcement of continuous disclosure obligations on the part of issuers, including the following three types of disclosure: (1) disclosing inside information as required by the relevant laws and regulations in a timely manner[28]; (2) disclosing necessary information to avoid false market[29]; (3) responding to enquiries from the Exchange regarding unusual fluctuations (price/turnover) of securities transactions of listed companies[30].

Other general matters that need to be disclosed include, but are not limited to:

- Financial result announcement/report (interim/annual);
- Amendments to the memorandum or articles of association;
- Change of directors or supervisors;
- Change of auditors and financial reporting year;
- Change of company secretary, share registrar and registered address;
- Change in share capital;
- Notice of shareholders' meeting.

3.2   Equity refinancing of listed companies

Upon listing on the Exchange, an issuer will have access to diversified and efficient channels for refinancing and increasing/reducing percentage interest of shareholders. In fact, post-listing equity refinancing channels for Hong Kong listed companies and their shareholders include, among others, post-listing issuance of new shares, rights issues, convertible/exchangeable bonds, equity warrants/options and block trades.

For many years, the amount of equity refinancing in the Hong Kong stock market has exceeded the amount of IPO funds raised, providing strong capital support for the sustainable development of companies listed in Hong Kong.

---

28 See Part XIVA "Disclosure of Inside Information" of the *Securities and Futures Ordinance* (Cap. 571) for details. As defined by the Ordinance, inside information "is not generally known to the persons who are accustomed or would be likely to deal in the listed securities of the corporation but would if generally known to them be likely to materially affect the price of the listed securities".

29 "False market" generally refers to a situation where there is material misinformation or materially incomplete information in the market which is compromising proper price discovery.

30 See Listing Rule 13.10 for details.

Figure 5. IPO funds raised and equity refinancing of listed companies at the SEHK
(2001 – Jun 2019)

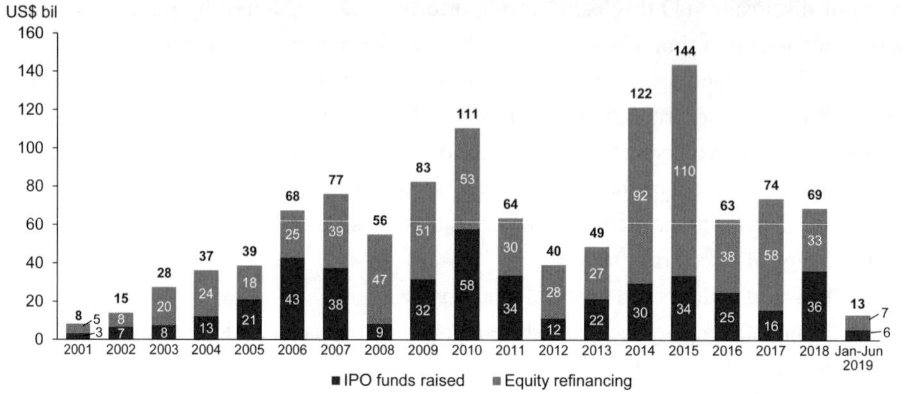

Source: HKEX.

3.2.1 Further equity issuance after listing

- It is the issuance of new shares by listed companies for financing purposes (not allowed within the first six months after initial offer[31]);
- The basic principle is to protect the pre-emptive rights of existing shareholders; the issuance of new shares is subject to prior approval by shareholders to prevent passive dilution of their interests;
- General mandates granted at annual general meetings allow listed companies in Hong Kong to freely and flexibly issue additional new shares or equity securities (usually not more than 20% of total share capital), which can be done speedily (normally completed overnight)[32];
- For additional issues for cash, the issue price under a general mandate must not be more than 20% below the market price.

3.2.2 Rights issues

- This is an offer by way of granting rights to existing shareholders, enabling them to subscribe securities in proportion to their existing holdings;

---

31 See Listing Rule 10.08 for details.
32 See Listing Rule 13.36(2) for details.

- The company should ensure that all shareholders are treated fair and equal, and should provide shareholders with sufficient information for making informed investment/ voting decisions;
- It is subject to approvals by the board of directors and shareholders at general meetings (if applicable);
- An announcement or listing document (if applicable) is required for listing by way of rights issue.

3.2.3 Convertible/exchangeable bonds

- Listed issuer and its shareholders may issue convertible/exchangeable bonds;
- The initial conversion price must not be lower than the benchmark price of the issuer's shares at the time of issue;
- Convertible bonds are issued by listed companies for financing;
- Exchangeable bonds are issued by shareholders with the tradable shares they hold in the listed company for capital replenishment (either the listed company they control or the listed company in which they hold a minority interest).

3.2.4 Equity warrants/options

- Equity warrants may only be issued (under general mandates) with a conversion price higher than the market price.

3.2.5 Block trades

- Shareholders may obtain financing by selling the stakes they hold in the listed company by way of block trades;
- Controlling shareholders are not allowed to reduce their percentage interest in the issuer within the first six months after the IPO; and from the 7th month to the 12th months after the IPO, the reduction of percentage interest is allowed only if it does not affect their controlling status.

Chapter 5

Reform and breakthrough of Hong Kong's listing regime

Chief China Economist's Office
and
Global Issuer Services
Hong Kong Exchanges and Clearing Limited

Summary

The Stock Exchange of Hong Kong Limited (SEHK, or the Exchange) consulted the market in February 2018 on proposed amendments to its Listing Rules to expand the original listing regime to facilitate the listing of companies from emerging and innovative sectors. In the consultation, a new concessionary route to secondary listing for large innovative companies already listed on qualified exchanges was also proposed. With broad market support, the relevant amendments to the Main Board Listing Rules were implemented on 30 April 2018.

This chapter presents, in a succinct way, the listing requirements stipulated in the respective newly introduced chapters of the Listing Rules for companies with weighted voting rights (WVR) structures, biotech companies and companies seeking a secondary listing in Hong Kong.

1 New chapter for companies with weighted voting rights (WVR) structures

1.1 Rationale and principles behind

The concept of proportionality between the voting power and equity interest of shareholders, commonly known as the "one-share, one-vote" principle, is an important aspect of investor protection. It helps align controlling shareholders' interests with those of other shareholders, and makes it possible for incumbent management to be removed, if they underperform, by those with the largest equity interest in the issuer.

The Exchange believes that the "one-share, one-vote" principle continues to be the optimum method of empowering shareholders and aligning their interests in a company. As such, companies wishing to list with a WVR structure shall be able to meet relevant qualifying requirements, and demonstrate the necessary characteristics of innovation and growth and the contribution of their proposed beneficiaries of WVR.

It is under this rationale that the Exchange launched a new Chapter 8A of the Listing Rules, allowing companies with WVR structures to list in Hong Kong subject to certain conditions. By 30 June 2019, two companies, Xiaomi-W (stock code: 01810) and Meituan-W (stock code: 03690), had successfully landed on Hong Kong's capital market under this chapter.

1.2 Listing qualifications and basic conditions

- Only applicable to new applicants; companies already listed are not allowed to apply for a change into WVR structures;
- A market capitalisation of at least HK$40 billion at the time of listing; or a market capitalisation of at least HK$10 billion at the time of listing and a revenue of at least HK$1 billion for the most recent audited financial year;
- The applicant shall be an innovative company;
- Successful business operation: the applicant must demonstrate a track record of high business growth, as can be objectively measured by operational metrics such as business operations, users, customers, revenue, profits or market values, and its high-growth trajectory is expected to continue;

- External validation: the applicant must have previously received meaningful third-party investment from at least one Sophisticated Investor[1].

1.3   Definition of innovative sectors[2]

The applicant must be an innovative company which is expected to possess more than one of the following characteristics:

(1)   The success of its operation is demonstrated to be attributable to the application, to the company's core business, of new technologies, innovations and/or new business models, which also serves to differentiate the company from existing competitors.

The applicant should elaborate on how its operations differ from conventional practices in its industry which sets it apart from peers. If the applicant's peers are employing similar technology/business model, the Exchange will take into account whether the applicant was the "first mover" in the industry by reference to the timeline of the implementation of its technology, innovations and/or business models compared to its closest peers.

(2)   Research and development (R&D) is a significant contributor of its expected value and constitutes a major activity and expense.

The applicant should, in addition to providing the amount of its R&D expenses during the track record period (both as a figure and as a percentage of revenue/total expenses), also explain how the R&D contributes value to the applicant. In this connection, the Exchange will examine whether the R&D expenses are capitalised as intangible assets in the accounts of the applicant as an indicator of the value generated through the R&D activities. Where a significant portion of the R&D expenses are not capitalised, the applicant should provide the reasons for this.

(3)   The success of its operation is demonstrated to be attributable to its unique business features or intellectual property, and/or it has an outsized market capitalisation or intangible asset value relative to its tangible asset value.

Providing a list of patents and trademarks alone is not sufficient to demonstrate this characteristic. The applicant should provide detailed explanation on how its intellectual properties enabled it to achieve business success.

---

1   An investor that the Exchange considers to be sophisticated by reference to factors such as net assets or assets under management, relevant investment experience, and the investor's knowledge and expertise in the relevant field.
2   See HKEX Guidance Letter HKEX-GL93-18 for details.

The Exchange recognises that what is considered "innovative" depends on the state of the industry(ies) and market(s) in which an applicant operates, and will change over time as technology, markets and industries develop and change. Accordingly, the fact that a particular company is qualified for listing with a WVR structure does not necessarily mean that another applicant with a similar technology, innovation or business model will also qualify for listing with a WVR structure. The Exchange will review the facts and circumstances of each case to determine if an applicant has met the relevant requirements. The superficial application of new technology to an otherwise conventional business will not be sufficient to demonstrate the characteristics set out in this section.

1.4   Restrictions on WVR beneficiaries

- Individuals only: materially responsible for the growth of the business (separate consultation will be carried out on corporate beneficiaries);
- Directors only: each WVR beneficiary shall be a director at listing and afterwards;
- Minimum shareholding: ≥10% at listing (collectively);
- Prohibition of shares transfer: natural sunset clause, allow trusts and legitimate tax planning.

1.5   Additional listing requirements and shareholder protection

- After listing, the WVR shares are not tradable, and their proportion should not be increased to above the proportion in issue at the time of listing;
- No increase in the ratio of WVR shares' voting rights after listing;
- Voting rights of WVR shares must not entitle the beneficiaries to more than ten times the voting power of ordinary shares, on any resolution tabled at the issuer's general meetings;
- Holders of ordinary shares must be entitled to cast at least 10% of the votes on resolutions at the issuer's general meetings;
- WVR beneficiaries must beneficially own collectively at least 10% of the applicant's underlying economic interest at the time of its initial listing.

The following material matters must be decided on a "one-share, one-vote" basis:
- Changes to constitutional documents;
- Appointment and removal of an independent non-executive director (INED);
- Appointment and removal of auditors;
- Variation of rights attached to any class of shares; and
- Voluntary winding-up of the listed issuer.

Enhanced disclosure and corporate governance:
- Warnings in listing documents and corporate communications;
- Marker "W" at the end of the stock name;
- Establish a Corporate Governance Committee to review, monitor and report on compliance with WVR safeguard measures;
- Ongoing appointment of a Compliance Advisor to provide advice to issuers on compliance with WVR safeguards and related rules;
- The Corporate Governance Committee must be comprised entirely of INEDs.

Constitutional backing and legal remedies:
- WVR safeguards must be incorporated into constitutional documents;
- WVR beneficiaries must give an undertaking to the issuer in a form acceptable to the Exchange to comply with WVR safeguards.

2 New chapter for biotech companies

2.1 Rationale and principles behind

In April 2018, the Exchange also introduced a new Chapter 18A "Biotech Companies" to its Main Board Listing Rules, to mainly accommodate biotech companies seeking to list on the Main Board but fail to meet the relevant "profit test" or "market capitalisation/revenue test" or "market capitalisation/revenue/cash flow test" under Chapter 8 of the Listing Rules, in a bid to further enhance the capabilities of the Hong Kong market to serve new-economy companies[3].

Under this guiding rationale, Hong Kong has become the second largest biotech listing venue in the world in just less than two years. Since the introduction of this new chapter up to 30 June 2019, a total of 14 biotech companies have been listed on the Main Board, raising funds of HK$47.5 billion in total. Of these companies, eight are pre-revenue biotech

---

3   It should be noted that biotech companies shall not apply for listing under Chapter 18A if they meet the relevant "profit test" or "market capitalisation/revenue test" or "market capitalisation/revenue/cash flow test" under Chapter 8. Listing applicants are encouraged to contact the Exchange early for confirmation on compliance/qualification.

companies raising funds of HK$23.5 billion in total. Besides, five more biotech companies have already submitted the A1 application form. Looking forward, Hong Kong's position as a leading capital market for biotech companies could be further strengthened.

	Stock short name	Stock code	Listing date	Principal business location
1	Ascletis-B	01672	01/08/2018	Hangzhou
2	BeiGene	06160	08/08/2018	Beijing
3	Hua Medicine-B	02552	14/09/2018	Shanghai
4	Innovent Bio-B	01801	31/10/2018	Suzhou
5	Junshi Bio-B	01877	24/12/2018	Shanghai
6	Cstone Pharma-B	02616	26/02/2019	Shanghai
7	CansinoBio-B	06185	28/03/2019	Tianjin
8	MabPharm-B	02181	31/05/2019	Taizhou

Table 1. New listings on the SEHK under the Biotech Chapter (End-Jun 2019)

Note: The affixed letter "B" to the stock name of BeiGene has been removed as a result of the approval granted by the Exchange on 28 June 2019 for the company's dis-application of Chapter18A due to its satisfaction of the market capitalisation/revenue test in Rule 8.05(3).
Source: HKEX.

2.2   Listing qualifications and basic conditions[4]

(1) The applicant must have developed at least one core product beyond the concept stage

A biotech product that is required by applicable laws, rules or regulations to be evaluated and approved by a Competent Authority[5] based on data derived from clinical trials (i.e. on human subjects) before it could be marketed and sold in the market regulated by that Competent Authority, is one of the most important bases on which a biotech company can apply for listing under Chapter 18A of the Listing Rules. Depending on the type of product, the Exchange would normally consider the following products have developed beyond the concept stage:

- **Pharmaceutical (small molecule drugs)** — It has passed Phase I clinical trials, or for products already approved (for example, the 505(b)(2) of FDA of the US), has successfully passed at least one clinical trial conducted on human subjects; and the relevant Competent Authority has no objection for it to commence Phase II (or subsequent) clinical trials.

---

4   See HKEX Guidance Letter HKEX-GL92-18 for details.
5   Competent Authority means the US Food and Drug Administration (FDA), the China Food and Drug Administration (CFDA), the European Medicines Agency (EMA) or any other authority recognised on a case by case basis.

- **Biologics** — In case of new biologic products, it has passed Phase I clinical trials, or for biosimilar, has passed at least one clinical trial conducted on human subjects; and the relevant Competent Authority has no objection for it to commence Phase II (or subsequent) clinical trials.
- **Medical devices (including diagnostic devices)** — The product is categorised as Class II (or equivalent) medical device (under the classification criteria of the relevant Competent Authority) or above; it has passed at least one clinical trial on human subjects; the Competent Authority has endorsed or not expressed objection for further clinical trials or the commencement of sales of the device.
- **Other biotech products** — This will be considered on a case by case basis; the applicant shall demonstrate that the relevant biotech product has been developed beyond the concept stage; the applicant shall have an appropriate framework or objective indicators for investors to make an informed investment decision.

(2) The applicant shall be primarily engaged in R&D for the purposes of developing its core product(s)
- It must have been engaged in the R&D of its core product(s) for a minimum of 12 months prior to listing;
- It must have, as its primary reason for listing, the raising of funds for R&D to bring its core product(s) to commercialisation;
- It must have durable patent(s), registered patent(s), patent application(s) and/or intellectual property in relation to its core product(s);
- If the applicant is engaged in the R&D of pharmaceutical (small molecule drugs) products or biologic products, it must demonstrate that it has a pipeline of these potential products.

(3) The applicant shall have recognition and secured investment from Sophisticated Investors

The applicant shall demonstrate that a reasonable degree of market acceptance exists for its R&D and biotech product. It must have previously received "meaningful investment" from at least one Sophisticated Investor at least six months before the date of the initial public offering (IPO) (which must remain invested in the applicant at IPO).

"Sophisticated Investor" refers to an investor that the Exchange considers to be sophisticated by reference to factors such as net assets or assets under management, relevant investment experience, and the investor's knowledge and expertise in the relevant field. Generally speaking, Sophisticated Investors include the following four types:
- A dedicated healthcare or biotech investment fund or an established fund with a division/department that specialises or focuses on investments in the biopharmaceutical sector;

- A major pharmaceutical/healthcare company;
- A venture capital fund of a major pharmaceutical/healthcare company; and
- An investor, investment fund or financial institution with minimum assets under management of HK$1 billion.

The Exchange considers an investment to be meaningful by reference to the nature of the investment, the amount invested, the size of the stake taken up and the timing of the investment. In general, there are three thresholds for a meaningful investment as follows:

- For an applicant with a market capitalisation between HK$1.5 billion to HK$3 billion, an investment of not less than 5% of the issued share capital of the applicant at the time of listing;
- For an applicant with a market capitalisation between HK$3 billion to HK$8 billion, an investment of not less than 3% of the issued share capital of the applicant at the time of listing; and
- For an applicant with a market capitalisation of more than HK$8 billion, an investment of not less than 1% of the issued share capital of the applicant at the time of listing.

(4) Additional listing requirements
- **Market capitalisation:** at least HK$1.5 billion at the time of listing;
- **Track record:** in its current line of business for at least two financial years, under substantially the same management;
- **Working capital:** the applicant shall ensure that it has sufficient working capital (after taking into account the IPO proceeds), as much as 125% of the applicant's costs for at least 12 months from the date of its prospectus, including general, administrative and operating costs and R&D costs.

2.3   Requirements for placement to cornerstone investors and original shareholders

Cornerstone investors and existing shareholders of a biotech company listed under Chapter 18A may only be allocated shares in the IPO if the company meets the additional minimum public float requirement at the time of listing of at least HK$375 million excluding shares allocated to them.

2.4   Special measures to manage risks

(1) Material changes in business activities
Without the prior consent of the Exchange, a biotech company listed under Chapter 18A must not effect any acquisition, disposal or other transaction or arrangement or a series

of acquisitions, disposals or other transactions or arrangements, which would result in a fundamental change in the principal business activities of the relevant issuer as described in its listing application document.

(2) Shorter delisting procedures

For issuers who have financial difficulties to an extent which seriously impairs their ability to continue their business or which has led to the suspension of some or all of their operations, and/or for issuers whose liabilities exceed their assets as at the balance sheet date[6], the Exchange will suspend the trading of, or even delist, their securities[7]. If the issuer fails to re-comply with the relevant requirements under Listing Rule 13.24 within 12 months, the Exchange will delist its securities.

(3) Special stock marker

The listed equity securities of a biotech company listed under Chapter 18A must have a stock name that ends with the marker "B". If a listed biotech company meets the relevant "profit test" or "market capitalisation/revenue test" or "market capitalisation/revenue/cash flow test" under Chapter 8, it can turn into a regular listed company upon the Exchange's approval, and its stock name will no longer bare the marker that ends with "B".

3 New chapter for secondary listing

3.1 Concepts and advantages of a secondary listing

In practice, many issuers listed overseas also wish to list in Hong Kong for liquidity and valuation support from the Hong Kong capital market. A primary listing in Hong Kong is one that results in the company becoming fully subject to the Listing Rules. A secondary listing, on the other hand, will mean that the company will be principally regulated by the rules and authorities of the jurisdiction where it is primary listed. The Exchange would also expect the dominant trading of the company's securities to be on the primary overseas exchange. On this basis, the Exchange exempts or waives certain requirements of the

---

6   See Listing Rule 13.24 for details.
7   See Listing Rule 6.01 for issues relating to delisting and suspension.

Listing Rules for overseas issuers with, or seeking, a secondary listing on it.

Therefore, for those issuers that are already listed overseas, application for a secondary listing in Hong Kong will be a most practical and least difficult option, compared to seeking a listing again in Hong Kong after being delisted from the overseas market, or seeking a dual primary listing in both markets. In particular, "China-concepts stocks" already listed overseas will have the following advantages from having a Hong Kong secondary listing:

- The US Generally Accepted Accounting Principles (GAAP) are acceptable for a secondary listing, so no extra auditing/re-auditing of the track record for three financial years and no reconciliations are needed, reducing the burden on the issuer in preparing the financial statements and relevant disclosure in the prospectus;
- Subject to conditions, a secondary listing applicant is entitled to a number of automatic and conditional waivers, including waivers of various continuing obligations of a primary listing on the Exchange (e.g. rules on notifiable transactions and connected transactions); certain qualified overseas listed issuers may submit its application on a confidential basis;
- A secondary listing could provide funding for the issuer's development through additional issuance of new shares; or the issuer can list by way of introduction, by way of sale of existing shares, or by way of repurchase and sale, without diluting existing shareholders' interests;
- Shares listed on the Exchange could serve as a valuation conduit between different trading venues, and are naturally closer to investors in Asian time zones, forming a full-day round circulation market for the securities, thus helping with more active trading in the shares;
- Given the convenience in capital operations in Hong Kong, a secondary listing in Hong Kong could provide better access to consumers in the Mainland and further improve brand recognition and presence.

Traditionally, the Exchange only accepts applications for a secondary listing from overseas issuers whose core businesses ("centres of gravity") are outside Greater China[8]. The primary purpose of this restriction is to prevent "regulatory arbitrage", i.e. companies with a centre of gravity in Greater China avoiding the full requirements of a primary listing by first listing on an overseas exchange and then a secondary listing in Hong Kong. In recent years, a number of large Mainland and non-Mainland companies from emerging and innovative sectors have been primary listed in the US or on other major international exchanges. Those from the Mainland sought these listings whilst the "centre of gravity"

---

8   For details, see HKEX's *Consultation Paper on A Listing Regime For Companies from Emerging and Innovative Sectors* published in February 2018.

restriction was in force and therefore did so for reasons other than "regulatory arbitrage". The application of the "centre of gravity" prohibition has prevented these Mainland companies from accessing Hong Kong investors via a secondary listing.

As such, the Exchange added a new Chapter 19C "Secondary Listings of Qualifying Issuers" to the Listing Rules since 30 April 2018 to attract fast-growing and innovative companies and, in particular, facilitate secondary listings in Hong Kong of applicants from Greater China.

3.2 Additional conditions and exceptions for a secondary listing on the Exchange by qualifying issuers under Chapter 19C

3.2.1 Qualifying issuers

There are three categories of "qualifying issuers" under Chapter 19C:
- **"Grandfathered Greater China Issuer":** an issuer with its centre of gravity in Greater China and was primary listed on a Qualifying Exchange[9] on or before 15 December 2017;
- **"Non-Grandfathered Greater China Issuer":** an issuer with its centre of gravity in Greater China and was primary listed on a Qualifying Exchange after 15 December 2017;
- **"Non-Greater China Issuer":** an issuer with its centre of gravity outside Greater China and is primary listed on a Qualifying Exchange.

3.2.2 The applicant shall be an innovative company

In addition to the general requirements for secondary listings, a qualifying issuer seeking a secondary listing under Chapter 19C shall be an innovative company. In general, an innovative company have two or more of the following characteristics:
- Applying new technologies, innovations or new business models, which also serve to differentiate the company from existing competitors;
- R&D is a significant contributor to its expected market value and a principal activity of the applicant and constitutes a major expense;
- The success of its operation is attributable to its unique business features or intellectual property; and/or

---

9   Qualifying Exchanges include The New York Stock Exchange LLC, Nasdaq Stock Market or the Main Market of the London Stock Exchange plc (and belonging to the UK Financial Conduct Authority's "Premium Listing" segment).

- It has an outsized market capitalisation or intangible asset value relative to its total tangible asset value.

The definition of "innovation" varies with the industry and market environment of the applicant, and will change with developments in technologies, market conditions and the sector concerned.

3.2.3   Grandfathered Greater China Issuers

Conditions and qualifications for listing:
- The issuer shall be primary listed on a Qualifying Exchange and have good regulatory compliance records for at least two full financial years;
- A market capitalisation of at least HK$40 billion at the time of listing; or a market capitalisation of at least HK$10 billion at the time of listing and a revenue of at least HK$1 billion for the most recent audited financial year;
- The issuer must demonstrate how the domestic laws, rules and regulations to which it is subject and its constitutional documents, in combination, provide protection equivalent to the key shareholder protections in Hong Kong (changes to constitutional documents may be required, as necessary).

    For example, super-majority vote of shareholders is required to approve certain matters (including variation of rights attached to any class of shares, material changes to constitutional documents and voluntary winding-up); appointment, removal and the remuneration of auditors require the approval of a majority of shareholders or other body(ies) independent of the board of directors.

Exceptions:
- As in the case of traditional non-innovative companies with a centre of gravity outside Greater China seeking a secondary listing in Hong Kong, Grandfathered Greater China issuers are waived from complying with a number of continuing obligations of a primary listing on the Exchange (e.g. notifiable transaction and connected transaction rules);
- No need to comply with requirements in the Listing Rules relating to the content and provisions of the constitutional documents;
- The issuer can secondary list with their existing WVR structures and variable interest entity (VIE) structures intact (if they have them);
- No need for the issuer to comply with WVR safeguards in the Listing Rules (other than those requiring disclosure);
- If the "bulk of trading" of the issuer's shares moves permanently to the Exchange's

markets[10], then the requirements in the Listing Rules relating to a primary listing, but not WVR safeguards (other than disclosure requirements), would apply.

3.2.4 Non-Grandfathered Greater China Issuers

Conditions and qualifications for listing:
- The issuer shall be primary listed on a Qualifying Exchange and have good regulatory compliance records for at least two full financial years;
- A market capitalisation of at least HK$40 billion at the time of listing; or a market capitalisation of at least HK$10 billion at the time of listing and a revenue of at least HK$1 billion for the most recent audited financial year;
- The issuer must demonstrate the shareholder protections in the overseas exchange on which it is primarily listed are at least equivalent to those provided in Hong Kong;
- The issuer must comply with requirements in the Listing Rules relating to the content and provisions of constitutional documents.

Exceptions:
- As in the case of traditional non-innovative companies with a centre of gravity outside Greater China seeking a secondary listing in Hong Kong, Non-Grandfathered Greater China issuers are waived from complying with a number of continuing obligations of a primary listing on the Exchange (e.g. notifiable transaction and connected transaction rules).

To deter "regulatory arbitrage":
- WVR structures and VIE structures (if they have them) must conform to all requirements in the Listing Rules relating to a primary listing;
- If the "bulk of trading" of the issuer's shares moves permanently to the Exchange's markets, then requirements in the Listing Rules relating to a primary listing would apply.

3.2.5 Non-Greater China Issuers

Conditions and qualifications for listing:
- The issuer shall be primary listed on a Qualifying Exchange and have good regulatory compliance records for at least two full financial years;
- Non-WVR issuers shall have a market capitalisation of at least HK$10 billion at the time of listing. WVR issuers shall have a market capitalisation of at least HK$40

---

10 The Exchange will consider this is the case if 55% or more of the global trading value (including that of depositary receipts) of the issuer's shares during the issuer's most recent financial year took place on the Exchange's markets.

billion at the time of listing; or a market capitalisation of at least HK$10 billion at the time of listing and a revenue of at least HK$1 billion for the most recent audited financial year;

- The issuer must demonstrate how the domestic laws, rules and regulations to which it is subject and its constitutional documents, in combination, provide protection equivalent to the key shareholder protections in Hong Kong (changes to constitutional documents may be required, as necessary).

Exceptions:

- As in the case of traditional non-innovative companies with a centre of gravity outside Greater China seeking a secondary listing in Hong Kong, Non-Greater China Issuers are waived from complying with a number of continuing obligations of a primary listing on the Exchange (e.g. notifiable transaction and connected transaction rules);
- The issuer can secondary list with their existing WVR structures and VIE structures intact (if they have them);
- No need for the issuer to comply with WVR safeguards in the Listing Rules (other than those requiring disclosure);
- Applicable regulatory requirements and exceptions remain unchanged if the "bulk of trading" of the issuer's shares moves permanently to the Exchange's markets.

Chapter 6

The practice of dual-class share structure in Hong Kong

Chief China Economist's Office

Hong Kong Exchanges and Clearing Limited

Summary

With an expanding number of growth companies in Internet, high-tech and biomedical research and development industries across the world, a listing regime for weighted voting rights (WVR) has been introduced or is being considered by major international financial markets, like the US, the UK and Singapore, to facilitate the listing and financing of such companies.

The primary concern about adopting a listing regime for WVR is the separation of management control and cash-flow rights ownership, which is expected to aggravate the corporation's agency problem and undermine the management's accountability to shareholders. However, a dual-class share structure (DCS structure) is conducive to a start-up's long-term development, especially if it is an innovative technology company with substantial initial investment, high uncertainties and high growth potential. To be specific, a DCS structure helps an innovative company build its long-term value, incentivises the founders to instill the company with greater innovation and more human capital, and forestalls hostile takeover attempts. To a certain extent, it is also seen as a self-protective measure taken by start-ups to avoid market short-term behaviour when there is an over-concentration of institutional investors in the financial market. Moreover, according to some empirical studies, corporate values were improved and agency costs were reduced after a DCS structure had been adopted.

Certainly, there is still much debate in theoretical and empirical studies as to whether a DCS structure incurs higher agency costs than a single-class share structure and is therefore less conducive to the protection of shareholders' rights. So how can a company with a DCS structure enhance internal supervision to ensure effective monitoring of its controlling shareholders? Several options have been suggested: (1) imposing suitable restrictions over the use of superior voting rights, including the cap of voting rights ratio of WVR shares relative to other ordinary shares, and a clear delineation of the applicable scope of superior voting rights; (2) establishing clear exit and transfer mechanisms for superior voting rights, including the commonly known "sunset clauses" and restrictions on the transfer of superior voting rights; (3) enhancing corporate governance and the parallel use of internal and external control mechanisms.

In April 2018, HKEX put forward new measures for allowing DCS structures while imposing control and restrictions as appropriate. Under the new measures, applicants are required to possess certain characteristics before they can list with WVR. The HKEX will reserve the right to reject an applicant on suitability grounds if its WVR structure is an extreme case of non-conformance with governance norms (for example, if the ordinary shares would carry no voting rights at all). HKEX also puts forward detailed investor protection measures to be applied to WVR companies after their listing. These include measures that restrict the power of WVR, protect the voting rights of non-WVR shareholders, and strengthen corporate governance and disclosure requirements. Issuers with WVR structures will be differentiated from others through a unique stock marker "W" after their stock name. In addition, WVR beneficiaries must be directors of the issuers to ensure they operate the companies with the obligations of a director as set out under relevant laws and regulations. The WVR attached to a WVR beneficiary's shares will lapse once the WVR beneficiary transfers the WVR shares to another person, or dies or is incapacitated, or ceases to be a director. WVR are therefore subject to natural sunset clauses and will not exist indefinitely.

Appropriate reforms in the listing regime with suitable listing rules will encourage the emergence of large innovative companies, contribute to the development of new-economy industries in the region, help upgrade the regional economy and expand its horizon. This is the kind of long-term positive impact that capital market reforms could have on the Hong Kong economy.

1 Application of weighted voting rights (WVR) in different countries

WVR refers to voting rights and other related rights enjoyed by certain shareholders that are disproportionate to the economic interest held by such shareholders in the company. The rights of such special shares take multiple forms — the holders of which may have no voting right, or they may have preferential voting rights, or enhanced or exclusive rights to elect directors. The most common share structure that adopts WVR encompasses two classes of shares — "class A" shares with one vote per share and "class B" shares with multiple votes per share. "Class B" shareholders are generally the founders, early partners, key strategic investors or senior management of the company. In practice, most innovative companies (exemplified by the share structures of Google and Baidu in the US), at the time of their initial public offers (IPO), list their A shares (one vote per share) on an exchange, and issue their B shares (multiple votes, usually 10 votes, per share) to the company's existing management. B shares may be converted into A shares on a one-to-one ratio, but A shares cannot be converted into B shares.

WVR structures, in essence, are dual-class share (DCS) structures under which founders can exercise effective control over a company with only a small percentage of shares with superior voting rights. Under such structures, founders of new-economy companies can continue to pursue innovation and maximisation of growth and company value for shareholders without the pressure from new investors. With an expanding number of growth companies in Internet, high-tech and biomedical research and development (R&D) industries, a listing regime for WVR has been introduced or is being considered by major international financial markets, like the US, the UK and Singapore, to facilitate the listing and financing of such companies.

1.1 Evolution and effects of WVR structures in the US

The US is one of the earliest countries where companies adopt a DCS structure. It is also a country in which such structure has operated relatively effectively. DCS structures dated back to 1898 when the International Silver Company issued 9 million preferential shares and 11 million ordinary shares with no voting rights. This was the first time in history when shareholdings were separated from voting rights. DCS structures became popular in the 1920s. Between 1927 and 1932, a total of 288 companies issued shares with no or limited voting rights[1]. Despite preference for a one-share-one-vote framework in the ensuing 40 years, some companies such as Ford continued to issue shares with different voting rights.

The rise of mergers and acquisitions (M&A) and fierce competition among exchanges in the 1980s substantially facilitated the adoption and use of DCS structures. At first, the three major securities exchanges in the US — New York Stock Exchange (NYSE), American Stock Exchange (AMEX) and NASDAQ — took different views towards these share structures. On one end of the spectrum, NYSE adhered to a strict policy of prohibition, demanding compulsory delisting of companies that sought to adopt WVR through share structuring. On the other end, NASDAQ had no restriction at all for WVR shares. In between was AMEX which allowed a conditional listing of dual-class ordinary shares. In 1984, NYSE suspended its policy to delist companies with a DCS structure and set up a special committee to evaluate its long-standing commitment to the one-share-one-vote principle. After a series of reform, NYSE, AMEX and NASDAQ accepted in 1994 the call of the US Securities and Exchange Commission (SEC) for a unified policy on the listing of companies with DCS structures. They agreed that, while a company must not reduce or restrict, through any action or share issuance, the voting rights of holders of ordinary shares that had been issued, a company issuing new shares might adopt a DCS structure[2]. With a unified policy on DCS structures, companies that adopted such a structure steadily increased in the US between 1994 and 1998. About 11.9% of IPOs and 24.9% of total funds raised by IPOs were attributed to such companies.

Since 2000, new listings have been dominated by high-tech Internet companies, and companies with DCS structures in the US notably increased. The DCS structure with which Google listed in 2004 was particularly popular in its industry. During the period from 2003 to before the financial tsunami in 2008, 64 out of 681 IPOs (9.4%) were companies that adopted a DCS structure, contributing to 20.8% of the total IPO funds raised. In the years after the financial tsunami up to 2013, 76 out of 461 IPOs (16.5%) were companies that adopted a DCS structure, contributing to 34.1% of the total IPO funds raised.[3] (see Table 1.)

1 Ashton, D. C. (1994). "Revisiting Dual-Class Stock", *John's Law Review*, Volume No. 68, p.863.

2 See Jiang Xiaomin (2015). "US' Dual-Class Share Structure: Development and Controversies" (《美國雙層股權結構：發展與爭論》), *Securities Market Herald*, September 2015.

3 Howell, J. W. (2017). "The survival of the US dual class share structure", *Journal of Corporate Finance*, Volume No. 44, pp.440-450.

| Table 1. The evolution of the use of dual-class share structures in the US since 1980s ||
Period	Application
1985	180 listed companies (2.8% of total number) adopted DCS structures
1988 — 1992	Companies with DCS structures accounted for 5.4% of IPOs
June 1994 — Sep 1998	Companies with DCS structures accounted for 11.9% of IPOs and 24.9% of IPO funds raised
2003 — 2008	Companies with DCS structures accounted for 9.4% of IPOs and 20.8% of IPO funds raised
2008 — 2013	Companies with DCS structures accounted for 16.5% of IPOs and 34.1% of IPO funds raised

Source: Howell, J. W. (2017). "The survival of the US dual class share structure", *Journal of Corporate Finance*, Volume No. 44, pp. 440-450.

DCS structures have been used extensively across a range of industries in the US. Figure 1 shows that 24 industries had more than 6% of the listed companies adopting DCS structures in 2010; and out of 44 industries of the listed companies in the US, only 6 did not have companies with DCS structures. Traditional industries such as machinery, retail and agriculture had 6% to 8% of companies with DCS structures. Companies in communications and printing and publishing industries adopted DCS structures most extensively (26.58% and 22.64% respectively). This shows the use of DCS structures is highly correlated with industry characteristics. The more information- and electronic-related an industry was, or the more a company required an organisational structure that conforms to the characteristics of the new economy, and the more receptive its investors were to DCS structures.

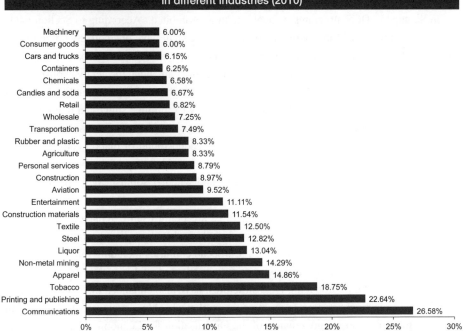

Figure 1. Percentage of US listed companies with DCS structures (more than 6%) in different industries (2010)

Industry	Percentage
Machinery	6.00%
Consumer goods	6.00%
Cars and trucks	6.15%
Containers	6.25%
Chemicals	6.58%
Candies and soda	6.67%
Retail	6.82%
Wholesale	7.25%
Transportation	7.49%
Rubber and plastic	8.33%
Agriculture	8.33%
Personal services	8.79%
Construction	8.97%
Aviation	9.52%
Entertainment	11.11%
Construction materials	11.54%
Textile	12.50%
Steel	12.82%
Liquor	13.04%
Non-metal mining	14.29%
Apparel	14.86%
Tobacco	18.75%
Printing and publishing	22.64%
Communications	26.58%

Source: Arugaslan, O., D. O. Cook & R. Kieschnick (2010). "On the decision to go public with dual class stock", *Journal of Corporate Finance*, Volume No. 16(2), pp.170-181.

1.2   Application of WVR structures in other countries

Dual-class share issuance fell in the UK in mid-1960s due to the widespread institutional participation in stock investment and the resultant rising demand for high corporate governance standards and shareholder protection, but the trend reversed in the 1990s. Table 2 shows that 23.9% of UK companies adopted DCS structures in 1996. Unlike the US, the UK currently allows only some sections of its market to list companies with a DCS structure. For Premium Listing (a section of the UK Main Board) which has stricter listing requirements, issuers have to comply with super-equivalent rules on information disclosure. For example, the listing applicant must be able to instil investor confidence by demonstrating an independent operation, a three-year track record with revenue, sufficient operating capital and a financial statement with unqualified opinions. Rules for Standard Listing (the other section of the UK Main Board) mainly apply to stocks, depositary receipts and bonds, and adopt the minimum requirements of the European Union

(EU) rather than the UK's super-equivalent rules. The UK listing rules amended in May 2014 provided that the listing regime for WVR only applies to Standard Listing.

In Singapore, DCS structures were at one time prohibited. According to Section 64(1) of the Company Act of Singapore, one ordinary share shall have one corresponding voting right with the exception of management shares issued by newspaper companies under the Newspaper and Printing Presses Act. In 2011, regulators amended the act by abolishing the restriction that one share could only confer one voting right and allowing public companies to issue WVR shares. In 2016, the Listings Advisory Committee suggested the Singapore Exchange (SGX) accept dual-class shares with appropriate protective measures to contain their risks. On 28 March 2018, SGX launched a second-round of market consultation on DCS structures, including seeking opinions on the consideration of the business models of DCS companies, event-based sunset clauses and whether sophisticated investors have participated in the company.

Dual-class or multiple-class share structures are being used extensively on a global basis. 30 out of the world's 46 largest national stock markets have listed companies that are adopting or had adopted a DCS structure[4]. European countries including Finland, Sweden, France and Ireland, in particular, have extensively adopted such structures (see Table 2). These countries, when introducing DCS structures, have also enhanced their corresponding shareholder protection measures to promote the market's healthy development. As stated in the principles of corporate governance of the Organisation for Economic Cooperation and Development (OECD), all investors should be able to obtain information about the rights attached to all series and classes of shares of a company before they purchase any shares of the company; and any changes in economic or voting rights should be subject to approval by shareholders of those classes of shares which are negatively affected, to ensure equitable treatment of all shareholders. Whether dual-class or multiple-class share structures should be accepted, therefore, depends to a certain extent on the investment knowledge of investors, the adequacy of information disclosure and the related control mechanisms. These are discussed in detail below.

---

4   Nenova, T. (2003). "The value of corporate voting rights and control: a cross-country analysis", *Journal of Financial Economics*, 2003, Volume No. 68, pp.325-351.

Table 2. Percentage of companies with DCS structures in different countries		
Country	Percentage	As of end of the year
Sweden	66.10%	1998
Switzerland	51.20%	1999
Italy	41.40%	1996
Finland	37.60%	1999
Ireland	28.10%	1999
UK	23.90%	1996
Australia	23.30%	1999
Germany	17.60%	1996
Norway	13.20%	1998
Canada	10.20%	1998
US	6.10%	2002
France	2.60%	1996

Source: Howell, J. W. (2017). "The survival of the US dual class share structure", *Journal of Corporate Finance*, Volume No. 44, pp.440-450.

2   How do WVR structures benefit companies?

The primary concern about adopting a WVR structure is the separation of management control and cash-flow rights ownership under such a structure. With superior voting rights, management may pursue personal gains not in the best interest of the company, giving rise to unfairness and aggravating the agency problem. Possible deeds include management's quest for private interest, excessive wages and fringe benefits, and their irresponsibility towards the company's major decisions[5].

WVR structures (or DCS structures) may also reduce management's accountability to shareholders. As most shareholders are holding inferior voting rights, they do not have the power to change the management even if it is performing poorly. It is also impossible for shareholders to accept a takeover offer that is opposed by the founders or management.

---

5   Howell, J. W. (2010). *The dual class stock structure in the United States.*

That should explain why DCS structures are not preferred by many investors (especially institutional investors)[6].

However, a DCS structure is critical to a start-up's long-term development, especially if it is an innovative technology company with substantial initial investment, many uncertainties and high growth potential. This is explained below.

(1) DCS structure helps an innovative company to build its long-term value.

With professional expertise and industry judgement, founders of innovative technology companies and their teams can make quick decisions in response to industry changes and high uncertainties in the external environment. This would significantly impact their companies' development. A DCS structure allows a founder to control a company through superior voting rights and focus on a company's long-term gain without being distracted by short-term share price movements. Therefore, DCS structures are vital to the company's implementation of its business model and long-term strategic planning.

Take the example of Facebook's acquisition of WhatsApp in February 2014. It took only 11 days for Facebook to take over the mobile social application, paying US$19 billion for this company with only 50+ staff members. The majority of market practitioners, including investors, considered the deal overpriced and not beneficial. Their pessimism caused a sharp plunge in Facebook's share price the day after the acquisition, driving down the company's market value by more than US$3 billion. Facebook, however, believed that WhatsApp, with its user coverage exceeding 90% in a number of countries/regions, would give it access to billions of active social media users. The deal would also bring in WhatsApp's brightest minds, removing a potential competitor and overcome Facebook's weakness in mobile social media platforms. The DCS structure, to a certain extent, allows Facebook's management to evade market pressure for short-term gains and to make the best decision for the company's long-term development, to expectedly generate sustainable investment returns to small and medium-sized investors. Within one year after the acquisition, Facebook stock reported a return that surpassed that of Google and the NASDAQ index. In almost three years after the acquisition, the cumulative return on Facebook stock exceeded that of Google and the NASDAQ index[7]. The case demonstrates that DCS structure is conducive to the decision-making process of innovative companies

---

6   Jarrell, G. A. & A. B. Poulsen (1988). "Dual-class recapitalizations as antitakeover mechanisms: The recent evidence", *Journal of Financial Economics*, Volume No. 20, pp.129-152.

7   Regarding the case of Facebook, see Li Haiying, Li Shuanghai & Bi Xiaofang (2017). "Protection of Small and Medium-Sized Investors under Dual-Class Share Structure: A Case Study of Facebook's Acquisition of WhatsApp" (《雙重股權結構下的中小投資者利益保護——基於 Facebook 收購 WhatsApp 的案例研究》), *China Industrial Economics*, 2017, Volume No. 1.

for increasing the company's long-term value.

(2)   A DCS structure effectively incentivises the founders to instill the company with greater innovation and more human capital.

Emerging companies are characterised by their ability to innovate, which to a large extent depends on the founders' creativity, their spiritual appeal and their insights on industry trends.   Any change in management might deprive the founders of returns on their intellectual property and human capital investment.   That would make it impossible for biomedical companies with long cycles and new-economy companies that require sophisticated expertise to accomplish their knowledge build-up and innovation.   Founders take more risks than external shareholders in pursuing investment objectives, resources and corporate development.   A DCS structure is in essence more like an incentive mechanism.   By protecting the interests of the founders, it helps the founders focus on continuous innovation, enhances the sense of belonging and cohesion within the company and enables the founders' team to create bigger values for the company[8].

(3)   A DCS structure forestalls hostile takeover attempts.

A hostile takeover generally refers to the acquisition of equity ownership in a target company without the consent of the target company's board of directors or without prior negotiation with existing shareholders of the target company, followed by a change of the target company's management and a takeover of the company's operational control.   A DCS structure confers critical trading rights on shareholders with superior voting rights.   This means that a bidder who successfully instigates a hostile takeover and obtains ordinary shares would still be unable to acquire sufficient decision making power to control a company or change its management.   A company that adopts a DCS structure is unlikely to be the target of a hostile takeover attempt.   That is why family businesses generally prefer these share structures.

(4)   A DCS structure may be seen as a self-protective measure taken by start-ups to evade market short-term behaviour given the over-concentration of institutional investors in the financial market.

Studies[9] show that primary voting rights are mostly held by fund managers who manage

8   He, L. (2008).   "Do founders matter? A study of executive compensation, governance structure and firm performance", *Journal of Business Venturing*, Volume No. 23(3), pp.257-279.

9   David Berger (2018).   *Why Dual-Class Stock: A Brief Response to Commissioners Jackson and Stein*, Wilson Sonsini Goodrich & Rosati, 22 February 2018.

other people's capital. These parties or institutional investors are generally motivated
by short-term gains and are concerned more about the short-term movements of share
prices. Excessive "financialisation" changes the nature of equity investment — hordes
of institutional investors who care for returns on investment dominate the stock market in
place of long-term investors that focus on a company's long-term healthy development.
DCS structures are a solution to help evade the negative impact of such problems on the
company.

(5) According to empirical studies, the adoption of a DCS structure led to improved corporate values and reduced agency costs[10].

For example, Dimitrov and Jain (2006)[11], based on a sample of 176 US companies,
found that a shift to a DCS structure from a single-class share structure resulted in an
impressive 23.11% return. For companies which newly issued shares with low voting
rights, the return was even greater. Jordan, Liu and Wu (2014)[12] found that companies with
DCS structures paid more cash dividends than companies that adopt a single-class share
structure, and regular dividend payouts were higher than special dividends and repurchases.
This indicates that despite the adoption of DCS structures, such companies did not neglect
the interests of external shareholders but fulfilled their undertaking by paying more
dividends. Howell (2017)[13] studied the survival time of companies with DCS structures,
and concluded that DCS structures are essential to the survival of an innovative company
since they allow a company to deliver good results by protecting its share ownership.
Based on the statistics of global Internet companies listed in the US, Shi Xiaojun and others
(2017)[14] found that DCS structures significantly motivated hi-tech companies to innovate.
This was especially the case where such companies were in developed countries with

---

10 Agency costs result from the separation of ownership and operational control. While owners and shareholders want
the management to run a company to maximise shareholders' returns, the management who are not shareholders
or who holds only a small percentage of shares often run the company based on their own interests. For example,
they may obtain additional benefits through in-service consumption at the expense of shareholders. The information
inequality between shareholders and management is also substantial. Management, as front-line operator with
knowledge of the company's cash flows, is more informed than shareholders. It is not easy for shareholders to
determine whether management's actions meet the objective of maximising shareholders' returns.

11 Dimitrov. V. & P. C. Jain (2006). "Recapitalization of one class of common stock into dual-class: Growth and long-run
stock returns", *Journal of Corporate Finance*, Volume No. 12(2), pp.342-366.

12 Jordan, B. D., M. H. Liu & Q. Wu (2014). "Corporate payout policy in dual-class firms", *Journal of Corporate Finance*,
Volume No. 26, pp.1-19.

13 Howell, J. W. (2017). "The survival of the US dual class share structure", *Journal of Corporate Finance*, Volume No.
44, pp.440-450.

14 Shi Xiaojun & Wang Aoren (2017). "Impact of the Specialty of Corporate Governance on Innovation: Global Evidence
from the Dual-class Structure of Internet Firms" (《獨特公司治理機制對企業創新的影響——來自互聯網公司雙層股權制
的全球證據》), *Economic Research Journal*, Volume No. 1, pp.149-164.

good external check-and-balance mechanisms and where the founders were the primary administrators.

3   Enhancement of regulation of companies with DCS structures

Certainly, there is still much debate in theoretical and empirical studies as to whether a DCS structure incurs higher agency costs than a single-class share structure and is therefore less conducive to the protection of shareholders' rights. Different samples and different development cycles of the companies under study will give different results.

While DCS structures could motivate founders and management and enhance a company's long-term value, they may also undermine internal governance, worsen asymmetric information and the agency problem and therefore reduce a company's value. So how can a company with a DCS structure effectively monitor its controlling shareholders? Different control mechanisms have been implemented worldwide, as discussed in the following sub-sections.

3.1   Imposing appropriate restrictions on the exercise of superior voting rights

Restrictions are mainly in the following two ways:

One way is to restrict the difference in voting rights — the voting rights attached to superior voting rights shares are capped at no more than ten times of the voting rights of the same number of ordinary shares. Although the US does not set a limit to this ratio of voting rights, the international practice is to restrict the voting right of one superior voting rights share to be no more than ten times the voting right of an ordinary share (a WVR ratio of 10:1). This ratio has been adopted by companies like Google, Facebook and Baidu (Table 3 sets out the WVR ratio and structure used by Mainland companies listed in the US). Other regions like Sweden and other European exchanges also require each share with superior voting rights to have a maximum voting right equal to that of 10 ordinary shares. When

SGX launched its consultation on the introduction of dual-class shares, it also proposed that each multiple-vote share's number of votes is to be capped at 10.

Table 3. WVR ratios and structures used by certain Mainland companies listed in the US				
Company	IPO date	Business	Share structure	Controlling shareholders
Baidu, Inc.	04/08/2005	Internet search engine	Class A (listed): 1 vote Class B (non-listed): 10 votes	Held by founders: • 15.9% of equity; • 53.5% of voting rights
Mindray Medical International Ltd.	25/09/2006	Development, manufacturing and marketing of medical devices worldwide	Class A (listed): 1 vote Class B (non-listed): 5 votes	Collectively held by all directors and executives: • 28.8% of equity; • 64.2% of voting rights
Shanda Games Ltd.	24/09/2009	Development and operation of online games	Class A (listed): 1 vote Class B (non-listed): 10 votes	Held by Shanda International: • 70.8% of equity; • 96.0% of voting rights
eCommerce China Dangdang Inc.	07/12/2010	Online B2C commerce platform	Class A (listed): 1 vote Class B (non-listed): 10 votes	Held by founders: • 35.3% of equity; • 83.3% of voting rights
Qihoo 360 Technology Co. Limited	29/03/2011	Internet and mobile security products	Class A (listed): 1 vote Class B (non-listed): 5 votes	Held collectively by all directors and executives (including two co-founders): • 40.4% of equity; • 64.9% of voting rights
Phoenix New Media Limited	12/05/2011	Media content provider	Class A (listed): 1 vote Class B (non-listed): 1.3 votes	Held by Phoenix Satellite TV: • 52.8% of equity; • 59.2% of voting rights
Youku Tudou Inc.	Youku and Tudou merged on 23/08/2012	Online video	Class A (listed): 1 vote Class B (non-listed): 4 votes	Held by founders: • 21.3% of equity; • 51.5% of voting rights
LightInTheBox Holding Co., Ltd.	06/06/2013	Global Internet retailer	One class of shares entitles the holder to one vote per share on most matters. Founders have three votes per share for voting on a change in control.	On change-of-control matters, founders have 43.0% of the voting rights with a holding of 20.1% in equity.
Autohome Inc.	10/12/2013	Online automobile sales	Class A (listed): one vote Class B (unlisted): one vote per share but carries up to 51% of voting rights if the controller's equity holding in the company is below 51% but above 39.3%	Held by Telstra: • 65.4% of equity; • 65.4% of voting rights

(continued)

Table 3. WVR ratios and structures used by certain Mainland companies listed in the US				
Company	IPO date	Business	Share structure	Controlling shareholders
iKang Healthcare Group, Inc.	08/04/2014	Private healthcare provider	Class A (listed): 1 vote Class C (non-listed): 15 votes	Held by founders: • 14.3% of equity; and • 35.9% of voting rights
JD.com	21/05/2014	Online direct sales	Class A (listed): 1 vote Class B (non-listed): 20 votes	Held by founders: • 20.7% of equity; • 83.7% of voting rights

Source: HKEX's WVR Concept Paper published in August 2014.

The second way is that, when designing its structure, a company should set out clearly the applicable scope of superior voting rights. On major matters involving the corporate operation and management by the company or the controlling shareholders (e.g. hostile takeovers), strategic decisions (e.g. corporate culture or business philosophy), national security and public interest, etc., the founders and certain shareholders can have additional voting rights. However, on matters directly relating to the legitimate personal interests of external investors (e.g. connected transactions or external guarantees or other major use of the company's properties, mandatory disclosure of core information, nomination of supervisors or independent directors etc.), the difference between superior voting rights and ordinary voting rights should be reduced or restored to one vote per share[15]. This enables ordinary shareholders to have a greater say in making decisions on major transactions and connected transactions. This will conform to the original purpose of a DCS structure, i.e. preventing a dilution of control while reinstating shareholders' oversight role over the company.

3.2　Clear exit and transfer mechanisms for superior voting rights

One automatically triggered mechanism is that when there is a transfer of the shares with superior voting rights, these shares will be automatically restored back into shares with ordinary voting rights. Ordinary shareholders accept the adoption of a WVR structure mainly because of their trust in the founders, including in their ability to innovate and managerial capability. When the founders or controlling shareholders leave the company or transfer their shares to a third party, it should be considered to go back to the original voting rights ratio as the company's control and operation has changed and the conditions for granting WVR no longer exist.

---

15 Cao Yang (2017). *Dual-class shares in China* — Introduction and System (《中國雙層股權結構的引入與規制》), December 2017.

In practice, different countries have different restrictions on the transfer of superior voting rights. In the US, superior voting rights are generally not tradable. Shares with superior voting rights are automatically converted into ordinary shares of one vote per share when they are transferred. The Toronto Stock Exchange in Canada requires companies with WVR structures to provide coat-tail protection to external shareholders, ensuring that bidders for shares with superior voting rights will also make an offer to holders of ordinary shares under the same conditions. The requirement will prevent internal shareholders of a company from selling their control in the company at a high premium at the expense of other shareholders. SGX proposed in its recent consultation paper that superior voting rights have to be converted into ordinary shares under specific conditions such as when a shareholder resigns as director or offloads his shares.

"Sunset clause" is another possible mechanism. Studies show that shares of companies with a DCS structure were traded at a premium shortly after their IPO, but such premium would disappear as the company became mature. This reflected that the costs and benefits of the DCS structure would evolve over the lifetime of a company — at the early stage, the protection of the founders' control is beneficial because the ability to innovate is vital to a company's competitiveness; but the effectiveness of DCS structures would need to be reconsidered when the company reaches a certain point in time (usually known as "sunset")[16]. Therefore, a company might impose certain restrictions on the conditions for the continued adoption of its DCS structure when it goes public, to reflect its management's willingness to return the voting rights to shareholders in some day.

Sunset clauses are in practice not common. Table 4 sets out the use of sunset clauses in some companies. Sunset clauses lay down the conditions for the restoration of superior voting rights into ordinary voting rights — either a minimum shareholding ratio for the founders, or a time limit (a certain number of years after listing) for the WVR structure.

Table 4. Triggers for sunset clauses used by some companies		
Company	Year of IPO	Triggers of sunset clauses
Groupon	2011	5 years after listing (became one vote per share in 2016)
Kayak Software	2012	7 years after listing
Yelp	2012	7 years or superclass falls below 10% of outstanding common
Workday	2012	20 years or superclass falls below 9% of outstanding common
Apptio	2016	7 years or superclass falls below 25% of outstanding common
Nutanix	2016	17 years after listing

---

16 Jackson, Robert J. (2018). *Perpetual Dual-Class Stock: The Case Against Corporate Royalty*, 15 February 2018.

(continued)

Table 4. Triggers for sunset clauses used by some companies		
Company	Year of IPO	Triggers of sunset clauses
Hamilton Lane	2017	10 years or founders and employees hold less than 25% of voting power
MuleSoft	2017	5 years after listing or when shares with superior voting rights are less than 15% of ordinary shares

Source: Council of Institutional Investors.

3.3   Enhancing corporate governance and the parallel use of internal and external controls

The agency problem of DCS structures can be addressed by internal measures, such as compulsory information disclose, including the disclosure of the WVR structure adopted and the associated risks. Others include the disclosure of the identities of WVR beneficiaries and the setup of incentives and penalty systems for controlling shareholders.

External control mechanisms can also impose control over corporate management. IPO pricing, for example, can be a market-based penalty mechanism. If investors perceive an agency problem in the management, the IPO price would have a big discount to compensate investors for their potential loss. The market would also drive the companies to select the appropriate share structures. Howell (2017)[17] found that 61 US companies with DCS structures had returned to a one-share-one-vote model, with positive market response. This reflects the capability of the market in exercising self-control.

DCS structures have pros and cons. On the one hand, the strategic vision and entrepreneurial spirit of the controlling founders drive a company's long-term development. On the other hand, their superior voting power impairs the interests of external shareholders and public investors. With good internal control and a flexible and suitable legal framework, a company can make the best share structure arrangement based on its own circumstances. A DCS structure will then be able to effectively deliver its comparative advantages.

---

17 Howell, J. W. (2017). "The survival of the US dual class share structure", *Journal of Corporate Finance*, Volume No. 44, pp.440-450.

4 Application and discussion of WVR in Hong Kong

In the past, five companies in Hong Kong had adopted DCS structures. Some of these have privatised or withdrawn their listings. Swire Pacific is currently the only listed company in Hong Kong that has class B shares in issue[18]. Hong Kong had banned the listing of companies with WVR in 1987 and adopted a listing regime that only accepted one vote per share — an arrangement under which each and every share enjoys the same voting rights to ensure proportionality between voting rights and equity holding, and equal treatment to all shareholders.

In 2004, the trend of listing with WVR structures began among innovative and technology companies, exemplified by Google in the US. Except for Twitter, most US technology companies were listed with a DCS structure. A considerable number of China-concept stocks listed in the US have a similar share structure. As of June 2017, 33 out of 116 (28%) Mainland companies listed in the US used WVR structures. Their market capitalisation reached US$561 billion, representing 84% of the market value of all Mainland companies listed in the US. 18 out of the 33 (55%, accounting for 84% of the market capitalisation) were innovative technology companies[19]. The US's embrace of DCS structures encouraged other countries to follow. Countries like the UK, Germany and Canada introduced WVR structures through introducing new listing boards or segmenting a listing board to serve the purpose.

In Hong Kong, there had been an extensive discussion of WVR structures in 2014 in response to Alibaba's listing demand. In 2017, HKEX launched a market consultation on the proposed New Board to explore new possibilities for WVR structures. In the consultation conclusions on DCS structures published in April 2018, measures were proposed to limit and control DCS structures. These include requiring applicants to possess certain characteristics before they can list shares with WVR. HKEX will also reserve the right to reject an applicant on suitability grounds if its WVR structure is an extreme case of non-conformance with governance norms (for example if the ordinary shares carry no voting rights at all). HKEX also puts forward detailed investor protection measures

---

18 Swire's class B shares have the same voting rights as class A shares, but their value is 1/5 that of class A shares.
19 See HKEX's *Concept Paper on New Board*, June 2017, on the HKEX website.

to be applied to WVR companies after their listing. These include measures that restrict the power of WVR, protect the voting rights of non-WVR shareholders, and strengthen corporate governance and disclosure requirements. Issuers with WVR structures will be differentiated from others through a unique stock marker "W" after their stock name. In addition, WVR beneficiaries must be directors of the issuers to ensure they operate the companies with the obligations of a director as set out under relevant laws and regulations. The WVR attached to a WVR beneficiary's shares will lapse once the WVR beneficiary transfers the WVR shares to another person, or dies or is incapacitated, or ceases to be a director. WVR are therefore subject to natural sunset clauses and will not exist indefinitely[20].

Through its WVR consultation in June 2015 and the New Board consultation in 2017, HKEX found that only a small number of respondents considered a mechanism for class actions a prerequisite for allowing WVR shares to be listed in Hong Kong. Market responses also indicated that most class actions in the US involved disclosure issues rather than the potential abuse of control under a DCS structure. Study findings have demonstrated that both the US and Hong Kong place a high priority on investor protection but achieve this goal in different ways. The US regime places greater emphasis on the ease by which shareholders can take private action to achieve redress for damages after abuse has occurred. The associated judicial costs involved are lower than in Hong Kong. In Hong Kong, greater reliance is placed on the Listing Rules to require disclosure and prevent the abuse of control before it occurs, and post-event legal action, involving listed companies, is primarily carried out on shareholders' behalf by the Securities and Futures Commission (SFC)[21].

As DCS structures continue to evolve and develop, non-typical DCS structures similar to Alibaba's "Chinese partnership" or DCS structures with Chinese characteristics may emerge. More mandatory disclosure or introducing whistle-blowing programmes (as the US has been doing) will keep these companies and their *de facto* controllers and management in check, and prevent fraud and insider dealing.

When HKEX planned to reform its listing regime in 2016, its major objective was to remove listing hurdles for high-growth companies invested by venture capital funds or pre-profit biomedical firms in response to the global rise of the new-economy sector, enabling them to list by using WVR structures. The listing of such international and Mainland companies in Hong Kong would create immense opportunities for Hong Kong and solidify its position as a global financial centre. Although "one-share-one-vote" has contributed

---

20 See *HKEX's consultation conclusions*, April 2018, published on the HKEX website.
21 See HKEX's *Weighted Voting Rights Concept Paper*, August 2014, published on the HKEX website.

fundamentally to investor protection in Hong Kong over the years, enterprise innovation and economic growth should not be constrained by the listing structure as market systems are further enhanced. Listing structures should be designed with flexibility and their effectiveness should be tested by the market. If innovative and technology companies do not have access to the public capital market, venture capitalists will not readily provide them with substantial funding, and SMEs will find it difficult to establish strategic international relationships. This will reduce the evolution of innovative companies and the formation of industry clusters.

Introducing appropriate listing criteria that suit the financial characteristics and investment risks of innovative and technology companies in their start-up stage will therefore be of significant help to the industries' development. More venture capital and private equity funds can then be directed into the industries and companies, making it possible for the emergence of large innovative companies, thereby stimulating the development of new-economy industries in the region, facilitating the upgrade of the regional economy and expanding its horizon. This is the kind of long-term positive effect that capital market reforms could have on the Hong Kong economy.

Chapter 7

Weighted voting rights:
Angel or evil to investors?

Chief China Economist's Office
Hong Kong Exchanges and Clearing Limited

Summary

Stocks with weighted voting rights (WVR) are usually known as dual-class stocks. A company with WVR structure issue at least two classes of shares with different per-share voting rights. While the founders and management usually own one class of shares with multiple votes per share, another class of shares with one vote per share are issued to general investors. New-economy innovative companies tend to adopt WVR structure in their initial public offerings in order to raise funds for business growth while the founders could maintain company control to pursue innovation with disproportionately diluted shareholdings. Mainland China, which is undergoing economic transformation to a new economy, has abundant potential supply of such companies. These WVR companies can choose to list on major overseas markets which accept the listing of WVR companies, such as New York, London, Hong Kong, Singapore and Tokyo.

In Hong Kong, the new listing regime for companies with WVR structure provides a set of safeguards for investors, which are more comprehensive and have even stricter regulatory requirements than other major global financial centres. The investor safeguards offered by the new regime include high entry requirement on market capitalisation, sunset provisions and enhanced corporate governance and disclosure requirements. These measures could not only effectively reduce agency costs, but also limit principal costs. Principal costs matter when investors do not have expertise on innovative sectors (principal competence cost) and when there are potential conflict of interests among investors (principal conflict cost). With appropriate safeguards, a WVR structure for company listing is not necessarily an evil to investors but can be an angel in that it could contribute positively to stock price and operational performance of a company. It is therefore potentially beneficial for long-term value creation.

In fact, international experience demonstrated a number of net benefits of WVR structure to investors for investing in companies with WVR structure. Firstly, the long-run share price returns of innovative companies with WVR structures in MSCI stock indices outperformed other index stocks across different regional and global indices. Secondly, there is no consistent evidence to support the potential adverse impact of WVR structure on companies' business performance; this is also valid in respect of US-listed Chinese companies with WVR structure. Besides, founders of WVR companies are likely to create more firm value as they can implement innovative ideas based on their expertise without much interference from investors. WVR companies could actually broaden investment opportunities to meet the demand of investors with different risk appetites. In this relation, the index company MSCI introduced new index series adjusted for voting rights to cater for investor preferences.

In conclusion, the listing of WVR companies is not a race to bottom, but widens the spectrum of investment opportunities. For an economy, the financial support to WVR companies contributes to its economic transition with new growth engines. For investors, they can choose to invest in trustworthy WVR companies, striking a balance between investment risk and opportunities.

1   An overview of global exchanges in listing WVR companies

Stocks with weighted voting rights (WVR) are usually known as dual-class stocks. Companies with WVR structure (referred to as "WVR companies" or "dual-class companies") issue at least two classes of shares with different voting rights — "class A" shares with one vote per share and "class B" shares with multiple votes per share. Holders of "class B" shares are insiders or managers of the company, including founders and directors in the management and the voting powers of their unlisted shares are disproportionately higher than their economic interests. This allows the class B shareholders to dominate in the decisions of corporate policies, e.g. nomination of individuals to its board of directors, share issuance and corporate actions[1] without the need to have a proportionately high shareholdings. In contrast, class A shares of the company with one vote per share are listed and are held by external investors. Shares with multiple voting rights are hereinafter referred to as "WVR shares" and shares with one vote per share are hereinafter referred to as "ordinary shares".

While "one-share-one-vote" (OSOV) is the default principle in local company laws of many countries, many stock exchanges provide the flexibility to list dual-class shares under different rules and standards[2]. Many of these exchanges are found in America and Europe but not in Asia until recently, with examples in Brazil, Canada, Denmark, France, Finland, Hong Kong, Italy, Japan, Russia, Singapore, Sweden, Switzerland, the UK and the US[3].

Innovative technology companies are the key contributors to the recent wave of initial public offerings (IPOs) of WVR companies (referred to as "WVR IPOs"). While technology companies have always been active in the IPO market, they increasingly prefer to adopt dual-class structures. Technology and innovative companies have strong funding needs, vis-à-vis traditional companies, to cope with their different term structure of revenues and investments. These companies would have high risks in their investments but would often have high growth potential. Their revenues would be highly volatile in the early stage

---

1   Corporate actions include dividends, right issues, stock splits, spin-offs and mergers and acquisitions.
2   See HKEX research report, "Listing regime reforms for dual-class share structure and biotech industry", published on the HKEX website, 24 April 2018.
3   Some of the examples are quoted in Tan, A. and B. Robertson "Why investors are fretting over dual-class shares", *Bloomberg QuickTake*, 10 July 2017.

of operation but the potential long-term growth could be extremely high. The new and innovative technology developments of these companies rely very much on the insights and capabilities of the founders. These companies tend to adopt WVR structure in their IPOs in order to raise funds for business growth while the founders could maintain company control to pursue innovation with disproportionately diluted shareholdings. Reasons for founders of innovative companies to adopt WVR structure may include facilitating the realisation of the company's long-term value, providing the incentives to founders for continuous inputs on innovative capacities and human resources to the company, shielding the threats of hostile takeover bids and protecting the emerging company from institutional investors' possible influence on company decisions with a short-term focus[4].

A listing regime that allows the listing of WVR companies would therefore meet the needs of technology and innovative companies. According to PwC, there were 100 technology IPOs globally raising $25.1 billion in 2017[5]. Another source showed that technology companies in the US accounted for about 36% of all US IPO activities and the share of WVR technology IPOs increased to a record high of 43% in 2017 and eased to 34% in 2018 (see Figure 1).

---

4 See HKEX research report, "Listing regime reforms for dual-class share structure and biotech industry", published on the HKEX website, 24 April 2018.
5 Source: PwC, "Global technology IPO review full-year and Q4 2017", 19 December 2017.

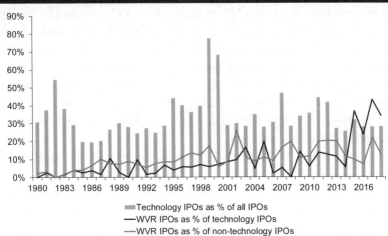

Figure 1. The share of technology IPOs in the US and the share of WVR IPOs (in number terms) in the technology and non-technology sectors (1980 – 2018)

Technology IPOs as % of all IPOs
WVR IPOs as % of technology IPOs
WVR IPOs as % of non-technology IPOs

Source: Ritter, J. R., "Initial public offerings: Updated statistics", manuscript, University of Florida, 31 December 2018 (https://site.warrington.ufl.edu/ritter/files/2019/01/IPOs2018Statistics_Dec.pdf).

These WVR technology IPOs include a number of technology giants from the Mainland. To welcome back the listing of these Mainland technology companies, Mainland China is considering ways to accept the listing of new-economy companies with WVR structure. In September 2018, the State Council clarified[6] that overseas-listed new-economy companies that have WVR structures or no profit yet could issue Chinese Depositary Receipt (CDR) in the onshore A-share market. On 5 November 2018, President Xi Jinping announced[7] the plan to launch the Science and Technology Innovation Board on the Shanghai Stock Exchange under a registration-based regime.   Based on the trial implementation measures published in March 2019[8], the IPOs of Chinese technology companies and technology companies with WVR structures will be accepted.

Currently, new-economy companies can choose to list in the major global financial centres of New York, London, Hong Kong, Singapore and Tokyo. The evolution of the listing regime for WVR companies varies across these markets. Below is an overview.

---

6   *Opinions of the State Council on Promoting the High-Quality Development of Innovation and Entrepreneurship and Creating an Upgraded Version of "Entrepreneurship and Innovation among All the People"* (《國務院關於推動創新創業高質量發展打造「雙創」升級版的意見》), 26 September 2018.

7   See "President Xi's keynote speech at the first China International Import Expo", *Xinhuanet*, 5 November 2018. (http://www.xinhuanet.com/world/ciie2018/jbhkms/index.htm)

8   *Measures for the Administration of the Registration of IPO Stocks on the Science and Technology Innovation Board (Trial implementation)* (《科創板首次公開發行股票註冊管理辦法 (試行)》), issued by the CSRC, 1 March 2019.

(1)   New York: The listing of dual-class stocks was once prohibited since 1926 and the New York Stock Exchange (NYSE) officially announced the ban in 1940[9]. In response to competition from other US exchanges, the NYSE proposed in January 1985 to relax the listing policies to allow dual-class structures. In 1994, the US exchanges implemented uniform listing requirements for dual-class stocks. However, from that date onwards, US stock exchanges voluntarily banned US companies from adopting a dual-class share structure after listing ("dual-class recapitalisations") given investors' concern over the potential for existing shareholders to be forced to give up their voting rights. In this relation, a legal research[10] suggested that dual-class recapitalisations could be beneficial to companies but only in early growth stages.

Currently, the IPOs of WVR companies have become increasingly popular in the US market. Empirical evidence showed that the share of WVR IPOs in the US increased to a record high of 28% in 2017[11] (see Figure 2). These WVR companies might have been attracted by the disclosure-based regime in the US under which companies are only subject to certain disclosure requirements on risk factors of the non-traditional governance structure[12].

---

9   See Appendix III in the HKEX consultation paper, *Concept Paper on Weighted Voting Rights*, HKEX website, August 2014.

10   Gilson, R. J. (1987) "Evaluating dual class common stock: The relevance of substitutes", *Virginia Law Review*, Vol. 73, pp.807-844.

11   Source: Ritter, J. R., "Initial public offerings: Updated statistics", manuscript, University of Florida, 31 December 2018. The dataset used in the research included only the IPOs with offer prices of at least US$5 and excluded American Depositary Receipts (ADRs), unit offers, closed-end funds, real estate investment trusts (REITs), natural resources limited partnerships, small best-effort offers, banks and savings and loans, and stocks not listed on the Center for Research in Security Prices (CRSP).

12   See Investor Advisory Committee of the US Securities and Exchange Commission (SEC), "Dual class and other entrenching governance structures in public companies", 27 February 2018.

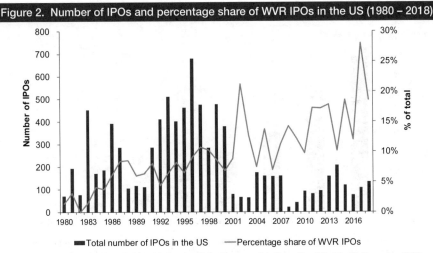

Figure 2. Number of IPOs and percentage share of WVR IPOs in the US (1980 – 2018)

■ Total number of IPOs in the US   ——Percentage share of WVR IPOs

Source: Ritter, J. R., "Initial public offerings: Updated statistics", manuscript, University of Florida, 31 December 2018 (https://site.warrington.ufl.edu/ritter/files/2019/01/IPOs2018Statistics_Dec.pdf).

(2)   London: The UK market only allows WVR IPOs under the rules of Standard Listing with minimum European Union (EU) requirements. For a listing on the Main board of the London Stock Exchange (LSE), a Standard Listing requires the applicant to meet only the minimum EU harmonisation standards while a Premium Listing requires the applicant to comply with the UK's super-equivalent rules which are stricter than the EU minimum requirements. Premium Listings may contribute to potentially lower cost of capital because of greater transparency. The Premium Listing principles have been tightened since 2014 to follow the proportionality of voting and equity interests[13], i.e. OSOV is required.

Empirical evidence showed that only 5% of listed companies had multiple voting rights in place in the UK market in 2007[14] (see Figure 3). A research paper[15] explained the two reasons behind this — the opposition from institutional investors and the elimination of pre-bid defences (e.g. through dual-class structure) in the takeover rules that force to adopt OSOV in case of a takeover bid.

---

13 See UK Financial Conduct Authority, "PS14/8: Response to CP13/15 — Enhancing the effectiveness of the listing regime", 3 August 2015.

14 Source: Shearman & Sterling LLP, *Proportionality between Ownership and Control in EU Listed Companies: Comparative Legal Study — Legal Study for Each Jurisdiction*, external study commissioned by the European Commission, 18 May 2007.

15 Huang, F. (2017) "Dual class shares around the top global financial centres", *Journal of Business Law*, Vol. 2, pp.137-154.

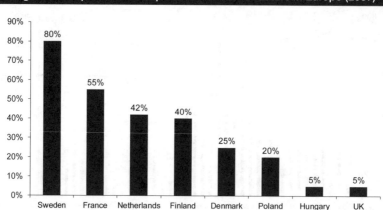

Figure 3. Proportion of companies with WVR structure in Europe (2007)

Source: Shearman & Sterling LLP, *Proportionality between Ownership and Control in EU Listed Companies: Comparative Legal Study — Legal Study for Each Jurisdiction*, external study commissioned by the European Commission, 18 May 2007.

(3)   **Hong Kong:** Hong Kong implemented a listing regime reform in April 2018 with the aim to become a listing hub for innovative companies. The Listing Rules of the Stock Exchange of Hong Kong (SEHK) are revised to allow Main board listing of innovative companies with WVR structures, biotech companies that cannot meet financial eligibility listing requirements and secondary listings of qualified overseas-listed companies. The reform revised the related listing rule (introduced in 1989) to allow deviation from the OSOV principle such that WVR companies may be listed. There were two IPO listings of Mainland technology giants with WVR structure during the first six months after the listing regime reform.

(4)   **Singapore:** The Singapore market extended the acceptance of dual-class structure from newspaper companies to other companies. The Singapore Exchange (SGX) used to ban listings of dual-class stocks except newspaper companies, which are mandated to issue two classes of shares whereby each management shareholder has 200 times the voting rights of an ordinary shareholder under Singapore Newspaper and Printing Presses Act. The Companies Act was amended in October 2014 to allow public companies to issue different classes of shares with either no voting rights or multiple voting rights. The SGX then reformed its listing regime and allow the listing of dual-class stocks from innovative sectors since 26 June 2018.

(5)   Tokyo: The Tokyo Stock Exchange (TSE) amended its listing rules in 2008[16] to relax the regulations on accepting WVR IPOs on a case-by-case basis but to continue to ban the issuance of a new share class for existing listed companies. Subsequently, the first WVR IPO on the TSE — Cyberdyne (a wearable robot developer) — was launched in March 2014. After the IPO, the TSE revised its listing rules in July 2014 to impose two additional requirements on the listing of dual-class stocks: (i) necessity and appropriateness for the use of a dual-class structure; and (ii) a sunset clause.

2   Hong Kong's new listing regime for WVR companies: Appropriate investor safeguards

Market demand for investing in innovative companies has driven the reform in the Hong Kong market's listing regime to accommodate the listing of ordinary share class of WVR companies (listed share classes of WVR companies are referred to as W-stocks). The Hong Kong Exchanges and Clearing Limited (HKEX) added a new chapter to the Main Board Listing Rules for WVR listings effective from 30 April 2018 after receiving strong support from stakeholders to the proposal in a consultation process. During the consultation process, concerns were expressed about the potential risks associated with investing in companies with WVR structure. The following sub-sections will discuss the concerns and how Hong Kong's new listing regime addresses them through introducing appropriate safeguards.

2.1   Founder manager as the key for success of WVR companies

The founders of innovative companies usually demonstrate that they are the most capable ones to develop and manage their companies. Adoption of the WVR structure is to ensure the absolute control of the company by the founder(s). The WVR structure can

---

16 See Tokyo Stock Exchange (TSE), "Listing system improvement FY2008", TSE website, 27 May 2008.

be a double-edged sword and the future of the company will hinge on the founder(s). If the founder(s) is/are visionary and consistently making right business decisions, it will be the fortune of the company, the blessing of investors. In this case, WVR preserves the entrepreneurship that supports the share price. On the contrary, if the founder(s) is/are not capable enough or unfortunately make(s) a mistake, this may be a potential hidden danger for the company and lead to an agency problem.

The "agency problem", or the potential misalignment of interests, between shareholders and managers is the primary concern of investors. It refers to the situation that the shareholders are interested in maximising the share value while the managers who operate the business are interested in maximising their pay and benefits. WVR companies' managers maintain the majority of control with a relatively small share of equity holdings. Other shareholders cannot exert their influence through voting against the managers even if they are not acting for the interests of shareholders. Therefore, proper investor safeguards should be imposed to limit the risk of agency cost for WVR companies. In the absence of appropriate safeguards, the agency problem may hurt investors' interests.

However, the benefits of WVR structure to managers are not necessarily achieved at the expense of investors' interest. On the one hand, investors are exposed to agency costs (or specifically agency conflict costs as explained above). On the other hand, academic literature[17] put forward the principal cost theory to explain how dual-class structure reduces agency competence costs, principal competence costs and principal conflict costs:

- **Agency competence costs:** The cost arising from honest mistakes by the management. The magnitude of the costs will vary. If the managers are intelligent, unbiased and informed, they are likely to make relatively few mistakes. For a WVR company, the risk of stepping down is relatively lower for the manager and he/she may be more willing to admit mistakes. These may avoid the risk of hidden mistakes which hurt the long-term growth.

- **Principal competence costs:** The cost arising from investors' mistakes due to the lack of expertise. The cost will be lower if the management (agency) already has the requisite expertise, which will enable the investors (principals) to reap the benefits of specialisation. For a WVR company, the manager usually demonstrates his/her capability to be positive for long-term growth (e.g. track records of successful visionary projects).

- **Principal conflict costs:** The costs arising from the conflict of interests among investors. The cost is likely higher when a company has multiple principals (investors)

---

17 Goshen, Z. and R. Squire. (2017) "Principal costs: A new theory for corporate law and governance", *Columbia Law Review*, Vol. 117, pp.767-796.

with conflict of interests (e.g. between activist investors with short-term focus and pension funds with long-term focus). For a WVR company, external investors have less impact and the potential conflict among them will be less likely to affect business decisions.

In other words, investors should consider the agency and principal costs together, i.e. total cost of control. Investors can benefit from the WVR structure if the total cost of control is smaller than single share class structure. These rely on whether a listing regime has appropriate safeguards to protect investors from agency costs.

2.2   Institutional investors' call for investor safeguards on WVR structure

Global institutional investors would ask for price discounts of W-stocks in case of the absence of appropriate safeguards. The European Commission conducted a survey[18] of 445 institutional investors worldwide with more than €4.9 trillion of assets under management (AUM) where European investors accounted for 13% of total AUM. 80% of the 445 institutional investors would expect a discount on the share prices of companies with control enhancing mechanisms (CEMs). It noted that there is no safeguard solution for CEMs that fits all, but more transparency on the existence and the impact of CEMs such as dual-class structure is preferred.

It happened in Europe that there was an increasing number of dual-class unifications (conversion of shares with multiple votes into OSOV shares) in early 2000s[19]. The same study also suggested that the dual-class companies should not be forced by law to switch to companies with single-class OSOV shares (referred to as "OSOV companies")[20].

In the US, the Securities and Exchange Commission (SEC) reviewed the listing regime and made recommendations in March 2017 on enhancing the disclosure requirements on dual-class structure and other entrenching governance structures[21]. Besides, the SEC Commissioner Robert Jackson advocated for sunset provisions of perpetual dual-class

---

18 Institutional Shareholder Services, Shearman & Sterling LLP and the European Corporate Governance Institute, *Report on the Proportionality Principle in the European Union*, external study commissioned by the European Commission, 2007.

19 Pajuste, A. (2005) "Determinants and consequences of the unification of dual-class shares", *European Central Bank working paper* No. 465, March 2005.

20 See also Lauterbach, B. and A. Pajuste (2015) "The long-term valuation effects of voluntary dual class share unifications", *Journal of Corporate Finance*, Vol. 31, pp.171-185.

21 See Investor Advisory Committee of the US SEC, "Dual class and other entrenching governance structures in public companies", 27 February 2018.

voting structure in February 2018 while recognising the benefits of these structures[22].

In Canada, the Canadian Coalition for Good Governance (CCGG)[23] admitted that there are advantages and disadvantages for adopting dual-class structures. They note that "it is important to encourage entrepreneurism in Canada and accordingly does not wish to hinder Canadian entrepreneurs from taking their companies public"[24]. As the listing regime and regulations in Canada did not change for more than 20 years, the CCGG published seven best practices for newly listed dual-class companies in 2013[25], which include the election of directors, putting a cap on voting right ratio, coat-tail[26] and sunset provisions, to protect investors' interests.

In Asia, market professionals pursue OSOV principles but ask companies for making adequate disclosure on the WVR structures in case these structures are legal. The CFA Institute[27] conducted a survey of 454 members on dual-class shares in April 2018. The survey results are similar to the CCGG's views in that they recognise both advantages and disadvantages of WVR structures[28]. The recognised advantages of a regime for the listing of dual-class structure companies include boosting attractiveness of the exchange and attracting the listing of companies from technology and other innovative sectors. The identified disadvantages include insufficient minority investor protection and skewed proportionality between ownership and control. The survey conclusion called for appropriate safeguards, including mandatory corporate governance measures, time-based sunset, limit on voting right ratio, coat-tail provision and conversion to OSOV shares in case of transfer of shares with multiple voting rights.

2.3 Absence of investor safeguards in the past for WVR structure in Hong Kong

The listing of companies with two share classes is not new to the Hong Kong market. During 1972 to 1973, seven companies issued and listed "B" shares[29] with lower

---

22 See "Perpetual dual-class stock: The case against corporate royalty", US SEC Commissioner's speech on 15 February 2018.

23 CCGG is a corporate governance organisation in Canada that is positioned to effect change as the voice of Canadian institutional shareholders.

24 See CCGG, "Dual class share policy", September 2013.

25 Ditto.

26 Coat-tail provision is a legal provision that allows the subordinate shareholder to participate equally in any formal bid to acquire multiple voting shares.

27 CFA Institute is a global association of investment professionals.

28 See CFA Institute, "Dual class shares and the need for safeguards", April 2018.

29 These "B" shares are different from the usual practice of class B shares with WVR in the US (e.g. class B shares are usually not listed).

denomination than "A" shares which had the same voting right per share. The underlying objectives were diverse, including fund-raising for the purchase of real estate or for expanding their businesses, and attracting a wider base of investors[30]. While six of them[31] had been either acquired or privatised, there is currently only one company — Swire Pacific — with "B" shares listed in Hong Kong.

At that time, there was no specific chapter in the Listing Rules for this kind of share issuance. At issuance, the company's existing listed shares became "A" shares and the company's "B" shares were offered to existing shareholders[32]. The "B" shares had the same voting rights per share as "A" shares but a lower denomination (either one-fifth or one-tenth of "A" shares) and a lower dividend entitlement. A controlling shareholder could spare more cash for takeover defence by selling their stakes in "A" shares and buying the same number of "B" shares to keep the same voting rights at lower prices[33]. In the case of Swire Pacific, each "B" share was issued with a nominal value one-fifth that of each "A" share (HK$0.12 vs HK$0.60) and dividends are paid by reference to these nominal values (even after the abolition of nominal value by law amendment in Hong Kong in 2014). If the control was being challenged, the "B" share would be traded at a premium over "A" shares. However, in absence of competition for control, the Swire Pacific "B" shares had been traded at a lower price, after adjustment for the degree of dividend entitlement, with an average discount of 12% to the "A" shares during 2000 to 2018 (see Figure 4), due largely to their low liquidity[34].

---

30 See Chapter 2 in HKEX consultation paper, *Concept Paper on Weighted Voting Rights*, HKEX website, August 2014.

31 These included five companies in the Wheelock Marden group — Wheelock Marden and Company Limited, Wheelock Maritime International Limited, Hong Kong Realty and Trust Company Limited, Realty Development Corporation Limited and Lane Crawford Limited. Another company with "B" shares was Local Property and Printing Company Limited. Source: HKEX's *Concept Paper on Weighted Voting Rights*, August 2014.

32 For example, Wheelock redesignated ordinary shares to A shares and issued B shares (1/10 of denomination of A shares) to existing shareholders in 1972 that shareholders could purchase one B share for holding every two A shares. (Source: Fung, B. Y. (2017) *History of Hong Kong Corporate Mergers and Acquisitions (Revised)* (《香港企業併購經典 (增訂版)》). Joint Publishing (HK) Company Limited.

33 See Chapter 26.3 in Chow, M. (2013) *Introduction to Corporate Finance Techniques II* (《財技密碼 (下篇)》), Hong Kong Mobile Financial Publication.

34 Ditto.

Figure 4. Price comparison of Swire Pacific "A" shares and "B" shares
(Jan 2000 – Dec 2018)

Note: The price of Swire Pacific "B" shares is multiplied by 5 for comparison purpose, as B shares' dividend
entitlement is one-fifth that of A shares.

Source: Bloomberg.

In 1987, a number of proposals were made by listed companies on the issuance of such "B" shares but their listings were prohibited by the stock exchange and the then securities regulator on the consideration of strong opposition by both Hong Kong and overseas brokers at that time. A review of the issue on "B" shares commissioned by the government in the same year noted that "there is a legitimate need for their continued availability in exceptional circumstances"[35]. The restriction on the listing of "B" shares was subsequently codified in Rule 8.11 of the Main Board Listing Rules in December 1989. Since then, there had been no new listings of "B" shares.

2.4   Appropriate investor safeguards under the new listing regime for WVR structure in Hong Kong

Hong Kong's new listing regime was designed to achieve a balance between investor protection and benefits to issuers. This gives the flexibility to provide more investment opportunities for investors with different risk appetites and to enable direct financing by innovative companies without diluting the control by their founders. To achieve these, the

---

35 Source: Companies Registry, *The Third Interim Report of the Standing Committee on Company Law Reform: B Shares*, July 1987.

only way is to introduce appropriate safeguards in the listing regime.

Contrary to the old regime, the listing of WVR stocks in Hong Kong is subject to strict requirements. In the past, there were no specific requirements on the listing of "B" shares. However, under the new listing regime, the chapter for WVR listings in the Main Board Listing Rules of the SEHK imposes a range of safeguards to address investors' potential concerns. These include higher entry requirements for a listing, limits on holdings and voting rights, sunset provisions and conditions on share transfers as well as enhanced corporate governance and disclosure requirements. In respect of these areas, Hong Kong's new listing regime for WVR companies provides more safeguards compared to exchanges in other major global financial centres (see Appendix). The characteristics of these safeguards are discussed below.

The entry requirement on market capitalisation for new WVR listings in Hong Kong is the highest among key global exchanges. The minimum market capitalisation is HK$40 billion (or HK$10 billion with at least HK$1 billion in revenue). High market capitalisation usually means that the company has developed to a reasonable scale and is not a highly risky start-up company. In the global market, the MSCI has included a number of technology giants in its global indices and the ones with dual-class structures usually have much higher market capitalisation than others in the same sector. As of 1 September 2017, MSCI ACWI includes 253 W-stocks and their market capitalisation accounted for 11.2% of the index total[36]. Of these, information technology sector had the largest share of 23.2% of the index's market capitalisation. The average market capitalisation of technology companies with WVR structures was the highest at US$72 billion (compared to the average of US$19 billion for all W-stocks in the index) (see Table 1).

Table 1. Average market capitalisation of W-stocks in MSCI ACWI (as of Sep 2017)			
Sector	Number of W-stocks	Total sector market capitalisation (US$ mil)	Average market capitalisation per stock (US$ mil)
Information technology	24	1,740,097	72,504
Health care	11	358,682	32,607
Financials	44	722,027	16,410
Consumer discretionary	61	870,424	14,269

---

36 Source: MSCI, *Should Equity Indexes Include Stocks of Companies with Share Classes having Unequal Voting Rights*, consultation discussion paper for the *Consultation on the Treatment of Unequal Voting Structure*, January 2018.

(continued)

Table 1. Average market capitalisation of W-stocks in MSCI ACWI (as of Sep 2017)			
Sector	Number of W-stocks	Total sector market capitalisation (US$ mil)	Average market capitalisation per stock (US$ mil)
Consumer staples	29	388,716	13,404
Real estate	6	70,459	11,743
Industrials	30	306,094	10,203
Telecommunication services	10	90,228	9,023
Materials	19	136,032	7,160
Energy	10	71,168	7,117
Utilities	9	28,126	3,125
All W-stocks in the MSCI ACWI	253	4,782,053	18,901

Source: Melas, D., "Putting the Spotlight on Spotify: Why have stocks with unequal voting rights outperformed?",
posted on MSCI blog, 3 April 2018.

Certain requirements on beneficial holders of WVR shares (referred to as "WVR beneficiaries") are imposed in Hong Kong to limit agency conflict costs. WVR beneficiaries are only limited to natural persons who are directors and materially responsible for the growth of the business. Other requirements on WVR beneficiaries are as follows.

- **Issuance of WVR shares of already-listed companies is not allowed.** After listing, WVR beneficiaries cannot increase the proportion of WVRs. In other words, the voting power of shareholders of ordinary shares (referred to as "non-WVR shareholders") will not be further disproportionately diluted. Non-WVR shareholders must always have at least 10% of total voting rights.
- **WVR shares are subject to a cap on voting right ratio.** The maximum ratio of voting rights between WVR shares and ordinary shares (or maximum voting differential) is 10 to 1 in Hong Kong, which is in line with many markets with dual-class stocks. A historical reason is that the cap for the ratio of 10 to 1 was recommended in the proposal of rules to relax the OSOV principle in 1985 by a subcommittee of the NYSE, although the NYSE did not adopt any caps on voting differential in the end. An extreme case is the IPO of Snap in the US that only issued non-voting shares to the public in March 2017. In terms of product nature, non-voting shares are similar to fixed-income or other financial products that the return is related to the company's performance but no impact on company decisions[37].

---

37 Source: CFA Institute, "Dual-class shares: The good, the bad, and the ugly", August 2018.

Natural sunset provisions of WVR structure are introduced in Hong Kong to mitigate potential agency conflict costs. The sunset provisions give a reasonable period of time for founders to invest in risky projects for long-term value creation. It is also fair to require conversion of the WVR shares into ordinary shares afterwards. Under Hong Kong's regime, the WVR assigned to the WVR shares is not time-based but subject to a natural sunset clause — WVR will cease when the beneficiary is dead or no longer a director or deemed to be incapacitated for performing duties as a director or no longer meet the requirements of a director, or upon the transfers of the WVR shares. In other words, their successors cannot inherit the super voting power. While the US market does not require any sunset clause for dual-class structure, the SEC Commissioner Robert Jackson alerted the market for the risk of "perpetual" dual-class stocks. He warned that perpetual dual-class structures will last forever and benefit visionary founders' successors who may not be visionary[38]. He highlighted that the median relative valuation[39] had worsened over time for perpetual dual-class stocks but improved slightly in longer term (7 years or later after IPO) for dual-class stocks with sunset provisions (see Figure 5).

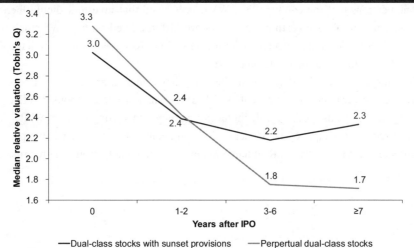

Figure 5. Relative valuation of dual-class companies in the US (IPOs during 2004 – 2018)

Source: "Perpetual dual-class stock: The case against corporate royalty", US SEC Commissioner's speech on 15 February 2018.

---

38 Source: "Perpetual dual-class stock: The case against corporate royalty", US SEC Commissioner's speech on 15 February 2018.

39 The firm's "relative valuation" is measured by Tobin's Q ratio. Tobin's Q ratio is the ratio between the market value of a company and the replacement cost of its assets.

WVR companies in Hong Kong are subject to enhanced corporate governance measures which would reduce agency competence costs. The multiple voting rights of WVR beneficiaries in Hong Kong are not applicable for certain resolutions for which OSOV will apply. These resolutions include changes to constitutional documents, appointment and removal of independent non-executive director(s) or auditors, variation of rights attached to any class of shares and voluntary winding-up of the listed issuer. A WVR company is required to have a corporate governance committee and an ongoing compliance advisor. These safeguards are incorporated into the companies' constitutional documents and WVR beneficiaries are required to provide an undertaking to the company to comply with the safeguards. Similar measures are also introduced in Singapore. In Japan, the articles of incorporation of Cyberdyne (a WVR company) require the conversion of WVR shares to ordinary shares if an offeror holds at least 75% of outstanding shares in a tender offer; that means OSOV will be applied in the case. Similarly, OSOV is applicable for takeover bids with 75% stake of WVR companies in the UK. In contrast, there are no specific enhanced corporate governance measures for WVR companies in the US market currently.

The enhanced disclosure requirements in Hong Kong help protect investors' interests. The new listing regime requires the inclusion of warnings in listing documents and communications that the company is a WVR issuer and that investors should therefore exercise caution when investing. W-stocks are differentiated with a "W" stock name marker. Among the major financial centres, only Hong Kong and Singapore have these unique stock name markers for alerting investors.

In summary, investor safeguards in the Hong Kong market are more comprehensive than those in other major financial centres. These measures would contribute to effectively limiting the total costs of control, including agency costs, for investors while WVR structure reduces principal costs. Under such a regime, investors are likely better protected for investing in stocks with WVR structure in pursuit of potentially promising returns.

3   WVR is not necessarily an evil but an angel to investors

Although appropriate safeguards are demanded for better protection of investor interests for investing in WVR companies, this should not be taken as a proof of WVR structure being an evil to investors. In fact, empirical evidence showed that WVR structure can be an angel to investors in that it contributes positively to price and operational performance of a company and is therefore good for long-term value creation. In the light of this, a listing regime with appropriate investor safeguards for WVR companies would enrich investors' choices while keeping investor protection in balance. Detailed discussion are given in sub-sections below.

3.1   Discount in prices but potential outperformance in returns

Given the concerns about the potential agency problem of WVR structure, general investors usually ask for a discount in ordinary share prices compared to the share prices of WVR shares to compensate the potential risks of managerial entrenchment and expropriation of firm resources as discussed in section 2. Empirical studies showed that listed shares with inferior voting rights tend to have price discounts. In the US, a study[40] found that the average discount of price-to-earnings ratio (PE ratio) was 17% for the five years following an IPO for 253 dual-class stocks during 1990 to 1998. Another study[41] estimated the value of control worth an average premium of 10%-14% of equity value of the company over the ordinary equity rights for privately negotiated transfers of 393 controlling blocks of shares in 39 countries during 1990 to 2000. A separate study found that the magnitudes of estimated premia for WVR shares relative to their corresponding ordinary shares were different across markets (see Figure 6).

---

40 Smart, S., R. Thirumalaib and C. Zutter (2008) "What's in a vote? The short- and long-run impact of dual-class equity on IPO firm values", *Journal of Accounting and Economics*, Vol. 45, No. 1, pp.94-115.
41 Dyck, A. and L. Zingales (2004) "Private benefits of control: An international comparison", *Journal of Finance*, Vol. 59, pp.537-600.

Figure 6. Estimated price premia for WVR shares in different countries (1997)

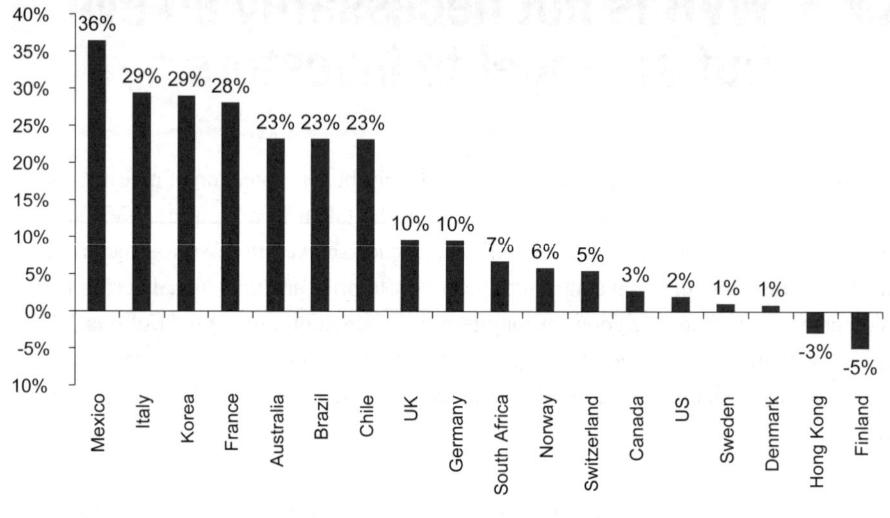

Note: The price premium was estimated as the value of control-block votes which was calculated from the prices of multiple- and limited-voting shares, adjusted to comprise 50% of the voting power, and scaled by the firm's market value.

Source: Nenova, T. (2003) "The value of corporate voting rights and control", *Journal of Financial Economics*, Vol. 68, pp.325-351.

However, a study[42] found that the long-term average return of W-stocks among the MSCI index constituents, in general, outperformed the rest in global and regional markets. As of September 2017, there were 2,493 constituents of MSCI's global benchmark ACWI and 243 of them were W-stocks of WVR companies[43]. During November 2007 to August 2017, the MSCI ACWI's W-stock constituents were found to have an annualised total return of about 7.2%, higher than the 4.5% for the overall index and the about 4.2% for the OSOV constituents. For the MSCI ACWI's regional indices of North America, Europe and Emerging Markets, the W-stock constituents also outperformed the index and OSOV constituents. The degree of active return[44] of W-stocks was found to be different across markets. (See Figure 7.)

---

42 Melas, D., "Putting the spotlight on Spotify: Why have stocks with unequal voting rights outperformed?", posted on MSCI blog, 3 April 2018.

43 The MSCI ACWI may include more than one share class of a WVR company, e.g. share class A (OSOV) and share class C (no voting right) of Google's parent company Alphabet are included in the ACWI but not the unlisted share class B which has multiple voting rights.

44 Active return is defined as the difference in returns between the share price and the benchmark index.

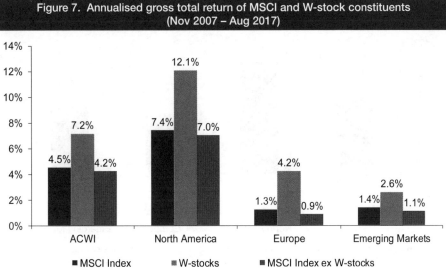

Figure 7. Annualised gross total return of MSCI and W-stock constituents (Nov 2007 – Aug 2017)

Source: Melas, D., "Putting the spotlight on Spotify: Why have stocks with unequal voting rights outperformed?", posted on MSCI blog, 3 April 2018.

In the same study, the active returns of W-stocks in MSCI indices were decomposed into returns contributed by different risk factors (see Figure 8). Industry factor gave positive returns (except in MSCI Emerging Markets Index) and this was in line with the strong performance of technology sector in the past decade. The contribution of other key risk factors (e.g. currency, country and market factors) did not give consistent contribution to their outperformance. In contrast, stock-specific risks accounted for a large proportion of active return except in Europe and these may include the founders' innovative visions and their choice of a WVR structure. The same study found that the W-stock constituents possessed the characteristics of relatively large market capitalisation, higher profitability but more volatile, higher asset growth but lower leverage and lower dividend yield, in comparison with OSOV stocks. It supported the assertion that a visionary founder of a company with WVR structure would tend to maximise long-term firm value through investing in risky projects.

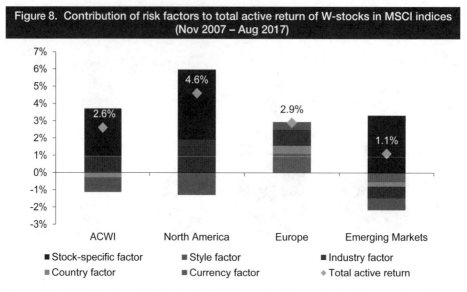

Figure 8. Contribution of risk factors to total active return of W-stocks in MSCI indices (Nov 2007 – Aug 2017)

Source: Melas, D., "Putting the spotlight on Spotify: Why have stocks with unequal voting rights outperformed?", posted on MSCI blog, 3 April 2018.

Some empirical studies also found excess returns in W-stocks. A thematic study[45] (referred to as the "D&J Study") looked into dual-class recapitalisation (i.e. an OSOV company became a WVR company by listing a share class of common stocks with limited or no voting rights) of 178 US-listed companies during the period from 1978 to 1998 and estimated the abnormal buy-and-hold price returns to be 5% and 23% in the existing OSOV share class respectively for the first year and the following fourth year after the announcement, compared to the returns of a matched sample without dual-class recapitalisation. Another empirical study[46] found better IPO returns of W-stocks during 2018H1 in the US — during the study period, the average of IPO returns (since the offer date) of W-stocks in the US were higher than those for all IPOs (see Figure 9). For sectoral comparison, the average IPO returns of W-stocks in the technology, media and telecommunication (TMT) sector outperformed the average for all IPOs in the sector; the TMT sector in turn had the highest average return among sectors during 2018H1. This reflects that W-stocks of companies in the innovative technology sector were well received by investors.

45 Dimitrov, V. and P. C. Jain (2006) "Recapitalisation of one class of common stock into dual-class: Growth and long-run stock returns", *Journal of Corporate Finance*, Vol. 12, pp.342-366.
46 Klausner, D., "Dual class IPOs are on the rise: Tech unicorns jump on board this new trend", PricewaterhouseCoopers' Deals Blog, 18 July 2018.

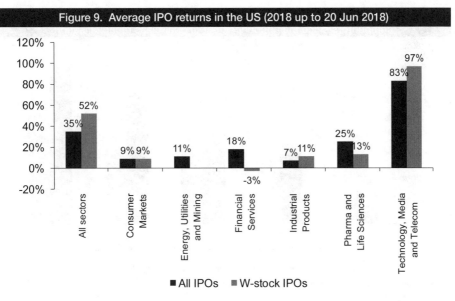

Figure 9.   Average IPO returns in the US (2018 up to 20 Jun 2018)

Note: "IPO return" of a stock is the return of the share price of the stock as of 20 June 2018 relative to the offer
price.   The dataset covers all IPOs in the US (excluding special purpose acquisition companies or SPACs)
during 1 January 2018 to 20 June 2018.

Source: Klausner, D., "Dual class IPOs are on the rise: Tech unicorns jump on board this new trend",
PricewaterhouseCoopers' Deals Blog, 18 July 2018.

3.2   Indifferent or better business performance for WVR companies

The price discount of W-stocks discussed in Section 3.1 above implies that W-stocks
tend to have lower PE ratios relative to their true value and the cost of equity funding
through a WVR structure may therefore be affected.   This may affect WVR companies'
leverage and cost of capital and hence their earnings, thereby impairing investment returns
to investors.   However, empirical evidence indicated that this might not be the case.

Results of the D&J Study quoted in section 3.1 showed that the income growth and
profitability of WVR companies in the US outperformed their non-WVR competitors
— WVR companies in the US had higher growth in sales, assets and operating income
compared to competitors in the same industry with similar size.   In particular, the operating
income of WVR companies grew by 72% in four years' time from the announcement of
dual-class recapitalisation, compared to 50% for their competitors (see Figure 10).   In
respect of profitability, the return on assets was found to be of no difference between WVR
companies and their competitors but the return on equity was higher for WVR companies
in the first few years following the recapitalisation.

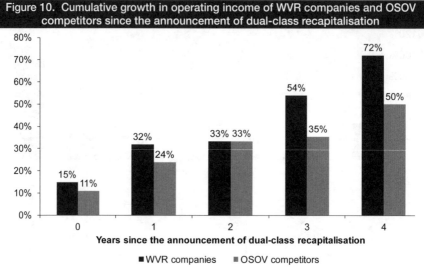

Figure 10. Cumulative growth in operating income of WVR companies and OSOV competitors since the announcement of dual-class recapitalisation

Note: "Year 0" refers to the year when the dual-class recapitalisation was announced.

Source: D&J study.

In Canada, WVR structures were not found to have negative impact on the companies' business performance either. An empirical study[47] examined the impact of WVR structure on profitability (return on assets), firm valuation (Tobin's Q), and takeover premium (cumulative abnormal return) for 10,366 firm-year observations of all Canadian listed companies during the period from 1996 to 2005. In the study, most of the controlling shareholders in Canadian WVR companies were still the founding family members and they still had concentrated ownership in the companies during the study period. Therefore, any expropriation of firm resources would have wealth effect on them too. The study results did not find significant positive or negative impact due to WVR structure.

3.3 Good for value creation

In addition to price discounts and business performance, investors may be concerned about the long-term valuation of WVR companies since the management may not act in line with the investors' interests to create shareholders' value. In fact, some external shareholders (e.g. activist investors) tend to have short-term focus and would like to exert pressure on business decisions of the companies they have invested in. For an OSOV

47 Jog, V., P. Zhu and S. Dutta (2010) "Impact of restricted voting share structure on firm value and performance", *Corporate Governance: An International Review*, Vol. 18, pp.415-437.

company, activist investors and certain kinds of institutional investors (e.g. hedge funds or other asset managers with short-term focus) can use their votes to alter business decisions to improve short-term share price performance (e.g. high dividend payout instead of re-investment) or to stop investments in risky projects that might create long-term values. On the contrary, for an innovative company with WVR structure, the visionary founders would be subject to less interference in their pursuit of risky projects in the hope of creating superior long-term value despite possible large fluctuations in short-term returns.

The potential conflict between visionary founders and investors is usually caused by asymmetric information. Investors usually have no expertise about the innovative sector, but the founder generally has better knowledge or an idiosyncratic vision for producing long-term superior returns, which will be shared on a pro rata basis between them and the investors. Therefore, certain long-term investors are willing to invest in WVR companies, with their degree of influence undermined. Although external shareholders have relatively low influence on business decisions of WVR companies, WVR structures of these companies reduce the chance of inappropriate business decisions forced to be made as a result of voting by investors without expertise (principal competence cost).

Some studies did find that the management of WVR companies could create extra value based on their expertise and reduce principal competence cost. One of such studies[48] examined the price premium of OSOV share class in case of takeover. The study worked on a sample of 142 companies with dual-class recapitalisation in the US during 1978 to 1998. 48 of them had significant decreases of insider ownership ("net-sell" companies) through the recapitalisation but insiders still owned about 39% of total shares. These net-sell companies became less risk averse while the ownership became more dispersed. The WVR structure enabled these companies to undertake high risk but value-enhancing projects. In fact, the return on assets of these companies rose significantly as a result of higher capital expenditure and higher leverage. 67 companies in the full sample of the study were subsequently acquired; among them were companies with both share classes being publicly traded. For these companies with dual-class shares listed, the average takeover premium (defined as the percentage difference between the reported acquisition price and the price four days prior to the announcement) was 53.4% for OSOV share class, which was higher than the 39.7% for WVR share class and the 36.1% for benchmark companies with single share class. The findings undermined the concerns on the potential impact of agency problem on value creation.

WVR structures also avoid principal conflict cost that comes from the potential conflict

---

48 Bauguess, S. W., M. B. Slovin and M. Sushka (2012) "Large shareholder diversification, corporate risk taking, and the benefits of changing to differential voting rights", *Journal of Banking and Finance*, Vol. 36, pp.1244-1253.

between long-term and short-term investors. Certain short-term investors (e.g. activist
investors) may tend to exercise control over companies they invest in. However, studies
found that activist investor behaviours are not always maximising firm value. According
to a McKinsey article[49], the number of activist campaigns against US companies increased
from 108 in 2010 to 280 in 2016. It highlighted that activist investors in the TMT sector
are mostly generalists that have no expertise in these industries. Although activist investors
intended to create values, the activist campaigns did not always increase the total returns to
shareholders ("total returns" is an overall measure of a company's operating performance,
market valuation of stocks, change in stock market expectations about the performance and
financial leverage[50]). The total returns to shareholders in excess of that in the corresponding
S&P sector index could be positive or negative during the first three years after activist
campaigns (see Figure 11).

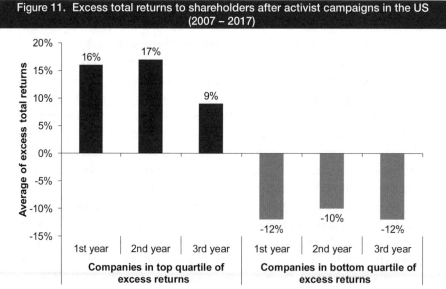

**Figure 11. Excess total returns to shareholders after activist campaigns in the US
(2007 – 2017)**

Note: The sample comprises 252 unique activist campaigns since 2007 across 151 companies (each with market
capitalisation over US$10 billion and revenues over US$1 billion) listed in the US that had the data of 3-year
total returns to shareholders available as of 15 March 2017. The excess return of a stock is calculated versus
the corresponding S&P sector index.

Source: Cyriac, J., S. Otto and D. Wells "The benefits of thinking like an activist investor", McKinsey website, June
2017.

---

49 Cyriac, J., S. Otto and D. Wells "The benefits of thinking like an activist investor", McKinsey website, June 2017.
50 See Deelder, B., M. H. Goedhart and A. Agrawal "A better way to understand TRS", McKinsey's article on its *Strategy and Corporate Finance* blog, July 2008.

3.4   Performance evidence from US-listed Chinese WVR companies

As of end-2018, there were 220 Chinese companies listed on US exchanges, a certain proportion of which were WVR companies. These include 28 companies with IPOs during 2018[51]. Six out of nine IPOs in the first quarter of 2018 (about 67% of the total) came from WVR companies[52]. As quoted in HKEX's concept paper on WVR structure[53] in 2014, about 30% of US-listed Chinese companies had WVR structures; in terms of cumulative amount of IPO funds raised, the share was 48% during January 1998 to May 2014. The paper also noted that 70% of US-listed Chinese WVR companies came from the information technology industry as of May 2014. Another source suggested that the IPOs of US-listed Chinese WVR companies raised US$34 billion during 2007-2017[54], compared to US$442 billion for all IPO funds raised in the US during the same period[55].

Given the possible differences in corporate culture, would Chinese WVR companies exhibit the same or different phenomenon compared to international experiences? Some empirical evidences showed that US-listed Chinese W-stocks did not underperform the stocks of OSOV companies (referred to as "OSOV stocks") in terms of stock price and business performance. A study[56] looked into 33 dual-class and 88 single-class Chinese companies listed on the US exchanges. In respect of price performance, the US-listed Chinese W-stocks performed less poorly in a bear market — the median abnormal return in the second year after IPO was a decline of 13% for W-stocks, compared to the decline of 38% for OSOV stocks (see Figure 12a). The differences in PE ratio and Tobin's Q ratio were not significant between WVR companies and OSOV companies. In respect of business performance, the difference in return on assets between WVR companies and OSOV companies was not statistically significant in the second year after IPO (see Figure 12b).

---

51 Source: Ritter, J. R., "Initial public offerings. Updated statistics", manuscript, University of Florida, 31 December 2018 (https://site.warrington.ufl.edu/ritter/files/2019/01/IPOs2018Statistics_Dec.pdf).

52 Source: Hu, K., "Chinese companies flooded into the U.S. IPO market in 2018", *Yahoo Finance*, 29 December 2018.

53 HKEX consultation paper, *Concept Paper – Weighted Voting Rights*, HKEX website, August 2014.

54 Source: Robertson, B. and A. Tan "Dual-class shares", *Bloomberg website*, 3 May 2018.

55 Source: World Federation of Exchanges.

56 Abdullah, J. Zhou and M. H. Shah (2017) "Effect of disproportional voting rights on firm's market performance: Evidence from Chinese firms cross-listed on US exchanges", *International Journal of Financial Studies*, Vol. 5, pp.1-11.

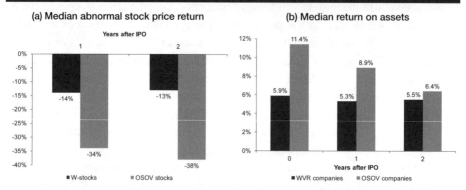

Figure 12. Comparison of stock price and business performances of US-listed Chinese companies (WVR versus OSOV)

Note: Abnormal return is defined as the difference between the actual rate of return in a specific period of time and the market return.

Source: Abdullah, J. Zhou and M. H. Shah (2017) "Effect of disproportional voting rights on firm's market performance: Evidence from Chinese firms cross-listed on US Exchanges", *International Journal of Financial Studies*, Vol. 5, pp.1-11.

3.5 Broadening the range of investment opportunities

Despite potential concerns of agency problem, WVR companies could be operationally efficient and likely to provide investors with upside potential as discussed above. While certain risk-averse investors may stay away from investments in WVR companies, other investors may be interested in taking the potential upside. The listing of WVR companies helps cater different investors' preferences. Investors interested in new-economy innovative sectors may be willing to take up shares with less voting rights than, but the same cash-flow rights as, the WVR shares. The upside opportunities offered by W-stocks of new-economy innovative companies may be appealing to even the sophisticated investors. An extreme case is the non-voting shares of Snap — the IPO was launched in the US in March 2017 with an oversubscription of more than ten times[57].

To meet the demand of different investors, the global index provider, MSCI, has resumed the inclusion of W-stocks in its indices. The MSCI once suspended the new additions of W-stocks as the Council of Institutional Investors (CII) [58] advocated for OSOV and pressured index providers to exclude W-stocks after the IPO of Snap's non-voting shares in March 2017. To reconsider the situation, the MSCI held a detailed consultation

---

57 Source: Reuters, "Snap tops expectations in pricing of long-awaited IPO", 1 March 2017.
58 CII is an industry organisation of US pension funds.

with market participants on W-stocks. The conclusion of MSCI's consultation was announced in October 2018 that supported the stakeholders to "find the right balance between investor views and comprehensive representation of the investable equity universe"[59]. Pursuant to the consultation conclusion, the MSCI has removed the restriction on new addition of W-stocks into its global benchmark from March 2019, leaving the methodology of existing indices unchanged[60], and launched new index series adjusted for voting rights in March 2019[61]. This is in line with an SEC Commissioner's comments on the investment value of WVR companies, "if we ban all dual-class companies from our major indices, Main Street investors may lose out on the chance to be a part of the growth of our most innovative companies"[62]. It echoed the case in Canada that the Canadian WVR companies are usually the largest companies and employers in key sectors[63]. Thus, it is hard to exclude them in investment portfolios.

4   Conclusion

Global stock markets in major financial centres have undertaken institutional reforms to allow the listing of companies with WVR structures. The main driver is the increasing demand from both the issuer side and the investor side. On the issuer side, visionary founders of innovative companies, including those from Mainland China, tend to prefer going public using a WVR structure to keep a dominating control on business decisions. On the investor side, the upside potential of investment in the new-economy sectors have become increasingly attractive, possibly outweighing the disadvantages of holding shares with inferior voting rights and the risks associated with a WVR structure. Hong Kong is among the major markets in the world that have reformed the listing regime to allow the listing of WVR companies.

---

59 See "MSCI will retain the MSCI Global Investable Market Indexes unchanged and launch a new index series reflecting the preferences of investors on unequal voting structures", MSCI's press statement on 30 October 2018.

60 See "Q&A on the results of the unequal voting structures consultation", MSCI, October 2018.

61 See "MSCI voting rights-adjusted indices methodology", MSCI, March 2019.

62 Source: "Perpetual dual-class stock: The case against corporate royalty", the US SEC Commissioner's speech on 15 February 2018.

63 Source: Allarire, Y., "The case for dual class of shares", the Institute for Governance of Private and Public Organisations website, 20 December 2018.

Launched in April 2018, Hong Kong's new listing regime includes a more comprehensive range of investor safeguards compared to other major markets. These safeguards could not only effectively reduce the potential agency problem of WVR structure, but also limit the principal costs associated with investors' lack of expertise and the conflict of interests among themselves. The new listing regime is therefore considered beneficial not only to issuers but also to investors.

International experience demonstrated a number of potential net benefits to investors for investing in WVR companies, including possible abnormal price returns and potential long-term value creation. The listing of WVR companies would broaden the spectrum of investment options to meet investors' wide range of needs and preferences. With the potential risks fully assessed, investment in stocks of WVR companies could be an angel to investors.

Appendix

Investor safeguards for the listing of WVR companies on exchanges in major financial centres

Safeguards	NYSE	LSE (under rules for Standard Listings)	TSE	SGX	HKEX
Minimum market capitalisation	US$40 million	£700,000 (~US$540,000)	25 billion yen for 1st section (~US$220 million)	S$300 million (~US$214 million)	HK$40 billion (~US$5.1 billion); or HK$10 billion (~US$1.3 billion) and HK$1 billion (~US$127 million) in revenue
Restriction to new issuers	Yes	No	Yes	No	Yes
Minimum equity threshold held by founders or other WVR beneficiaries as percentage of total issued share capital	No	No	No	None	10%
Maximum voting differentials	No	No	No	Yes (10:1)	Yes (10:1)
Restriction to particular industries	No	No	No	Yes	"Innovative" companies only
Sunset provisions	No	No	Natural sunset provisions	Natural sunset provisions	Natural sunset provisions
Automatic conversion to OSOV shares on retirement/incapacity/ death of founder	No	No	Yes	No	Yes
Automatic conversion to OSOV shares on share transfers	No	No	Yes	No	Yes
Enhanced corporate governance measures	No	Yes	Yes	Yes	Yes
Unique stock name marker	No	No	No	Yes	Yes

Source: Exchanges' websites on listing criteria; CFA Institute, "Dual-class shares: The good, the bad, and the ugly", August 2018; and Toshima, K. (2014) "Cyberdyne's dual-class IPO", *International Financial Law Review*, Vol. 33, pp.43-45.

Glossary

OSOV
: One share with one vote: the OSOV rule (or proportionality principle) refers to equal voting rights for each share, along with the same fraction of cash-flow rights (e.g. dividend entitlement).

OSOV company
: One-share-one-vote company: a company with a single class of shares that follows OSOV rule.

OSOV shares/stocks
: One-share-one-vote shares/stocks: the share class of a company that follows OSOV rule, while the issuing company may have a single class or multiple classes of shares.

WVR
: Weighted voting right: a WVR share structure deviates from the OSOV rule in that a share class has multiple voting rights per share in contrast to OSOV shares.

WVR beneficiary
: Weighted voting right beneficiary: the holder of shares with multiple voting rights per share in contrast to OSOV shares.

WVR company
: Weighted voting right company: a company with at least two share classes, one of which is the class of WVR shares.

WVR shares
: Weighted voting right shares: shares with multiple voting rights per share in contrast to OSOV shares.

W-stocks
: The share classes of WVR companies listed on an exchange which are not the WVR shares. In Hong Kong, they are the OSOV shares.

Chapter 8

Overseas bond markets support the development of Chinese new-economy enterprises

Adrian KHOO

Managing Director, Capital Markets Origination and Co-Head of
Asia Debt Origination, Citigroup Global Markets Asia Limited

JIANG Guorong

Managing Director, Chairman and Head of China Corporate and
Investment Banking, Citigroup Global Markets Asia Limited

Katherine FANG

Managing Director, China Investment Banking
Citigroup Global Markets Asia Limited

MAO Junxiang, David

Managing Director, Capital Markets Origination
Citigroup Global Markets Asia Limited

Summary

Over the past decade, Chinese companies are becoming more international-supported by the robust growth of the Chinese economy and the gradual opening-up of its financial sector. As pioneers and crucial players in the Chinese economy, Chinese new-economy enterprises are inevitably attracting attention from global investors. As the largest capital market internationally by fund size, the overseas bond market plays a very important role in this process. This chapter will briefly discuss the trends of debt capital market, examine how Chinese issuers obtained recognition over the past ten years and discuss on how to leverage the offshore bond markets for further development.

1 Evolution of bond issuances by Chinese issuers in overseas bond markets (2009-2019)

With rapid domestic economic development and active overseas business expansions, offshore bond financing has become the new trend for Chinese enterprises. Chinese issuers saw an annual increase in issuance volume since 2010, gradually building up their market share in international bond markets and becoming key players in the market. As shown in Figure 1, the total issuance volume is US$3.3 billion in 2009. This number peaked in 2017, reaching US$217.2 billion, which is more than 65 times of the volume in 2009. Although the issuance volume dropped in 2018, it still maintained a high level of US$175.5 billion (see Figure 1). The issuance volume reached US$136.7 billion in 2019 up to July, with projected full-year volume higher than the peak in 2017. Undoubtedly, Chinese issuers have played pivotal roles in credit supply in the Asia-Pacific region (see Figure 2).

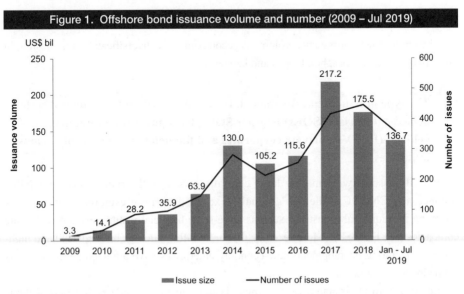

Figure 1. Offshore bond issuance volume and number (2009 – Jul 2019)

Source: Dealogic.

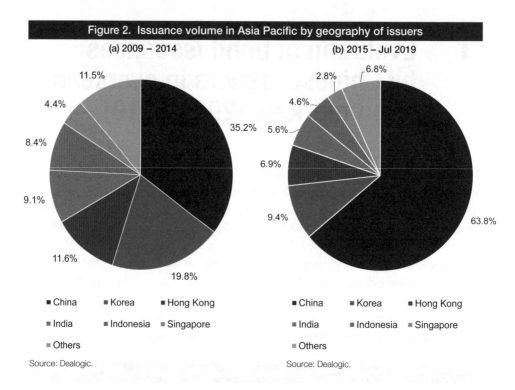

Figure 2. Issuance volume in Asia Pacific by geography of issuers

(a) 2009 – 2014   (b) 2015 – Jul 2019

China   Korea   Hong Kong

India   Indonesia   Singapore

Others

Source: Dealogic.

The rapid growth of issuance volume is accompanied by diversification, manifesting in the variety of issuer types, bond types and key terms.

(1)   The type of issuers has diversified, from a few core centrally-owned state-owned enterprises (SOEs) to other SOEs, local government financing vehicles (LGFVs), private corporates and financial institutions of various industries

Initially, issuers were mainly large centrally-owned SOEs represented by CNPC, Sinopec and CNOOC. Then, it gradually extended to more established companies operating in traditional industries, such as large-scale financial institutions and real estate companies (see Figure 3). More recently, driven by information technology (IT) and high-end manufacturing development, new-economy enterprises have attracted strong interests from global investors.

Looking into all issuances by Chinese issuers up to July 2019, there are 7 deals with issuance volume above US$5 billion, of which three were issued by new-economy enterprises. Notably, Alibaba's inaugural issuance in 2014, which raised US$8 billion, set the highest record. A series of such issuances marked the rising recognition of China's new-economy enterprises in the overseas bond market. China's new-economy enterprises did not enter the international bond markets until late 2011, when Tencent, the first publicly recognised Chinese new-economy issuer, completed its inaugural offering of a 5-year fixed-rate note with a size of US$600 million. In the subsequent 8 years, Tencent completed 6 public offshore bond issuances with increasing size and variety, updating its strategies and credit highlights to investors. In 2014, Tencent issued its first dual-tranche bond of different tenors totalling US$2.5 billon. In 2018, with a size upgrade, Tencent offered 4 tranches of different tenors in total of US$5 billion and included a floating-rate note for the first time. The floating-rate structure met investors' asset allocation appetite in the rising interest rate environment back then, and thus received strong interest from US investors with an over-subscription rate of 9.6 times. In 2019, Tencent, as a mature issuer who attracted extensive interest in the international market, successfully priced a US$6 billion bond with 5 tranches — Tencent's largest offering as of July 2019 — including 5/7/10/30-year fixed-rate notes and a 5-year floating-rate note. Subscriptions amounted to US$27 bill ion. In the past 8 years, Tencent's issuer rating has been upgraded from Baa1/BBB+/-- to A1/A+/A+, reflecting the company's growth and market recognition.

Figure 3. Chinese enterprises' offshore bond issuance volume by industry (2009 – Jul 2019)

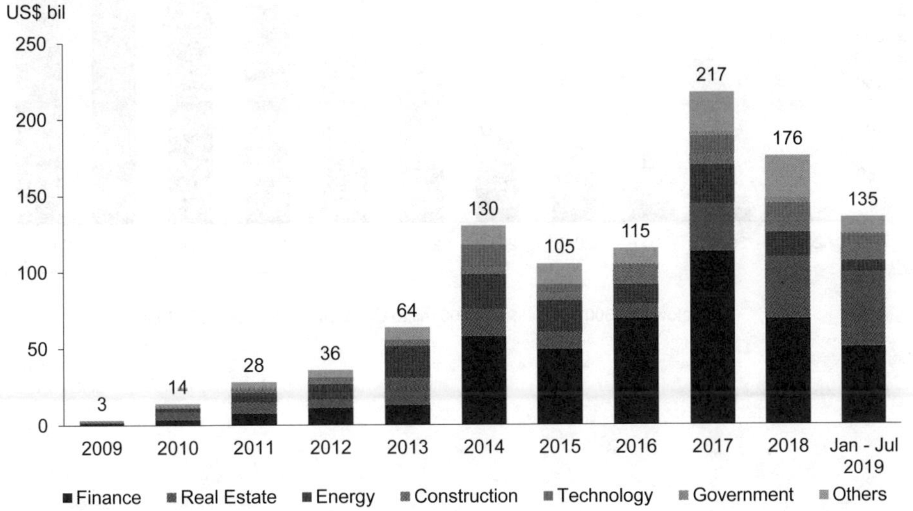

Source: Dealogic.

(2) Deal size has increased significantly

Before 2009, Chinese offshore bond issuers were mainly sovereigns or governments. There were few corporate bond issuances of over US$500 million. Nowadays, with rising global recognition of Chinese issuers, issuances valued more than US$2 or US$3 billion are more common (see Figure 4). In the last 5 years, with the rapid growth of new-economy enterprises, high-quality issuers have completed a number of symbolic offshore offerings (see Table 1). In 2014, Alibaba completed its US$8 billion inaugural issuance of fixed-rate and floating-rate bonds through 6 tranches of different tenors. In 2017, it completed another US$7 billion offering with 5 tenors (5.5/10/20/30/40). Meanwhile, Tencent made 5 issues of US dollar (USD) bonds since its inaugural issuance in 2011, from a vanilla 5-year single tranche of US$600 million to an aggregate of US$6 billion with a tenor ranging from 5 years to 30 years.

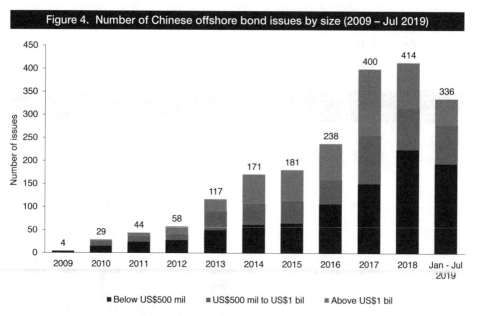

Figure 4. Number of Chinese offshore bond issues by size (2009 – Jul 2019)

Source: Dealogic.

Table 1. Chinese offshore corporate bond issues with size of US$1.5 billion or above (up to Jul 2019)			
Pricing date	Issuer	Issuer ratings (Moody's/S&P/Fitch)	Deal size (US$ bil)
20/11/2014	Alibaba Group Holding Ltd.	A1/A+/A+	8.0
29/11/2017	Alibaba Group Holding Ltd.	A1/A+/A+	7.0
03/04/2019	Tencent Holdings Ltd.	A1/A+/A+	6.0
11/01/2018	Tencent Holdings Ltd.	A2/A+/A+	5.0
22/04/2014	Tencent Holdings Ltd.	A3/A-/--	2.5
29/04/2016	Huawei Investment & Holding Co., Ltd.	--/--/--	2.0
03/02/2015	Tencent Holdings Ltd.	A3/A-/--	2.0
25/01/2018	Tsinghua Unic Ltd.	--/--/--	1.9
22/03/2018	Baidu Inc.	A3/--/A	1.5
28/06/2017	Baidu Inc.	A3/--/A	1.5
14/02/2017	Huawei Investment & Holding Co., Ltd.	--/--/--	1.5
29/04/2014	Lenovo Group Ltd.	--/--/--	1.5

Source: Dealogic.

(3)   Issuance has more tenor options

At the beginning, simple 5/10-year fixed-rate notes dominated Chinese issuances in overseas bond markets. However, tenors began to vary with issuers' diversified funding needs. The market began to see tenors of 365 days, 20/30/40 years, and perpetuity, for both fixed-rate and floating-rate notes (see Figure 5). Investment-grade issuers inclined to multi-tenor. For example, Alibaba included a 3-year floating-rate note and 3/5/7/10/20-year fixed-rate notes in its US$8 billion offering in 2014. In 2017, Alibaba extended the long end of its issuance to 40 years. Similarly, in 2019, Tencent issued a 5-year floating-rate note and 5/7/10/30-year fixed-rate notes in its US$6 billion offering. In 2017, Lenovo issued a perpetual note which was successfully classified as equity in terms of accounting treatment.

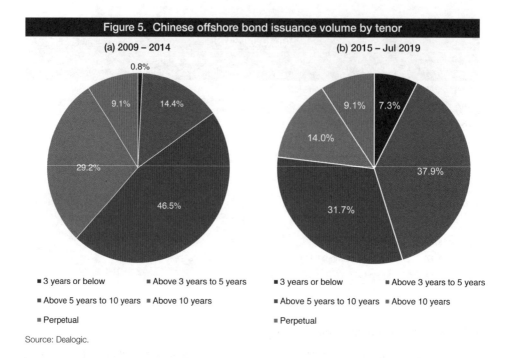

Figure 5. Chinese offshore bond issuance volume by tenor

(a) 2009 – 2014

(b) 2015 – Jul 2019

Source: Dealogic.

(4) The issuance structure is more sophisticated

For example, Lenovo issued a US$1 billion perpetual note in 2017 and managed to recognise it as equity for accounting treatment, which lower the leverage of the company. In 2018, Lenovo conducted a tender offer for its existing notes that would mature in 2019 and 2022 with a total size of US$2 billion and then issued a US$750 million bond under its MTN programme to finance the tender offer. Thus, Lenovo minimised its refinancing risk, extended the average maturity of existing debt and secured a lower interest rate under the prevailing favourable market conditions.

(5) Issuance currency is more diversified

Issuance currency has evolved from single tranche USD to a more diversified combination of euro (EUR), British pound (GBP), offshore Renminbi (CNH), Hong Kong dollar (HKD), Australian dollar (AUD), Japanese yen (JPY), Singaporean dollar (SGD), etc. While USD has been dominating the market so far, accounting for 87% of issuance on average (see Figure 6), non-USD currency offerings such as EUR and CNH offerings are also gaining momentum under different market conditions (see Figure 7). The low absolute yields in these markets have provided issuers with a wider choice of financing. In particular, bonds denominated in a local currency serve as a natural hedge for issuers with revenue in the same currency.

Figure 6. Chinese offshore bond issuance volume by currency (2009 – Jul 2019)

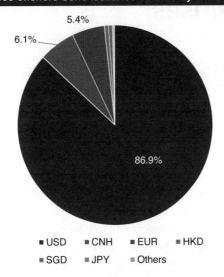

5.4%

6.1%

86.9%

■ USD ■ CNH ■ EUR ■ HKD
■ SGD ■ JPY ■ Others

Source: Dealogic.

Figure 7. Volume of Chinese offshore bonds in CNH and EUR (2009 – Jul 2019)

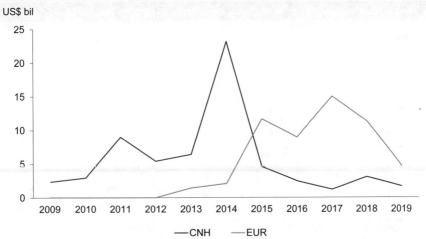

US$ bil

—CNH   —EUR

Source: Dealogic.

2 Benefits of offshore bond fund-raising to new-economy enterprises

Over the past decade, Chinese issuers benefit from the development of the offshore bond market in terms of broadening financing channels, reducing refinancing risks, optimising capital structure and deepening internationalisation.

(1) Lower cost of financing

The year 2009 was a turning point for global macroeconomic environment. Under extensive loosening of monetary policies, the benchmark rates of USD and other international currencies have been at historical lows, which brought down the cost of corporate debt financing to an unprecedentedly low level. Besides, a number of market factors such as abundant global liquidity, demand for high-quality assets and higher level of global recognition of Chinese enterprises led to the continuous tightening of credit spreads (see Figure 8). This provides an ideal external macro environment for Chinese issuers in the offshore bond market.

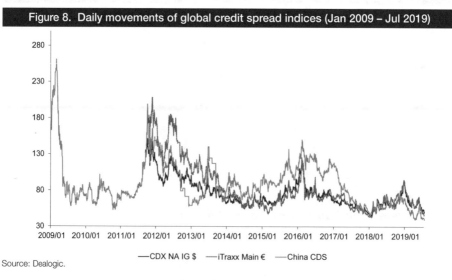

Figure 8. Daily movements of global credit spread indices (Jan 2009 – Jul 2019)

Source: Dealogic.

(2) Larger sizes and longer tenors

Compared to the onshore bond market (investors dominated by banks), investors in the offshore bond markets are more diversified, including high-quality global long-only funds, hedge funds, banks, insurance companies and pension funds, generating robust fund flows. These global long-only funds, insurance companies and pension funds are more interested in bonds with long tenors, which supports bond offerings with larger size and longer tenors. One of the examples is the Alibaba's 2017 issuance, in which the company successfully priced a US$1 billion 40-year bond. For this tranche, the majority of investors came from Europe and the US and a great portion of allocations went to global funds and insurance companies.

(3) Flexible use of proceeds

There are fewer restrictions for issuers on issuing overseas bonds than on issuing domestic bonds. Greater flexibility is allowed in terms of issuance windows and the use of proceeds, helping Chinese issuers to effectively obtain sufficient capital in time for overseas business expansion. The senior management of Alibaba, for instance, expressed in the investor meetings that the proceeds from the offshore USD bonds would assist its acquisition of overseas assets in retail and logistics in order to support its global strategy. In addition, SF Express completed its inaugural public USD bond issuance in 2018, which facilitated its overseas investment in a bunch of physical assets.

(4) Efficient and flexible execution process

Overseas bond issuance can be executed in a highly efficient manner. A fund-raising channel is built once the first issuance is completed such that the execution for subsequent issuances becomes simple and the timeline can be further shortened. Additionally, the flexibility of execution windows effectively helps companies to diversify the source of financing under various market conditions. Such a complement to domestic financing channels reduces reliance on traditional onshore funding channels, largely alleviating refinancing risks.

(5) Greater recognition by overseas investors

The overseas bond market has a very broad investor base and generally has a deeper level of understanding of the high-tech, Internet, and other new-economy industries. Therefore, overseas investors have a stronger interest in subscribing the bonds. Many

European and US investors have a dedicated research team specialised in specific sectors in the new-economy industry. For example, Weibo has recently completed the issuance of its first USD-dominated bond. As a Chinese social platform unicorn, the scarcity of such bond issuance has attracted a large number of international investors to participate.

(6) Effectively improved business management ability and operation transparency

In the process of public fund-raising, through preparing documents, obtaining international ratings and meeting with international investors, issuers could effectively improve their business management ability and transparency — successful issuance indicates successful marketing, which is of great help to enterprises in further developing international businesses, overseas investment and financing and establishing a trustworthy international image. For more active and frequent issuers in the market, each issuance is an excellent opportunity to update their business status and future development strategies for international investors, so as to timely showcase their latest credit highlights to achieve lower financing costs. For example, in 2017, through effective communication with investors, Alibaba priced US$7 billion multi-tranche bonds in its second issuance, where all tranches realised negative "new issue concession" (NIC), i.e. spread at issue lower than the secondary trading levels of the company's existing bonds. The issuance redefined Alibaba's credit curve and reflected investors' full recognition of Alibaba's strengthened credit.

3 Case studies of offshore bond financing by new-economy enterprises[1]

Chinese new-economy enterprises offer an attractive investment opportunity to overseas investors. Compared to their state-owned counterparts, new-economy enterprises are

---

1 The information about the transactions discussed in the case studies in this section is publicly available.

generally more competitive, transparent, financially robust, and technologically advanced. Investors, particularly from developed markets, are more familiar with sectors like IT and high-tech manufacturing, and therefore have higher interest and stronger risk appetite to invest in these sectors. The supply of Chinese bonds, on the other hand, is still dominated by issuers from traditional sectors, adding to the investment value of new-economy enterprises due to scarcity. These factors combined highlight the huge investor demand for China's new-economy issuers.

The international bond market opens the gate to promising growth for high-quality new-economy issuers. Through issuing bonds offshore, qualified issuers are able to achieve financing with a lower cost, larger size, longer tenor, and in multiple currencies. This complements their existing onshore financing channels and supports their strategic overseas expansion. Successful examples range from Internet giants such as Alibaba, Tencent, Baidu, JD.com and Weibo, to emerging industry leaders such as Lenovo, Sunny Optical, and SF Express. Among them, we have chosen to focus on three typical new-economy issuers, each with their own characteristics, and made their stories into the following case studies.

3.1 Alibaba's offshore bond offerings

As a pioneer in China's IT and high-tech sectors, Alibaba Group made a successful debut in the international bond market with a US$8 billion six-tranche offering (including 3/5/7/10/20-year tranches) on 20 November 2014. This landmark transaction was record-breaking on multiple fronts and placed itself in the centre of global investors' attention — it was the largest debut international bond offering at that time, and Asia's largest ever international bond offering with the biggest number of tenor tranches. This deal came shortly after Alibaba's blockbuster Initial Public Offering (IPO) on the New York Stock Exchange (NYSE) on 19 September 2014, which made history by becoming the largest IPO of all time opened on the NYSE. With full recognition by global investors, Alibaba soon worked on obtaining its initial credit ratings from major international rating agencies to lay the groundwork for its proposed bond offering. All of the three agencies granted Alibaba ratings equal to China's sovereign ratings at A1 (Moody's) / A+ (S&P) / A+ (Fitch), which was unprecedented for a Chinese private company.

As soon as the preparation was completed, Alibaba kicked off its debut offshore bond issuance. Since the company had just finished its IPO a short while back, only simple adjustment and update were required based on existing documentation. The execution timetable was shortened to less than six weeks before everything was nearly finalised. After the mandate announcement, Alibaba sent out two teams to go on a three-day global roadshow across five international financial centres, namely Hong Kong, Singapore, London, Boston and New York, meeting with more than 300 investors through one-on-one meetings and global investor calls to broadcast Alibaba's credit story and present its business performance and market status. The marketing campaign was well received with positive feedbacks. On the day of bookbuilding, the subscription ratio reached 6.9 times. The total order book size was only second to that for Verizon's US$49 billion issuance in September 2013. This had marked the second largest reported orderbook for a USD issue in the world as of that date and was the largest among Asian issuers.

Riding on such enthusiasm, the issuer was able to price the bonds in its 5-year and 10-year tranches at among the tightest spreads seen for any international issuance from China. The tight pricing was a testament to the global recognition of the company's credit strength and successfully positioned Alibaba among US blue-chip technology giants like Amazon, Cisco, and Oracle. Impressively, a significant portion of the final allocation went to US investors, demonstrating the strong interest and support shown by US investors who have been closely following the development of Chinese IT and high-tech companies like Alibaba. By tapping into the international bond market, Alibaba effectively broadened its investor base and further elevated its brand image through the global marketing campaign. Large volume of long-term funding was secured at record-low costs, which helped the company diversify financing channels and optimise its capital structure, thus laying a solid foundation for its subsequent overseas expansion.

Three years later in November 2017, as Alibaba deepened its international presence and expanded its global footprint, the company priced its second US$7 billion bond offering at even lower costs, representing the largest Asian corporate issuance since its debut offering in 2014. Notably, this transaction made a breakthrough in terms of tenor — the offering was composed of 5.5/10/20/30/40-year tranches, among which the longer-dated 40-year tranche had been the first seen among Asian issuers in two decades. New pricing benchmarks were established for Alibaba as negative NIC was achieved across all tranches. The issuer successfully compressed the reoffer spreads by at least 20 basis points (bps) below the final pricings of the first offer in November 2014, further narrowing its pricing differential relative to major global technology peers. This deal again demonstrated how Alibaba used offshore bond financing to its great advantage to strategically lower its financing costs and optimise its capital structure.

3.2   Sunny Optical's offshore bond offering

Regarded as the leader in high-end optical components manufacturing, the Hong Kong-listed Sunny Optical Technology successfully completed its inaugural US$600 million

5-year Regulation S (Reg S) bond offering in January 2018. The deal recorded the lowest reoffer spread from an Asia ex-Japan inaugural BBB area-rated corporate issuer as of that date. Prior to the issuance, drawing strength from its leading industry position, low debt leverage and decent profitability, Sunny Optical had secured a Baa2 investment-grade rating from Moody's, an outstanding result among peers. The rating was proved to be instrumental in forming foundations for a successful bond issuance and set an important benchmark for its industry peers. Concurrently, the company was able to capture an optimal market window to launch the deal in early 2018, when its stock price had surged by 170% over the preceding 12 months driven by strong investor demand. Following the announcement, the company commenced a roadshow interfacing with around 160 investors in major financial hubs including Hong Kong, Singapore and London. The roadshow team was welcomed with enthusiasm and many meetings featured more than 20 participants at one time (usually attended by three or four investors). Despite the sporadic volatility on the pricing day, the deal was well received by the market, witnessing a peak orderbook at US$9.1 billion or oversubscription rate of 15.2 times. This was the highest oversubscription rate amongst all Asian USD bond offerings. On the back of the strong orderbook, the company managed to price its bonds with 35 bps tightening from the Initial Price Guidance (IPG). Moreover, the bond tightened further by more than 10 bps in the secondary market on its first trading day, reflecting the investor demand. This inaugural offshore bond offering helped Sunny Optical secure record-low financing cost, diversify funding sources, optimise capital structure, and establish its position in the international capital markets.

3.3   Lenovo's tender offer and new issuance

Listed in Hong Kong, Lenovo Group has a diversified overseas business portfolio with revenue and assets denominated in USD. Through raising funds in USD, the group can naturally hedge its foreign exchange rate risks. In 2014, Lenovo witnessed significant improvement in its financial results and a successful inaugural US$1.5 billion 5-year USD bond offering overseas. The bond offering was the largest ever single-tranche senior Regulation S-only bond offering in Asia at that time. Even without credit ratings, this issuance was highly sought after by investors, garnering a high-quality orderbook at US$8 billion or 5.3 times' oversubscription rate. The proceeds would be used for supporting the group's business expansion and strategic development. The company went on to acquire Motorola and IBM's x86 server business. However, the post-acquisition integration did not generate revenue growth as promising as expected. Faced with increasing pressure from debt servicing, Lenovo decided to actively manage its liability through launching a tender offer for the bonds due 2019 and refinance through a new 5-year bond offering. This was a strategic move taking advantage of the flexibility permitted and product diversity offered in the offshore bond market. This facilitated Lenovo's capital restructuring by extending its debt maturity and minimised its refinancing risk in the short-term. The tender offer was a success receiving over US$1.02 billion tender instructions from bondholders. At the same time, Lenovo priced its US$750 million 5-year new bond, proceeds from which financed the cash tender offer. This liability management exercise helped the group to effectively extend the maturity of its liabilities without raising the overall leverage, which in turn lowered its refinancing cost.

4 Conclusion

In conclusion, the offshore bond market provides attractive opportunities for new-economy issuers, who can draw experiences from many successful examples as listed above. Through issuing bonds offshore, qualified issuers are able to achieve financing with a lower cost, larger size, longer tenor, and in multiple currencies. This complements their existing onshore financing channels and supports their strategic overseas expansion. Overseas investors have tremendous interest in the top players in China's IT and high-tech manufacturing sectors. Bonds issuance by these companies are, however, still relatively limited in volume and infrequent compared to the market scale, underlining their premium value due to scarcity. We are strongly confident that with the further development of China's new-economy industries, more qualified Chinese issuers will become active participants to tap into the offshore bond market. In this context, before taking up the opportunity, companies are advised to fully consider the potential risks and challenges based on their specific situations and thoroughly consult and discuss with financial institutions and legal counsels with relevant expertise.

Note: The following team in Citigroup Global Markets Asia Limited also contributed to the preparation of this article —
Melody ZHANG, Director, Capital Markets Origination; Natalie NI, Vice President, Capital Markets Origination; Amy ZHANG, Vice President, China Investment Banking; WU Tianhao, Analyst, Capital Markets Origination; Jasmin Qi, Analyst, Capital Markets Origination; Skye HAN, Analyst, China Investment Banking.

Chapter 9

New-economy companies' overseas mergers and acquisitions and financing

David WU

Vice Chairman and Financial Services Industry Leader, Deloitte China

Sitao XU

Chief Economist, Head of Deloitte Research and Partner, Deloitte China

Yun YU

Risk Advisory and Treasury Service Partner, Deloitte China

Nina ZHOU

Tax Service and Global Infrastructure Services Centre Partner, Deloitte China

Summary

In recent years, China's new-economy companies are the rising stars in the digital transformation of the global economic system. Their globalisation activities will become more and more active in the future. Overseas mergers and acquisitions (M&A) play an important role in promoting globalisation in various ways. They can help new-economy companies rapidly expand overseas markets and absorb advanced foreign technologies. In order to raise enough M&A funds, new-economy companies often need financing support. However, many of them are still in the early stage of development, so it is difficult for them to obtain traditional bank credit. Overseas financing is therefore a more convenient channel for them to grow. As an international financial centre, Hong Kong can make use of its institutional and channel advantages to provide a platform for the Mainland new-economy companies to absorb overseas funds, including attracting support from international banks, listing in Hong Kong or issuing bonds, to help them develop globally.

1 Trends and characteristics of overseas M&A of China's new-economy companies

1.1   Overall slowdown in overseas M&A

With the outbreak of the financial crisis in 2008, undervalued international assets have become the target of overseas "bargain-hunting" of Chinese enterprises, and overseas M&A of Chinese enterprises have entered a rapid growth stage. Until recent years, policy and legal restrictions have gradually become stricter, and overseas M&A have gradually returned to rationality. Since November 2016, the Chinese government has issued policies to strengthen foreign exchange control, strictly control capital outflow[1] in succession, and has strengthened the foreign exchange review of large overseas M&A transactions. This has led to a sharp decline in overseas M&A cases and in transaction amounts of Chinese enterprises since 2017 with a continued slowdown trend. According to the Ministry of Commerce, the total value of overseas M&A transactions of Chinese enterprises in the first half of 2019 was only US$16.95 billion, dropped 72% compared with the corresponding period in 2018.

1.2   Overseas M&A of the new-economy companies will remain active

Despite the overall slowdown in the trend of M&A, the Belt and Road Initiative (BRI) still provides new opportunities of overseas M&A for Chinese enterprises. Especially at present, the new-economy forms, represented by the Internet, knowledge economy and high and innovative technology, have developed rapidly to meet the needs of the consumers. The overseas M&A of China's new-economy companies represented by "unicorn companies"

---

1   In November 2016, the State Administration of Foreign Exchange (SAFE) required the outward remittance under the capital account of funds of US$5 million or above to be submitted to SAFE for review and approval; in January 2017, SAFE issued the *Notice on Further Promoting the Reform of Foreign Exchange Administration and Enhancing the Verification of True Compliance*; and in July 2017, the Ministry of Commerce promulgated the *Decision on Revising the Interim Administrative Measures for the Record-filing of the Incorporation and Change of Foreign-funded Enterprises.*

have attracted wide attention.

The business models and products of new-economy enterprises are based on technological innovation and application. In the early stage of development, these enterprises were facing fierce competition. Enterprises needed to invest a lot of capital resources to establish a leading position in the market. Therefore, acquiring overseas technology and market through M&A has become an ideal strategic layout. Practices show that a myriad of new-economy enterprises have developed rapidly in a few years, and have grown into "unicorn companies". From 2018 to the first half of 2019, among the Chinese enterprises undertaking large overseas M&A, new-economy companies have performed well, such as Tsinghua Unigroup's US$2.6 billion acquisition of Linxens SA, the world's largest maker of smart security chip modules, and the social media platform YY's acquisition of the entire stake in BIGO, a Singaporean video social platform, for US$1,453 million.

1.3 Characteristics of overseas M&A of new-economy companies

1.3.1 M&A subject: Active private enterprises and the increase in the number of forms of cooperation among diversified shareholders

State-owned enterprises are the traditional main subjects of M&A who play a leading role and are the "main force" in the construction of BRI. Their advantages are concentrated in large-scale traditional industries such as energy and infrastructure. New-economy companies represent innovation and competitiveness with their advantages shown in digital technology, communications, media, Internet, biomedicine, intelligent manufacturing and cultural tourism. Private enterprises and financial investors represented by venture capital and private equity funds are the main forms undertaking overseas M&A by new-economy companies, such as Huawei, Alibaba, Fosun and other large corporate groups. Moreover, to reduce the risk, the M&A subject enterprises usually look for other partners to set up joint ventures to implement the M&A. As a result, more and more diversified shareholder cooperation models will be adopted.

1.3.2 Target industries: Emerging industries with high technology content and high added value

Driven by the trend of domestic economic transformation and upgrading, overseas M&A of Chinese enterprises tends to focus on new industries with high technology content and high added value, such as Internet, TMT[2], new retail[3], biomedicine, intelligent manufacturing, etc. According to the data of Thomson Reuters, the industries of Chinese new-economy enterprises with more overseas M&A were high-tech, accounting for 20% of the total, followed by medical care and consumption-related industries which had relatively high percentage shares in 2018 (see Figure 1). Recent representative acquisitions include Tencent's US$8.6 billion acquisition of 84.3% of Supercell, a Finnish mobile games developer, and Shanghai RAAS' announcement of its US$5.4 billion acquisition of GDS, the world's leading blood testing company.

Figure 1. Distribution of Chinese companies undertaking overseas M&A by industry (2018)

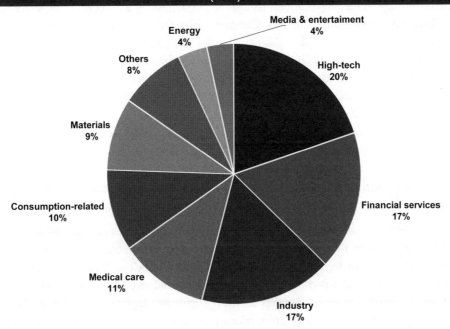

Source: Thomson Reuters, Deloitte Research.

---

2 TMT refers to Science and Technology, Media and Telecommunications.
3 New retail refers to the new retail model of deep integration of online services, offline experience and modern logistics in which individuals and enterprises rely on the Internet to upgrade and transform the production, circulation and sales processes of goods through the use of big data and artificial intelligence, and then reshape the business structure and ecosphere.

1.3.3 Target area: European and American advantages have declined and the proportion of Asia-Pacific region has increased

For a long time, the preferred areas of overseas M&A of Chinese enterprises have been in Europe and the United States (US). However, since 2017, the number of M&A in Europe and in the US has declined dramatically. M&A in the Asia-Pacific region accounted for half of the total[4]. The growth in M&A of Southeast Asian enterprises is particularly prominent. Such change in major M&A region is mainly due to two factors. Firstly, Europe and the US set restrictions on direct investments and M&A, and trade protectionism is on the rise. For example, US President Donald Trump announced the *Foreign Investment Risk Review Modernisation Act* in June 2018, while in September 2017, the European Union announced the proposal for a regulation on establishing a framework for screening of foreign direct investments into the European Union (EU), which was formally adopted in March 2019. These acts take strict measures for foreign investments in high-tech fields, making Europe and the US no longer the preferred investment target regions for Chinese companies. Secondly, the BRI stimulated Chinese enterprises to increase their business layout in ASEAN[5] countries and other related places. For example, in recent years, Chinese e-commerce companies especially favoured Southeast Asia for their overseas acquisitions. Alibaba bought Indonesia's online shopping company Tokopedia and Singapore's e-commerce platform Lazada in 2017 and 2018 respectively.

1.3.4 Purposes of M&A: Acquiring key capabilities and developing overseas markets

On the one hand, M&A can directly acquire key capabilities and intangible assets of the target enterprise. These include research and development (R&D) capabilities, trademarks, goodwill, technology, management, sales channels, etc. It can also achieve the optimisation of business portfolio through certain cross-industry M&A. The development of new-economy companies depends on R&D talents and high-tech investment. New-economy companies therefore often take overseas companies with many patents or cutting-edge technologies as the targets of M&A, hoping to achieve rapid breakthroughs in key technologies and to enhance R&D capabilities. For example, in April 2017, Truking Technology merged with Romaco Group, the world's leading supplier of solid pharmaceutical equipment, which greatly increased the production and plant size of Truking

---

4   Deloitte, *Embarking on the Journey to a New Era of Opening-up — 2018 Deloitte Outbound Investment and Operation Guide for Chinese Business.*
5   ASEAN is the Association of Southeast Asian Nations.

Technology.

On the other hand, cross-border M&A is the preferred way for multinational companies to speed up market expansion and enter the target market in a short time because it does not need a long construction period to do so. Moreover, when the domestic market is close to saturation and lacks growth space, new-economy companies can exploit the international market through M&A. Overseas M&A and transnational operations can help companies gain pre-empt rights in undeveloped overseas markets more quickly. For example, in January 2018, Didi Chuxing acquired "99", the largest transportation platform in Brazil, precisely for the purpose of expanding the transportation market of the company in Brazil and even in Latin America.

2 Challenges and problems of overseas M&A of new-economy companies

2.1 Risk and regulation are the main challenges

A large number of empirical examples have proved that the main reasons for the failure of overseas M&A of many enterprises are the lack of risk preparation in advance, poor response to risks in the event, and inadequate review of risks for improvement afterwards. Against the background of the BRI, the line of thought about risk management should be updated comprehensively. Most of the BRI countries have complex situations. It would not be enough to analyse the risks at one point in time, in a single place or in a single dimension because it needs to consider multiple factors, including the risk of policy change, the risk of economic instability, the risk of legal differences, the risk of interest rate and exchange rate, the financial risk, and the risk of cultural differences.

Domestic and foreign regulatory requirements and law enforcement are becoming increasingly stringent, especially for the review and approval of foreign investments in Europe and in the US. Enterprises have become more and more aware of the importance of understanding the regulatory environment in advance and carrying out operations in compliance with the rules and regulations.

2.2   Problems run through the whole life cycle of M&A

Affected by many factors, overseas M&A activities of Chinese enterprises are not smooth sailing. Throughout the whole life cycle of overseas M&A, the problem lies in three stages: pre-M&A, M&A in-process and post-M&A integration.

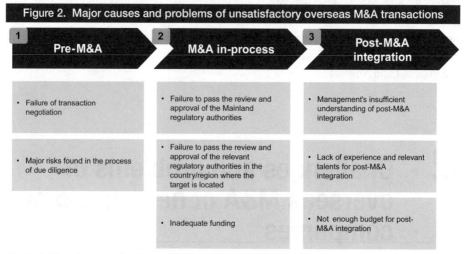

Figure 2. Major causes and problems of unsatisfactory overseas M&A transactions

Source: Deloitte's *Survey on the Present Situation of Overseas M&A and Integration of Chinese Enterprises*, Deloitte Research.

2.2.1   The problems in the whole life cycle of M&A

Deloitte's survey results[6] show that before M&A, the failure of transaction negotiation and the discovery of significant risks in the process of due diligence leading to the termination of the transaction are the two main reasons which account for 24% and 22% of the failed M&A cases respectively. Chinese investors are often weak in the reasonable planning of projects, or lack understanding of the materials required by the seller. This may lead to project abortion due to procedural reasons rather than the price quotation.

During the M&A process, failure to pass the approval by domestic and foreign regulatory authorities and inadequate funding are the main factors of transaction termination. Firstly, overseas investments of Chinese enterprises must be approved by government departments including the Ministry of Commerce, the National Development

---

6   See Deloitte, "Active M&A, lagging integration: A survey of overseas mergers & acquisitions and post-mergers integration of Chinese enterprises", an article posted on Deloitte's website.

and Reform Commission (NDRC) and the State Administration of Foreign Exchange (SAFE). The long approval cycle and the rigorous approval process may lead to missing the best timing for M&A. Secondly, in order to increase the control of domestic technology, the target country of M&A will also conduct a strict review of cross-border M&A, especially in respect of overseas M&A of high-tech enterprises which reside in countries or regions that give priority to intellectual property protection. Another problem is the lack of M&A funds of new-economy companies or the withdrawal of partners that lead to the termination of transactions.

In the process of post-merger integration, many cases of integration fail to achieve satisfactory results and the realisation of the expected transaction value. There are multiple reasons: first, the management is not well aware of the need for post-merger integration; second, the company lacks integration experience and relevant talents for post-M&A integration; third, some investment institutions are short of budget to carry out the relevant work.

2.2.2 Urgent need to solve the financing problem

Some of the above problems can be solved by means of third-party professional service agencies (in areas such as transaction negotiation, due diligence, regulatory approval), or by enterprises themselves (such as by focusing on post-merger integration, strategic optimisation and talent cultivation). However, capital inadequacy is a thornier problem, because many new-economy companies are in the early stage of development such that they lack assets for mortgage loans and their operation is unstable. It is difficult for them to reach the threshold of bank credit access. Moreover, the Mainland's domestic multi-level capital market is still being developed, and enterprises cannot get access to the financing and loan support in need.

In the face of such funding plight, Hong Kong can become a financing platform for overseas M&A of new-economy companies to solve their financial difficulties. With the establishment of the Mainland-Hong Kong Mutual Market Access programme, Mainland new-economy companies can get access more easily to overseas financing through Hong Kong, which in turn can leverage on its advantages in respect of systems, channels and talent to serve the overseas M&A of new-economy companies. Related service providers in Hong Kong may build a financing platform for enterprises to help them design multi-dimensional financing channels and connect with the relevant financial and asset management institutions to meet the needs for overseas operation funds and to reduce financing cost.

3 Analysis of the advantages of Hong Kong as an overseas M&A platform and its flexible usage

Over the years, Hong Kong has ranked 3rd in the world according to GFCI (Global Financial Centres Index)[7]. As the largest international financial centre in Asia and the world's largest offshore Renminbi (RMB) market, Hong Kong has naturally become a platform where Mainland companies of diverse economic types face up to the risks and challenges in their overseas M&A. It has not only close ties with the Mainland, but also prominent advantages in cross-border financing and institutional environment.

3.1 Close ties with Mainland China

Hong Kong, adjacent to Shenzhen, stands at the core of the Guangdong-Hong Kong-Macao Greater Bay Area. Hong Kong provides an exclusive option of geographical location for the Mainland enterprises that intend to expand their business in overseas markets, let alone its advantages in language and culture that are self-evident.

Since China's economic reform and opening-up, Hong Kong has witnessed increasing economic exchanges with the Mainland. The total volume of trade between the two places expanded by 65 times[8] during the period of 1979 to 1997. Through frequent trade contacts, the Mainland enterprises have been gradually taking the advantages of Hong Kong as a global free-trade port. The return of the sovereignty of Hong Kong has led to increasing tied economic cooperation between the two places after signing the framework agreement of "Mainland and Hong Kong Closer Economic Partnership Arrangement" (CEPA) by both sides. With the advantage of Hong Kong as an international financial centre getting more prominent, more Mainland enterprises go to Hong Kong for financing. The number of H shares that got listed in Hong Kong increased from 26 in January 1997 to 267 by the end of 2018[9].

---

7   "Global Financial Centres Index" is the most authoritative ranking index of international financial centres and is jointly compiled by the British think tank, Z/Yen Group, and China Development Institute (CDI).
8   Source: The Census and Statistics Department of the Hong Kong government.
9   Source: China Securities Regulatory Commission.

Recently, Hong Kong and the Mainland have gradually developed into a mutual capital market through the Mutual Market Access schemes. Shanghai-Hong Kong Stock Connect and Shenzhen-Hong Kong Stock Connect were launched in 2014 and 2016 respectively, leading to the rapid increase in the turnover value of cross-border stock trading. "Bond Connect" that went live in 2017 have attracted over a thousand foreign institutional investors in two years' time[10].

The Mutual Market Access model has further cemented the connection between Hong Kong and the Mainland, created more and more global investment opportunities for the Mainland investors and, to a certain extent, promoted the overseas investments and M&A by the Mainland enterprises.

3.2   Financing advantages of Hong Kong

Based on the close ties between Hong Kong and the Mainland, Hong Kong has many notable advantages in the process of overseas M&A by the Mainland new-economy enterprises. The most important of which is its ability to provide the Mainland enterprises with adequate, diversified, low-cost and convenient financing.

In the actual business process, we have observed that the Mainland enterprises generally tend to use the financing mode of bank loans in their overseas M&A, in the hope of receiving the loan support of large international commercial banks to ensure timely completion of the M&A. However, according to an incomplete statistics of the China Banking and Insurance Regulatory Commission (CBIRC), private enterprises accounted for only 25%[11] of the outstanding loans in the Mainland banking industry. Most of them are private new-economy enterprises which are difficult to obtain financing through loans. The banking industry of Hong Kong can better satisfy the M&A needs of new-economy enterprises. As of July 2019, there were 194 recognised banking institutions in Hong Kong, including the affiliates set up by 70 of the world's 100 largest banks[12]. More than 40% of the loans granted to the Mainland enterprises in Hong Kong went to private enterprises. (See Figure 3).

---

10 Source: Xinhua News Agency (http://hm.people.com.cn/n1/2019/0704/c42272-31213073.html).

11 The report on Guo Shuqing, Chairman of the CBIRC, answering reporters' questions on the CBIRC's website: http://www.cbirc.gov.cn/chinese/home/docView/7F7EDCDDD5A04396A00E8E23F8E2E813.html.

12 Source: Hong Kong Monetary Authority (https://www.hkma.gov.hk/gb_chi/key-functions/banking-stability/banking-policy-and-supervision/three-tier-banking-system.shtml).

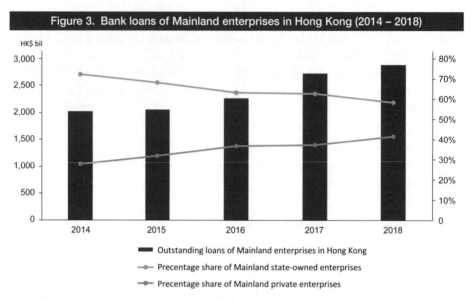

Figure 3. Bank loans of Mainland enterprises in Hong Kong (2014 – 2018)

Source: *Semi-annual Report on Monetary and Financial Stability* (2014-2019), Hong Kong Monetary Authority.

In addition to bank loans, Hong Kong also has advantages in equity and bond financing. According to the statistical data of Refinitiv, in 2018, Hong Kong Exchanges and Clearing Limited (HKEX) achieved the highest amount of funds raised by initial public offering (IPO) in 8 years. There were 208 newly listed companies with a total amount of IPO funds raised of HK$286.6 billion, 95% of which were contributed by the Mainland enterprises[13]. Following the implementation of the new listing regime, 28 new-economy companies, including unicorn enterprises like MI and Meituan, went to list in Hong Kong[14]. In respect of bond issuance, Hong Kong offered even better funding. According to the statistics of the World Federation of Exchanges, the funds raised on the HKEX bond market far exceeded that on the Shenzhen Stock Exchange in 2018 (see Figure 4).

---

13 Source: Deloitte, *IPO Market Review of 2018 and Prospect Report of 2019 for Mainland China and Hong Kong.*
14 Source: https://tech.sina.com.cn/roll/2019-01-05/doc-ihqfskcn4174725.shtml.

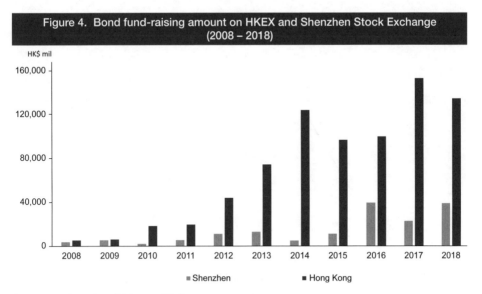

Figure 4. Bond fund-raising amount on HKEX and Shenzhen Stock Exchange (2008 – 2018)

Source: World Federation of Exchanges, Wind.

Furthermore, under the pegged exchange rate system of Hong Kong, which is operated by the Hong Kong Currency Board, Hong Kong features lower financing cost and risks. The Hong Kong dollar (HKD) exchange rate is stabilised at HK$7.8 for US$1 through an automatic interest rate adjustment mechanism, meaning that interest rate is fully market-based in Hong Kong. The pegged exchange rate system has maintained the stability of the financial market in Hong Kong, while the fully market-based interest rate enables enterprises to enjoy low-cost financing (see Figure 5).

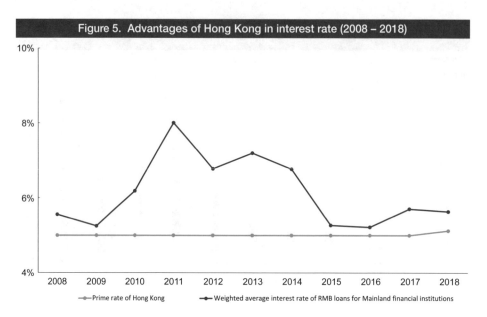

Figure 5. Advantages of Hong Kong in interest rate (2008 – 2018)

Note: The interest rate of Mainland is that applicable in the fourth quarter of the relevant year.
Source: Hong Kong Monetary Authority, People's Bank of China, Wind.

3.3   Advantages of Hong Kong in institutional environment

The Mainland implements strict regulatory measures on the M&A loans of commercial banks and the flow of foreign exchange funds, which highlights the advantages of Hong Kong in its institutional environment. In February 2015, the China Banking Regulatory Commission (CBRC) (currently the CBIRC) revised the *Guide to Risk Management of the M&A Loans of Commercial Banks*, indicating that it would strictly supervise M&A loans through a rigorous system of statistics, summarisation and analysis. After 2016, a number of ministries, including the SAFE, jointly imposed restrictions on the overseas investments of the Mainland enterprises. In this regard, Hong Kong can provide the Mainland enterprises with greater flexibility and convenience.

Additionally, under the support of the basic national policy of "One Country, Two Systems", Hong Kong has maintained its original common law system after the return of its sovereignty. In the actual operation process of cross-border M&A, most transaction documents apply the framework of common law. Once any legal dispute takes place, the judgment handed down in Hong Kong is valid in all other jurisdictions in the world. Besides, Hong Kong has a large number of professional service agencies and related high-end talents, e.g. lawyers, accountants, investment banks and consultants, thus enhancing the ability of Hong Kong to support cross-border M&A.

Lastly, the financial infrastructure established in Hong Kong has also provided great convenience for the financing of the Mainland enterprises. The Hong Kong Monetary Authority has invested heavily in payment systems (e.g. the Real Time Gross Settlement (RTGS) system), bond settlement systems (e.g. the Central Moneymarkets Unit (CMU) system) and the networking system, resulting in the increasing improvement of the financial infrastructure of Hong Kong.

In fact, with the advancement of RMB internationalisation and the increasing number of Chinese enterprises "going global", Hong Kong is also facing challenges from other Asian financial markets. Singapore, with a high degree of economic freedom, breadth and depth of transactions in foreign exchange, futures and other financial derivatives, and a robust management of fiscal and foreign exchange reserves, has gradually attracted the Mainland investors. Abu Dhabi in the Middle East is also actively building up a high-tech and digital financial industry chain, and has been working together with a number of Mainland agencies such as the NDRC to establish a regional investment and financing centre for the Mainland enterprises in the BRI regions. However, Hong Kong boasts more outstanding advantages in the short term.

3.4   Flexible use of the Hong Kong platform in cross-border M&A

Most Mainland enterprises hope to take full leverage on the advantages of Hong Kong in their overseas M&A process. A common mode is that a Mainland enterprise registers a platform company in Hong Kong or completes the financing activity through a subsidiary listed in Hong Kong, utilising the flexible financial tools and the capital market of Hong Kong. In recent practices, more and more companies choose to set up a corporate treasury centre ("CTC") in Hong Kong in order to maintain financing efficiency and to make full use of the capital advantages of Hong Kong.

3.4.1   CTC in Hong Kong contributes to the allocation and integration of global funds

Take the example of an enterprise which is a large financial service group of comprehensive global businesses covering forfaiting, international factoring, asset management and over-the-counter transactions. With business expansion into a number of financial centres around the globe, it would be hard to continue to use the method of direct overseas investments using domestic funds. Moreover, the decentralised management of domestic and foreign funds has lowered the utilisation efficiency of funds. Therefore, there is an urgent need of setting up an offshore CTC to manage its operation and trading funds and achieve effective funds operation and transactions through funding mismatch management, position management and active management of foreign exchange.

3.4.2 CTC plan and benefits

In practice, a professional service team (referred to as the Professional Team) in Hong Kong assisted the enterprise in establishing its Hong Kong CTC with an investment and financing platform, an offshore fund centralisation platform, an offshore cash pool and a fund risks intelligent platform. This helped the enterprise to complete the integration and optimisation of its global fund resources allocation.

- **Select the site for the overseas CTC:** In the light of the prevailing conditions and requirements of the enterprise, the Professional Team and the enterprise, after discussion, decided on the establishment of an overseas CTC to plan the enterprise's functions of investment, financing, centralised management of funds and control over fund risks. Finally, Hong Kong was selected to be the site for setting up the CTC after comparing Hong Kong and Singapore in respect of their capital market activities, financial regulation and offshore RMB settlement.

- **Design the investment architecture and optimise the layout of overseas reinvestment:** Using its Hong Kong CTC as a holding company of its overseas businesses, the enterprise re-adjusted its equity structure and optimised its investment and dividend allocations. Meanwhile, the relatively lax policy supervision environment of Hong Kong was leveraged on to optimise its investment portfolio.

- **Design the financing plan and support its implementation:** Using the Hong Kong CTC as the financing platform, the Professional Team helped the enterprise to design multi-dimensional financing channels and contact the relevant financial and asset management institutions to meet the funding needs for overseas operations and to reduce financing cost.

- **Build a funds pool to achieve the physical and virtual centralised management of funds:** The enterprise used the relatively lax exchange rate policy and monetary policy in Hong Kong and selected a partner bank for establishing a cross-border funds pool of multiple currencies to maximise the capital centralisation on a global scale.

4   Conclusion

The increasing uncertainty of global geopolitics, the rise of trade protectionism and the trend of tightening regulation and laws in countries around the world will take a toll on overseas M&A of Mainland enterprises. Nevertheless, some optimistic and positive factors support the expectation that overseas M&A of new-economy companies will remain active.

Firstly, the global economic system is undergoing a digitalisation reform and the field of digital economy will be the forerunner of a new round of industry reforms. Most Chinese companies of new-economy types have already become global leaders in the field. To maintain their own advantages, they will actively expand their business into overseas markets through direct investments or cross-border M&A.

Secondly, Mainland China is striving to achieve "high-quality development" by enthusiastically carrying out supply-side structural reform, promoting the upgrade of the consumption market, improving business environment and opening-up at an even higher level. These measures have ensured the vigour of the Chinese economy and will play a positive role in the development of new-economy companies.

Thirdly, the outstanding China-US trade friction may have dampened the enthusiasm of new-economy enterprises for seeking M&A in the technical fields in North America. However, under the BRI, M&A activities will become enlivened gradually in Europe, Israel, India and the Southeast Asia regions in the future.

Hong Kong should fully seize this opportunity to play well its own role of the "largest international financial centre in Asia" and join the Mainland in achieving win-win economic development. This requires Hong Kong to actively conduct reforms, fully utilise its own advantages in the offshore RMB market, hold onto the opportunity of development in the Greater Bay Area, enhance the Mutual Market Access programme with the Mainland, maximise its advantages of "international financial centre" and avoid the risk of being marginalised.

Note: Thanks to Deloitte Research Director Lydia CHEN, Deloitte China Risk Advisory and Treasury Services Director David WONG, Deloitte Research Public Policy Manager Yu ZHAO and Deloitte Research Financial Services Manager Annie ZHOU for their participation.

Part 3

The Connectivity platform for onshore and offshore cross-border asset allocation

Chapter 10

China A shares inclusion into MSCI indices:
The journey and impact

Zhen WEI

Head of China Research

MSCI

Summary

The partial inclusion of China A shares in the MSCI ACWI and Emerging Markets Indices in 2018 and the increase to the weighting of this class of shares in 2019 have prompted investors to reconsider the appropriate allocation to China A shares in their equity portfolios.

In February 2019, MSCI announced the raising of the inclusion factor of China A shares from 5% to 20% by the end of 2019. This reaffirmed that the domestic China A shares market has become more accessible to global investors. As of 30 May 2019, Chinese equities accounted for 31% of the MSCI Emerging Markets Index by free float-adjusted market capitalisation. If the weight of A shares continues to increase, many long-term investors may want to put a greater focus on the China-allocation decision in their investment policy.

1　The world comes to China: MSCI inclusion journey

China's stock markets grew up with the expansion of the "Reform and Opening Up" policy. After 20 years of developing the market mechanisms since China's two key stock exchanges and the China Securities Regulatory Commission (CSRC) were founded in early 1990, the combined market value of stocks listed on the Shanghai and Shenzhen Stock Exchanges has become the second largest in the world.

Despite this success, the A shares market has often been criticised since inception. From investors' perspective, there are several issues that could seriously delay the domestic stock market's further development and pose obstacles to a well-functioning financial market. For example, the A shares market was long dominated by individual investors characterised by an investment philosophy not based on an understanding of global economy and by high levels of speculation. After the reform of the exchange rate regime in 2015, the Chinese Renminbi switched from a unilateral appreciation trend to a two-way fluctuation mode, which exerted capital outflow pressure.

International investors began meaningfully investing in China's onshore stock market after the Qualified Foreign Institutional Investor (QFII) scheme was introduced in 2002. The opening-up of financial markets, however, was relatively slow at the beginning. In 2013, when MSCI first considered adding China A shares to its mainstream indices, global investors were unenthusiastic. Few were familiar with the Chinese domestic equity market even though the MSCI China Index (covering other share classes and Chinese stocks listed overseas) already represented the largest country by market capitalisation in the MSCI Emerging Markets Index. In general, global investors were unfamiliar with this market because they were unable to access it.

China had cautiously opened up its domestic A shares market to the world through efforts targeting institutional investors: the QFII and Renminbi Qualified Foreign Institutional Investor (RQFII) schemes[1]. While these programmes attracted some early capital, they failed to provide the minimum level of market accessibility needed by institutional investors.

1　The QFII scheme was introduced in 2002, allowing foreign investor's direct access to China's capital market. The RQFII scheme was introduced in 2011, allowing the use of Renminbi funds raised in eligible locations outside of Mainland China to invest in the domestic securities market.

Thus, the journey to China inclusion into indices was far from smooth. Different market accessibility issues were raised by investors during the past MSCI consultations from 2014 to 2016. In addition to capital mobility restrictions and unequal market access under the qualified investor schemes, investors were uneasy about issues such as uncertainty of capital gains tax (raised in 2014 consultation), questionable beneficial ownership under the early phase of the Stock Connect scheme (raised in 2015 consultation)[2], widespread voluntary stock trading suspensions (raised in 2016 consultation) and pre-approval restrictions on launching financial products (raised in 2016 consultation). Despite numerous setbacks, Chinese policymakers and regulators made progress in addressing investor concerns. Finally, when the Stock Connect scheme was expanded to include Shenzhen-traded shares in December 2016, it provided the breakthrough that global investors were seeking — a market that afforded open accessibility to all investors.

Since MSCI announced in June 2017 that it would partially include China A shares in several of its key indices, we have seen increased interest in investing in Chinese stocks by global investors as market accessibility has improved:

- The opening of Special Segregated Accounts (SPSA[3]), a proxy of international investors' readiness and appetite to trade A shares, has surpassed 8,000 accounts in May 2019, up from less than 1,700 in June 2016.
- The total portfolio value settled through the Northbound Stock Connect reached RMB 910 billion (US$132 billion) in February 2019[4].
- More than 1,800 China A shares including those being added to the MSCI Emerging Markets Index are held by Stock Connect investors. Stock Connect is rapidly establishing itself as the primary channel to access A-shares.

Various efforts by Chinese regulators aim to avoid accessibility issues. For example, in 2018, the People's Bank of China (PBoC) quadrupled the Northbound Stock Connect daily limit to RMB 52 billion (US$8.3 billion) and the Hong Kong Monetary Authority (HKMA) sought to ensure ample liquidity in the offshore RMB (CNH) market with a series of supporting measures.

---

2   The Stock Connect scheme allows international and Mainland Chinese investors to trade securities in each other's markets through the home exchanges in the Mainland and Hong Kong.

3   See HKEX's website on the definition of SPSA (https://www.hkex.com.hk/Services/Settlement-and-Depository/Special-Segregated-Account-Services?sc_lang=en) and related statistics.

4   Source: HKEX's website.

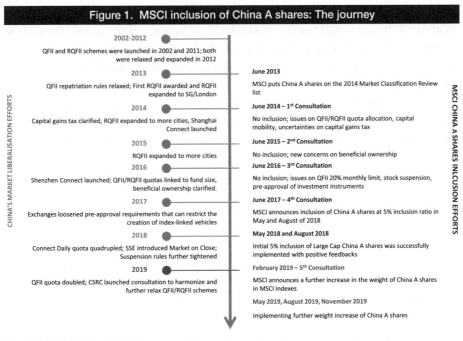

Figure 1. MSCI inclusion of China A shares: The journey

Source: Related official and MSCI announcement.

2 Potential impacts on China's financial markets and asset management industry

Historically, international investors' ownership of local stocks has increased after inclusion in the MSCI Emerging Markets Index[5].

When considering the potential impact of partial inclusion of A shares into MSCI Emerging Markets Index, we should also consider some special conditions. First, the

---

5 Michelle Liu, "Bringing Their 'A' Game", Lazard Asset Management, February 2019.

Chinese Renminbi has been added to the currency basket of the Special Drawing Rights (SDR) of the International Monetary Fund (IMF), and its international adoption is touching the levels of developed country currencies; second, the market size and liquidity of China A shares are far beyond that of any other emerging markets (EMs) at their first-time inclusion; third, the ratio of non-Chinese holdings of A shares at this point of time is much lower than that of other countries at their initial inclusion; fourth, besides the A shares market, there are plenty of other investment opportunities to Renminbi assets, such as the bond market, futures market and private funds. With these specialties, the path of the internationalisation of China A shares market is so far a unique one.

The total market value of China A shares is now the second largest in the world, behind only the US stock market (based on calculation by MSCI Equity Database). If calculated by investable free-float market value, however, the size of China A shares lags that of some developed countries, such as Japan and the UK. As China reforms its stock markets, improvements in free-float may help further increase the interest of institutional investors, as well as opening up room for further development of domestic asset managers.

Emerging-market fund managers affiliated with international asset managers have long and deep experience with the international investment process. Their advantages include a deep understanding and participation in the international asset owners' investment process, and experience with portfolio management and risk evaluation tools. Their investment philosophy reflects asset owners' investment values in the portfolio management process, such as a specific requirement in respect of target companies' environment, social and corporate governance (ESG) aspects.

However, there are few emerging-market fund managers who have long-term investment experience in A shares and have consistently good track records. The "scarcity" of A shares portfolio management experience creates opportunities for Chinese local fund managers. Their advantages may include a deep understanding of Chinese economic and industry cycles, close to Mainland policies, industry trend and company insights, and also having a hands-on experience with A shares portfolio management.

3   Potential impacts on the investment process

For international institutional investors using the MSCI Emerging Markets Index as a policy benchmark, how to respond to the inclusion of A shares may depend on whether investors already have exposure to A shares. For global institutional investors without current A shares exposure, they may address their policy allocation to emerging markets, including China. Such a re-evaluation may help reflect their long-term risk and return preferences more accurately.

Global institutional investors who already have an off-benchmark allocation to China A shares face potential benchmark misalignment with their existing A shares allocations. In addition, they may consider whether to retain or switch their A shares specialist allocation to a generalist emerging-market or global manager. In this process, China A shares specialist managers may have an advantage because of their longer track records and because not all emerging-market and global managers may be ready to manage China A shares.

The challenge for A shares specialist managers, however, will be to demonstrate that their investment processes are aligned with investor beliefs and that their performance is stable and persistent. While generalist managers may have less experience investing in China A shares, they may be able to point to better conformity with asset owner beliefs and benchmarks, and less volatility in their ability to deliver excess returns over time.

With or without a current A shares exposure, global institutional investors may want to examine how ready their current emerging-market and global equity managers are to handle the inclusion of A shares in their benchmarks. In particular, it helps to understand if managers have a clear insight on the pros and cons of different access paths, sources of risk and return (including differing sector compositions), macro drivers, the premium or discount between A shares and H shares, style factors and other idiosyncratic drivers of China A stocks.

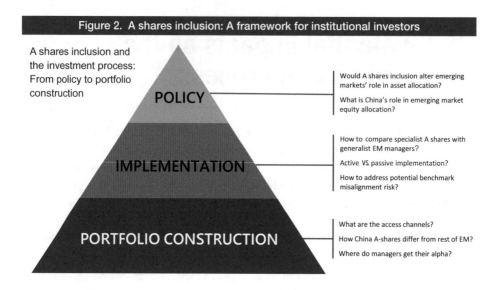

Figure 2. A shares inclusion: A framework for institutional investors

A shares inclusion and the investment process: From policy to portfolio construction

POLICY

Would A shares inclusion alter emerging markets' role in asset allocation?

What is China's role in emerging market equity allocation?

IMPLEMENTATION

How to compare specialist A shares with generalist EM managers?

Active VS. passive implementation?

How to address potential benchmark misalignment risk?

PORTFOLIO CONSTRUCTION

What are the access channels?

How China A-shares differ from rest of EM?

Where do managers get their alpha?

4   Weighting China in equity portfolios

Investors make strategic allocations to the equity asset class for a variety of reasons. While a significant number look to capture the long-term growth premium historically attached to the cash flow of equity portfolios, many also allocate to equities to meet future liabilities, to obtain the desired mix of reserve currencies, to enhance liquidity and transparency for the total portfolio or to provide diversification to other asset classes.

Return-driven investors often look at the relative size of the economies that underlie listed equity performance. Based on the IMF's latest forecasts, China is set to contribute 19% of global gross domestic product (GDP) by 2024, not far behind the US's 22% contribution[6].

6   IMF, *World Economic Outlook Reports*, April 2019.

China's purchasing power overtook the US in 2014 and is forecast to be 1.5 times the size of the US by 2024[7]. This drastic shift in the balance of economic power may have significant implications for investors' equity portfolios if they seek exposure from financial assets that may benefit from this growth trend.

A direct result of China's growth has been that many countries around the world are now exposed to its economy via increasing cross-border trade and investments. Demand has risen for assets linked to Renminbi-denominated cash flows, and there is also greater recognition of China's technology and infrastructure capabilities.

For example, as the Renminbi plays a more significant role as an alternative reserve currency to the US dollar, the euro and the Japanese yen, central banks and monetary authorities have increased their allocations to Chinese debt and equity. Liability-driven investors, such as pension funds, who see the future of their domestic economies closely tied to China, could, in some cases, reassess the pros and cons of having an explicit strategic allocation to Renminbi-denominated financial assets.

4.1   Have global and emerging-market investors caught up with this fast-changing reality?

Evidence suggests that investors have only started to reconsider the place of China in their equity allocations, particularly since June 2017, when MSCI announced its decision to partially include China A shares in the MSCI Emerging Markets Index and the MSCI ACWI Index, among other indices.

However, global funds remained underweight in Chinese stocks compared to the MSCI ACWI Index. China's role in EM allocations has been considerably higher, as it has been the largest emerging-market country by market capitalisation since 2007. On average, EM mutual funds allocated about a quarter of their portfolios to China in 2018, though still lagging the index on average[8].

Whether in global or emerging-market funds, this consistent underweight may be due to various investors' concerns, such as China's macroeconomic risks, lower familiarity with the universe of stocks and limited access to risk-management tools. As China's capital markets and financial infrastructure evolve, investors are likely to consider alternatives to the role of Chinese equities in their broader global equity opportunity set.

---

7   Ibid.
8   According to Cambridge Associates, institutional active EM funds had a higher allocation to China than EM mutual funds, but this still fell short of the MSCI Emerging Markets Index.

4.2   Ways to slice the world

Traditionally, many institutional investors have used market-capitalisation-weighted (cap-weighted) indices to help inform their asset allocation decisions. Given cap-weighted indices are designed to reflect the performance of the underlying investable opportunity set, they can serve as a neutral starting point for capital allocation decisions.

China currently represents 16% of global GDP or economic exposure weight but makes up only 4% of the cap-weighted index[9].

This discrepancy in weights mainly reflects two issues: publicly-held companies in less-developed capital markets typically have lower levels of free-floating shares, and China A shares are still not fully represented in the index.

In the context of emerging-market allocations, however, China's current index weight does not significantly trail its GDP or economic exposure weight.

4.3   Investment conviction

Under a classic asset allocation framework, investors typically have two objectives in mind when investing in an asset class or segment. The first is to improve the expected return of the portfolio, while the second is to improve portfolio diversification. In other words, the choice is fundamentally driven by the investor's return objectives and risk tolerance, although practical considerations including liability obligations, accessibility constraints and regulatory restrictions may also influence the decision.

Improving the expected return may require a high level of conviction about the potential of the opportunity set, given that no investor can claim to predict the future. Institutional investors, however, have long considered emerging-market equities, of which China is now the biggest component, as an important segment in their global equity portfolios.

Despite the already large weight of China in the MSCI Emerging Markets Index, some investors sought early exposure to China A shares, taking the view that the restricted domestic Chinese equity market would one day become more open to international investors and potentially become a significant part of their portfolio. Such investors have traditionally taken an off-benchmark bet by investing in dedicated QFII/RQFII China mandates.

Assuming all else remains constant, their approach to allocating to China may not necessarily change in the event of a full inclusion of China A shares in the Index. Given

---

9   As of 28 February 2019; calculated based on the MSCI ACWI Index and MSCI Economic Exposure Security Data
    module.

this logic, these investors are likely to continue to overweight China and pursue a strategy that works for them.

Benchmark-sensitive investors, on the other hand, may choose to take a more conservative approach when it comes to off-benchmark allocation. These investors may continue to approach the new opportunity set prudently while improving their understanding of the market.

5   A framework for policy benchmarks

When considering how to allocate to China, global investors commonly choose between a dedicated allocation to China or letting the China exposure be determined by its weight in the benchmark index, such as the MSCI Emerging Markets Index.

Some investors with a positive long-term conviction about China have chosen to pursue a dedicated China allocation, thus compensating for the country's lower weight in the benchmark, while benchmark-oriented investors have preferred to follow changes to indices as they occur. These different approaches partly reflect investors' different levels of experience and varying outlooks when investing in Chinese equities. In addition, these approaches may evolve if more A shares are added to the benchmark index over time.

The approach to the China allocation may be determined either at the policy level, strategic asset allocation level or portfolio implementation level. Here, we address the question in the context of the policy benchmark, recognising that this yardstick has the largest impact on the overall investment process. In addition, most investors have discretion and flexibility over their asset allocation and implementation decisions.

Some investors may believe that China will grow in size and importance within the global equities framework, while others may not. They may have varying approaches to how they treat China within their broader global equity allocation.

Investors with a neutral or a mildly positive view on China, but who do not need to act urgently, could choose to grow their China exposure organically by simply sticking with their current policy benchmark — basically, by doing nothing. Any increase in A shares to the benchmark index theoretically would result in a larger China allocation as these investors would seek to remain neutral (or close to neutral) to the benchmark.

Investors with positive convictions about the long-term prospects for China's economic growth or a bullish view of China's equity markets may wish to raise their exposures quickly. This implies a higher China exposure than the global equity benchmark. Investors considering this may select different routes, depending on how much they care about deviating from the policy benchmark:

- More benchmark-sensitive investors may consider an integrated approach to an index equity allocation, with a benchmark reflecting a higher weight for China;
- Less benchmark-sensitive investors may simply raise their exposure via a dedicated allocation to China or, more narrowly, to China A shares only.

Investors who are less certain about the implications of adding China A exposure to an existing emerging-market portfolio may consider delaying the decision to add China A shares' weight.

Some investors might be less convinced of the prospects for China's growing role in the global economy or the long-term trajectory of its equity market. In this case, a complete "carve out" solution may be more appropriate.

Figure 3. Example of a typical framework for policy benchmark configuration

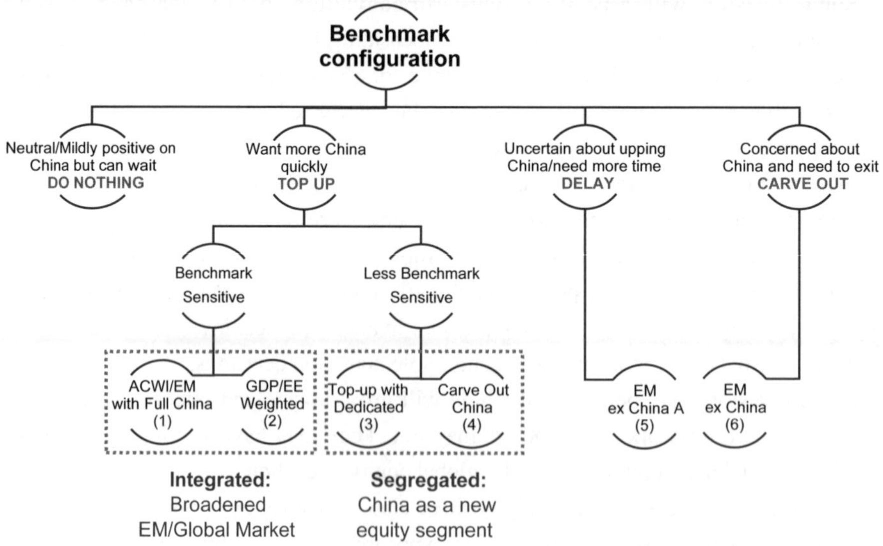

5.1   Integrated versus dedicated China approaches

Choosing which of the approaches previously described depends, to a large degree, on the investors' objectives and constraints. There is no "one-size-fits-all" answer.

With integrated solutions, investors may be able to maintain continuity with the existing investment process and governance structure by fine-tuning the weight of China in a broader equity portfolio. Turning to dedicated approaches, investors who opt for a "top-up" approach to the China allocation may have greater flexibility to reflect their views on China. However, introducing China as a dedicated new equity segment is likely to mean a change in the investment process with potential implications on the governance structure.

Table 1. Integrated versus dedicated China approaches		
Integrated	Dedicated (China as a new equity segment)	
Emerging markets/ACWI with neutral or higher China weights	Top up: Dedicated allocation to China strategies	Carve out: China + EM (ex China) configuration
Pros	• Continuity with existing investment process and governance structure • Little interruption in manager selection or quantitative investment strategy • No need to spend dedicated resources on China	• Flexibility in configuring China's weight in policy and tactical asset allocation decisions • Historic risk premium for being in a less efficient market • Possibly no impact on existing EM/ACWI policy	• Additional precision in policy configuration and flexibility in tactical asset allocation decisions • Historic risk premium for being in a less efficient market
Cons	• Less flexible in tactical asset allocation • Shortage of managers who can manage global equities with significantly higher China weight	• May need to change investment process and governance structure • Need to spend dedicated resources on China • May need to re-deploy existing allocation to fund new China strategy	• Possible disruption to investment process and governance structure • Shortage of managers who can manage EM equities (excluding China) • Higher exposure to China market-timing risk

Note: Using MSCI Index as the benchmark index as example.

5.2   MSCI China index options

Investors have traditionally used the MSCI China Index to define the opportunity set for dedicated China strategies. With the partial inclusion of China A shares in MSCI indices, these constituents are now incorporated in the MSCI China A Index and MSCI China All Shares Index, among others.

5.3 Opportunity set: Capacity, breadth and factor premium

Configuring policy benchmarks often requires a detailed assessment of the chosen benchmark's implications on broader asset allocation decisions. Specifically, investors may want to evaluate:

- **Asset class investability:** What is the market capitalisation and liquidity of the benchmark index?
- **Market breadth and alpha opportunity:** How many investable securities are included? How big are the potential alpha (i.e. the excess return with reference to a benchmark index) and factor premium opportunities?
- **Investment structure availability:** How many investment structures are available for the opportunity set? How large are they? What is the trend in new structures?

In other words, besides investment conviction and their merits for asset allocation, benchmark configuration decisions may also take into account of the current and future availability of investment structures at scale.

An integrated China universe as represented by the MSCI China All Shares Index (including onshore A shares and Chinese companies listed abroad) has exhibited broad capacity and diversity. However, an integrated EM benchmark including China has offered even greater investment capacity. Investors may account for these types of consideration when selecting a policy benchmark.

We find the largest return dispersion (the return difference between the top performers and the bottom performers, indicating market breadth) occurred in the MSCI China A and MSCI China All Shares Indices, followed by the MSCI Emerging Markets Index and MSCI Asia ex Japan Index. Return dispersion of constituent stocks in the US and other developed markets was lower than emerging markets and Asia ex Japan market over the past decade.

While not indicative of future results, factor-based quantitative strategies delivered better relative performance in the Chinese equity markets than elsewhere in the world[10]. In addition, integrating ESG factor historically enhanced return more in the China, emerging-market and Asia ex Japan universes than in developed markets[11].

---

10 We explained potential reasons for the better returns of factor-based strategies in China A shares in a blog post: "Can your investment strategy work with China A shares?" on the MSCI website, 27 June 2018.

11 Based on annualised active return (%) of MSCI ESG Leaders Indices from December 2008 to April 2019.

6   Conclusion: China and the future of equity allocations

China's growing economic strength and improving market accessibility may potentially bring significant investment opportunities, as well as challenges, to global investors. As MSCI partially included China A shares into the MSCI Emerging Markets and MSCI ACWI Indices, understanding China's role in equity allocation may be crucial to making sound investment policy decisions.

Given investors' wide variety of starting points and constraints, no universal allocation solution can address all the issues investors may face. Investors with a more strategic approach to equity allocation may want to look beyond near-term market and macroeconomic volatilities when assessing how to allocate to Chinese equities. In their assessment, investors may want to calibrate two main points: (1) the appropriate size of their China allocation, both within the global equity and EM allocation, and (2) how to implement this allocation.

On the first point, investors may need to assess how each scenario could meet their own investment objectives, align with their segmentation of the global equity opportunity set and provide a match with their investment convictions.

For the second point, investors are likely to face a choice between two broad configuration types — either an integrated approach that incorporates China within the benchmark index opportunity set, or a dedicated approach where China is considered to be a separate equity segment. Under the dedicated approach, there are different implications depending on whether the investor decides, for example, to "carve out" China from the emerging-market portfolio, to pursue an integrated "China All Shares" approach, or to take a specialist China A shares focus. When deciding on the approach to take, points for consideration will likely include investment convictions, potential asset allocation merits within the total portfolio and the expected availability of various investment structures.

No matter what choice an investor might make, the importance of understanding China's role in equity asset allocations cannot be underestimated.

Chapter 11

"Belt and Road" investment opportunities offered by Stock Connect companies

Chief China Economist's Office
Hong Kong Exchanges and Clearing Limited

Summary

The "Belt and Road" Initiative (BRI) was initiated by the Chinese President Xi Jinping in 2013, with the visions and action plan released in 2015. The BRI is expected to help promote the economic prosperity of the countries along the Belt and Road (B&R) (referred to as the B&R countries) and regional economic cooperation, strengthen exchanges and mutual learning between different civilisations, and promote world peace and development. More than 100 countries are involved in the BRI, with over 60 B&R countries, through various forms of cooperation. Over the five years after the announcement of the BRI, significant progress has been achieved in respect of economic and trade cooperation, investments and services.

The BRI has opened up abundant business opportunities and possibly high potential return on investments along the B&R countries to worldwide enterprises across all industry sectors. For providing insights on investment opportunities to capital market participants, the China Exchanges Services Company Ltd. (CESC) conducted its B&R Survey to reveal the trend of B&R business participation by listed companies in the Mainland and Hong Kong. The Survey revealed that (1) there has been continuous increase in B&R participation and growth in B&R businesses of the listed companies; (2) participation has been increasingly active particularly in engineering contracts and insurance businesses; (3) Mainland companies have assumed dominance in non-financial businesses and have a growing share in financial businesses; (4) participation by companies in B&R businesses have become more diverse across industry sectors; and (5) the geographical distribution of B&R businesses have also become increasingly diverse.

A number of B&R indices have been developed to track these B&R opportunities. In particular, the CES Belt and Road Index is the first and only Mainland-Hong Kong cross-border benchmark for B&R investment, with index constituents being listed stocks tradable through the Mainland-Hong Kong Mutual Market Access pilot programme, i.e. the Stock Connect schemes. Investment tools such as exchange traded funds could be developed based on the index to facilitate investors to reap the B&R investment opportunities across the border.

1 The Belt and Road Initiative

1.1 Initiation and significance

The Belt and Road Initiative (BRI) was firstly raised by the Chinese President Xi Jinping in 2013 during his visit to Asian countries. The initiative consists of the Silk Road Economic Belt (SREB) and the 21st-Century Maritime Silk Road (MSR). The SREB runs through Central Asia, West Asia, the Middle East to Europe, with extension to South Asia and Southeast Asia. The MSR runs through Southeast Asia, Oceania and North Africa. In a subsequent event in 2013, the Chinese Premier Li Keqiang emphasised the need to build the MSR oriented toward the Association of South East Asian Nations (ASEAN)[1]. In March 2015, the Chinese Government issued the foundation document on the visions and actions on the BRI (the BRI Action Plan)[2], stating that accelerating the building of Belt and Road (B&R) can help promote the economic prosperity of the countries along B&R and regional economic cooperation, strengthen exchanges and mutual learning between different civilisations, and promote world peace and development. According to the B&R Action Plan, the BRI is aimed at promoting orderly and free flow of economic factors, highly efficient allocation of resources and deep integration of markets; encouraging the countries along B&R to achieve economic policy coordination and carry out broader and more in-depth regional cooperation of higher standards; and jointly creating an open, inclusive and balanced regional economic cooperation architecture that benefits all (see Section 1.2).

The BRI is said to involve over 100 countries — over 60 along B&R[3] and many others participating through joining the Asian Infrastructure Investment Bank (AIIB)[4], developing

---

1 ASEAN consists of the ten nations of Brunei, Cambodia, Indonesia, Laos, Malaysia, Myanmar, the Philippines, Singapore, Thailand and Vietnam.

2 *Vision and Actions on Jointly Building Silk Road Economic Belt and 21ˢᵗ-Century Maritime Silk Road* (《推動共建絲綢之路經濟帶和 21 世紀海上絲綢之路的願景與行動》), jointly issued by the National Development and Reform Commission, the Ministry of Foreign Affairs and the Ministry of Commerce of the People's Republic of China (PRC), with State Council authorisation, March 2015.

3 The report, *Industrial Cooperation Between Countries Along the Belt and Road* (《「一帶一路」沿線國家產業合作報告》), released by the China International Trade Institute in August 2015 identified 65 countries along B&R that will be participating in the initiative.

4 AIIB is a multilateral development bank headquartered in Beijing, which is promoted by China to be established to support the BRI, with a mission to improve social and economic outcomes in Asia and beyond. As of 8 October 2018, AIIB had 44 Regional Members, 24 Non-Regional Members and 19 Prospective Members from around the world (source: AIIB website, viewed on 12 December 2018).

transport infrastructure in collaboration with China, or through other forms of cooperation[5]. A series of infrastructure projects, such as highways and railways, in countries along B&R (B&R countries) have been initiated and underway. China has been actively participating in these projects, both in financing and in the construction work. At the Belt and Road Forum for International Cooperation in May 2017, the Chinese President reaffirmed China's commitment to globalisation and free trade by promising a US$124 billion investment into countries along B&R.

Many of the B&R countries lack the financial capacity to develop the infrastructure themselves either by public or private means. Backed by China in the B&R projects, these countries will benefit from the resultant enhanced infrastructure establishments, foreign investment and the trade and economic activities so stimulated. China itself has committed to further opening up its market with greater international cooperation.

The identified B&R countries had a combined share of about 65% of the world's population, 40% of the world's gross domestic products (GDP), three quarters of global energy resources and one-quarter of global trade in goods and services[6]. The BRI is expected to create abundant job and trade opportunities for people and enterprises in the region.

1.2   Multilateral cooperation priorities and initiatives

According to the BRI Action Plan, the BRI upholds the principles of peaceful co-existence[7]; it is open for cooperation, harmonious and inclusive, follows market operation and seeks mutual benefit of all parties involved. Five key areas are highlighted to be of cooperative or "connectivity" priorities, briefed below:

- **Policy coordination** — promote inter-governmental cooperation, building a multi-level inter-governmental macro policy exchange and communication mechanism, reaching new cooperation consensus;
- **Facilities connectivity** — form an infrastructure network connecting all sub-regions in Asia, and between Asia, Europe and Africa step-by-step;
- **Unimpeded trade** — with investment and trade cooperation, improve investment

---

5   The report, *The Belt and Road Initiative: 65 Countries and Beyond*, Fung Business Intelligence Centre, May 2016, identified 48 such countries.

6   Sources: "China says it's building the New Silk Road. Here are five things to know ahead of a key summit", *Time*, 12 May 2017; website of the Singapore Government (https://www.iesingapore.gov.sg).

7   There are five principles of peaceful co-existence: mutual respect for each other's sovereignty and territorial integrity, mutual non-aggression, mutual non-interference in each other's internal affairs, equality and mutual benefit, and peaceful co-existence.

and trade facilitation, and removing investment and trade barriers for the creation of a sound business environment within the region and in all related countries;

- **Financial integration** — deepen financial cooperation, make more efforts in building a currency stability system, investment and financing system and credit information system in Asia;
- **People-to-people bond** — carry forward the spirit of friendly cooperation of the Silk Road by promoting extensive cultural and academic exchanges, personnel exchanges and cooperation, media cooperation, youth and women exchanges and volunteer services.

The BRI has received strong positive responses from the B&R countries since its announcement. Over the five years after the announcement of the BRI, China has signed 118 cooperation agreements with 103 countries, regions and international organisations on developing the B&R and signed or upgraded five free trade agreements with 13 B&R countries by August 2018. The initiative's scope has been thereby extended from the Eurasian continent to Africa, Latin America and the Caribbean, and the South Pacific region.[8]

Apart from infrastructure projects, BRI business opportunities involve also trade and investment, and financial and professional services — in terms of finance, some institutions estimated that BRI projects will require up to US$5-6 trillion of financing during the next 15 years[9]. B&R countries reported to have partnered with China in respect of infrastructure projects include, among others, Sri Lanka, Malaysia, the Philippines, Pakistan, Belgium, Ethiopia, Kenya, Cambodia and Mozambique[10]. Total trade between China and B&R countries over the past five years has exceeded US$5 trillion, with an annual average growth rate of 1.1%, and 82 overseas economic and trade cooperation zones have been established in these countries, with an accumulated investment of US$28.9 billion[11]. Countries and regions like the UK, Singapore and Hong Kong are also actively grasping the opportunities in offering financial and professional services including banking and insurance, legal, engineering, city planning, etc. for the BRI.

Driven by the BRI, Chinese enterprises were reported to have increased their businesses

8 Sources: "China sees trade, investment growth with B&R countries", *Xinhua*, 27 August 2018, "Belt and Road Initiative in 5 years", *Xinhua*, 27 August 2018, viewed on the website of the State Council Information Office (SCIO) of the PRC.

9 HSBC gave an estimate of US$6 trillion while PricewaterhouseCoopers gave an estimate of US$5 trillion (source: "BRI helps companies build global bridges of connectivity", *China Daily*, 14 November 2018).

10 Source: Various media reports.

11 Source: "B&R countries eye production capacity cooperation with China", *Xinhua*, 22 October 2018, viewed on the SCIO website.

and investments in the B&R countries, achieving a win-win outcome with their B&R business partners. BRI benefits enterprises in the industries of construction, port operations, cargo/shipping and logistics, energy, trading and financial services[12]. For example, for the trade and cargo industries, the shipping time has been significantly reduced as a result of railway and ocean cargo routes and port facilities established under BRI[13]. In 2018 up to October, new investments by Chinese enterprises of US$11.9 billion were recorded in 55 B&R countries, a year-on-year increase of 6.4%[14]. Chinese enterprises signed new construction contracts along B&R worth US$57.11 billion in 2018 up to July[15]. Chinese banking institutions are also taking important roles in B&R financing. For example, Bank of China's Singapore branch was reported to have raised an accumulated sum of US$16 billion for financing BRI projects[16].

The BRI, therefore, has opened up abundant business opportunities and possibly high potential return on investments along the B&R countries to enterprises in the Mainland, Hong Kong and across the world.

2   The CESC's B&R Survey

In view of the significance of the BRI in the global business and investment world, China Exchanges Services Company Ltd. (CESC), a joint venture of Hong Kong Exchanges and Clearing Limited (HKEX), the Shanghai Stock Exchange (SSE) and the Shenzhen Stock Exchange (SZSE), initiated the B&R Survey to collect data and information from listed companies about their scope and extent of business participation in the BRI. The survey is conducted on a regular basis, with an attempt to keep track of the changes in the capital market landscape along with the development of the BRI. The survey results are expected to provide insights to capital market participants in the following ways:

---

12 As reported by various media reports.

13 These include, among many others, the China-Singapore rail-sea transit route, the Zhengzhou-Liege (Belgium) cargo train route, Lanzhou-Islamabad (Pakistan) rail and road cargo service, the China-Europe Land-Sea Express Route and the Sino-Euro Cargo Railway program.

14 Source: 〈前 10 個月我國企業對「一帶一路」沿線國家新增投資 119 億美元〉, the SCIO website, 15 November 2018.

15 Source: "Belt and Road Initiative in 5 years", *Xinhua*, 27 August 2018, viewed on the SCIO website.

16 Source: "China, Singapore explore more potentials for BRI cooperation", *Xinhua*, 12 November 2018, viewed on the SCIO website.

(1) Revealing the current characteristics and development trends in B&R economic activities;

(2) Revealing the different business models of listed companies in their product and service markets in capturing BRI opportunities; and therefore

(3) Revealing the investment opportunities in the capital market offered by Mainland- and Hong Kong-listed companies participating in BRI.

Target respondents of the survey cover listed companies on the SSE, the SZSE and the HKEX with their stocks being eligible securities for trading under the Mainland-Hong Kong Mutual Market Access pilot programme — the Stock Connect schemes[17]. The first survey was conducted in 2017 (the 2017 Survey) to collect data for the financial years of 2014 to 2016 while the latest survey was conducted in 2018 (the 2018 Survey) to collect data for the financial year of 2017. The 2017 Survey and the 2018 Survey covered 1,931 and 1,939 listed companies respectively in the target population, achieving respective response rates of 67% (1,286 valid responses) and 66% (1,274 valid responses).

The measures of BRI participation of listed companies are identified with reference to the five key areas of connectivity among B&R countries (see Section 1.2 above). Three of the five key areas are considered to be quantifiable — "facilities connectivity", "unimpeded trade" and "financial integration". The following six B&R business participation indicators are identified:

(1) **Goods/services sales amount** — Amount of trade exports, selling goods and/or providing services domestically, etc. in order to generate revenue from the B&R countries;

(2) **Goods/services purchases amount** — Amount of trade imports, buying goods and/or receiving services domestically, etc. for production or resale purpose from the B&R countries;

(3) **Newly signed engineering contract value** — For infrastructure design and/or construction, civil engineering, and equipment manufacturing, etc. offered to the B&R countries;

(4) **New investment amount** — Capital expenditure and/or equity investment in the B&R countries;

(5) **Financing/loan amount (for financial institutions only)** — Amount of direct financing, bond issuance and/or lending business in the B&R countries;

(6) **Insured amount (for financial institutions only)** — Insured amount in the

---

17 Consisting of the Shanghai-Hong Kong Stock Connect (Shanghai Connect) and the Shenzhen-Hong Kong Stock Connect (Shenzhen Connect).

insurance services and underwriting of customers' insurance applications in the
B&R countries.

Provided with the reference list of B&R countries[18], the survey respondents were asked
to report the amounts of the respective six indicators listed above which are applicable
to them, together with any other information on B&R-related businesses. As there is no
definite geographical boundary of the B&R, the list of B&R countries in the survey series
may expand over the years[19].

Section 3 below gives the trends of B&R participation of the responding companies in
the key areas of infrastructure, trade and investment, and financial services as revealed by
the survey results.

3   B&R participation by listed companies[20]

3.1   Overall B&R participation

Current status and future plans on B&R participation

In the 2018 Survey (for financial year 2017), the majority of the responding companies
(52%) have participated in B&R business (referred to as "B&R companies" hereinafter),
showing an increase in participation over the financial years from 2014 to 2017 (42%,
44% and 46% in 2014, 2015 and 2016 respectively). Of the B&R companies in the 2018
Survey, 60% intended to increase their B&R business and 33% intended to maintain it in
the coming three years. For companies without B&R business, 29% of them intended to

---

18 See Appendix 1 for the reference list of B&R countries for the 2018 Survey. Hong Kong, which is a Special
    Administrative Region of the PRC, was not included in the list.

19 Compared to the 2017 Survey, nine countries were newly added in the 2018 Survey — Korea in Northeast Asia; New
    Zealand in Oceania; Ethiopia, Madagascar, Morocco and South Africa in Africa; Austria in East Europe; Panama, and
    Trinidad and Tobago in America.

20 For details, please see the 2018 Survey report released by the CESC on its website on 11 December 2018 (https://
    www.ces.com).

get involved in the coming three years. Overall, 94% of the responding companies intended to increase or maintain their B&R business participation. (See Figure 1.)

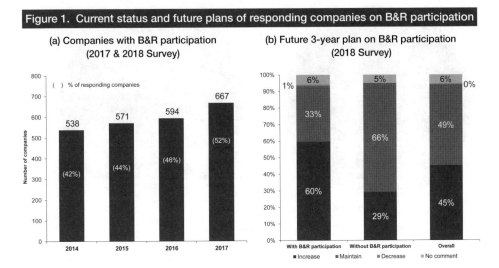

Figure 1. Current status and future plans of responding companies on B&R participation

(a) Companies with B&R participation
(2017 & 2018 Survey)

(b) Future 3-year plan on B&R participation
(2018 Survey)

B&R companies by type and industry (2018 Survey)

The B&R companies in response to the 2018 Survey were predominantly Mainland companies[21] (98%) — 36% were state-owned enterprises (SOEs)[22] and 62% were non-SOEs. In respect of distribution by industry, companies in the Industrials sector were of the largest proportion (28%), followed by the Materials sector (18%), Consumer Discretionary (15%) and Information Technology (IT) sector (14%). Companies in these four top industry sectors constituted in aggregate 75% of the overall sample of the B&R companies in response to the 2018 Survey. (See Figure 2.)

---

21 "Mainland company" is defined as a company which meets any of the following criteria: (1) the entity is incorporated in Mainland China; (2) the operation centre of the company is in Mainland China; and (3) at least 50% of the company's revenue is originated from Mainland China.

22 SOEs comprise (1) "local SOEs" which are beneficially controlled by local bureaus of the State-Owned Assets Supervision and Administration Commission (SASAC) or local governments or other local SOEs; and (2) "central SOEs", which are beneficially controlled by SASAC under the State Council or by the Ministry of Finance.

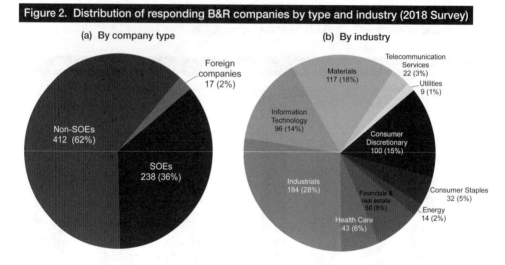

Figure 2. Distribution of responding B&R companies by type and industry (2018 Survey)

(a) By company type

(b) By industry

Note: The base sample consisted of 667 listed companies with B&R business.

3.2 The growth in B&R businesses

Based on the results of the 2017 Survey and the 2018 Survey covering the financial years of 2014 to 2017 of the responding companies, apparent trends of growth in the various types of B&R business measured were observed[23] — positive compound annual growth rates (CAGR) were recorded during the period of 2014 to 2017. When businesses in newly added countries were included, the B&R business type recorded the biggest CAGR was new engineering contracts (63%), followed by insurance (14%) and financing/loans (10%). Even when newly added countries were excluded, the CAGR for new engineering contracts was still at a high level of 60%, while that for insurance and financing/loans were 10% and 9% respectively. (See Figure 3.)

In dollar terms (including newly added countries), the amounts of B&R insurance and financing/loans of responding companies reached RMB 4,036.6 billion and RMB 2,546.1 billion respectively in 2017. In particular, the insurance amount recorded a significant growth of 30% over 2016. Comparatively, new B&R investments amounted to only RMB 94.2 billion in 2017. The amounts of B&R purchases and sales were almost the same

---

23 Although the sizes of the respective responding samples to the two surveys are different, they are close to each other
     in terms of actual number (1,286 and 1,274 respectively) and response rate (67% and 66% respectively) and their
     results are therefore considered to be comparable.

in 2017 — RMB 1,113.7 billion and RMB 1,110.9 billion respectively, while new B&R engineering contracts amount was RMB 680.6 billion, recording an increase of over three times from the RMB 158.4 billion three years ago or doubled the RMB 327.7 billion in 2016.

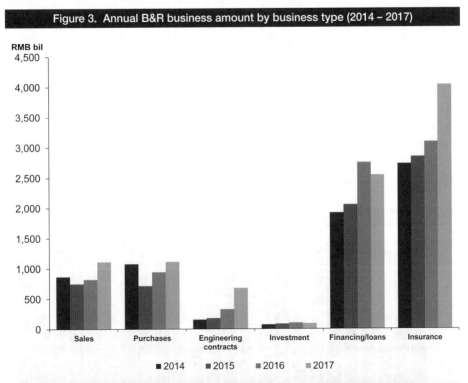

Figure 3. Annual B&R business amount by business type (2014 – 2017)

Include/exclude newly added countries	CAGR (2014 – 2017)					
	Sales	Purchases	Engineering contracts	Investment	Financing / loans	Insurance
Inclusion	9%	1%	63%	7%	10%	14%
Exclusion	5%	-1%	60%	6%	9%	10%

3.3 B&R participation by type and industry of listed companies

By company type

Mainland companies (SOEs and non-SOEs) had the dominant share in non-financial B&R businesses among responding companies — in 2017, they shared 83%, 93% and 100% in sales, purchases and new engineering contracts respectively, and 76% in respect of new investments (see Figure 4). For the business types of sales, purchases and new engineering contracts, SOEs played a much more significant role than non-SOEs (65%, 82% and 97% respectively in 2017). On the other hand, non-SOEs had a rapidly growing share in new investments over the years, from 6% in 2014 to 39% in 2017.

Financial business types had been dominated by foreign companies, but Mainland companies were slowly catching up. The share of foreign companies in financing/loans and insurance were respectively 67% and 79% in 2017, down from respectively 98% and 88% in 2014. (See Figure 4.)

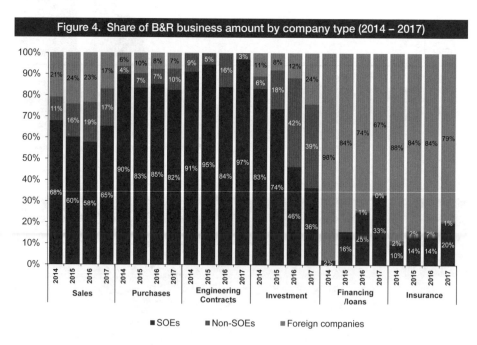

Figure 4. Share of B&R business amount by company type (2014 – 2017)

Note: Percentages may not add up to 100% due to rounding.

By industry sector

Companies in the Energy sector, which constituted only 2% in number of the responding B&R companies in the 2018 Survey, played a dominant role in B&R purchases and sales — constituting the majority share of 68% and 33% respectively in 2017, albeit showing a decline from 85% and 52% respectively in 2014. Companies in the Industrials sector showed increasing shares in all four non-financial B&R business types — from 76% in 2014 to 99% in 2017 for new engineering contracts, from 13% in 2014 to 25% in 2017 for sales, from 6% in 2014 to 11% in 2017 for purchases and from 8% in 2014 to 17% in 2017 for new investments. Notably, companies in the Finance sector had relatively high shares in new investments (23% in 2017) and sales (12% in 2017). Note that only companies in the Finance sector, by nature of their business, provide B&R financing/loans and insurance services. (See Figure 5.)

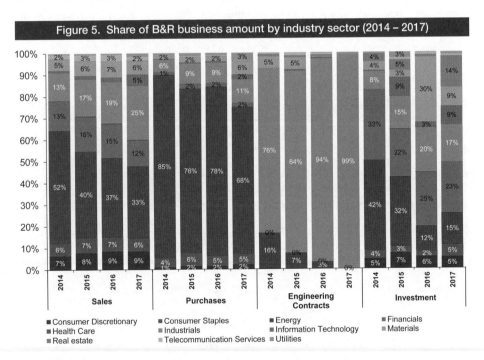

Figure 5. Share of B&R business amount by industry sector (2014 – 2017)

Note: Percentages may not add up to 100% due to rounding.

3.4 B&R businesses by geographical region and country

Figure 6 shows the distribution of B&R businesses of responding companies[24] by geographical region in 2016 and 2017 in terms of value, with the percentage shares shown. In the 2018 Survey (for collecting data on the financial year of 2017), nine countries in five geographical regions were newly added[25] such that Africa, Oceania and America were newly added regions compared to the 2017 Survey (for collecting data on the financial years of 2014 to 2016).

The responding B&R companies had their B&R businesses relatively concentrated in Southeast Asia — in 2017, the percentage shares were 59% for sales, 42% for purchases, 50% for new engineering contracts, 55% for new investments, 45% for financing/loans, and especially high for insurance (78%). The Middle East region constituted a significant share in purchases (31% in 2017) and financing/loans (36% in 2017), while South Asia constituted a relatively high share in new engineering contracts (18% in 2017, same as the Middle East) and in sales (11% in 2017, compared to 9% for the Middle East).

---

24 A total of 512 and 657 companies provided answers to the questions on country breakdown in the 2017 Survey and 2018 Survey respectively. These responding companies' B&R businesses measured by the survey amounted, in aggregate, to RMB 6,700 billion and RMB 8,409 billion in the financial years of 2016 and 2017 respectively, representing a coverage of 83% and 88% of the total B&R business in value terms in the respective years of the overall sample of responding B&R companies to the survey in the two years.

25 See Appendix 1 for the list of B&R countries in the 2018 Survey and footnote 19 above for the changes.

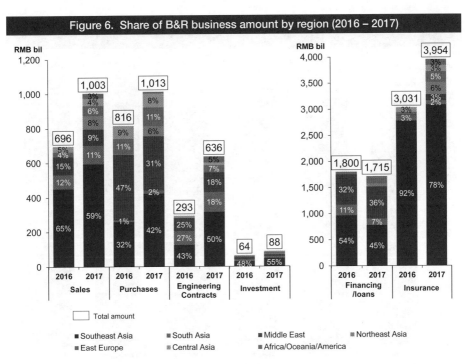

Figure 6. Share of B&R business amount by region (2016 – 2017)

Notes: The base samples consisted of 512 and 657 companies in the 2017 Survey and 2018 Survey respectively, which provided the geographical breakdown of their B&R businesses.

Percentages may not add up to 100% due to rounding.

Southeast Asia, the dominating region in B&R businesses, recorded in 2017 the strongest annual growth in new engineering contracts (153% in value terms) and also considerable annual growth in purchases (64%), investment (58%) and sales (32%). On the contrary, the B&R businesses in the Middle East suffered an annual decline of 17% in purchases and 11% in sales, but managed to have a 51% annual growth in new engineering contracts in 2017. The growth in B&R businesses in Northeast Asia in 2017 (which constituted an overall 5% in 2017, compared to 0% in 2016) was mainly attributed to the newly added country, Korea.

Although the top three regions constituted about 80% or more in each business type in 2017, the concentration had declined compared to 2016 and that B&R businesses of responding companies had achieved a wider distribution among different geographical regions in 2017. (See Figure 7.)

Figure 7. Percentage share of top three regions for each B&R business type (2016 & 2017)

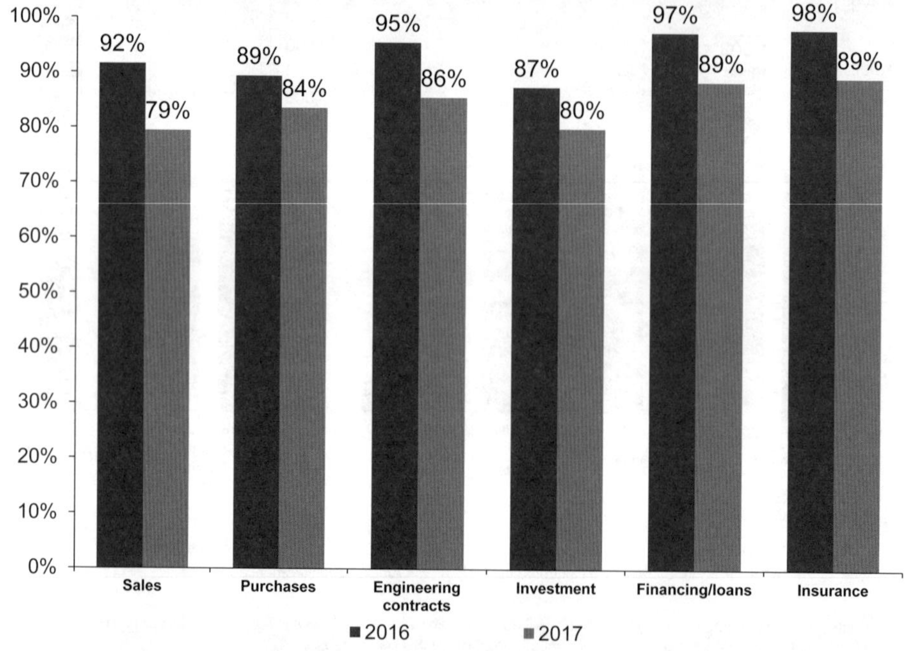

Financial year	Sales	Purchases	Engineering contracts	Investment	Financing / loans	Insurance
2016	SE Asia (65%) Middle East (15%) South Asia (12%)	Middle East (47%) SE Asia (32%) East Europe (11%)	SE Asia (48%) South Asia (27%) Middle East (25%)	SE Asia (48%) Middle East (24%) South Asia (15%)	SE Asia (54%) Middle East (32%) South Asia (11%)	SE Asia (92%) South Asia (3%) Central Asia (3%)
2017	SE Asia (59%) South Asia (11%) Middle East (9%)	SE Asia (42%) Middle East (31%) East Europe (11%)	SE Asia (50%) South Asia (18%) Middle East (18%)	SE Asia (55%) Middle East (13%) East Europe (12%)	SE Asia (45%) Middle East (36%) South Asia (7%)	SE Asia (78%) NE Asia (6%) East Europe (5%)

Figure 8 shows the top five countries by percentage share in each B&R business type of the responding B&R companies for the financial year of 2017. In particular, Singapore ranked top in five of the six identified B&R business types. Malaysia assumed the top rank in new engineering contracts.

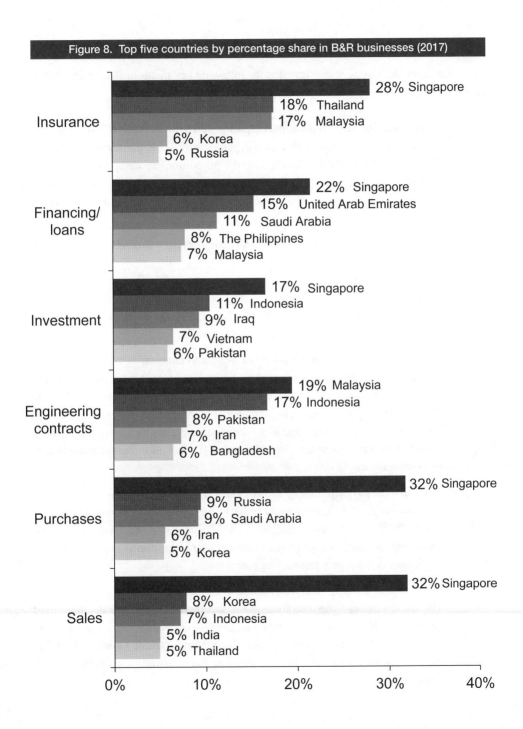

Figure 8. Top five countries by percentage share in B&R businesses (2017)

3.5 Key trends from survey results

In summary, the survey results revealed the following key trends of B&R participation by listed companies surveyed:

(1) Continuous increase in B&R participation and growth in B&R businesses

There were a growing number of companies with B&R participation over the financial years of 2014 to 2017, with the proportion increased from being the minority in prior years to being the majority in 2017. Significant growth in business amounts was also recorded over the years in various business types measured by the survey.

(2) Increasing active participation in B&R engineering contracts and insurance businesses

Among the business types measured, new engineering contracts and insurance business showed particularly high growth in 2017 (over 100% and 30% respectively compared to that in 2016). This reflects the continuous significance of infrastructure developments in the B&R countries.

(3) The significance of participation from Mainland companies

Mainland companies have dominated the businesses of B&R sales, purchases, new engineering contracts and new investments. The Mainland companies (mainly the SOEs) are also gaining increasing and relatively more significant market shares in the financial business (financing/loans and insurance), in which foreign companies have had the majority share. Notably, non-SOEs had an increased share in new investments.

(4) More diverse participation by industry sector in B&R businesses

The dominance of the Energy sector in B&R sales, purchases and new investments has declined, giving way to increasing participation from other industry sectors. Nevertheless, companies in the Industrials sector continued to have business dominance in B&R new engineering contracts.

(5) More diverse geographical distribution of B&R businesses

The B&R business concentration in the top three regions had declined over the years. The top three countries with the biggest market share varied across different B&R business types. Among the countries outside Greater China that were assessed by the survey, Singapore ranked top in five of the six B&R business types measured in 2017.

4   The CES B&R Index and other B&R indices

In June 2017, HKEX appointed CESC to explore a B&R index in response to the market's interest in a barometer for the BRI. The CES Belt and Road Index (CES B&R Index) was therefore developed.

The CES B&R Index is the first cross-border B&R Index in the offshore market, with the objective to reflect the price performance of companies which demonstrate substantial participation in the BRI.

The constituents of the CES B&R Index are selected from the stocks eligible for Northbound Trading or Southbound Trading of Stock Connect (excluding those which are labelled as "ST" or "ST*" on the SSE or the SZSE[26]) based on their degrees of B&R participation in respect of the six business areas as measured in the B&R Survey[27]. That is, the CES B&R Index consists of constituent companies listed in the Mainland exchanges and HKEX. The composition of the index would reflect the progress of the BRI in the various economic sectors participated in by listed companies tradable through Stock Connect. The index benchmarks the performance of B&R businesses-related stocks.

Apart from the CES B&R Index, there are three major Mainland onshore B&R indices comprising A shares of companies engaged in industries related to the BRI theme. These are:

(1) **The CSI One Belt & One Road Index** (referred to as the "CSI B&R Index" hereinafter[28]) launched by the China Securities Index Co., Ltd. (CSI) on 16 February 2015;

(2) **The SSE One Belt & One Road Index** (referred to as the "SSE B&R Index" hereinafter[29]) launched by the CSI for the SSE on 24 June 2015; and

---

26 Stocks with "ST" or "ST*" label are stocks under "special treatment", which are a kind of risk alert label for the attention of investors.

27 See Appendix 2 on the methodology of the CESC B&R Index.

28 The official English short name of the index is CSI OBOR Index. It consists of the most representative A shares listed on the SSE or the SZSE from five B&R-related industries: Infrastructure Construction, Traffic & Transportation, Equipment Manufacturing, Electric Power & Communication, and Resource Development. (Source: CSI website, accessed on 10 December 2018.)

29 The official English short name of the index is SSE OBOR Index. It consists of the most representative A shares listed on the SSE from the same five B&R-related industries as the CSI OBOR Index. (Source: CSI website, accessed on 10 December 2018.)

(3)   **The CNI Silk Road Economic Belt Index** (referred to as the "CNI B&R Index" hereinafter[30]) launched by the Shenzhen Securities Information Company Limited (SSIC) on 8 January 2015.

In addition, there is at least a global B&R-related index, the **MSCI Global China Infrastructure Exposure Index** (referred to as the MSCI B&R Index), which tracks listed companies within developed, emerging and frontier markets with high revenue exposure to China infrastructure development from within a specified set of industries related to B&R. China has the highest country weighting in the index (43.5%), followed by Singapore (10.2%), Malaysia (7.2%), Russia (6.6%) and Thailand (6.1%), as of end-November 2018.[31]

Figure 9 shows the industry composition of the five B&R indices. The CES B&R Index gives a relatively wide distribution across industry sectors compared to the CSI, SSE and MSCI B&R indices. The CNI B&R Index gives a comparably wide industry distribution as the CES B&R Index but its constituents are selected based on place of incorporation of the companies rather than their substantive B&R business relations.

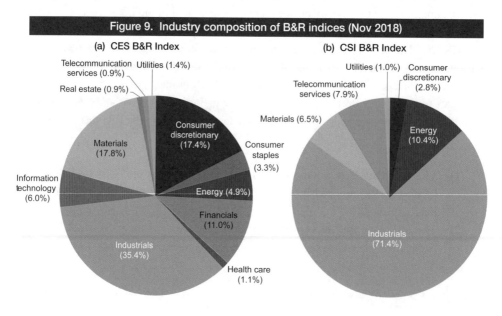

Figure 9. Industry composition of B&R indices (Nov 2018)

(a) CES B&R Index

(b) CSI B&R Index

---

30 The official English short name of the index is New Silk Road Index. It consists of companies with A shares listed on the SSE or the SZSE which are incorporated in the five provinces of Shaanxi, Gansu, Ningxia, Qinghai and Xinjiang (these are economic regions along the B&R). (Source: SSIC website, accessed on 10 December 2018.)

31 Source: The index factsheet (November 2018) on the MSCI website, accessed on 10 December 2018.

(continued)

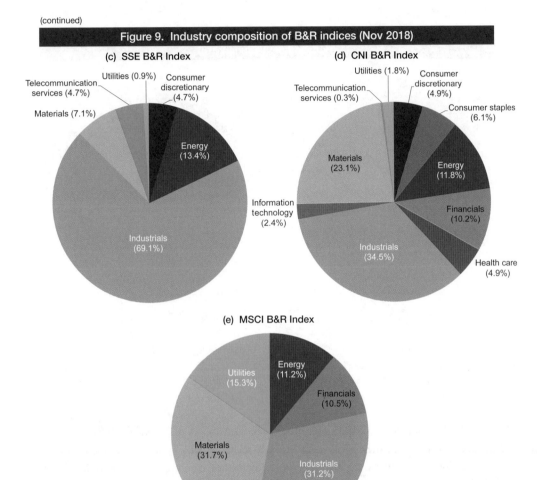

Figure 9. Industry composition of B&R indices (Nov 2018)

(c) SSE B&R Index

(d) CNI B&R Index

(e) MSCI B&R Index

Note: Percentages may not add up to 100% due to rounding.

Source: CESC for CES B&R Index; index factsheets from the respective index companies' websites.

Figure 10 shows the relative performance of the four B&R indices comprising constituent companies listed on the Hong Kong and/or the Mainland markets for the period July 2015 to October 2018[32].

32 The MSCI B&R Index is excluded as the index data series is only dated back to August 2017.

Figure 10. Daily closings of the B&R indices (30 Jun 2015 – 31 Oct 2018)
(Rebased on 30 June 2015)

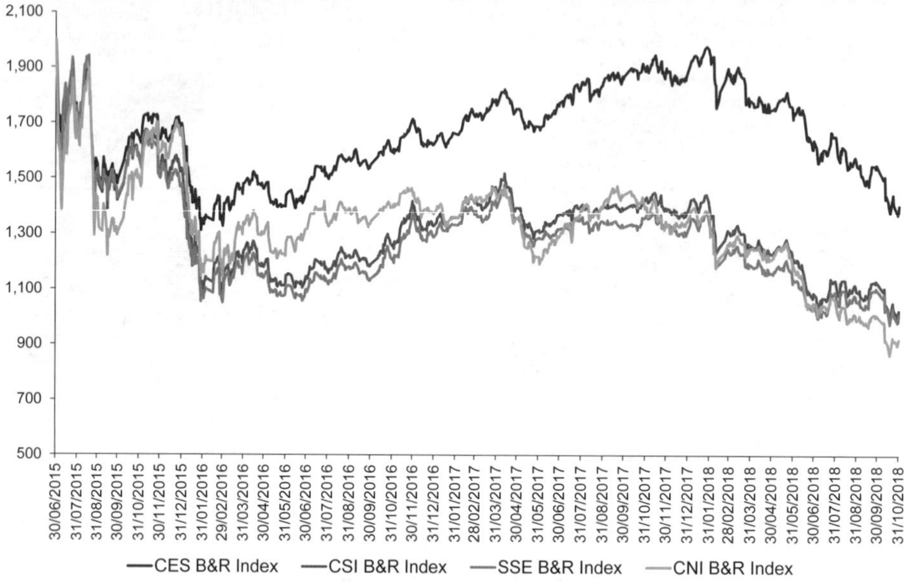

Note: The rebased date (30 June 2015) is the base date of the CES B&R Index. The base point for all index series
        in the chart is set to 2000 points on the rebased date.

Source: Wind.

With the availability of B&R indices of specialised scopes of coverage, investors would be able to benchmark their investment in B&R-related stocks by making reference to the suitable index. Furthermore, thematic investment tools such as exchange traded funds (ETFs) based on these B&R indices could be developed to facilitate participation by retail and institutional investors. In particular, the CES B&R Index is the only Mainland-Hong Kong cross-border B&R index with the underlying stocks tradable by both Mainland and Hong Kong investors through the Stock Connect scheme. Investment tools based on this index help provide B&R investment opportunities across the border.

5   Conclusion

The BRI is considered a significant driver of multilateral economic cooperation and development, providing abundant business opportunities and possibly high potential return on investments along the B&R countries to worldwide enterprises across all industry sectors. The CESC's B&R Survey provides insights on investment opportunities to capital market participants by revealing the trend of B&R business participation by listed companies in the Mainland and Hong Kong. These opportunities are being tracked by related B&R indices. In particular, the CES B&R Index is the first and only Mainland-Hong Kong cross-border benchmark for B&R investment. Investment tools such as ETFs could be developed based on the index to facilitate investors to reap the B&R investment opportunities across the border.

Appendix 1

List of Belt and Road countries in the CESC 2018 B&R Survey

Region	Countries
Africa	Ethiopia, Madagascar, Morocco, South Africa
America	Panama, Trinidad and Tobago
Oceania	New Zealand
East Europe	Albania, Austria, Belarus, Bosnia and Herzegovina, Bulgaria, Croatia, Czech Republic, Estonia, Hungary, Latvia, Lithuania, Macedonia, Moldova, Montenegro, Poland, Romania, Russia, Serbia, Slovakia, Slovenia, Turkey, Ukraine
Middle Asia	Afghanistan, Kazakhstan, Kyrgyzstan, Tajikistan, Turkmenistan, Uzbekistan
Middle East	Armenia, Azerbaijan, Bahrain, Egypt, Georgia, Iran, Iraq, Israel, Jordan, Kuwait, Lebanon, Oman, Palestine, Qatar, Saudi Arabia, Syria, United Arab Emirates, Yemen
Northeast Asia	Mongolia, Korea
South Asia	Bangladesh, Bhutan, India, Maldives, Nepal, Pakistan, Sri Lanka
Southeast Asia	Brunei, Cambodia, Indonesia, Laos, Malaysia, Myanmar, the Philippines, Singapore, Thailand, Democratic Republic of Timor-Leste, Vietnam

Source: CESC 2018 B&R Survey report.

Appendix 2

Methodology of the CES B&R Index

Feature	Details
Universe of underlying stocks	• Stocks eligible for Stock Connect, which includes both "Northbound Trading" and "Southbound Trading". • Excluding ST or ST* stocks listed on the SSE or the SZSE.
Selection criteria	• **Negative EPS Screen:** Exclude companies with negative earnings per share (EPS) in the most recent annual report. • **Participation Screen:** (1) Each company will be ranked based on measures of six factors — sales, purchases, new orders in construction, investments, financing and insurance associated with the BRI. (2) For each factor, top 1% of companies in terms of amount involved will immediately become index constituents. The remaining companies will be assigned scores equivalent to the cumulative probabilities of cumulative distribution function of the respective t-distribution for each factor. (3) The companies will be ranked according to its highest score. The top-ranked companies* will then be selected for addition to the index so that the number of constituents equal to 100. * For companies with both listed A shares in the Mainland and H shares in Hong Kong, the share class with a higher 1-year average daily turnover value (ADT) will be selected.
No. of constituents	100
Calculation methodology	Equally-weighted
Review frequency	Yearly
Rebalance frequency	Half-yearly
End-of-day index (currency)	CNY (onshore Renminbi)
Base value	2000
Base date	30 June 2015
Dissemination	Once a day, after the close of trading

Source: CESC.

Chapter 12

Investment benchmark for Asia's biotechnology sector: CES HK Biotechnology Index

Chief China Economist's Office
Hong Kong Exchanges and Clearing Limited
and
China Exchanges Services Company Limited

Summary

Global economic development, population growth and an aging society have given new impetus to the biomedical industry. Compared to traditional sectors, biomedical companies are characterised by having substantial investment, high-value outputs and high risks, and being technology-intensive. They usually adopt equity financing rather than debt financing as an important source of financing during their growth period. Financing by different methods at different stages of development of a biomedical company according to development characteristics provide strong financial support for the sector's development. Major global stock markets have in recent years developed new listing rules specifically for the biotechnology (biotech) sector to facilitate effective allocation of venture capitals and private equity funds, completing the "input-output" cycle of capital deployment and optimise resources allocation.

The Mainland's biomedical industry thrives largely because of policy support, increasing capital inputs, accelerating industry consolidation and other favourable factors. In April 2018, HKEX amended its Listing Rules and opened up a new listing channel for pre-revenue or pre-profit biotech companies in Asia, creating a sound environment for the financing of, and investment in, this dynamic and promising sector.

Relevant indices and financial products have been rolled out across major global capital markets to broaden investor access to the biomedical sector. The most representative ones are the NASDAQ Biotechnology Index and the S&P Biotechnology Select Industry Index, and a series of exchange traded funds (ETFs) based on these indices. In the Asian market, the CES HK Biotechnology Index ("CES HK Biotech" for short) tracks biotech companies listed according to Hong Kong's new listing rules and Hong Kong-listed biomedical companies which are in relatively mature development stage. The index gives a comprehensive picture of the performance of Hong Kong's biomedical sector and serves as an industry investment benchmark.

The launch of CES HK Biotech will further increase the diversity of the Hong Kong capital market. The use of biotechnology by index constituent stocks in therapies for human illnesses represents the future development direction of biotechnology in Asia. This is of high significance for China to strengthen its competitiveness in biomedical innovation.

1 The capital market as the booster of biotech medical industry

1.1   The recent rapid growth of biotech medical industry

Global economic development, population growth and an aging society have given new impetus to the biomedical industry. The global pharmaceutical market exhibited much stronger growth than the overall economic growth. Genetics, molecular biology and biochemistry continue to make technological breakthroughs. Given technological advancement and the increasing demand, the biomedical sector enters into a golden age of development. According to the EvaluatePharma 2017 report, the global prescription drug market will see a compound annual growth rate (CAGR) of 6.5% in five years and be expected to reach US$1.06 trillion in 2022. The biologics segment, in particular, will develop most rapidly. By 2022, 52% of the top 100 best-selling pharmaceutical products would be biotech products. The share of biomedical industry in the entire pharmaceutical market would increase from 17% in 2008 to 30% (US$326 billion) in 2022[1].

The biomedical sector, due to its speedy growth, has gradually become the new engine of economic growth. In the US, the powerhouse for the biomedical industry, the industry's total production accounts for about 17% of the gross domestic product (GDP) and biomedical clusters have emerged in cities like Boston and Los Angeles. In developed countries like Japan and Germany, the share of the healthcare industry in GDP also exceeds 10%, making the industry a key engine of social and economic growth[2]. Strong growth was also seen in China's medical industry. The Mainland's pharmaceutical market expanded at a CAGR of 15% between 2011 and 2016[3]. In the Outline of the Healthy China 2030 Plan released in 2016, the development of China with a healthy population has become a national strategy. Mainland healthcare service is estimated to reach RMB 16 trillion by 2030[4]. With the wide application of frontier biotechnology (or biotech for short) in the medical field, it is foreseeable that biomedicine will gradually become the fastest growing and most technology-intensive involved sub-sector in the healthcare industry, and that

---

1   Source: EvaluatePharma, *World Preview 2017*, June 2017.
2   Source: "Healthcare industry is key engine of national economic growth" (〈發展健康產業是引領我國經濟增長的重要動力〉), *Xinhuanet*, 20 July 2017.
3   Source: McKinsey, *CPA-McKinsey China Hospital Pharmaceuticals Report: An In-depth Perspective*, August 2017.
4   See the State Council's *Outline of the Healthy China 2030 Plan* (《健康中國 2030 規劃綱要》), October 2016.

some remarkably innovative companies will emerge in this field to become a new driver for national economic growth and industry innovation.

1.2 Equity financing providing the necessary financial resources to biomedical companies

Firstly, equity financing (rather than debt financing) is an important mode of financing for biomedical companies in their growth stage.

Compared to traditional sectors, biomedical companies are characterised by having substantial investment, high-value outputs and high risks, and being technology-intensive. A new drug has to go through clinical trials at multiple stages in its production cycle before launch, where it is tested for safety, curative effects, hazards and adverse reactions. It also needs to be approved by regulators before it is released to the market. According to a report on clinical drug development success rates issued by the Biotechnology Innovation Organisation (BIO) in the US[5], the likelihood of a drug candidate in Phase I clinical trials receiving final approval by the US FDA[6] is merely 10%. Most candidates fail to advance from Phase II to Phase III, rendering a failure in the entire research and development (R&D) process. Therefore, according to the experience of some large overseas biomedical companies, investing in a biomedical company has been long-term, costly and high risky. It required an average funding of US$250 million, but a product took about eight to ten years from concept to launch, with an average annual loss of US$30 million[7].

The huge uncertainties facing drug development imply that the biomedical industry has very high entry barriers, and that venture capital and equity financing would be the key channels for meeting the financing needs of these companies. Different modes of financing may be adopted by a biomedical company at its different stages of development according to its development characteristics, thereby providing strong financial support for the sector's development. In recent years, new listing regimes have been developed in the capital market specifically for the listing of biomedical companies to facilitate effective allocation of venture capitals and private equity funds, thereby completing the "input-output" cycle of capital deployment and optimise resources allocation.

---

5 BIO, *Clinical Development Success Rates 2006-2015*, May 2016.

6 US Food and Drug Administration (FDA) is the highest law enforcement agency authorised by the US Congress that specialises in food and drug management.

7 Source: R&D-based Pharmaceutical Association Committee under the China Association of Enterprises with Foreign Investment, et. al. "Facilitating the design and implementation of the clinical research framework, deepening the formation of the medical innovation ecosystem" (〈推動臨床研究體系設計與實施，深化醫藥創新生態系統構建〉), December 2017.

Secondly, as a result of intensive competition in drug research and development, the global biomedical industry has become more concentrated. Major pharmaceutical companies need to acquire massive capital from the capital market in order to maintain their market dominance.

Examples are Pfizer's acquisition of Warner Lambert in 2000 and Pharmacia in 2003 and Aventis's merger with Sanofi in 2004. In 2017, there were more than 400 mergers and acquisitions (M&As) in the global biomedical industry, involving more than US$180 billion. Most of the deals were acquisitions of small and medium-sized enterprises (SMEs) by large enterprises to obtain emerging technologies for market expansion[8]. Companies can use funds obtained through venture capital financing and equity financing to industrialise their products, promote R&D innovation among peers, and speed up growth. Table 1 presents the list of major pharmaceutical companies in the world and their businesses.

Table 1. The world's top 10 pharmaceutical companies and their major biologics products			
Company	Pharmaceutical business revenue in 2017 (USD mil)	R&D expenses (USD mil)	Principal business
Pfizer	52,540	7,657	A research-based pharmaceutical company dedicated to a wide range of therapeutic fields for health purposes and the prevention and treatment of diseases, providing medicine in the fields of cardiology, oncology and immunology.
Roche	44,368	10,392	A R&D-focused healthcare company that develops, manufactures and delivers innovative therapies as well as diagnostic instruments and tests. Pharmaceutical products cover oncology, diabetes, ophthalmology, neuroscience, immunology, infectious diseases, etc.
Sanofi	36,663	6,697	French pharmaceutical company providing prescription and over-the-counter drugs for the central nervous system, cardiovascular diseases, oncology and diabetes, general medicine and vaccines.
Johnson & Johnson	36,256	10,554	Providing drugs for hepatitis C (HCV), HIV/AIDS, and digestive ailments.
MSD	35,390	10,000	Research-based pharmaceutical company offering drugs for oncology, fertility, neurodegenerative diseases and endocrinology.
Novartis	33,000	8,972	Innovative biotherapies and drugs in a variety of therapeutic areas such as oncology, cardiology, neuroscience, immunology, ophthalmic care, generics and biosimilar.

---

8   Source: hsmap（火石創造）ed. (2018). *Blue Book on the Development of Biomedical Industry in China (2017)*（《中國生物醫藥產業發展藍皮書 2017》）, August 2018.

(continued)

Table 1. The world's top 10 pharmaceutical companies and their major biologics products			
Company	Pharmaceutical business revenue in 2017 (USD mil)	R&D expenses (USD mil)	Principal business
AbbVie	28,216	4,982	Research-based biopharmaceutical company that mainly develops and markets drugs for rheumatism, gastroenterology, dermatology, oncology, virology, neurological diseases, metabolic diseases, etc.
Gilead	25,662	3,374	Research-based biopharmaceutical company that specialises in virology, hepatology, hematology, oncology, cardiovascular diseases, inflammation, respiratory diseases, etc.
GSK	24,038	6,235	Research-based biopharmaceutical company that provides drugs for cardiovascular diseases, gynecology, diabetes, anti-infection, skin diseases, respiratory diseases etc.
Amgen	22,849	3,562	Pharmaceutical company headquartered in California, US, specialising in the discovery and development of innovative biologics and small molecular drugs.

Source: Igeahub, *Top 10 Pharmaceutical Companies in 2018*.

Thirdly, cancer immunotherapy is a major field where biotechnology is applied in the pharmaceutical industry. Most of the innovative biomedical companies in this area raised capital through listing on the exchange market to support drug development for cancer immunotherapy.

Representative drugs for immunotherapy are PD-1 and PD-L1 antibodies. There are five types of antibody drug in the world, including Bristol-Myers Squibb's Opdivo and MSD's Keytruda. Opdivo's global revenue increased from US$20 million in 2014 to US$5,753 million in 2017, with a CAGR exceeding 560%. Keytruda's global revenue increased from US$55 million in 2014 to US$3,809 million in 2017, with a CAGR exceeding 310%[9]. These became the best-selling drugs in the biomedical sector. The sales performance encourages biomedical companies to increase their R&D in this segment and offers a safeguard for biomedical companies' profit after listing on the exchange market. Effective support offered by the capital market to biotech companies would boost the emergence of more cancer treatment drugs.

---

9   Source: Frost & Sullivan.

1.3   Recent reforms in the global capital market driving further development of the biomedical industry

Biotech companies which had been barred from traditional capital markets are attracted to list on major securities markets which have undertaken market rules reforms in recent years targeting the listing of biotech companies. NASDAQ in US is the key market for the listing of global biomedical companies. It has a flexible listing regime under which companies may list on various boards according to their financial performance in terms of profits, net assets, total assets and market size. In April 2012, the US promulgated the Jumpstart Our Business Startups Act (JOBS Act) to support the equity financing and listing of start-ups and innovative SMEs, especially biotech companies. It facilitates the refinancing and supports the development and industrialisation of innovative biomedical products. As of February 2019, about 745 health-related companies were listed on NASDAQ, of which 171 were biomedical companies with a total market capitalisation of US$490.7 billion[10]. Several Chinese medical companies have also listed in the US (see Table 2).

Table 2. Chinese biomedical enterprises listed on NASDAQ in recent years		
Company	Listing year	Company profile
BeiGene	2016	Specialised in innovative molecular targeted drugs and immune pharmaceuticals for cancer treatment, listed on NASDAQ in February 2016.
Hutchison MediPharma	2016	Specialised in the discovery and development of innovative therapies for cancer and auto-immune diseases; listed on AIM of London Stock Exchange in the UK in May 2006 and obtained a secondary listing on NASDAQ in March 2016.
Zai Lab	2017	Mainly engaged in R&D for drugs that treat cancer, auto-immune and infectious diseases; listed on NASDAQ in September 2017.

Source: Compiled according to public information.

Amendments of main board listing rules have been continued in other stock exchanges as well to promote biotech industry development. Since 1993, the London Stock Exchange (LSE) in the UK undertook a series of institutional reforms to its main board and launched the Alternative Investment Market (AIM) in 1995 to speed up the growth of the UK biotech industry. In 2014, the UK topped other countries in Europe in terms of the number and value of new venture capital investments in the biotech industry. In 2016, a total of 11 health-related companies were newly listed on the main board and AIM of the LSE. Among them was the biomedical company, ConvaTec, which raised GBP 1,465 million of

---

10 Source: NASDAQ's website, as of 25 February 2019.

funds — the largest case of initial public offer (IPO) by a European medical company in nearly two decades[11].

In an attempt to solve the financing needs of high-growth technology companies, the Frankfurt Stock Exchange (FWB) had established outside its main board a new market (Neuer Markt). In 2003, the FWB sought to reshape its stock market into segments that adopt different disclosure standards[12]; and in 2005, it set up a junior board for SMEs. Such innovative moves facilitated biotech R&D and corporate developments, and accelerated Germany's biotech industry development. Germany now leads other European countries in new drug R&D, accounting for over 40% of drugs produced in Europe[13]. In January 2018, the FWB listed its first Mainland biotech company[14].

2 New growth impetus for Asian biomedical industry

2.1 Policy support for Mainland biomedical sector and industry consolidation

The biomedical industry in the Mainland thrives owing largely to the support of favourable policy reforms that have created new development opportunities for the industry. The speedy implementation of national strategies such as the "13th Five-Year Plan" and "Healthy China 2030 Plan", the acceleration of reforms undertaken by the State Food and Drug Administration (SFDA) and other regulatory authorities, and the reforms in launching new version of the medical insurance catalogue, standardised evaluation, priority review, and fast-track approval of innovative drugs. The government continues to increase support for autonomous innovation with the focus moving from generic drugs during the 11th

---

11 Source: Beyond Laboratory.

12 This included splitting the market into two independent boards (Prime Standard and General Standard) and creating new industry sector indices. Only issuers listed on Prime Standard are eligible for admission into FWB indices; enterprises of high market capitalisation and turnover value are included in DAX; SMEs of traditional industries are included in MDAX and SDAX; SMEs of technology industries are included in TecDAX.

13 Source: "Research Report on China's Medical Industry Development and Investment Strategy 2017-2022" (《2017-2022 年中國醫藥工業行業市場行情動態與投資戰略研究報告》), on Zhiyan.org.

14 The company is Beroni Group Ltd.

Five-Year Plan period and autonomous innovation during the 12th Five-Year Plan period to the current national blueprint for technological innovation in the 13th Five-Year Plan. The strong government support has promoted the rapid growth of Chinese biomedical companies and provided critical support for nourishing innovative capabilities conducive to the establishment of a healthy nation.

With increasing capital inputs, there has been accelerated consolidation within the Mainland biomedical industry. Biomedical industry clusters formed around the Yangtze River Delta, the Pearl River Delta and Bohai Bay gradually generate industry agglomeration effect. Shenzhen's biomedical industry had a value exceeding RMB 200 billion in 2016. There are now key innovative companies in the city, such as BGI, Mindray and Beike Biotech[15]. In the first 10 months of 2018, 357 M&As valued at more than RMB 120 billion were completed among Mainland medical and healthcare companies, a year-on-year increase of almost 20% (see examples in Table 3). Among them, biomedical M&As accounted for three fifths of the total in number terms. More than 240 cases (or 49%) were related to biopharmaceutical investment[16]. In particular, increasing industry consolidation was seen in the areas of cancer treatments and drugs, products and services for gene diagnosis and treatment, and implanted interventionist medical products. This further improves the R&D capabilities for innovative biomedical products.

Table 3. Certain M&A cases of Mainland biomedical companies			
Year	Acquirer	Company acquired	Amount/Equity holding	Field
2014	MicroPort	Wright	US$290 million	Joint implant
2014	PW MedTech	TianXinFu	RMB 800 million/100%	Orthopedic implanted device
2015	Anke Biotechnology	AGCU ScieTech	RMB 450 million/100%	Genetic testing
2015	Shanghai Yiliao	Zhonggu Shengwu	RMB 270 million/67.5%	Anti-cancer intermediate drug
2016	Nanjing Cenbest	Sinocord	RMB 3.4 billion/76%	Hematopoietic stem cells
2016	Bai Hua Cun	Huawei Medicine	RMB 1,954 million/100%	Anti-cancer drug R&D
2017	SanPower Group	Findgene	RMB 680 million	Genetic testing
2018	Huaxi Holdings, Hicin Pharmaceutical	NMS	US$369 million	Anti-cancer drug R&D
2018	Huadong Medicine	UK Sinclair	GBP 169 million/100%	Biotechnology

Source: *China Biomedical Industry Development Blue Paper 2017*（《中國生物醫藥產業發展藍皮書 2017》）；
GBI SOURCE Database.

---

15 Source: "Shenzhen's biotech industry value exceeds RMB 200 billion"（〈深圳生物產業規模超 2,000 億元〉）, *China Economic Daily*, 31 March 2017.
16 Source: zyzhan.com.

2.2 Better supports from the Hong Kong capital market for the development of biomedical companies in the region

In April 2018, HKEX amended its Listing Rules to open up a new listing channel for pre-revenue or pre-profit biotech companies. After the implementation of the new listing rules, in addition to the listing of companies with revenue and profit pursuant to the pre-existing listing rules, companies without operating income are now able to access the Hong Kong capital market upon satisfying the requirements under the new listing rules.

The Hong Kong new Listing Rules in 2018 recognise China Food and Drug Administration (CFDA) as a regulator qualified to assess biotech products — putting it on par with the US' FDA and the European Medicine Agency (EMA). This is conducive to the application and promotion of Chinese standards in the international market.

The exit channel provided by the listing platform of HKEX may help attract more venture capital to the high-risk and high-return biotech field. Compared to investors in overseas markets, investors in the Hong Kong market would have a better understanding of Mainland regulations and market conditions, allowing them to better evaluate the investment risks in Mainland biotech companies. On the other hand, Mainland investors can buy biotech stocks listed in Hong Kong through Stock Connect, thereby helping to create a sound investor base for biotech companies and a good environment for financing of, and investment in, these companies.

3   Biotech indices and related products in the capital market

3.1 Major biotech indices in the US capital market

Being in the most dynamic and the most promising industry sub-sector, biomedical companies have attracted much attention from the market. In order to facilitate investor participation in this market, related indices and financial products have been launched in major capital markets in the world. The most representative ones are: the S&P Biotechnology Select Industry Index launched by Standard & Poor's (S&P) in January 2006, which has 119 constituent stocks and a total market capitalisation of US$7,078 million[17]; and the NASDAQ Biotechnology Index (NBI) launched in 1993, which has 221 constituent stocks[18] and is now a major sector index that covers NASDAQ-listed companies specialised in biotech and gene pharmaceuticals. Over the past decade, both indices had a total return two to three times that of the S&P 500 index (see Figure 1 and Table 4)[19].

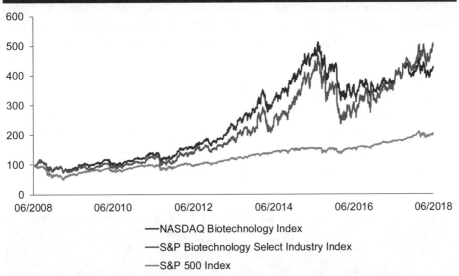

Figure 1. Comparison of daily movements of major US biotech indices with S&P 500 (6 Jun 2008 − 8 Jun 2018) (Rebased on 6 Jun 2008)

——NASDAQ Biotechnology Index

——S&P Biotechnology Select Industry Index

——S&P 500 Index

Source: Bloomberg.

Table 4. Performance comparison of major US biotech indices with S&P 500				
Index	3-year return	5-year return	10-year return	PE (times)
NASDAQ Biotechnology Index	34.75%	85.17%	325.96%	48.03
S&P Biotechnology Select Industry Index	110.81%	162.50%	398.97%	-13.02
S&P 500 Index	42.56%	69.10%	104.24%	20.28

Source: Bloomberg. Data for returns were as of 6 June 2018 and data for PE were as of 29 January 2019.

---

17 Source: S&P's website, as of 31 January 2019.

18 Source: NASDAQ's website, as of 25 February 2019.

19 Past performance is not an indicator of future performance.

3.2 ETF products based on biotechnology indices

The US has the world's largest market of biotechnology exchange-traded funds (ETFs), having a total of 18 biotech ETFs available with the total assets under management (AUM) amounted to about US$19 billion (see Table 5). All of the underlying stocks of these ETFs are listed in the US. These ETFs mainly track three major US biomedical indices: the NASDAQ Biotechnology Index, the S&P Biotechnology Select Industry Index and the New York Stock Exchange's biotechnology index. Among them, iShares NASDAQ Biotechnology ETF had the largest AUM (US$9.6 billion), tracking the NASDAQ Biotechnology Index. There are also six leveraged and inverse biotech ETFs which further broaden the range of US biotech index products. In Europe, two biotech ETFs are listed respectively in London and Frankfurt, the larger (AUM of US$490 million) of which was launched by Invesco and tracks the NASDAQ Biotechnology Index.

In the Mainland, there are six biotechnology index funds but no biotech ETFs. Four of these funds track the Mainland A-share biotechnology indices and two track the US biotechnology indices. Three of the six funds are structured funds. There are currently no biotechnology index futures or other related derivatives available for trading in the Mainland. (See Table 6.)

Table 5. Major biotechnology index ETF products listed in the US and Europe					
Stock code	Product name	Product nature	Issuer	Underlying index	Total asset value* (USD mil)
US-listed					
IBB	iShares NASDAQ Biotechnology ETF	Traditional	BlackRock iShares	NASDAQ Biotechnology Index	9,563.22
XBI	SPDR S&P Biotech ETF	Traditional	State Street Global Advisor	S&P Biotechnology Select Industry Index	5,386.64
FBT	First Trust Amex Biotechnology Index ETF	Traditional	First Trust	NYSE Arca Biotechnology Index	1,728.72
BBH	VanEck Vectors Biotech ETF	Traditional	VanEck	MVIS US Listed Biotech 25 Index	430.79
LABU	Direxion Daily S&P Biotech Bull 3x Shares	Leveraged (300%)	Direxion	S&P Biotechnology Select Industry Index	411.66
BIB	ProShares Ultra NASDAQ Biotechnology ETF	Leveraged (200%)	ProShares	NASDAQ Biotechnology Index	363.94
PBE	Invesco Dynamic Biotechnology & Genome ETF	Traditional	Invesco	Dynamic Biotechnology & Genome Intellidex Index	280.12
ARKG	ARK Genomic Revolution Multi-Sector ETF	Traditional	ARK Investment Management	Nil	226.11

(continued)

Table 5. Major biotechnology index ETF products listed in the US and Europe					
Stock code	Product name	Product nature	Issuer	Underlying index	Total asset value* (USD mil)
SBIO	ALPS Medical Breakthroughs ETF	Traditional	ALPS	Poliwogg Medical Breakthroughs Index	224.71
BBC	Virtus LifeSci Biotech Clinical Trials ETF	Traditional	Virtus	LifeSci Biotechnology Clinical Trials Index	79.18
LABD	Direxion Daily S&P Biotech Bear 3x Shares	Inverse (-300%)	Direxion	S&P Biotechnology Select Industry Index	77.47
CNCR	Loncar Cancer Immunotherapy ETF	Traditional	Exchange Traded Concepts	Loncar Cancer Immunotherapy Index	59.67
BTEC	Principal Healthcare Innovators Index ETF	Traditional	Principal Financial Group	NASDAQ U.S. Health Care Innovators Index	54.62
UBIO	ProShares UltraPro NASDAQ Biotechnology ETF	Leveraged (300%)	ProShares	NASDAQ Biotechnology Index	37.64
BBP	Virtus LifeSci Biotech Products ETF	Traditional	Virtus	LifeSci Biotechnology Clinical Trials Index	34.36
BIS	ProShares UltraShort NASDAQ Biotechnology ETF	Inverse (-200%)	ProShares	NASDAQ Biotechnology Index	28.97
IEIH	iShares Evolved U.S. Innovative Healthcare ETF	Traditional	BlackRock iShares	Nil	5.15
ZBIO	ProShares UltraPro Short NASDAQ Biotechnology ETF	Inverse (-300%)	ProShares	NASDAQ Biotechnology Index	2.99
Europe-listed					
SBIO	Invesco NASDAQ Biotech UCITS ETF	Traditional	Invesco	NASDAQ Biotechnology Index	490.65
A2DWAW	iShares NASDAQ US Biotechnology UCITS ETF	Traditional	BlackRock iShares	NASDAQ Biotechnology Index	14.36

* As of 10 July 2018.

Source: Bloomberg and websites of the issuers.

Table 6. Biotechnology index funds in China		
Product name	Underlying index	Total asset value* (RMB mil)
SWSMU CSI SWS Health Care Index Structured Fund	CSI SWS Health Care Index	853.53
China Merchants CNI Biomedicine Index Leverage Fund	CNI Biomedicine Index	394.56
GF NASDAQ Biotechnology Index Launched Type Securities Investment Fund	NASDAQ Biotechnology Index	359.15
E Fund Biotech Index Graded Securities Investment Fund	CSI WIND Biotechnology Index	210.00

(continued)

Table 6. Biotechnology index funds in China		
Product name	Underlying index	Total asset value* (RMB mil)
E Fund's S&P Biotech Index Securities Investment Fund	S&P Biotechnology Select Industry Index	78.50
China Universal CSI Biotechnology Theme Index Launched Fund	CSI Biotechnology Thematic Index	77.50

* As of 10 July 2018.

Source: Bloomberg and websites of the issuers.

3.3   Asia-based CES HK Biotechnology Index

Unlike in the US and Europe, the biotech industry in Asia is in its infancy. There are few large biopharmaceutical companies in the region while the world's top 10 medical companies, such as Roche and Novartis (see Table 1 above), are in Europe and the US. Neither does Asia have world class biotech R&D centres such as the Sanger Institute in Cambridge in the UK, which can commercialise biomedical findings. Hence, biotech R&D clusters cannot easily be formed in Asia to attract knowledge and talents. Asian investors and analysts also lack adequate experience and expertise to assess pharmaceutical companies. Hong Kong's new listing rules will no doubt attract more quality biotech companies to the region and instill new energy into the region's biomedical industry.

Increasing demand will be seen in the capital market for benchmarks that track and reflect developments of the biotechnology sector. Biotech stocks are prone to be volatile as most of the products of biomedical companies are still at R&D stage and their stock prices are to a large extent subject to the progress of clinical trials. Therefore, there is a need for an industry benchmark to help investors diversify the risk of investment in individual biotech stocks.

For this purpose, China Exchanges Services Company Limited (CESC), a joint venture of the three stock exchanges in Shanghai, Shenzhen and Hong Kong, launched the CES HK Biotechnology Index ("CES HK Biotech" with the index code of CESHKB) on 14 November 2018 as the benchmark that measures the performance of Hong Kong-listed biotech stocks. The methodology and performance of the index are described below.

(1) Selection of constituents

The base date of the index is 12 December 2014. The base point is 2000. The index universe for CES HK Biotech is comprised of common stocks with primary or secondary listing on the Main Board of the SEHK. On 11 April 2019, the index had 17 constituents.

The constituents are selected according to the following selection criteria:

- To be eligible for inclusion, a company must be classified as either "Biotechnology Company" according to the Industry Classification, or "Biotechnology Company" listed under Chapter 18A of the SEHK Main Board Listing Rules; and
- The stock must attain a daily average total market capitalisation of at least HK$1.5 billion in the most recent year.

In order to include as many suitable biotech stocks as possible, the fast entry rule will be applied to a newly listed stock which meets the selection criteria after the close of its 10th trading day if the number of constituents falls below 20. For newly listed stocks, the daily average total market capitalisation is derived from the data between the 4th trading day and the review cut-off date for assessing the stock's eligibility.

(2) Index calculation

The index value of CES HK Biotech is calculated based on the free float-adjusted weighted market capitalisation of the constituents. The bigger the free float market capitalisation, the greater the impact of the constituent is on the index. Changes in the share price of a constituent will be directly reflected in the movement of the index. When the share price of a constituent rises, the market capitalisation of the constituent also rises and so does the weight of the constituent. Conversely, the constituent's market capitalisation as well as the weight of the constituent fall when the share price of a constituent falls. The calculation is as follows:

Calculation formula

$$\text{Current index} = \frac{\text{current adjusted market capitalisation of constituents}}{\text{divisor}} \times 2000$$

Where adjusted market capitalisation = Σ (price $\times$ adjusted number of shares $\times$ weight factor)

Determination of the divisor

The initial value of the divisor is the total market capitalisation of the constituents on 12 December 2014. To ensure the continuity of the index, when the list of constituents or their share structure changes, or when there is a change in the market capitalisation of the constituents due to non-trading factors, the divisor will be revised according to the index maintenance rule.

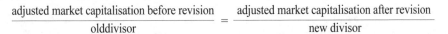

$$\frac{\text{adjusted market capitalisation before revision}}{\text{old divisor}} = \frac{\text{adjusted market capitalisation after revision}}{\text{new divisor}}$$

Where adjusted market capitalisation after revision = adjusted market capitalisation before revision +/- the increase/decrease in adjusted market capitalisation.

Determination of the share price

The share price (X) of each constituent is determined according to the following principle[20]:

If there is no transaction that day, X = reference opening price;

otherwise X = the last traded price

Adjusted number of shares

Adjusted number of shares = Total number of issued shares × inclusion factor

The inclusion factor is a ratio applied to the total number of issued shares based on the ratio of the free-float shares to the total issued share capital (free float ratio[21]). To reflect the change in the actual number of free-float shares in the market, restricted shares and non-tradable shares held for strategic and other reasons are excluded from index calculation. The remaining shares are the free-float shares or simply called free float. The inclusion factor of each constituent is calculated based on the free float ratio to ensure the number of shares of each constituent is relatively stable for calculating the index.

Weight adjustment

When calculating CES HK Biotech, the weight factor is set between 0 and 1 so that no constituent accounts for more than 10%. However, if the number of constituents is less than 20, a 15% cap will be applied to their weights. When the number of constituents is less than 8, a 25% cap will be applied. When the number of constituents is less than 5, the constituents will be equally weighted. Weight factors will be adjusted along with changes to the list of constituents at periodic reviews, with the same implementation date. The weight factor generally remains the same until the next periodic review[22].

(3) Index performance and outlook

Based on data as of 6 March 2019, WuXi Biologics had the highest weight (16.30%) among constituents of CES HK Biotech; Innovent Biologics had a weight of 13.46%; Sino Biopharmaceutical, 3SBio and Genscript Biotech each had a weight of over 10%[23]. The index consists of biotech companies listed under the new listing rules as well as relatively more developed companies in the industry, providing a comprehensive picture of the whole sector to investors. Derivative products on the index would also facilitate investors to

---

20 Real-time stock prices of index constituents are obtained from the trading data disseminated by HKEX via various channels. The real-time index is calculated during the trading hours of HKEX.

21 Free float ratio = free float shares / total number of issued shares.

22 For the index calculation methodology, see "Calculation and Maintenance Methodology" (https://www.cesc.com/en/ Index/Hong-Kong-Overseas/Ces-Hk-Biotech.html).

23 Source: CESC.

diversify the investment risk stemmed from individual biotech stocks.

The annualised volatility of CES HK Biotech was close to that of the NASDAQ Biotechnology Index and higher than that of the NASDAQ Composite Index. Its annualised return was 42.41% in the past three years, surpassing the performance of the Hang Seng Index and the NASDAQ Biotechnology Index in the same period (see Figure 2 and Table 7).

Figure 2. Comparison of CES HK Biotech with Hang Seng Index and NASDAQ Biotechnology Index (12 Dec 2014 — 6 Mar 2019) (Rebased on 12 Dec 2012)

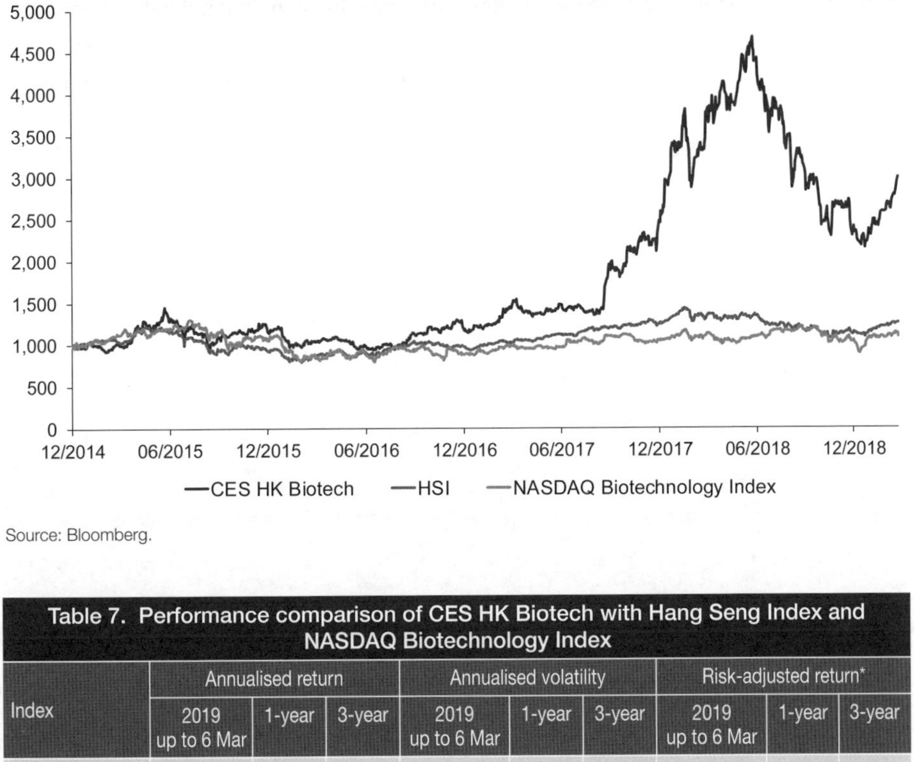

Source: Bloomberg.

Table 7. Performance comparison of CES HK Biotech with Hang Seng Index and NASDAQ Biotechnology Index

Index	Annualised return			Annualised volatility			Risk-adjusted return*		
	2019 up to 6 Mar	1-year	3-year	2019 up to 6 Mar	1-year	3-year	2019 up to 6 Mar	1-year	3-year
CES HK Biotech	30.00%	-10.03%	42.41%	28.52%	41.84%	35.43%	1.05	-0.24	1.20
HSI	12.35%	-2.84%	12.90%	15.48%	18.64%	15.97%	0.80	-0.15	0.81
NASDAQ Biotechnology Index	13.51%	-0.93%	7.85%	24.09%	24.69%	31.10%	0.56	-0.04	0.25

* Risk-adjusted return is calculated by dividing the annualised return by the annualised volatility.

Source: Calculation based on daily closing data from Bloomberg, as of 6 March 2019.

4 Conclusion

China's biotechnology sector has entered a stage of growth with a projected long period of development. HKEX introduced new listing rules that suits the special financial characteristics and investment risks of these companies in their start-up stage (no profit or revenue for a long time before and after listing) will help channel more venture capital and private equity funds into the industry and companies. This will be conducive to the emergence of large innovative biotech companies, thereby stimulating the development of such core industries in the region and facilitating the upgrade of the regional economy.

The launch of CES HK Biotech will further increase the diversity of the Hong Kong capital market. The use of biotechnology by index constituent stocks in therapies for human illnesses represents the future direction of biotechnology in Asia. This is of high significance for China to strengthen its competitiveness in biomedical innovation.

Appendix

CES HK Biotechnology Index Constituents and their profile

Stock code	Company name	Listing date	Return on assets in 2018 (%)	Return on equity in 2018 (%)	Earnings per share in 2018		Income growth in 2018 (%)	Debt to equity ratio in 2018 (%)
					Annual growth (%)	Value (HKD)		
775	CK Life Sciences	16/07/2002	2.50	5.92	1.86	0.03	11.50	108.98
1035	BBI Life Sciences	30/12/2014	8.24	10.82	22.88	0.17	25.78	0.86
1061	Essex Bio-Technology	27/06/2001	17.71	27.70	36.13	0.41	30.78	5.92
1177	Sino Bio-pharmaceutical	29/09/2000	25.59	47.09	274.67	0.87	40.96	9.09
1530	3SBio	11/06/2015	9.26	15.95	35.14	0.59	22.75	36.99
1548	Genscript Biotech	30/12/2015	2.99	6.09	-22.37	0.09	51.34	2.13
1672	Ascletis Pharma	01/08/2018	-0.32	-0.36	86.54	-0.01	212.63	0.00
1801	Innovent Biologics	31/10/2018	-134.69	-512.60	-926.38	-20.44	-48.88	18.88
1877	Junshi Biosciences	24/12/2018	-26.16	-28.56	-123.29	-1.41	-18.64	0.00
2269	Wuxi Biologics	13/06/2017	8.86	10.49	116.67	0.62	56.56	0.00
2359	WuXi AppTec	13/12/2018	12.83	18.81	70.23	2.64	23.85	0.85
2552	Hua Medicine	14/09/2018	-415.05	-1602.61	-1221.06	-11.94	274.07	0.00
2616	CStone Pharmaceuticals	26/02/2019	-158.92	-370.68	N/A	-12.28	N/A	0.00
6118	Austar Lifesciences	07/11/2014	0.01	0.02	-175.00	0.00	49.30	4.16
6160	BeiGene	08/08/2018	-41.03	-56.00	-416.67	-7.29	-22.17	2.82
6185	Cansino Biologics	28/03/2019	-16.64	-24.92	N/A	N/A	N/A	29.86
6826	Shanghai Haohai Biological Technology	30/04/2015	9.74	12.17	11.16	3.07	14.94	0.96

N/A: Not applicable.

Note: Constituent list is as of 11 April 2019.

Source: CESC website for constituent list; Bloomberg and Wind for stock data, retrieved on 4 April 2019.

Chapter 13

Rising demand for block trading in the Mainland and Hong Kong securities markets

Chief China Economist's Office
Hong Kong Exchanges and Clearing Limited

Summary

Block trading mechanism of equities is important to minimise price impact and avoid unwanted signalling of large-sized transactions, which is detrimental to best execution by brokers. It provides an efficient way to buy or sell large blocks of shares in a single transaction with more certainty in transaction price and timing than trading on the auction market. Various block trading mechanisms prevail in global markets. These include off-exchange negotiated trades with post-trade reporting to an exchange, separate off-exchange trading platforms or dark pools, or specialised on-exchange block trading mechanism. Despite the differences in block trading mechanisms, there is a growing trend in block trading on major stock exchanges as hinted by the growth in negotiated trades.

The growth of the asset management industry is a primary driver for block trading. For an active asset manager, the price impact of lit orders of on-exchange large-sized transactions erodes the active returns. For a passive asset manager, the costs of rebalancing on index stocks can be significant and much higher than the management fees received. In addition, creation/redemption of fund units sometimes involves large-sized transactions within a short time frame. Demand also comes from other trading purposes: mergers and acquisitions, share repurchases as well as transactions by private equities and major shareholders. The development of equity derivatives and structured products may increase demand for block trading to enhance efficiency of shares delivery at settlement. Besides, findings of empirical studies showed that negotiated trades (or the "upstairs market") effectively provide extra liquidity and reduce price impact, particularly for stocks with low liquidity in the on-exchange limit-order book.

In the Mainland market, the Shanghai and Shenzhen stock exchanges provide dedicated platforms for block trading. In contrast to world major markets, block trading of A shares in the Mainland has been dominated by sell-initiated trades — mainly the placement of shares by major shareholders after the lock-up period of their substantial shareholdings, rather than by trades addressing the demand from asset management business. This may explain why the average transaction size of negotiated trades in the Mainland has been much larger than those in major stock markets. Nevertheless, the situation is expected to change as the access by global investors to the A-share market has been expanded through Northbound Stock Connect. The asset management of offshore institutional investors in A shares is expected to grow further after the inclusion of A shares into global indices and the quadrupling of daily quota of net purchases under Northbound Stock Connect. The increased global institutional participation is expected to increase the demand for block trading of A shares.

In Hong Kong, the exchange's manual trade mechanism accommodates the reporting by Exchange Participants of block trades through private negotiations between brokers, internal crossing or dark pools. The turnover of negotiated trades reached the highest in 2018. Evidenced by the relatively high average transaction size, negotiated trades in Hong Kong are believed to comprise mostly of block trades. The growth in block trading could, to a certain extent, be attributable to the growth in asset and wealth management businesses in Hong Kong. Moreover, Mainland investors have become increasingly interested in diversifying their portfolios through mutual funds with exposures in Hong Kong stocks. The potential demand for block trading from Mainland funds through Southbound Stock Connect is expected to grow further.

Given the growing potential demand for cross-border block trading and that Mainland and Hong Kong cross-border trading is now conducted largely via the Stock Connect schemes, it is worth considering the enhancement of Mainland-Hong Kong market connectivity with block trading facilities under appropriate arrangements and controllable environment, so as to meet investors' demand. Cross-border market liquidity can thereby be further enhanced and investors can benefit from even better execution.

1 The significance of block trading mechanism in equity trading

1.1 What is a block trade?

A block trade refers to a purchase or sale of a large amount of securities in a single transaction. Block trades are often conducted not on the open auction market in order to lower transaction cost or price impact, i.e. to prevent the price to be driven up for purchases or down for sales, by avoiding the disclosure of large-sized orders to the open market, such that a "best execution" price could be attained. Stock exchanges which have block trading mechanism impose different thresholds on the number of shares or transacted amount for block trades. They also implement different modes of block trading.

Block trades are usually executed through off-exchange methods. These include negotiation between brokers who then report the trade to the stock exchange (negotiated trades), as well as intransparent trading on alternative trading venues (or dark pools). The advantage of off-exchange execution is pre-trade anonymity as other investors cannot see the large orders in bid-ask queues on the exchange. This avoids price movements in a direction unfavourable to the buyer/seller that placed the large order before its execution. Besides, large orders may be able to match off-exchange with other reserved or hidden liquidity which do not place their orders on exchange.

There are two common modes of block trading:

- **Negotiated trades** are also called "upstairs market" where brokers make private large order negotiations on behalf of clients and the resultant transaction will be reported to a stock exchange for central clearing for mitigating counterparty risks. These trades are mainly liquidity-motivated with the aim to reduce execution costs.
- **Dark pool** block trading has become increasingly popular. While the anonymity of investor identity and trade size avoids unwanted signalling that would have price impact, investors are concerned about the potential trade price deviation between dark pools and exchanges, due to the information asymmetry of investors[1], and the risk of front-running by the dark-pool operator. The use of dark pools should be accompanied by proper regulations and investor safeguards.

---

1 To limit the price difference between trading on dark pools and exchanges, the US regulator adopted regulations (e.g. Regulation NMS Rule 611) in 2005 to ensure that orders are executed at the best displayed prices or better.

On-exchange (referred to as "downstairs market") block trades usually have high price impact. It is because the lit orders may send an information signal to the market that attracts more trades in the same direction. An empirical study[2] estimated the asymmetric price impacts of on-exchange block trades in major markets — during 1999 to 2002, the block trade price was about 0.16%-0.34% higher than the open price (in terms of ask quotes) for block purchases but 0.32%-0.42% lower than the open price (in terms of bid quotes) for block sales. Some may use algorithmic trading of smaller trades to execute a block order (e.g. trade strategies using volume-weighted average price (VWAP) or time-weighted average price (TWAP)), but this may add uncertainty to execution cost. The use of on-exchange block trades may be enhanced by exploring ways to avoid information leakage (e.g. introducing hidden order type).

These modes of block trade execution contribute to minimising price impacts and uncertainty of the transaction price.

1.2   The underlying demand for block trading in global markets

The growth of asset management is a fundamental driver for higher demand for block trading by buy-side institutions. As the assets under management (AUM) of investment funds continue to grow, the total transaction amount for purchases and sales will increase. According to PricewaterhouseCoopers (PwC), the global AUM of the asset and wealth management industry were US$84.9 trillion in 2016 and were expected to grow across all regions and reach US$111.2 trillion in 2020 and US$145.4 trillion in 2025 (see Figure 1). Mainland China is one of the fastest growing markets — its total AUM reached RMB 48 trillion (US$7.4 trillion) in 2016 and was expected to double to about RMB 90 trillion (US$14 trillion) in 2022[3]. Hong Kong's asset and wealth management industry reached an AUM of HK$23.9 trillion (US$3.1 trillion) in 2018[4].

2   Frino, A, D. Gerace and A. Lepone. (2007) "Block trades and associated price impact: International evidence on the two asysmmetries", *Papers for European Financial Management Association 2007 Annual Meetings.*

3   Source: Oliver Wyman. (2018) "Global asset managers in China: Riding the waves of reform", Oliver Wyman's website, March 2018.

4   Source: The Securities and Futures Commission (SFC)'s survey report. "Asset and Wealth Management Activities Survey 2018", published on the SFC's website, July 2019. (Referred to as the "SFC Asset Management Survey 2018".)

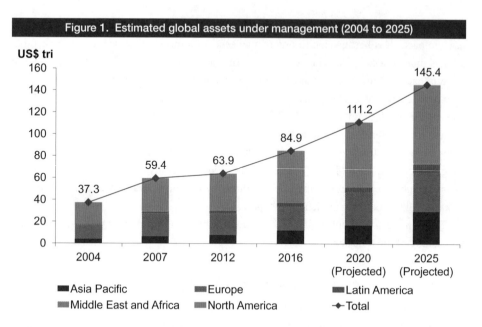

Figure 1. Estimated global assets under management (2004 to 2025)

Source: PwC. (2017) "Asset & wealth management revolution: Embracing exponential change", PwC website, 30
October 2017.

Block trading mechanism is important to minimise the price impact for both active and passive asset management. The absence of block trading mechanism on exchanges may increase the cost of execution significantly.

- **For active asset management**, asset managers apply various investment strategies with the aim to outperform benchmark indices for active returns ("alpha" returns). Active investment strategies accounted for 71% of global AUM in 2016[5]. An empirical study[6] looked into 26 active equity managers in Australia and found that the price impact of round-trip transactions (a pair of purchase and sale) were significant compared to their active returns (0.27% versus 0.92%). In addition, the redemptions of fund units may become intense during market downturn that the fund needs to sell large blocks of securities to meet the redemption needs. This explains why some funds adopt "swing pricing" that allows the fund manager to adjust the fund's net asset value (NAV) per share in order to pass on the estimated trading costs associated

---

5   Source: PwC. (2017) "Asset & wealth management revolution: Embracing exponential change", PwC website, 30
    October 2017. (Referred to as the "PwC 2017 Paper".)
6   Gallagher D. R. and A. Looi. (2003) "An examination of the market impact cost of active Australian equity managers",
    SIRCA Research Report, Sydney.

with redemptions and subscriptions to investors[7].

• **For passive asset management**, the asset managers track the performance of benchmark indices. Passive investment strategies were found to account for 17% of global AUM in 2016[8]. A research paper[9] noted that index trackers accounted for more than 10% of market capitalisation for S&P 500 (US large-cap index) stocks and about 6% for Russell 2000 (US small-cap index) stocks in 2000. The same study found that the market impact of index rebalancing activities of passive funds in the US could be significant (up to 168 basis points (bps)), compared to the annual management fee of about 10 bps for index funds (see Table 1). The price impact could be even more significant during the last few minutes before market close or at the closing auctions of trading days of index rebalancing if there is no block trading mechanism in place. As for creation and redemption of units of exchange traded funds (ETFs), similar to mutual funds, a large amount of securities may be traded within a short time frame for arbitrage to minimise the premium/discount of the ETFs or liquidation to meet redemption needs[10].

Table 1. Estimated market impact of index rebalancing in the US			
Index	Ratio of market capitalisation of additions and deletions to the index total	Abnormal returns of additions/deletions	Impact on returns
S&P 500	5.0%	5%	25 bps
S&P 1500	3.4%	7%	25 bps
Russell 2000	16.8%	10%	168 bps
Russell 3000	1.8%	10%	18 bps
NASDAQ 100	5.6%	3%	17 bps
Dow Jones TMI	1.8%	0%	0 bps

Source: Quinn and Wang 2003 Paper.

In addition to long-only asset management, some sophisticated professional investors are interested in equity structured products (e.g. equity-linked instruments or ELI) or derivatives. These products are usually settled through physical delivery. For example,

7   See Arnold and Porter Kaye Scholer LLP. (2016) "SEC adopts new rules: Fund liquidity, reporting and disclosure and 'swing pricing' ", Lexology.com, 6 December 2016.
8   Source: PwC 2017 paper.
9   Quinn, J. and F. Wang. (2003) "The impact of adds and deletes on the returns of stock indexes", Applied Finance Project, Haas MFE Program, University of California Berkeley. (Referred to as the "Quinn & Wang 2003 Paper".)
10 Source: Dickson, J. M. and J. J. Rowley. (2014) "Best practices for ETF trading: Seven rules of the road", Vanguard research, September 2014.

an investor invested in a bull equity-linked note (ELN) where the investor is required to purchase the shares at an agreed strike price when the share price at the settlement date falls below an airbag level that is lower than the agreed strike price. The transaction amount at settlement on behalf of investors can be high, especially for cases in which the products are created for only a few high-net-worth individuals. If investors need to purchase a large amount of shares on-exchange for settlement of the ELN, it may send a wrong signal to the market and increase the stock price volatility. In this case, anonymous block trading is a way to provide higher certainty of transaction prices for ELN settlement, which will be beneficial to investors and the overall market. It would also support the development of various structured products and derivatives to meet the demand of asset management.

Further to asset management, corporate actions also contribute to the demand for block trading mechanism. Block purchases can be used by companies for repurchases to support the share price. Some research[11] noted the benefits of privately-negotiated repurchases over on-exchange repurchases — the company can repurchase a large amount of shares through block trading within a short period of time and the company can avoid the administrative expenses associated with approaching a large number of shareholders.

Another demand for block trading is from mergers and acquisitions (M&A). Negotiated block purchase is one of the ways for an acquirer to gain control of a large fraction of a company's shares and the blocks are typically traded at a premium[12]. A research paper[13] explained that the premium consists of benefits to managers and owners and found that the trades for M&As with subsequent changes in top management had higher premia among 756 negotiated block trades during 1987 to 2002. The paper also noted that the transaction prices of block trades may serve as a reference for the bid price on large shareholders' stakes. This highlights the importance of block trading on price discovery among institutional investors.

Moreover, private equities and major shareholders would benefit from block trading. Private equity sponsors (or financial sponsors) acquire companies with a view towards eventual sale or public offering to potential investors. Block trading is one of the ways to trade share stakes that could provide higher certainty of execution price and shorter period of execution for buyers and sellers of share blocks. According to Dealogic, sponsor-related

---

11 See, for example, Atkins, P. and P. Korff. (2013) "Questions surrounding share repurchases", *Harvard Law School Forum on Corporate Governance and Financial Regulation*, 14 March 2013.

12 See, for example, Bittlingmayer, G. (1999) "The market for corporate control (including takeovers)", *Encyclopedia of Law and Economics, Vol. III — The Regulation of Contracts*, Boudewijn Bouckaert and Gerritt De Geest, Edward Elgar and the University of Ghent.

13 Hwang, J. H. (2004) "Whose private benefits of control — Owners or managers?", unpublished working paper, Indiana University, Bloomington.

block trades in the US rose from US$28.6 billion during 2005 to 2010 to US$170.8 billion during 2011 to 2016[14]. Similar case happened in Asia Pacific: Naspers owned 33% stake of Tencent at its start-up stage since 2001 and sold HK$9.8 billion or 2% stake of Tencent in a block trade in Hong Kong at a discount of 7.8% to the prevailing market prices[15]. This was the largest block trade in Asia Pacific on record[16]. During the bull market in 2015, some major shareholders traded their share blocks in the Hong Kong market and the discounts were just around 3%-4%, less than the normal level of about 10%[17].

In summary, block trades of equities are used in the global markets by buy-side and sell-side institutional investors with benefits in terms of minimising price impact and certainty of timing and execution costs. Moreover, these benefits are not at the expenses of other investors on the exchange (downstairs) market. These are discussed in the next sub-section.

1.3   International experience on the complementary nature of block trading

Some investors are concerned about fairness, for example, whether the upstairs market provides or takes liquidity from the downstairs market and whether brokers will front-run their clients (like trades in dark pools). The international experience may shed light on these.

Empirical evidence showed that the upstairs market for block trading is a complement and not a substitute of the downstairs markets for small orders. Although there are worries that the fragmentation of markets will lead to order substitution, the upstairs market was said to add extra liquidity and trading from order creation — literature[18] highlighted that some traders in the upstairs market wish to transact a large block of shares at lower costs (liquidity motivated) but do not want to reveal their full orders or information to the downstairs market. These trades can only be effected in the upstairs market but not in the downstairs market.

An exchange example is the Euronext Paris. The exchange allows block trades through

---

14 Source: "Building up US ECM a block at a time", Dealogic website, 17 October 2017.

15 Source: "Tencent drops a further $22 billion after Naspers trims stake", *Bloomberg* website, 23 March 2018.

16 Source: Ditto.

17 Source: "DEALTALK: Hong Kong stock surge spurs block trade boom; China bank deals seen", *Reuters* website, 21 April 2015.

18 See Burdett, K. and M. O'Hara. (1987) "Building blocks: An introduction to block trading", *Journal of Banking and Finance*, Vol. 11, pp.193-212, and Seppi, D. J. (1990) "Equilibrium block trading and asymmetric information", *Journal of Finance*, Vol. 45 (March 1990), pp.73-94.

negotiations, VWAP rules and exercising options and derivatives. A study[19] looked into the data during January to March 2007 and found that stocks with active upstairs markets were more liquid (lower bid-ask spread) and had a larger trading volume on the downstairs market, compared to other stocks. The study also found that low liquidity (high bid-ask spread) on the downstairs market does not induce investors to route their orders to the upstairs market.

Another example is the Toronto Stock Exchange (TSX). The exchange has a separate system for orders in the upstairs market but immediately disclosed the trades to the market after transaction conclusion. Those orders in the upstairs market are required to be priced at least as favourable as those available in the downstairs market at the same time. A study[20] looked into the data of June 1997 on 5,840 upstairs trades (about 55.5% of trading volume) reported to the TSX. The study concluded that the upstairs market does not "cannibalise or free ride off" the downstairs market. The upstairs market offers price improvement over the limit orders available in the downstairs market for one-quarter of the block trades. It also found that a trade is more likely to be executed upstairs when there is a wider bid-ask spread, and when there is less depth on the opposite side of the limit order book, in the downstairs market.

Evidence showed that block trades in the upstairs market are liquidity-motivated and attracts unexpressed liquidity. A study[21] looked into 92,170 block trades of French stocks during April 1997 to March 1998 and found that 67% of these trades (compared to 20%-30% in the New York Stock Exchange (NYSE)) were conducted in the upstairs market. The findings suggested that there was order creation from unexpressed liquidity and showed that the actual execution costs in the upstairs market was only 35% of displayed and unexpressed liquidity in the downstairs market. Similarly, another study[22] on 21,077 block trades on the NYSE supported that block trades in the upstairs market were liquidity-motivated and these trades will not occur without the upstairs market.

Upstairs brokers not only intermediate block trades but also play an important role to uphold fairness to investors. A study[23] examined 69,449 block trades on the Australian

---

19 See Lefebvre, J. (2010) "Block trades and market liquidity on Euronext Paris", Working Paper, Paris December 2010 Finance Meeting EUROFIDAI — AFF.

20 Smith, B. F., D. A. S. Turnbull and B. W. White. (2001) "Upstairs market for principal and agency trades: Analysis of adverse information and price effects", *Journal of Finance*, Vol. 56, Issue 5, pp.1723-1746.

21 See Bessembinder, H. and K. Venkataraman. (2004) "Does an electronic stock exchange need an upstairs market?", *Journal of Financial Economics*, Vol. 73, Issue 1, pp.3-36.

22 See Madhavan, A. and M. Cheng. (1997) "In search of liquidity: Block trades in the upstairs and downstairs Markets", *The Review of Financial Studies*, Vol. 10, Issue 1, pp.175-203.

23 Fong, K. Y. L., A. Madhavan, P. L. Swan and F. D. Foster. (2004) "Upstairs, downstairs: Does the upstairs market harm the downstairs?", unpublished working paper, University of New South Wales.

Securities Exchange (ASX) during 1993 to 1998 and found no evidence of higher market impact or wider bid-ask spread of the downstairs market due to the existence of the upstairs market. Besides, the paper suggested lowering or cancelling the threshold for block trades at AUS$1 million to facilitate block trades for small-sized stocks so as to let more block traders enjoy the benefits of lower execution costs. The paper highlighted the importance of brokers in the upstairs market to filter out "front-running" trades and search for counterparties. In respect of the role of filtering "front-running" trades, although upstairs brokers know the identities of the counterparties, they are willing to trade off short-term information gains from principal trades for the relationships with long-term customers who pay more for agency trades[24]. In respect of the role of searching for counterparties, a paper[25] cited an example of routing large orders from the limit order book on the Paris Bourse to the dealer market in London, which is partly attributable to lower costs and relative ease of search for counterparties in London's deep dealer market.

These study findings provide empirical support to the complementary nature of the upstairs (off-exchange) market to the downstairs (on-exchange) market.

1.4   Global growth trend of block trading on the upstairs market

The turnover of negotiated trades in the upstairs market has increased across many major exchanges in recent years. Statistics of the World Federation of Exchanges (WFE) on negotiated trades (negotiated deals and reported trades[26]) on an exchange are used as the proxy for block trading in the upstairs market. The statistics showed that the average daily turnover value of negotiated trades during 2016 to 2018 was higher than that during 2013 to 2015 for major exchanges in the US, Europe and Asia (see Figure 2). The proportion of negotiated trades in total turnover value on the exchange was higher in the US and Europe than in Asia. For the US, larger transaction size contributed to the increase (the average deal size of negotiated trades rose from US$9,473 during 2013 to 2015 to US$13,830 during 2016 to 2018). For Europe, the increase may contribute to a reshuffling in the

---

24 See Aitken, M. J., G. Garvey and P. L. Swan. (1995) "How brokers facilitate trade for long-term clients in competitive securities markets", *The Journal of Business*, Vol. 68 (1), January 1995, pp.1-33.

25 De Jong, F., T. Nijman and A. Roell. (1995) "A comparison of the cost of trading French stocks on the Paris Bourse and on SEAQ International", *European Economic Review*, Vol. 39, pp.1277-2301.

26 Negotiated deals are trades confirmed through a system managed (directly or indirectly) by the exchange, where both seller and buyer agree on the transaction (price and quantity). This system checks automatically if the transaction is compliant with the exchange rules, including most often the consistency with electronic order book price. Reported trades are trades reported through a Trade Reporting Facility (TRF) where only one counterparty provides information on the trade and the TRF offers dissemination services at the request of the reporting trader. The other counterparty could use this facility if reporting is mandatory. Note that data across different exchanges may not be fully comparable due to different trading and reporting rules.

composition of block trading among exchanges and other trading venues. (See Appendix 1 on the recent development of alternative trading venues in Europe in the light of regulatory changes.) For Hong Kong, the turnover of negotiated trades has been relatively stable in the two periods. It is worth noting that the trading value of the negotiated trades on the Shenzhen Stock Exchange (SZSE) during 2016 to 2018 increased by about 3.1 times from the level during 2013 to 2015 while that on the Shanghai Stock Exchange (SSE) rose by 30% across the same periods.

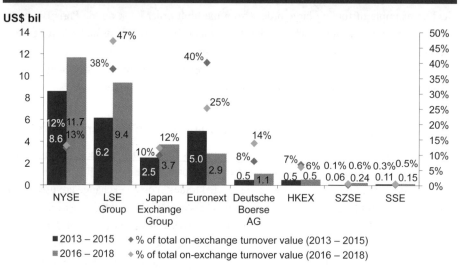

Figure 2. Average daily turnover value of negotiated trades of major stock exchanges (2013 – 2015 and 2016 – 2018)

Note: The numbers may include not only placements or upstairs block trades but also other types of negotiated deals, and are therefore not entirely comparable across exchanges because of different reporting rules and calculation methods.

Source: Calculated based on the statistics available on the WFE website.

Table 2 shows the larger average transaction size of negotiated trades vis-à-vis order-book trades on selected key exchanges, including Hong Kong Exchanges and Clearing Limited (HKEX). The average transaction size of negotiated trades on the SSE and the SZSE was about US$5-6 million during 2016 to 2018, which were significantly higher than other major stock exchanges and were over 1,800 times of the order-book trades on the respective exchanges. The large average transaction size of negotiated trades on the SSE and the SZSE hinted at the potentially strong demand for block trading of A shares. However, block trading accounted for only a very small proportion (less than 1%) of total on-exchange trading in the Mainland, compared to other exchanges (for example, 6% on HKEX, 13% on NYSE and 47% on the London Stock Exchange (LSE) Group during 2016 to 2018, see Figure 2). The contributing factors to this and the characteristics of Mainland block trading activities are discussed in Section 2.

Table 2. Average transaction size on major exchanges by execution method (2016 – 2018)			
Exchange	Negotiated trades (US$)	On-exchange electronic order book (US$)	Ratio (times)
SSE	6,105,183	3,007	2,030.4
SZSE	5,262,434	2,905	1,811.6
Euronext	1,582,698	8,669	182.6
LSE Group	74,354	7,241	10.3
Deutsche Boerse AG	64,358	11,192	5.8
HKEX	17,400	7,634	2.3
NYSE	13,830	10,530	1.3

Note: Japan Stock Exchange Group is excluded since the data on number of transactions is not available.

Source: Calculated based on the statistics available on the WFE website.

2   Block trading in Mainland China

2.1   Block trading mechanism for A shares in Mainland China

Block trading of A shares in Mainland China is subject to the exchanges' trading rules[27]. There are limits on the transaction size and trade price. The minimum transaction size of A-share block trades is 300,000 shares or RMB 2 million. For block trades of state-owned A shares, the major shareholders are subject to additional requirements, which include the approvals required for share sales exceeding the thresholds based on the ratio of shareholding and the minimum transaction price which is set at the higher of volume-weighted average of closing prices during the last 30 trading days or the net asset value

---

27 Sources of information on the Mainland block trading mechanism are these versions of the trading rules: *Trading Rules of the Shanghai Stock Exchange* (《上海證券交易所交易規則》), issued by the SSE, 6 August 2018; *Trading Rules of the Shenzhen Stock Exchange* (《深圳證券交易所交易規則》), issued by the SZSE, 30 September 2016.

based on the latest audited annual report[28].

The market prices of A shares will exclude block trades. The prices of block trades will not be used for the calculation of indices and the settlement prices of derivatives such as index futures. Nevertheless, block trade turnover is included in the calculation of securities turnover at the end of the trading day.

The SSE and the SZSE provide dedicated platforms for block trading. The block trading platforms on the SSE and the SZSE are available only to onshore institutions (exchange members or accredited investors[29]). Qualified Foreign Institutional Investors (QFII) and Renminbi Qualified Foreign Institutional Investors (RQFII) can participate through their onshore brokers who are exchange members[30]. Users can submit orders or trades to the block trading platform. The information of block trades, including stock code and name, price and volume of transactions and names of buyers and sellers, will be disclosed at the end of the trading day on the respective websites of the stock exchanges.

The submission types of order/trade allowed by the SSE/SZSE are:

- **Intent orders** show the intention of block purchases or block sales. These can be submitted during 09:30-11:30 and 13:00-15:30 on the SSE and during 09:15-11:30 and 13:00-15:30 on the SZSE.

- **Fixed-price orders** are orders with the price fixed at either the market closing price or VWAP[31]. These orders can be submitted during 15:00-15:30 on the SSE and during 15:05-15:30 on the SZSE. The submitted orders will be matched according to time priority.

- **Execution trades** are reported negotiated block trades. Both the buyer and the seller are required to report the block trade. Input of execution trades on equities takes

---

28 Source: Measures for the *Supervision and Administration of State-owned Equities of Listed Companies* (《上市公司國有股權監督管理辦法》), issued by the State-owned Assets Supervision and Administration Commission (SASAC), 16 May 2018.

29 Accredited investors include Social Security Fund, pension funds and charity funds, investment plans registered with the Asset Management Association of China (AMAC), fund managers of private funds and investors that meet the China Securities Regulatory Commission (CSRC)'s requirement on accredited investors. See details in the *Interim Measures for the Supervision and Administration of Privately-Offered Investment Funds* (《私募投資基金監督管理暫行辦法》), issued by the CSRC, 21 August 2014.

30 See the *SSE Detailed Implementation Rules on Securities Trading of QFIIs and RQFIIs* (《上海證券交易所合格境外機構投資者和人民幣合格境外機構投資者證券交易實施細則》), issued by the SSE, 19 March 2014 and the *SZSE Detailed Implementation Rules on Securities Trading of QFIIs and RQFIIs* (《深圳證券交易所合格境外機構投資者和人民幣合格境外機構投資者證券交易實施細則》), issued by the SZSE, 25 April 2014.

31 Fixed-price orders at VWAP are allowed on the SZSE but these are suspended on the SSE at the moment according to the *Notice on Revising the Trading Rules of the Shanghai Stock Exchange* (《關於修訂《上海證券交易所交易規則》的通知》), issued by the SSE, 6 August 2018.

place during 15:00-15:30 on both exchanges[32].

2.2   Block trading activities

Despite the given discount of 30% to the exchange handling fees for block trading of A shares[33], the annual turnover value of negotiated trades (comprising mostly of block trades) accounted for only about 0.1%-0.6% of the market total during 2012 to 2018 (see Figure 2). Among the factors are the relatively low institutional participation and the liquidity constrained by the block trading mechanism and related regulations.

Firstly, although the Mainland market allows the direct access by virtually all kinds of onshore institutional investor, and the indirect access by QFIIs/RQFIIs, to the block trading platforms, the majority of offshore institutional investors actually do not have the QFII/RQFII status but invest in A shares through the Northbound Stock Connect schemes. Moreover, the onshore institutional investor base is still small — the institutional trading value accounted for a small share of about 10%-20% of total market trading value[34], which is lower than in other major markets (e.g. 55% of total trading value in Hong Kong)[35]. The volume of block trades generated by the asset management needs of these onshore institutional investors, not to say by QFIIs/RQFIIs, would therefore be relatively low. The restricted access of other offshore institutions would further limit the number of counterparties on the block trading platform and affect the timing and pricing of block trades. Secondly, the liquidity is constrained because, as reflected by offshore market participants, both money and shares are locked up in the block trading platforms until the block trades are reported and disclosed. Thirdly, block trading of state-owned shares is subject to even more restrictions as discussed in Section 2.1. All these contribute to the relatively low market share of block trading in Mainland China.

Nevertheless, block trading on the Mainland exchanges appeared to have picked up in recent years. The turnover of negotiated trades rose to a peak of RMB 736 billion in 2016 after the stock index reached a recent high in 2015 (see Figure 3).

---

32 The input of execution trades during 09:30-11:30, 13:00-15:00 and 16:00-17:00, which was previously allowed, is suspended on the SSE at the moment. See the *Notice on Revising the Trading Rules of the Shanghai Stock Exchange* (《關於修訂《上海證券交易所交易規則》的通知》), issued by the SSE, 6 August 2018.

33 See the details of trading fees on the SSE's and the SZSE's websites.

34 Source: "Comparison between A shares and US stocks: Investor structure and trading pattern"(〈A 股與美股對比：投資者結構及交易特徵〉), Sina's website, 9 June 2019.

35 Source: HKEX survey, "Cash Market Transaction Survey 2018", published on the HKEX website, July 2019.

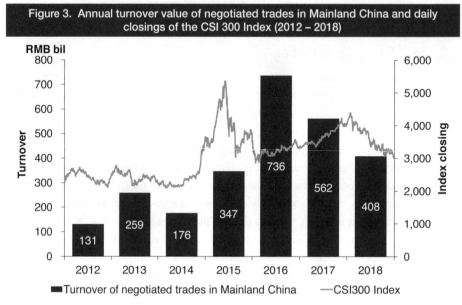

Figure 3. Annual turnover value of negotiated trades in Mainland China and daily closings of the CSI 300 Index (2012 – 2018)

Source: WFE for turnover value of negotiated trades; Bloomberg for stock index data.

Block trading is one of the main ways to buy and sell non-tradable shares after the expiry of trading restrictions (unlocking)[36]. When these shares become tradable, major shareholders can sell their stakes to investors in the market. Block trades of A shares, therefore, have been dominated by sell-initiated trades, which may add pressure to price volatility. These sell-initiated block trades are usually traded at a discount to market prices. It is observed that the share of block trading traded at discounted prices contributed about 74%-89% of total block trading turnover in recent years (see Figure 4).

---

36 Source: "While unlocking of non-tradable shares under new regulations on cutting stakes is approaching, private equity investment institutions actively participate in block trading" (〈減持新規限售股解禁期至　私募積極介入大宗交易業務〉), *China Fund* (《中國基金報》), 4 June 2018.

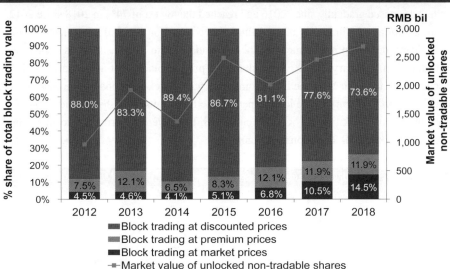

Figure 4. Composition of block trades in Mainland China and market value of unlocked non-tradable shares (2012 – 2018)

Block trading at discounted prices
Block trading at premium prices
Block trading at market prices
Market value of unlocked non-tradable shares

Source: eastmoney.com.

To promote more balanced trading behaviour and a transparent market, the China Securities Regulatory Commission (CSRC) issued guidelines in July 2015, January 2016 and May 2017[37] for substantial shareholders and senior management of listed companies, as well as private placement funds[38] on selling their stakes. According to the latest guidelines in 2017, block sales of shares by the aforesaid stakeholders are subject to certain restrictions. For those who hold more than 5% of a listed company's stakes or have acquired shares through private placement, the sales of non-tradable shares shall not exceed 50% of their total holdings in a 12-month period after unlocking. The shares transferred through block trading shall not exceed 2% of the company's total shares in 90 days, and the

---

37 These are: *CSRC Announcement [2015] No. 18 — Substantial Shareholders, Directors, Supervisors and Senior Executives of Listed Companies Shall Not Reduce Shareholding Within Six Months* (《中證監公告 [2015]18 號 —— 上市公司大股東及董事、監事、高級管理人員 6 個月內不得減持》), *CSRC Announcement [2016] No. 1 — Certain Provisions on Shareholding Reduction by Substantial Shareholders, Directors, Supervisors and Senior Executives of Listed Companies* (《中證監公告 [2016]1 號 —— 上市公司大股東、董監高減持股份的若干規定》), and *CSRC Announcement [2017] No. 9 — Certain Provisions on the Reduction of Shares Held in a Listed Company by Substantial Shareholders, Directors, Supervisors and Senior Executives of Listed Companies* (《中證監公告 [2017]9 號 —— 上市公司大股東、董監高減持股份的若干規定》), issued by the CSRC.

38 Private placement funds are equity mutual funds dedicated to acquire shares by private placement of shares.

buyers are subject to a lock-in period of six months[39]. Owing to the various restrictions, the proportion of block trades effected at discounted prices (a proxy of sell-initiated block trades) decreased gradually since 2016 and reached the lowest of 74% in 2018 since 2012.

In addition, certain equity mutual funds have relied on block trading mechanism to meet their redemption requirements. These include private placement funds, which were introduced in February 2015[40]. The shares in private placement are usually sold at a discount to market prices by the listed companies with a lock-in period of one year or three years[41]. Upon the expiry of the lock-in period, the fund can sell its shares held through the open market or block trading[42]. As of end-2018, there were 53 private placement funds with an AUM of RMB 28 billion (see Figure 5).

Figure 5. The AUM and number of private placement funds in Mainland China at quarter-ends (2015Q2 – 2018Q4)

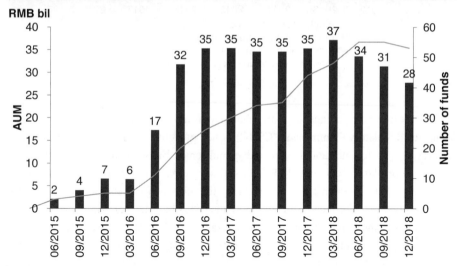

Source: Wind.

---

39 *SSE's Detailed Implementation Rules on the Reduction of Shares Held in a Listed Company by Substantial Shareholders, Directors, Supervisors and Senior Executives of Listed Companies, issued by the SSE, 27 May 2017; SZSE's Detailed Implementation Rules on the Reduction of Shares Held in a Listed Company by Substantial Shareholders, Directors, Supervisors and Senior Executives of Listed Companies,* issued by the SZSE, 27 May 2017.

40 Source: "UBS SDIC RUILI transform from private placement funds into LOF, stay or go?"(〈轉型 LOF，國投瑞利告別定增，該走還是留？〉), *National Business Daily*(《每日經濟新聞》), 8 August 2016.

41 Source: "Analysis on investment method and source of return for the private placement market" (〈定增市場的投資方法及收益來源解析〉), *Gelonghui*, 30 November 2017.

42 Source: "New trends of private placement funds: Block trading becomes new 'stairs' for selling stakes" (〈定增基金真會玩：大宗交易成減持新「梯子」〉), *The Economic Observer*(《經濟觀察報》), 16 September 2017.

2.3   Potential demand for block trading of A shares via Northbound Stock Connect

The investor base of A shares has been expanded through the Northbound Stock Connect scheme to include a more extensive global institutional investor (and retail investor) population than QFIIs and RQFIIs. The scheme has become the most preferred channel to trade A shares in 2018 (see Figure 7). These investors may potentially have a high demand for block trading of A shares.

Figure 6.   Year-end total foreign investments in Mainland onshore stocks (2013 – 2018)

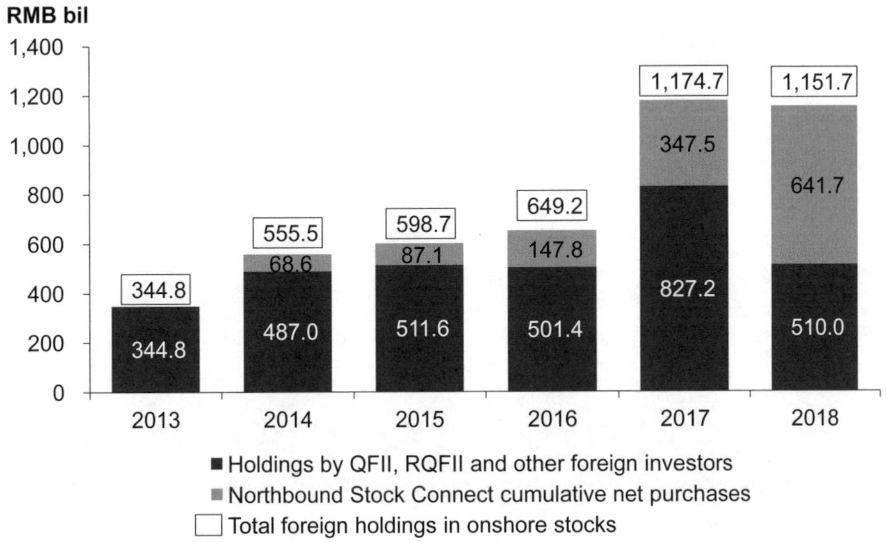

Note: The amount of holdings by QFIIs, RQFIIs and others is estimated as the difference between the total foreign holdings in onshore stocks and the cumulative net purchases via Northbound Stock Connect.
Source: HKEX for the cumulative net purchases via Northbound Stock Connect; Wind for the total foreign holdings in onshore stocks.

Statistics showed that more and more institutions participate in Northbound Stock Connect — the number of Special Segregated Accounts (SPSAs)[43] increased from about

---

43 An SPSA is used by an institution (under an investor identification number) to fulfill the pre-trade checking requirement for trading through Northbound Stock Connect. One SPSA can execute orders through up to 20 Exchange Participants (EPs). The EP should have prior arrangements to confirm that the fund manager client has the authority to aggregate SPSA orders across SPSAs and decide the appropriation allocation and it complies with applicable laws, rules and regulations without any misappropriation of client assets. (See *Frequently Asked Questions* (question 1.48.1) issued by HKEX on its website, 3 September 2018.)

1,700 in June 2017 to 6,363 in September 2018 [44] and further to over 8,200 in July 2019[45]. This is partly attributable to the daily quota of net purchase through Northbound Stock Connect being quadrupled from RMB 13 billion to RMB 52 billion (the same quota applies to Shanghai Connect and Shenzhen Connect) since May 2018 ahead of the inclusion of A shares in major global indices. Currently block trading arrangement through Northbound Stock Connect is only available to certain cases of non-trade transfers[46]. However, if an institution needs to execute a large Northbound trading order for other purposes, it can only break it down into smaller orders to be traded on the exchange. This is evidenced by the average transaction size of Northbound trading (see Figure 7) being comparable to that of Mainland's on-exchange electronic order book at around RMB 20,000 (see Table 2).

---

44 See "Consultation on further weight increase of China A shares in the MSCI indexes", MSCI's website on index consultations, December 2018.

45 Source: HKEX.

46 These include stock borrowing and lending, rectification of erroneous trades, post-trade allocation of shares from aggregated orders to funds or sub-funds in different accounts by fund managers and a few other cases. See "Information Book for Market Participants" for Stock Connect (Section 3.37 Non-trade Transfer), HKEX website, 15 February 2019.

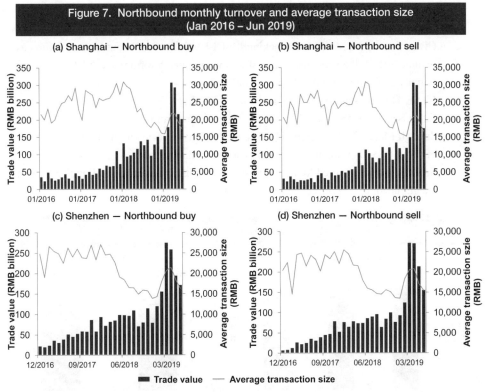

Figure 7. Northbound monthly turnover and average transaction size (Jan 2016 – Jun 2019)

Note: Shenzhen Connect was launched in December 2016.

Source: HKEX.

The turnover of A shares through Northbound Stock Connect is expected to grow further along with the increasing offshore passive investment in A shares, which is boosted by the inclusion of A shares into global equity indices. With the expected further expansion of global index baskets covering A shares and the growth in associated ETFs, the demand for block trading of A shares by global investors will increase in the events of index rebalancing or creation/redemption of fund units:

- **MSCI** started to include A shares into its global index series, including Emerging Markets Index (EM Index), since June 2018 with an initial partial inclusion factor of 5% (or about 0.8% by weighting as of May 2018) in two phases through August 2018. The inclusion factor will gradually increase to 100% (or about 16.2% by weighting)[47].

---

[47] Source: Chia, C. P., "The world comes to China", a blog post on MSCI's website, 23 May 2018

In February 2019, MSCI decided to increase the inclusion factor from 5% to 20% for large-cap and mid-cap stocks in phases throughout 2019 and 2020[48] (Figure 8a). Given that the AUM of funds tracking MSCI EM Index was about US$1.8 trillion as of June 2018[49], the investment in A shares would increase from about US$20 billion initially to about US$300 billion after full inclusion.

- **FTSE Russell** started Phase 1 inclusion of A shares into the FTSE Global Equity Index Series (GEIS), including the Emerging Index, from June 2019 with 3 tranches through March 2020 (Figure 8b). The inclusion is expected to attract initial inflows of about US$10 billion into A shares by index tracking funds[50].
- **S&P Dow Jones** started to include A shares into its global benchmark indices with a weight factor of 25% effective from September 2019[51].

Figure 8. Projected weightings of Chinese stocks in MSCI EM Index and FTSE Emerging Index after inclusion of A shares

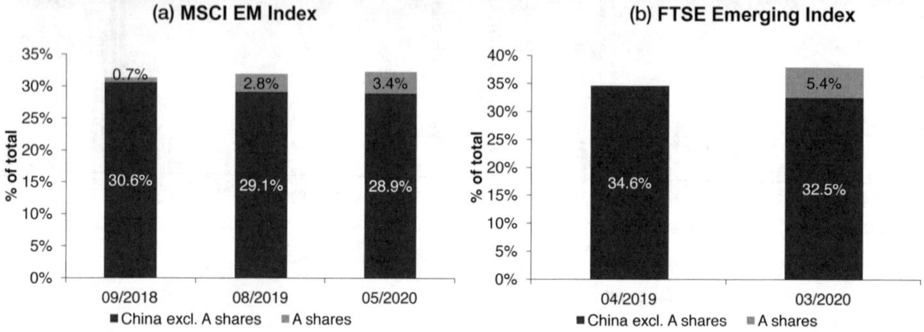

(a) MSCI EM Index

(b) FTSE Emerging Index

Note: For MSCI EM Index, the weightings were calculated based on share prices as of 22 January 2019. For FTSE Emerging (large- and mid-cap) Index, the weightings after 3 tranches of Phase 1 in March 2020 were calculated based on share prices as of 21 June 2019.

Source: "Conclusion of consultation on further weight increase of China A shares in the MSCI indexes", Index Consultations webpage on MSCI's website, February 2019; "China A shares inclusion — Seven key points", a blog post on FTSE Russell's website, 24 June 2019.

The increased popularity of structured products would be another driver of the demand

---

48 Source: "Conclusion of consultation on further weight increase of China A shares in the MSCI indexes", Index Consultations webpage on MSCI's website, February 2019.

49 Source: "MSCI — A leader in equity indexes", MSCI's website, 2019.

50 Source: "FTSE Russell promotes China A shares to emerging market status", FTSE Russell Press Release, 26 September 2018.

51 Source: "S&P Dow Jones Indices' 2018 country classification consultation results", index announcement of S&P Dow Jones Indices, 5 December 2018.

for Northbound block trading. A study[52] reported a gradual growth of the sales of retail structured products in Asia, despite tighter regulations imposed after the global financial crisis during 2007-2008. The study noted that the outstanding sales volume for retail structured products offered in major Asian markets (including Mainland China, Hong Kong, Japan, South Korea, and Singapore) exceeded US$750 billion as of end-2015, representing a compound annual growth rate of 4% since 2012. An expected growth of offshore equity structured products and derivatives with A shares as the underlying assets will drive up the demand for block trading in order to enhance the settlement efficiency of these products (see Section 1.2).

In a nutshell, the potential demand for block trading of A shares is expected to grow with a more extensive A-share institutional investor base in Hong Kong and their increased asset allocation to A shares.

3   Block trading in Hong Kong

3.1   Manual trade mechanism in Hong Kong

The Hong Kong market provides a manual trade mechanism which accommodates block trades of Hong Kong stocks. A manual trade is a trade concluded by private negotiation between Exchange Participants (EPs) but not by public auction and order matching on the Orion Trading Platform — Securities Market (OTP-C) of the Stock Exchange of Hong Kong (SEHK), the securities market operating subsidiary of HKEX. Manual trades are subject to requirements of trade reporting to the SEHK. There are two types of manual trade — non-direct business transactions (trade type "M") and direct business transactions (trade type "X", sometimes called "crossing transactions"). A Type M transaction refers to a trade executed by two different EPs. A Type X transaction refers to a trade between different investors executed by the same EP. Direct business transactions also cover transactions in dark pools or Alternative Liquidity Pools (ALPs) operated in Hong Kong. Block trades in Hong Kong can be executed as either type of manual trade.

Manual trades in Hong Kong are flexible in terms of order size and trading hours but

---

52 Wu, A. and C. Pitts (2017) "Asian structured products", *Research Foundation Briefs*, Vol. 3, Issue 6, August 2017.

subject to limits on order price for crossing transactions. There is no explicit threshold of transaction size for manual trades (e.g. applicable for a one-share trade). However, a large order with a size above 3,000 board lots[53] during pre-opening and Continuous Trading Session (CTS) can only be executed as a manual trade because it exceeds the maximum order size for automatic order matching on OTP-C. Manual trades can be conducted any time, including the pre-opening session, CTS, Closing Auction Session (CAS) and after trading hours (after-hours), which are subject to exchange rules of reporting time for manual trades (see the paragraph below). Manual trades that are crossing transactions during CTS are subject to upper and lower price limits of 24 tick sizes. In other words, the price must be within the range of the lowest of 24 spreads below the previous closing price, the lowest bid and the lowest ask price up to the time of the transaction on the day and the highest of 24 spreads above the previous closing price, the highest bid and the highest ask price up to the time of the transaction on the day[54].

For trade transparency, timely trade reporting is required for manual trades in Hong Kong. For manual trades not executed on dark pools, EPs are required to report the trade within 15 minutes of trade execution on the trading day (T day) and not later than 9:45 a.m. on the following trading day (T+1 day) for after-hours manual trades. For manual trades executed on dark pools[55], EPs are required to report them within 1 minute of trade execution. For a direct manual trade (i.e. a crossing transaction), including transactions on dark pools, the EP is required to report and review the details of the manual trade. For a non-direct manual trade, the selling EP is required to report it while the buying EP is required to review the reported details of the trade within the trading session of the reporting day. Every EP is required to perform day-end review of both sales and purchases journals including manual trades.

The prices of manual trades are not be used for determining a stock's market price for dissemination and index calculation purposes while the turnover of manual trades is included in calculating the market turnover. For constituents of Hang Seng Composite LargeCap, MidCap and SmallCap indices and H shares which have corresponding A shares listed on a Mainland exchange, the closing prices are determined through the auction

---

53 A board lot is the fixed number of shares as the trading unit for a stock, which may differ across stocks.
54 See SEHK's *Rules of the Exchange*, Chapter 5 Trading.
55 According to a report by the SFC (*Report on the Thematic Review of Alternative Liquidity Pools in Hong Kong*, 9 April 2018), there are 16 dark pools operated in Hong Kong, which are mainly brokers' or banks' internal crossing engines. The buy and sell orders received by the same EP in the dark pool are matched internally (internalisation) before sending to the exchange market. Based on HKEX's data, the turnover of dark pools operated by EPs was about HK$15-25 billion (or 1.0%-1.7% of total market turnover) during October 2016 to September 2017. The regulators are aware of the potential investor protection issues relating to dark pools and therefore have banned the access of retail investors to dark pools since December 2015.

scheme during CAS. The closing prices of other stocks[56] are determined by taking the median of 5 snapshots on the nominal prices at 15-second interval starting from 3:59:00 p.m of the CTS until they are added into CAS effective from 8 October 2019.

The different trading arrangements for block trades on the SSE, the SZSE and HKEX are presented in Appendix 2.

3.2   Block trading activities of Hong Kong stocks

The WFE statistics on negotiated trades are used as the proxy of block trades on HKEX. The statistics reflect manual trade[57] in which block trades accounted for the majority share. The turnover value of negotiated trades rose to a record high in 2018 when the Hang Seng Index (HSI) reached a record high in the same year (see Figure 9). The turnover of negotiated trades and the performance of the HSI appeared to have similar trends. This reflects, to a certain extent, that block trading in Hong Kong tends to be more active during times of market rally and less active during a market downturn. The phenomenon is different from the prevalence of sell-initiated block trading in the Mainland.

---

56 Effective from 8 October 2019, the list of securities covered by CAS will be expanded to include all equities, including depositary receipts, investment companies, preference shares and stapled securities, as well as funds (including ETFs and real estate investment trusts (REITs)). Source: "HKEX to expand closing auction session to all equities and funds", HKEX News Release, 5 July 2019.

57 Given no volume threshold, manual trades in Hong Kong cover not only placement and upstairs block trades, but also some trades with much smaller order sizes — odd lots (transacted quantities of less than one board lot) and special lots (transacted quantities not equal to a whole number of board lots). HKEX's data showed that placement of shares (a major kind of block trades) accounted for about 4% of the total market turnover, whereas all negotiated trades accounted for about 6% of the total market turnover.

Figure 9. Annual turnover of negotiated trades in Hong Kong and daily closings of Hang Sang Index (2007 – 2018)

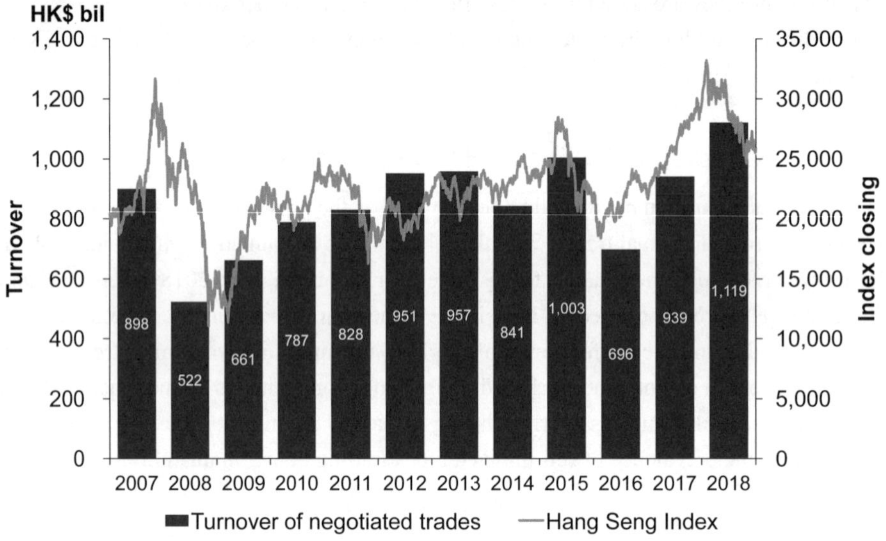

Source: WFE for turnover of negotiated trades; Bloomberg for index data.

The demand for block trading of Hong Kong stocks for purposes discussed in Section 1 above is supported by the growth of asset and wealth management businesses in Hong Kong, which reached HK$23,955 billion as of end-2018[58]. These include asset management and fund advisory businesses, which reached HK$16,447 billion in 2018. These institutional investors include active and passive asset managers who may trade blocks of shares for purposes as mentioned in Section 1.2. Stocks listed in Hong Kong involved in block trading would include H shares, the liquidity of which would be boosted by the "H share full circulation" pilot scheme. The CSRC announced the launch of a pilot scheme in December 2017[59] to implement "full circulation" of H shares through the conversion of non-tradable shares held by Mainland shareholders into H-shares for selected H-share companies. Subsequently, three Chinese companies were approved for H-share full circulation during 2018. The non-tradable shares held by major shareholders and senior management of H-share companies upon H-share full circulation would become tradable

---

58 Source: The SFC Asset Management Survey 2018.

59 "CSRC deepen the reform of offshore listing regime"(〈中國證監會深化境外上市制度改革〉), CSRC News Release, 29 December 2017.

in Hong Kong. Given the smooth operation of the pilot scheme, the CSRC and relevant authorities are exploring to expand the scope of H-share full circulation[60]. This may lead to the rising demand for the buying and selling of large blocks of H shares originally held by the Mainland major shareholders.

2.3   Potential demand for block trading via Southbound Stock Connect

Continual growth in cross-border asset management covering Hong Kong-listed stocks is expected upon increased Mainland-Hong Kong connectivity. Similar to Northbound Stock Connect, the daily quota of net purchases is expanded four times from RMB 10.5 billion to RMB 42 billion in May 2018 for Southbound Stock Connect (the same quota applies to Shanghai Connect and Shenzhen Connect). The cumulative Southbound net purchases of Hong Kong stocks have been growing — the amount reached HK$657.1 billion for Shanghai Connect and HK$223.1 billion for Shenzhen Connect as of end-June 2019 (see Figure 10). During the first half of 2019, the average daily turnover (ADT) (buy and sell included) through Southbound Stock Connect was about HK$7.1 billion and HK$4.0 billion respectively for Shanghai Connect and Shenzhen Connect. The turnover accounted for about 5%-6% of total market turnover in Hong Kong. (See Figure 11.)

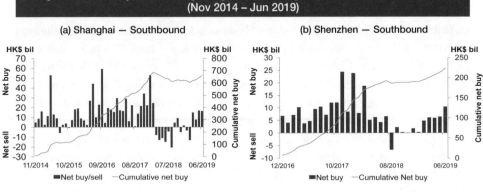

Figure 10. Net buy/sell of Hong Kong stocks through Southbound Stock Connect (Nov 2014 – Jun 2019)

Source: HKEX.

---

60 Source: "CSRC spokesman's press Q&A on further expansion of opening capital markets" (〈證監會有關負責人就進一步擴大資本市場對外開放答記者問〉), the CSRC's website, 14 June 2019.

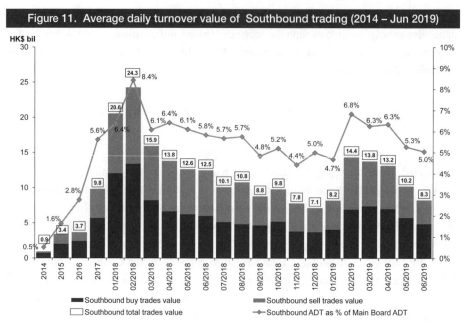

Figure 11. Average daily turnover value of Southbound trading (2014 – Jun 2019)

Note: Data starts from 17 Nov 2014 when Shanghai Connect was launched. Shenzhen Connect is included since
         launch on 5 Dec 2016.
Source: HKEX.

The growth in Southbound trading activities is partly attributable to the growing participation of Mainland mutual funds. These include Qualified Domestic Institutional Investors (QDII) funds and funds that invest in Hong Kong stocks through Southbound Stock Connect (referred to as "Mainland Connect Funds")[61]. As of end-2018, the number of Mainland Connect Funds reached a record high of 375, with total net assets of RMB 219 billion in Hong Kong stocks (see Figure 12). The active management and creation/ redemption of fund units may involve large-sized transactions within a short time frame, which increase the demand for block trading (see Section 1.2). As block trading is currently not available under Southbound Stock Connect, these mutual funds may execute large orders by algorithmic trading of small trades. This is evidenced by the average transaction size through Southbound trading (see Figures 13) being comparable to that on the on-exchange electronic order book in Hong Kong at around HK$60,000 (see Table 2 in Section 1.4).

---

61 See "Earnings reports of SH-HK-SZ Connect Funds: Limited development of QDII funds and the rise of SH-HK-SZ
    Connect Funds"(〈滬港深基金業績報告：QDII 基金發展受限　滬港深基金崛起〉), jrj.com.cn, 19 January 2018.

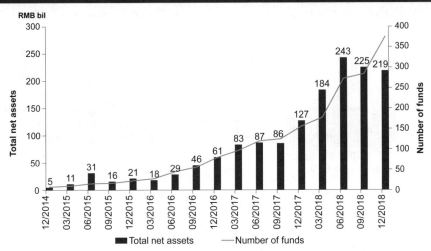

Figure 12. Mainland Connect Funds – Total net assets and number of funds at quarter-ends (2014Q4 – 2018Q4)

Note: Only the funds reported quarterly total net assets are included. Some funds have more than one tranche.

Source: Wind.

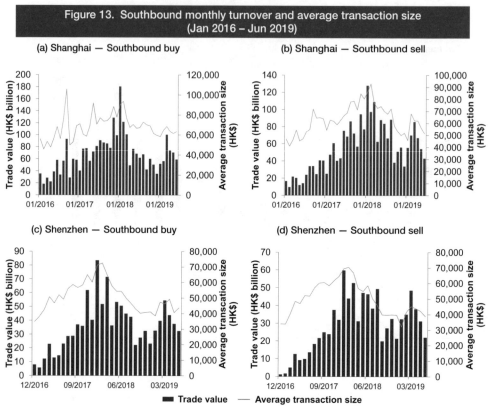

Figure 13. Southbound monthly turnover and average transaction size
(Jan 2016 – Jun 2019)

Note: Shenzhen Connect was launched in December 2016.
Source: HKEX.

4   Conclusion

Block trading mechanism of equities is important to minimise the price impact of large-sized transactions and avoid unwanted signalling to the market. It provides an efficient way to buy or sell large blocks of shares in a single transaction with more certainty in timing and transaction price than trading on the auction market. The growth of global asset management has been driving up the demand for block trading. Such demand arises from index rebalancing and the creations and redemptions of fund units, as well as the development of structured products and equity derivatives, which require block trading for efficient settlement of securities.

In both Mainland China and Hong Kong, the stock exchanges provide special mechanisms to facilitate block trading or the reporting of negotiated trades (which include block trades). In the Mainland, block trading has been dominated by sell-initiated trades, which are mainly the placement of shares by major shareholders after the lock-up period of their substantial shareholdings. This is in contrary to block trading driven more by asset management activities as in developed markets like Hong Kong. Nevertheless, market-opening factors, including the enhanced access by global investors to the A-share market through Northbound Stock Connect, the expanded daily quota of Northbound net purchases, as well as the expected increase in global passive investment in A shares as a result of the inclusion of A shares into global indices, would all contribute to the rising demand for block trading in the Mainland for asset management purposes, like that in the global markets.

The Hong Kong market, in parallel, also faces the rising demand for block trading. Driving factors include the growth of asset management businesses in Hong Kong, the growing number and size of Mainland funds investing in Hong Kong stocks and the potential expansion of H-share full circulation scheme that may lead to the buying and selling of large blocks of H shares originally held by the Mainland major shareholders of H-share companies.

Given the growing potential demand for cross-border block trading and that Mainland and Hong Kong cross-border trading is now conducted largely via the Stock Connect schemes, it is worth considering the enhancement of Mainland-Hong Kong market connectivity with block trading facilities under appropriate arrangements and a controllable environment to meet investors' demand. Cross-border market liquidity can thereby be further enhanced and investors can benefit from better execution.

Appendix 1

Recent development of alternative trading venues in Europe

European stock exchanges have launched block trading facilities to tap into dark-pool liquidity. The growth of activities on these facilities is prominent because of regulatory developments — the second version of the Markets in Financial Instruments Directive (MiFID II). In Europe, a dark pool can be exempted for pre-trade transparency by meeting one of the four waivers[62], including on negotiated trade, reference price, order management facility and large-in-scale (Table A1).

Table A1. MiFID pre-trade transparency waivers for trading venues	
Type of waiver	Description
Negotiated Trade Waiver (NTW)	This allows trades to be negotiated off order book between two parties provided that it is within the VWAP spread of the trade (i.e. the low and high VWAP of a trade), or the price of the trade is subject to conditions other than the current market price.
Reference Price Waiver (RPW)	This allows the matching of orders at an external reference source at the midpoint of the BID and ASK of, say, the primary venue.
Order Management Facility Waiver (OMF)	This allows a venue to hold orders in its order management system pending execution and disclosure (e.g. iceberg orders or stop-loss, etc.).
Large-in-Scale (LIS)	This waiver is to allow for large orders at any price or mechanism to forgo the pre-trade transparency requirement. The trade has to be over a certain level (which varies per size up to 500,000 Euros for large caps).

Source: European Securities and Markets Authority (ESMA).

Since 2018, MiFID II has banned the use of broker crossing networks (which accounted for 4%-5% of European equities trading)[63]. Double volume caps (DVC)[64] are introduced to limit the dark trading under the reference price waiver (RPW) and the negotiated transaction waiver (NTW) — 4% of the total volume of trading in an instrument per trading venue and 8% of the total in that instrument across all Europe Union (EU) trading venues. Once these thresholds are passed, trading venues are required to suspend the dark trading of the equity instrument for a six-month period. The waiver on large-in-scale (LIS) has

---

62 Source: "Waivers from Pre-trade Transparency", European Securities and Markets Authority (ESMA), 20 June 2016.
63 Source: Puaar, A., J. Schack and A. Kemmsies, "Let there be light: Rosenblatt's Monthly Dark Liquidity Tracker — European Edition", Rosenblatt Securities' market structure analysis, 21 September 2018.
64 See Double volume cap mechanism webpage on ESMA's website, viewed on 2 August 2019.

become more important for dark pools and the thresholds (minimum size) of large-in-scale orders increase with increasing average daily turnover of the securities.

While some banks set up systematic internalisers (SI) to fill the gap[65], European stock exchanges have launched their dark-pool facilities for the large-in-scale trades. These include Turquoise Plato (co-operation between the London Stock Exchange and Plato Partnership) for European and US stocks launched in September 2016 and Euronext Block (partnership between Euronext and AX Trading) for European stocks launched in July 2017. Block trading on these facilities recorded daily record highs in recent two years — €653 million on 31 January 2019 on Turquoise Plato Block Discovery (a platform of Turquoise Plato for matching block trade indications or conditional orders)[66] and €24 million on 1 August 2018 on Euronext Block[67].

---

65 See Hadfield, W., "These are the two words traders will be paying attention when Europe's New Rules kick in", *Bloomberg*, 18 December 2017.

66 Source: *Turquoise Plato Block Discovery* webpage on LSE Group's website, viewed on 31 July 2019.

67 Source: Taylor, V., "Euronext Block continues to grow: Record trades and planned improvements", a post on Linkedin's website, 4 September 2018.

Appendix 2

Block trading arrangements in the Mainland and Hong Kong

Feature	Shanghai Stock Exchange	Shenzhen Stock Exchange	HKEX
Execution method	Dedicated platform for trade intention, private negotiations and fixed-price auctions		Private negotiations (No dedicated platform)
Thresholds on quantity	300,000 shares or RMB 2 million		No
Trade hours	Intent order: 09:30-11:30; 13:00-15:30 Execution trade: 15:00-15:30 Fixed-price order: 15:00-15:30	Negotiated trade (Intent order): 09:15-11:30; 13:00-15:30 Negotiated trade confirmation (Execution trade): 15:00-15:30 Fixed-price order: 15:05-15:30	Any time
Limits on transaction price	Within ±10% of previous close (For ST and *ST stocks: within ±5% of previous close or RMB 0.01 if previous close of the A share is below RMB 0.1)	Within ±10% of previous close (For ST and *ST stocks: within ±5% of previous close)	Within 24 tick sizes above the day-high and below the day-low up to the time of transaction during CTS of the trading day
Impact on index calculation and settlement of derivatives	Not included in the calculation of prices for stock indices or settlement prices of equity derivatives but included in the securities' turnover		
Reporting requirement	For negotiated block trades, both buyers and sellers are required to report the trade. The reporting is done through the submissions according to the trading hours above.		For negotiated trades, only sellers (or the EP for a crossing transaction) are required to report the trade Must be reported within 15 minutes on T day (or not later than 9:45 on T+1 day for after-hours trades) for trades concluded in non-dark pools and within 1 minute on T day for trades concluded in dark pools

Note: Stocks with "ST" or "ST*" label are stocks under "special treatment", which are a kind of risk alert label for the attention of investors.

Source: HKEX's website and Chapter 5 of *Rules of the SEHK*; *Trading Rules of the SSE* (《上海證券交易所交易規則》), issued by the SSE, 6 August 2018; *Trading Rules of the SZSE* (《深圳證券交易所交易規則》), issued by the SZSE, 30 September 2016.

Chapter 14

Hong Kong's ETF market as a door to global investment

Chief China Economist's Office

Hong Kong Exchanges and Clearing Limited

Summary

Exchange traded funds (ETFs) have become increasingly popular in global markets. These products are more cost-effective, convenient and transparent than traditional mutual funds, leading to rapid growth in both the global number of ETFs and their assets under management (AUM). The growth is accompanied by innovation and diversification of products as well as increased use of fixed-income ETFs. While the growth in the US and European ETF markets continued, the ETF growth momentum in Asia Pacific is remarkable. Hong Kong has been one of the leading ETF markets in Asia Pacific while the Mainland ETF market is growing significantly in recent years to be the second largest in Asia Pacific upon the Mainland investors' increasing acceptance to passive investments.

In respect of passive investments, the Mainland investors can currently invest only in ETFs listed in the Mainland. Equity ETFs on a large variety of Mainland indices have dominated the market. Money market ETFs have become attractive owing to better yields than deposit rates. In addition to ETFs, listed open-ended funds (LOFs), which are traded like ETFs, provide convenient investment choices of active returns. However, ETFs and LOFs on global equity indices have been subject to time delay to receive cash in case of redemptions in the primary market. Besides, Mainland-listed funds are traded in the Renminbi (RMB) only. These hint at the potential demand of the Mainland investors for access to the Hong Kong ETF market for global asset allocation and international currency exposure.

The Hong Kong ETF market is a door to global investment. It offers products on diversified asset classes from global markets and a well-established institutional investor base to support market liquidity. The Hong Kong ETF market offers a high degree of global equity market exposures, including Hong Kong stocks, Mainland A shares, Asia-Pacific and other overseas equities. It also covers asset classes of fixed income and currency as well as commodities in the global markets. In addition, a number of ETFs on global indices are traded in multiple currencies, including the Hong Kong dollar (HKD), the RMB and/or the US dollar (USD), which can fit the diverse needs of global investors. ETF investors in Hong Kong can also enjoy the high market depth, competitive tax savings as well as an ecosystem of associated risk management tools. The Hong Kong ETF market is also attractive to issuers who can enjoy a relatively market-friendly regulatory environment, a broad base of global institutional investors and an effective market making mechanism that ensures secondary market liquidity and facilitates arbitrage activities in the primary market. Besides, ETF issuers in Hong Kong can benefit from the geographical advantage of "Trade Asia in Asia" for hedging risks in the underlying markets for their Asian products.

Hong Kong is well-positioned to be Asia's ETF marketplace for the issuance and trading of ETFs. The Hong Kong ETF market can not only meet the needs of global investors, but also the demand by the Mainland investors potentially. Given the relatively low institutional participation for supporting ETF primary market activities and the limited ETF coverage of global assets in the Mainland market, the Hong Kong ETF market with diverse products and investor base would be attractive to the Mainland investors who have shown growing interests in ETF investments. Conversely, the large variety of ETFs on Mainland assets in the Mainland market would also be attractive to global investors in Hong Kong. The potential mutual access of the Mainland and Hong Kong ETF markets[1] would facilitate a more balanced mix of investors in the ETF primary and secondary markets on both sides, accompanied by a widened spectrum of ETF products. Further development in Mainland-Hong Kong market connectivity in respect of the ETF segment will therefore be conducive to the mutual growth of the two markets.

---

1   Subject to regulatory approval.

1   ETFs have become attractive to global investors

An exchange traded fund (ETF) is an open-ended fund[2] traded like a stock that tracks the performance of the underlying assets. The world's first ETF was launched in Canada in 1990, tracking a benchmark equity index. It was followed by the US in 1993[3], Japan in 1995, Hong Kong in 1999 and Europe in 2000. Subsequently, the global ETF market has experienced significant growth. An increasing number of ETFs are listed and traded in global markets and the coverage of underlying assets is extended from equities to other asset classes, including bonds, commodities and baskets of assets. In particular, the growth in Asia has accelerated in recent years.

Hong Kong has been one of the leading ETF markets in Asia. It was the first mover to issue and trade ETF in Asia Pacific (excluding Japan) back in 1999 when the Tracker Fund (tracking the Hang Seng Index (HSI)) was launched. Hong Kong was also the first in the world to offer ETFs on underlying assets in the Mainland or on Renminbi (RMB) assets. These include the first A-share ETF[4] launched in 2004 (a synthetic ETF[5] tracking the FTSE China A50 Index), the first physical A-share ETF (tracking an A-share index through RMB Qualified Foreign Institutional Investor (RQFII) scheme with dual counters traded in both Hong Kong dollars and RMB) launched in 2012[6], the first offshore RMB bond ETF in Asia launched in June 2013 and the first ETF tracking the onshore bond market in Mainland China launched in February 2014.

The Hong Kong ETF market has now developed into a platform of diversified asset

---

2   An open-ended fund is an investment fund that can issue and redeem units of the fund. Investors can buy the fund units directly from the fund manager.

3   In the US, the first ETF (traded on the New York Stock Exchange (NYSE)) was launched in 1993, tracking the S&P 500 Index. Prior to this, the American Stock Exchange and the Philadelphia Stock Exchange offered "Index Participation Shares" (like futures that have long and short positions for every share) in May 1989 that allowed investors to buy or sell an interest in an equity market without having to purchase the individual stocks. Owing to lawsuits by US regulators, the federal court ruled that the shares were actually a futures contract (i.e. a derivative product) and therefore could not trade on a stock exchange. (Source: Foucher, I. and K. Gray (2014) "Exchange-traded funds: Evolution of benefits, vulnerabilities and risks", *Bank of Canada's Financial System Review*, December 2014 issue, pp.37-46.)

4   See Dennison, T. (2018), "International investing and the importance of breaking the country box", *Invest Outside the Box: Understanding Different Asset Classes and Strategies*, Palgrave Macmillan, pp.228-278.

5   A synthetic ETF is an ETF replicating the performance of an underlying index using derivatives and swaps rather than using underlying physical securities as in the case of a physical ETF.

6   Source: "First RQFII A-share ETF to debut at HKEX", HKEX's News Releases, 16 July 2012.

classes from global markets traded in multiple currencies to meet the needs of global investors. The market can serve as a regional issuing and trading hub of ETFs to meet the growing market demand of global and Asia-Pacific investors, particularly from Mainland China.

ETFs have become increasingly popular as an asset class not only in Hong Kong, but also in global markets. Why are ETFs so popular in global and regional markets? How does product innovation contribute to this? These are discussed in the following subsections.

1.1   Why are ETFs more popular than mutual funds?

Prior to the launch of ETFs, mutual funds have been a popular choice of investment funds since their first launch in 1920s[7]. However, the growth in demand for ETFs has been higher than that in mutual funds in recent years, evidenced by the investment flows during market correction in late 2018. It was reported that the redemption of mutual funds reached US$56.2 billion during the week ended 19 December 2018 which was the highest since October 2008. On the contrary, ETFs received an inflow of US$25.2 billion during the same week[8].

Compared to traditional mutual funds, ETFs not only provide for creation and redemption activities, but also allow secondary trading activities on a stock exchange, contributing to higher liquidity. It is convenient to trade ETFs in the secondary market when the exchange market is open. Besides, market makers of ETFs are obliged to provide bid-ask quotes to ensure the secondary market liquidity. The number of ETF units available for trading in the secondary market is determined by the creation or redemption of ETF units in the primary market between participating dealers[9] and ETF issuers. The arbitrage activities between primary and secondary markets, based on the demand and supply, help mitigate the premium over or discount to the net asset value (NAV)[10] of an ETF.

In respect of costs to the issuer, the average expense ratio[11] of ETFs in the US was

---

7   Source: Divakaran, S., S. Buxton, St. Giles, M. Valentine, and S. Atamuratova (2015) "Mutual funds in developing markets: Addressing challenges to growth", World Bank's Working Paper, No. 100976.

8   Source: Stein, C., "Fund investors pull $56 billion in biggest exit since 2008", *Bloomberg News*, 27 December 2018.

9   Participating dealers (sometimes called authorised participants) are the institutional investors who place orders of creation or redemption of ETF units to the ETF issuer directly in the primary market. Some participating dealers are market makers.

10 NAV of an ETF (per unit) represents the market value of all the underlying securities held by the ETF (such as shares or bonds and cash) minus any liabilities (such as management fees and administrative expenses) and divided by the number of fund units outstanding. See "ETF pricing and valuations", Blackrock's website, viewed on 7 August 2019.

11 Expense ratio of a fund is calculated by dividing the total operating costs of the fund by the average value of fund assets under management during a specific period (usually a year).

0.23% in 2016, compared to 0.73% for index-tracking mutual funds and 1.45% for actively managed mutual funds[12]. Investors can therefore enjoy lower fund management fees for investments in ETFs than investments in mutual funds. Certain global investors also enjoy lower taxes for investments in ETFs and these include the withholding tax advantage of 15% of Irish Undertakings Collective Investment in Transferable Securities (UCITS) ETFs over mutual funds when investing in the US[13].

In respect of transparency, the information of an ETF is usually readily available on its own website that includes timely disclosure of the full portfolios (usually on a daily basis).

1.2 Overview of global and regional ETF markets

The number of ETFs in the global markets reached a record high of 6,310 as at the end of 2018 (5,707 at end-2017), with assets under management (AUM) of US$4.66 trillion (US$4.69 trillion at end-2017) (see Figure 1). The compound annual growth rate (CAGR) in AUM was about 21% during 2009 to 2018.

---

12 Source: "ETFs vs. mutual funds: Cost comparison", Fidelity's website, June 2017.
13 Source: "ETFs 2018: Opportunities and obstacles for active ETFs", Ernst and Young's website, December 2018.
    UCITS ETFs are ETFs domiciled in European markets and can be sold to global investors; these ETFs are subject to the UCITS regulation, which is a harmonised regulatory framework for European markets.

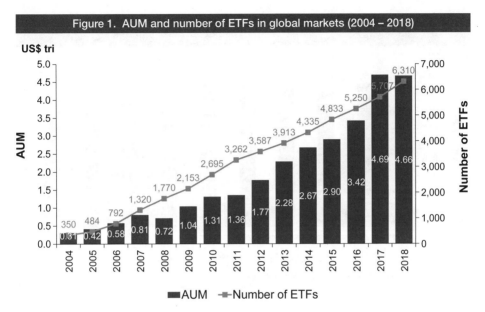

Figure 1. AUM and number of ETFs in global markets (2004 – 2018)

Source: "ETFGI reports ETFs and ETPs listed globally gather net inflows of US$76.24 billion during December 2018, 2nd Highest on record", ETFGI press release on *etfgi.com*, 9 January 2019.

The popularity of ETFs is attributable to the growth of passive investment that tracks the performance of indices. The share of passive investment was estimated to be about a quarter of open-ended funds in 2018 and was expected to increase to about 31% by end-2020 (see Figure 2). Another source noted that the percentage shares of active and passive equity funds in the US total equity funds was the closest to each other in a decade as of end-November 2018 — about 52% and 48% respectively[14]. A key driving factor is the long-run outperformance of passive investment because of their lower costs[15] — more than 80% of US equity actively-managed mutual funds underperform their respective benchmark indices in the 10-year horizon ending 2018[16].

---

14 Source: Stein, C. "Shift from active to passive approaches tipping point in 2019", *Bloomberg News*, 31 December 2018.

15 These costs include management fees, bid-ask spreads, administrative costs, commissions, market impacts and, where applicable, taxes.

16 Source: Rowley Jr. J. J., D. J. Walker, and C. Zhu, "The case for low-cost index-fund investing", *Vanguard Research*, April 2019.

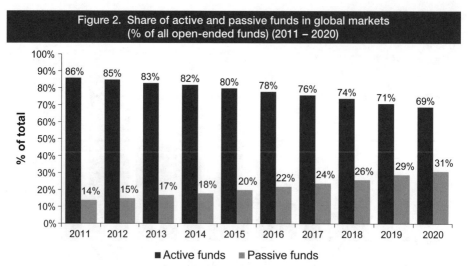

Figure 2. Share of active and passive funds in global markets (% of all open-ended funds) (2011 – 2020)

Note: The figures after 2016 are not actual statistics but estimates by Ernst and Young.
Source: "Global ETF Research 2017 — Reshaping around the investor", published on Ernst and Young's website, 2017.

Equities have dominated the underlying asset classes of ETFs. Among exchange traded products (ETPs)[17], which are mainly ETFs, equities accounted for about 77% of total AUM as of end-2018, amounting to US$3.7 trillion (see Figure 3). Of these, the exposure to US equities was the largest, amounting to US$2.1 trillion as of end-2018. For fixed-income ETPs, both the percentage share and the amount of AUM rose steadily in recent years, reaching about 18% of the total AUM of all ETPs and US$0.9 trillion respectively as of end-2018. Commodities and others accounted for about 4% of the total AUM of all ETPs, amounting about US$0.2 trillion.

---

17 ETPs comprise ETFs and other ETPs, including exchange traded vehicles (ETVs), exchange traded notes (ETNs) and certificates. Based on the data of ETFGI, ETFs accounted for 89%-98% of AUM of ETPs at year-ends between 2004 to 2018.

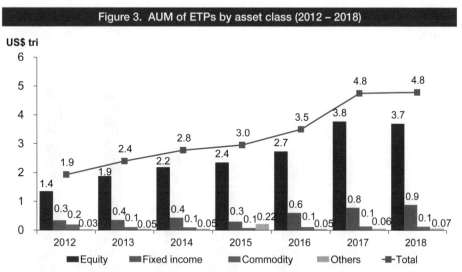

Figure 3. AUM of ETPs by asset class (2012 – 2018)

Note: "Others" include currency, money market and alternatives (e.g. volatility).

Source: "Blackrock Global ETP landscape: Industry highlights", December issues of 2012 to 2018, published on Blackrock's website.

Although the US ETF market remained the largest among major ETF markets (see Figure 4), the 6-year CAGR during 2012 to 2018 of ETP AUM was the highest in Asia-Pacific markets at 26.1%, compared to 16.5% for the US market and 13.4% for European markets.

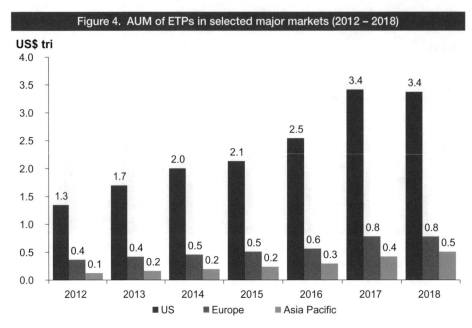

Figure 4. AUM of ETPs in selected major markets (2012 – 2018)

Source: "Blackrock Global ETP landscape: Industry highlights", December issues of 2012 to 2018, published Blackrock's website.

In terms of secondary market trading, the annual total turnover of ETFs on global exchanges has been increasing and reached a record high of US$24.1 trillion in 2018 (see Figure 5). The American exchanges dominated the turnover with a 6-year CAGR at 13.9% during 2012 to 2018, led by the US exchanges. The total turnover of ETFs in the Asia-Pacific region surpassed the combined ETF turnover in the regions of Europe, Africa and the Middle East since 2015. It reached a record high of US$1.7 trillion in 2018, with a 6-year CAGR of 31.2% during 2012 to 2018.

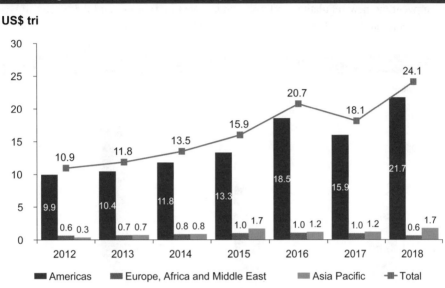

Figure 5. Total turnover of ETFs by regional exchanges (2012 – 2018)

Source: World Federation of Exchanges (WFE) statistics on its website.

In the fastest growing Asia-Pacific region, the Japanese ETF market is the largest in terms of AUM, accounting for 62% of the total in the region as of end-2017 while each of the Mainland and Hong Kong markets accounted for 9% of the regional total respectively[18]. Another source noted that the AUM of ETFs in the Japanese market reached US$309 billion as of end-June 2018 — the largest in Asia Pacific, followed by US$37 billion in the Mainland market, US$36 billion in the Korean market and US$34 billion in the Hong Kong market[19].

Despite the weak equity market performance in 2018, the AUM of Asian-Pacific equity ETPs continued to increase (see Figure 6). For fixed-income ETPs, the AUM increased by 1.5 times in 2018. For commodity ETPs, the AUM was relatively stable. The growth in AUM of other asset classes was attributed mostly to money-market ETFs. In addition to the trend of growing asset management, government policy incentives have also contributed to the growth of a number of Asia-Pacific ETF markets. Examples are the central bank's

---

18 Source: "APAC ETF: Finding the tipping point", presentation slides of Broadridge Insights at Investment Management Association of Singapore (IMAS) Lunchtime Talk Series — APAC ETF Opportunities: The Next Growth Frontier, 13 February 2018.
19 Source: "Are China's ETFs taking off?", Fund Selector Asia's website, 12 July 2018.

asset purchase programmes in Japan[20], the surge of bond ETF purchases by insurers to circumvent the new regulation on foreign investment in Taiwan[21] and the pursuit of cost-effective investments after the ban of "conflicted remuneration structures" in Australia[22].

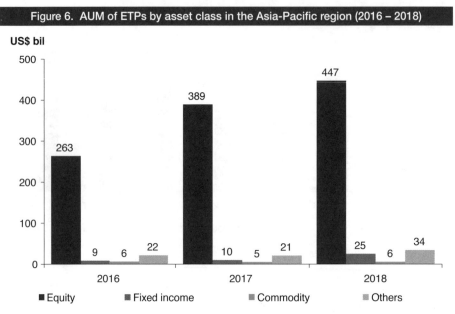

Figure 6. AUM of ETPs by asset class in the Asia-Pacific region (2016 – 2018)

US$ bil

Source: "Blackrock Global ETP landscape: Industry highlights", December issues of 2016 to 2018, published Blackrock's website.

1.3 Recent product developments to support the growth of ETF markets

Continued product innovation is a key to meet the evolving demand of institutional and retail investors. ETFs can fill the gaps of investment choices along the efficient frontier

---

20 ETF serves as a policy tool in the central bank's asset purchase programmes. See: Petrov, A. (State Street Global Advisors), "ETFs in Monetary Policy — Case Study: Bank of Japan", November 2017.

21 Taiwan's insurance companies purchases ETFs to circumvent the new regulation of a cap on foreign currency investment. See: Herrero, A. C. and G. Ng "Taiwan: Big data show positive sentiment for a new type of overseas assets — Bond ETFs", published on Natixis' website, 12 March 2019.

22 The Australian regulator has banned "conflicted remuneration structures", including commission, of financial advisors since July 2013, leading to higher incentives to promote cost-effective investments like ETFs. See: *Regulatory Guide 246 — Conflicted remuneration*, Australian Securities & Investments Commission (ASIC), March 2013.

for asset allocation. For example, the Hong Kong market offers ETFs with different risk-return profiles (see Figure 7). These investment choices would attract new investors while providing alternative tools to existing investors for diversification.

Figure 7. Scatter plot of 3-year return versus 3-year volatility for ETFs in Hong Kong (2016 – 2018)

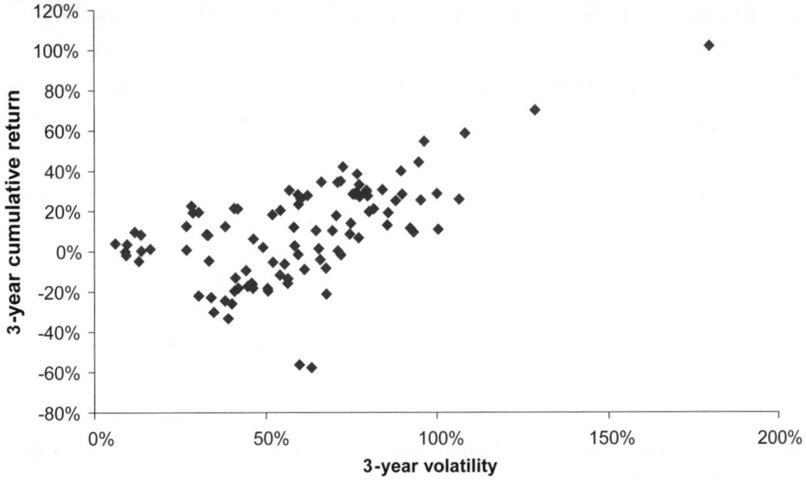

Note: The sample covers only the ETFs listed in Hong Kong during January 2016 to December 2018. The cumulative return is the price return during the period adjusted for dividend payout. The volatility is calculated as the difference between the highest price and the lowest price during the 3-year period divided by the closing price at the end of 2018.
Source: Wind.

A number of product innovations in the global ETF market are observed in recent years that have broadened global investment opportunities:

- **Smart beta ETF:** Tracks the performance of indices with adjustments in investment components based on certain macroeconomic or style factors. These factors include size, value, momentum and quality (a composite of fundamental indicators), etc. The number of newly issued smart beta ETFs kept increasing during 2013 to 2017, with their global AUM risen at a CAGR of 29% from US$280 billion in 2012 to US$999 billion in 2017[23]. Another source noted that there were 1,298 smart beta ETPs with a total AUM of US$618 billion as of end-2018[24].

---

23 Source: "Five-year trends and outlook for smart beta", FTSE Russell's website, 5 July 2018.
24 Source: "ETFGI reports Smart Beta ETFs and ETPs listed globally gather net inflows of US$12.42 billion during December 2018", ETFGI press release on etfgi.com, 30 January 2019.

- **Actively managed ETF (or active ETF):** Tracks an underlying benchmark index, but deviation from changes in sector allocations, timing of trades or from the underlying index itself is allowed, which can potentially generate higher returns. There were 773 active ETFs globally with an AUM reaching a record high of US$109.4 billion as of end-November 2018 and the AUM is expected to grow to US$217 billion by 2020[25].

- **The rise of artificial intelligence (AI):** The first ETF tracking a global index on robotics and AI stocks was launched in the US in 2013. The annualised cumulative 3-year return of the ETF was 16.7% as of 9 July 2019, which outperformed the 14.1% for the S&P 500 Total Return Index[26]. A number of major ETF managers issued similar ETFs since 2016[27]. The first AI-powered ETF tracks the performance of 30-70 stocks or real estate investment trusts (REITs) in the US market with high probabilities of outperformance over the next 12 months, based on the interpretation of market information by the AI system. The ETF was launched in the US market on 18 October 2017. The first-year performance of 11.8% outperformed the 8.1% for the S&P 500 Index during the same period and was reportedly better than those for 87% of all actively managed funds[28]. The same ETF issuer launched another ETF in the US in June 2018, which covers a portfolio of 80-250 stocks in global markets with the support of the same AI system.

- **Leveraged and inverse (L&I) product:** Delivers a multiple or the opposite of the return of the underlying index with the use of futures or swaps. The number of L&I products in global markets rose from 608 as of end-2013 to 901 as of end-June 2018, and the AUM of these products rose from US$53.2 billion to US$79.9 billion during the same period[29].

- **Environmental, Social and Governance (ESG) investing:** Supports the sustainable development of an economy but the definition or framework of ESG assessment varies across asset managers. A survey in 2018 found that about 99% of institutional investors in the Greater China region considered that ESG factors are important[30]. The number of ESG ETFs rose from 47 as of end-2013 to 208 as of end-2018 and

---

25 Source: "Active ETFs — The next step in Asia's ETF innovation", *Risk.net*, 12 February 2019.

26 The cumulative 3-year returns of the US-listed ETF on robotics and AI stocks (with ticker of ROBO) and S&P 500 total return index are available at Marketwatch.com. The figures were as of 9 July 2019.

27 Source: Lim, J., "Robotics and A.I. ETFs: Which will win the battle?", POEMS website of Phillip Securities Pte., Ltd., 20 July 2018.

28 Source: Ponczek, S., "Robot ETF leaves pros in dust, scoring wind on small-cap fliers", Bloomberg News, 19 October 2018.

29 Source: "ETFGI reports that assets invested in leveraged and inverse ETFs and ETPs listed globally reached $79.9 billion at the end of June 2018", ETFGI press release on etfgi.com, 31 July 2018.

30 Source: "Ready to grow: ETF opportunities in Greater China", Brown Brothers Harriman's website, April 2018.

their AUM rose from US$5 billion to US$23 billion during the same period[31].

- **Fixed-income ETF:** The AUM of fixed-income ETPs rose from US$780 billion as of end-2017 to US$882 billion as of end-2018[32]. Of this, the AUM of ETPs with exposures to US treasuries rose from US$55 billion to US$130 billion. The AUM of ETPs with exposures to emerging markets' debts (excluding corporate bonds) rose from US$51 billion to US$58 billion during the same period. This may be driven by the inclusion of Chinese bonds into global bond indices, which is facilitated by the onshore access through Bond Connect. Such growth in fixed-income ETFs may signal investors' increasing interests in debt instruments. A survey on global institutional investors showed that 60% of respondents considered bond ETFs as an alternative vehicle for fixed-income exposure[33]. One underlying reason for the increase in the interest in bond ETFs may be the higher capital efficiency than buying the underlying bonds[34].

- **Cross-listing of ETFs:** Cross-listing (or mutual recognition) of ETFs broadens the range of asset classes available to investors in a single market. In Europe, an ETF is often listed across multiple European exchanges to accommodate investors' preference to trade ETFs and settle the transactions on their home exchanges. There were 2,260 ETPs listed in Europe with 7,278 listings on 27 exchanges as of end-2017, compared with 2,116 ETPs listed in the US on 4 exchanges[35]. In Hong Kong, 24% of ETFs were cross-listed from other jurisdictions as of end-June 2017[36].

One of the key reasons for product developments of ETFs is to meet the diversification needs of institutional investors. According to a survey in 2017, 51% of global institutions (vs. 39% of Asian institutions) rated asset allocation (through core and tactical strategies) as the most important factor for ETF purchases[37]. Other uses of ETFs found by the survey include risk management or hedging, rebalancing/transitional allocation, interim beta for transition, cash equitisation for reducing idle cash, "liquidity sleeves" for providing liquidity and risk management (e.g. through L&I products).

---

31 Source: Wilson, D, "China, Hong Kong must look offshore for ESG ETFs", published on Fund Selector Asia's website, 31 January 2019.

32 Source: "Blackrock Global ETP landscape: Industry highlights", issues in December 2017 and December 2018, Blackrock's website.

33 Source: McCollum, A., "Institutions turn to ETFs for bond market liquidity", issued by Greenwich Associates, 18 September 2018.

34 See "Growing an institutional footprint in Asia's ETF market", *Risk.net*, 19 November 2018.

35 Source: Crigger, L., "New light shines on Europe ETF trading", *ETF.com*, 18 April 2018.

36 Source: Securities and Futures Commission (SFC) research paper, "Hong Kong market and topical issues in the ETF space", published on the SFC's website, 16 January 2018. (Referred to as the "SFC ETF Paper 2018".)

37 Source: "Global trends in institutional ETF trading", survey report commissioned by Jane Street, *Risk.net*, 5 October 2017. The survey covers responses from 210 institutions, which comprise 86 from the US, 79 from Europe and 45 from Asia.

2 The growth of the Mainland and Hong Kong ETF markets can be complementary to each other

The growth of Asia's ETF market is expected to continue. The parallel growth in the Mainland and Hong Kong ETF markets will be complementary to each other and is expected to accelerate on further increase in market connectivity. A study predicted that the percentage share in the total AUM of Asia-Pacific ETFs will rise from 9% in 2017 to 19% by 2025 for Mainland ETFs and from 9% to 18% for Hong Kong, but fell from 62% to 40% for Japanese ETFs in the same period[38]. The same study expected that about 75% of new net ETF inflows into Asia-Pacific markets will come from Japan, Mainland China and Hong Kong during 2018 to 2025.

2.1 Strong diversification demand from the Mainland investors

The Mainland ETF market has been growing rapidly since the first ETF was listed on the Shanghai Stock Exchange (SSE) in February 2005, which tracked the performance of the SSE 50 Index (an index for large-cap stocks listed on the SSE). Both the AUM and the number of ETFs in the Mainland reached record highs as of the end of the first quarter of 2019 (2019Q1) — 205 ETFs with a total AUM of RMB 583.1 billion (see Figure 8). These comprised 146 ETFs on the SSE and 59 ETFs on the Shenzhen Stock Exchange (SZSE).

---

38 Source: "APAC ETF: Finding the Tipping Point", presentation slides of Broadridge Insights at Investment Management Association of Singapore (IMAS) Lunchtime Talk Series — APAC ETF Opportunities: The Next Growth Frontier, 13 February 2018.

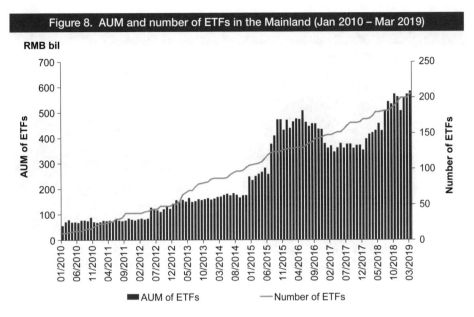

Figure 8. AUM and number of ETFs in the Mainland (Jan 2010 – Mar 2019)

Note: The AUM refers to the outstanding amount of ETFs at the China Securities Depository and Clearing Corporation (CSDC).

Source: Wind.

In terms of number of ETFs as of end-2019Q1, equity ETFs on Mainland indices dominated — 151 in number or 74% of the total, while the number of cross-border ETFs on foreign equity indices was only 13 (6% of the total) (see Figure 9a). For the ETFs on onshore stocks, the underlying indices cover different combinations of smart beta (e.g. size, value, growth or dividend), industry sector, ownership and markets (e.g. Shanghai, Shenzhen or cross-markets) for Mainland equity exposure. As for the 13 cross-border ETFs, nine of them track Hong Kong equity indices (including 4 ETFs on H shares), three of them track US equity indices and one tracks German stocks. There were 37 ETFs investing in the Mainland fixed-income market — 27 money market ETFs[39] (13% of the total) and 10 bond ETFs (5% of the total). Of the 4 commodity ETFs, all were gold ETFs (2% of the total), tracking gold contracts traded on the Shanghai Gold Exchange.

---

39 Money market ETFs invest in short-term fixed-income investments that include short-term government bonds, commercial papers, repurchase instruments (repos), bank acceptances and negotiable certificates of deposit.

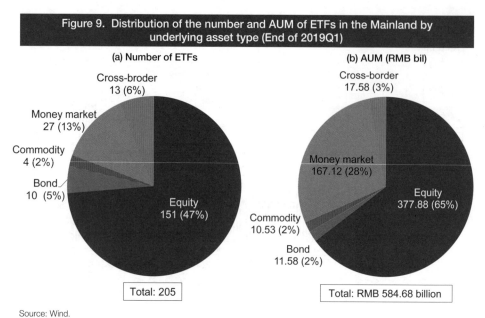

Figure 9. Distribution of the number and AUM of ETFs in the Mainland by underlying asset type (End of 2019Q1)

Source: Wind.

In terms of AUM as of end-2019Q1, money-market ETFs accounted for 28% of the total, second only to the 65% for equity ETFs (see Figure 9b). Money market ETFs are attractive to investors because they are very liquid and usually have higher yield from money-market funds than from 1-year fixed deposits at banks[40]. For money market ETFs in the Mainland, the proceeds from the sales in the secondary market can be used for purchasing securities on the same trading day (T+0) or can be withdrawn on the next trading day (T+1), while creation and redemption of money market ETF units in the primary market follow a T+0 settlement cycle on the SZSE or a T+2 settlement cycle on the SSE[41]. Money market ETFs therefore have become an alternative means of asset allocation for enhancing yields and diversification. Separately, there are on-exchange money market funds (non-ETFs) on the SSE available only for primary market transactions, which follow a T+1 settlement cycle for creation/redemption[42] (compared to T+0 or T+2 for money

---

40 See "What are the advantages of money-market ETFs?" (〈貨幣 ETF 基金的優勢有哪些？〉), *Touzi.com*, 26 April 2016; "An article to understand money market ETFs" (〈一文看懂場內貨幣基金〉), *Sina finance* website, 5 September 2018.

41 For the money market ETFs on the SZSE, the newly created ETF units can be redeemed in the primary market or sold in the secondary market on the same trading day (T+0). For money market ETFs on the SSE, newly created ETF units can be redeemed in the primary market or sold in the secondary market on the third trading day (T+2).

42 According the statistics available in Wind, there were 12 money market funds for primary market transactions only on the SSE, with a total AUM of RMB 21.9 billion, as of end-2018.

market ETFs). However, all of the listed funds in the Mainland are traded in the RMB at the moment. In other words, the Mainland investors cannot diversify currency exposure by investing in ETFs in the Mainland.

ETFs in the Mainland mainly cover domestic asset classes (94% of the total). In the light of this, the Mainland exchanges are exploring ways to broaden the exposure to global asset classes, including cooperation with global exchanges. The SSE and the Tokyo Stock Exchange under the Japan Exchange Group (JPX) cross-listed the first batch of ETFs with underlying assets on each other's market on 25 June 2019 under the China-Japan ETF Connectivity Scheme. Under the scheme, a Japanese or Chinese ETF issuer develops a different class of fund units called "feeder ETF" that invests in at least 90% of a target ETF listed on the other exchange, subject to quotas under the Qualified Foreign Institutional Investor (QFII) and Qualified Domestic Institutional Investor (QDII) schemes. The target ETF should have been listed for more than one year and should have enough liquidity in the home market. For the first batch of four ETFs cross-listed on the SSE, three of them track Nikkei 225 index and one tracks the Tokyo Stock Price Index (TOPIX). They amounted to a total AUM of RMB 509 million as of 21 June 2019[43]. The average daily turnover of these ETFs was RMB 161.8 million during the first week and fell to RMB 65.1 million during the second week[44].

In the Mainland listed fund market, listed open-ended funds (LOFs) are listed and traded on exchanges like ETFs. Similar to ETFs, LOFs are liquid and convenient tools for investing in different underlying asset classes, including stocks, bonds, mixed strategy (with both stocks and bonds), commodities and cross-border investments (through QDIIs). LOFs usually adopt active strategies with the aim to outperform benchmarks and hence have higher management expenses than ETFs.

As of end-2019Q1, there were 298 LOFs (74 on the SSE and 224 on the SZSE) listed in the Mainland with a total AUM of about RMB 358.3 billion[45]. LOFs with mixed strategy on onshore stocks and bonds accounted for the highest proportion by number (36% of the total) and by AUM (68% of the total) (see Figure 10), partly because these funds offer different risk-return profiles from ETFs in the Mainland. Like ETFs in the Mainland, onshore asset classes dominated — 89% by number and 98% by AUM of LOFs.

---

43 See "First batch of China-Japan cross-listed ETFs start trading today" (〈首批 4 隻中日互通 ETF 今日上市〉), *Xinhuanet. com*, 22 April, 2019.

44 Source: Wind.

45 Source: Wind.

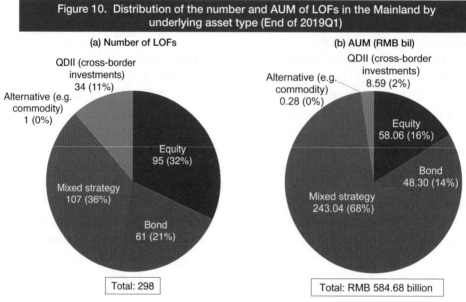

Figure 10. Distribution of the number and AUM of LOFs in the Mainland by underlying asset type (End of 2019Q1)

Source: Wind.

In addition to ETFs and LOFs, Mainland-listed funds also include structured funds[46]. The mother fund units are divided into sub-units (structured funds) with different expected risks and incomes, among which part or all of the fund unit types may be traded. The creation and redemption can be conducted only for mother units but not for sub-units. The mother fund units of structured funds on the SSE are LOFs[47] and those on the SZSE are not listed and traded on the exchange. To protect retail investors, entry barriers are imposed since May 2017 for an investor to trade structured funds. These include a minimum asset value of RMB 300,000 during the past 20 trading days and the requirement to sign a risk disclosure statement with the broker. In contrast, there are no minimum requirements on assets for investors to trade ETFs and LOFs. As of end-2018, there were 122 structured

---

46 There is another fund type — closed-end fund. However, as a result of the reform of the Mainland fund industry, closed-end funds have diminished over the years. As of 7 August 2019, there were 2 closed-end funds traded on exchanges (one on each of the SSE and the SZSE). Source: Lists of funds on SSE's website and SZSE's website, viewed on 7 August 2019.

47 The mother fund units can be formed by the combination of sub-units on the same trading day of purchase for redemption on the same trading day. See "Guidelines on business management of structured funds" (《分级基金业务管理指引》), issued by the SSE and the SZSE respectively on 26 November 2016; "Can I buy mother funds of structured funds? How can I buy them?" (〈分级基金母基金能买吗？ 分级基金母基金怎麽買？〉), PPmoney（網貸), 28 April 2018.

funds (each with A class and B class of sub-units), with a total AUM of RMB 111.0 billion. Similar to ETFs and LOFs, the investment scope of structured funds is mainly confined to onshore stocks and bonds.

In terms of trading value, ETFs have dominated the Mainland listed fund market. The average daily turnover value (ADT) of ETFs on the Mainland exchanges rose to a record high of RMB 18.4 billion in 2018 — RMB 6.8 billion on the SSE and RMB 11.5 billion on the SZSE. This was much higher than the corresponding figures of RMB 158.5 million for LOFs (RMB 15.7 million on the SSE and RMB 142.8 million on the SZSE) and RMB 1.1 billion for structured funds (RMB 21.1 million on the SSE and RMB 1.1 billion on the SZSE)[48] (see Figure 11). The ADT of ETFs was about 5% of the ADT of onshore stocks during 2018[49] (see Figure 12). Margin trading and short selling have been allowed for eligible ETFs since 2011[50].

The SZSE is in the process of revising its securities clearing and settlement arrangements since April 2019[51], the settlement cycle will be standardised for A shares, ETFs and LOFs listed on the SZSE, which will become the same as that for the same security types listed on the SSE — securities settlement will be on the same trading day (T+0) and money settlement will be on the next trading day (T+1). Before the revision, the settlement of ETFs on the SZSE followed delivery versus payment (DVP) on T+1 — both securities and money settlement were on T+1. Besides, cross-market ETFs tracking A shares on the SSE and the SZSE, before the corresponding rules revisions, could not be purchased and redeemed for A shares to be sold on T+0, but this will be allowed upon the implementation of the revised rules. Investors may then purchase and sell ETFs more frequently on T+0, hence supporting the secondary market liquidity of ETFs.

---

48 Source: SSE's and SZSE's monthly statistics for ADT of LOFs and structured funds, on the respective exchanges' websites.

49 The average daily turnover of ETFs was estimated from the WFE statistics on total ETF turnover per exchange. The onshore stock market turnover is available from Wind.

50 The eligibility criteria of ETFs for margin financing and securities lending (which enables short selling) include being traded for at least 5 trading days after listing, an average AUM of at least RMB 500 million in the past 5 trading days and at least 2,000 holders of the fund. See the SSE's and the SZSE's related rules revisions on margin financing and securities lending (《融資融券交易實施細則（2019 年修訂）》), issued by the SSE and the SZSE respectively on 9 August 2019.

51 *Notice on technical preparation on the trading and settlement mechanism for ETFs on SZSE* (《關於做好深市 ETF 交易結算模式調整相關技術準備的通知》), issued by the SZSE on 8 April 2019; *Notice on the Detailed Implementation Rules on the Trading, Creation and Redemption of Securities Investment Funds on the SZSE (2019 revision for consultation)* (《關於就《深圳證券交易所證券投資基金交易和申購贖回實施細則（2019 年修訂徵求意見稿）》公開徵求意見的通知》), issued by the SZSE on 23 August 2019.

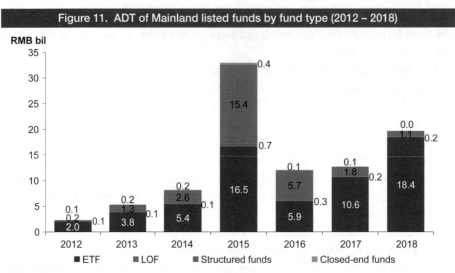

Figure 11. ADT of Mainland listed funds by fund type (2012 – 2018)

Source: SSE and SZSE.

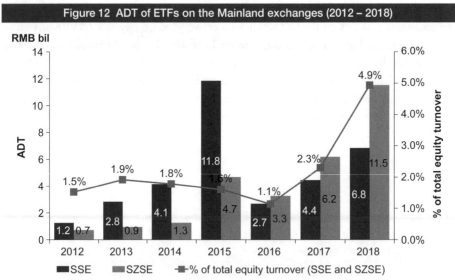

Figure 12 ADT of ETFs on the Mainland exchanges (2012 – 2018)

Source: WFE statistics on its website and Wind.

The primary market is another source of liquidity. For A-share ETFs listed in the Mainland (including cross-market ETFs), arbitrage activities between the primary market and the secondary market will be allowed on the same trading day (T+0) upon the implementation of the SZSE's revised rules — newly created ETF units can be sold in the secondary market on the date of creation and the redeemed securities from cancelled ETF units can be sold in the secondary market on the date of redemption. This will be conducive to the price discovery of these ETFs given that short selling activities in ETFs are restricted. However, the settlement cycles are still different across different types of ETF and between ETFs and other listed funds in the primary market. For LOFs, the creation and redemption of fund units are cash-settled only and therefore may involve additional costs of buying and selling underlying securities. Like LOFs, the creation and redemption of cross-border ETF units are also cash-settled only. Noteworthily, for cross-border ETFs and QDII LOFs, it takes longer time to receive the cash — 4 to 7 days for cross-border ETFs and about 10 days for QDII LOFs[52].

To summarise, the Mainland investors may demand investment diversification through ETFs. Rising interests in ETFs by the Mainland investors were observed from the growing market size and trading activities of the Mainland ETF market. However, Mainland-listed funds are dominated by funds on domestic assets (mainly equity) while the supply of listed funds on global assets was limited. The time delay to receive cash from the redemption of listed funds on global assets may constrain the liquidity of these funds. In the light of these, it would be beneficial to the Mainland investors if they have access to the Hong Kong ETF market which provides more varieties of global asset classes (see Section 2.2).

2.2   Hong Kong ETF market as a platform for global investments

In two decades' time, the Hong Kong ETF market has developed into an investment platform for global investors to get exposure to multiple asset classes of domestic and global equities, fixed income and commodities traded in different currencies (see Figure 13). ETFs in Hong Kong are not only traded in HKD (63% of the total number of ETFs and L&I products[53]), but also in US dollars (USD) (14% of the total) and RMB (23% of the total).

Types of ETF in Hong Kong[54] include:

---

52 Sources: Li, C., Z. Song and B. Cheng (Hwabao Securities), "Detailed explanation on trading mechanism and net asset valuation of cross-border ETFs" (〈跨境 ETF 交易機制及淨值計算詳解〉), *xueqiu.com*, 20 November 2015; "Settlement rules on cross-border LOFs" (〈跨境 LOF 基金的交收規則〉), Guotai Junan's website.

53 Source: "List of securities" webpage for ETPs on HKEX's website. As of June 2019.

54 All the statistics on the different types of ETF presented in this part are sourced from "ETF perspective", June 2019 issue, published on HKEX's website, unless otherwise stated.

- **Equity ETFs:** Cover at least 80% of total exposure of global equity markets[55], including ETFs tracking Mainland China A shares, Hong Kong equities, Asia-Pacific equities and other overseas equities. As of the end of the first half of 2019 (2019H1), there were 90 equity ETFs with an AUM of HK$279.7 billion (86.0% of the total).
  - ○ Hong Kong equity ETFs track the Hang Seng Index and Hong Kong indices, some with smart beta factors;
  - ○ A-share ETFs track not only the Mainland flagship indices, but also A-share indices compiled by global index providers such as MSCI and FTSE;
  - ○ Asia-Pacific equity ETFs track equity indices in the markets of India, Indonesia, Japan, Korea, Malaysia, the Philippines, Singapore, Taiwan and Vietnam;
  - ○ Other overseas equity ETFs track US equities, European equities and equities in some emerging markets (e.g. Brazil and Russia). These include the world's first ETF tracking the NASDAQ Overseas China New Economy Companies Top 50 Index (covering Mainland new economies listed in Hong Kong and the US) launched in September 2018.
- **L&I products:** Deliver up to ±2 times performance on Hong Kong and overseas equity indices. As of end-2019H1, there were 22 L&I products with an AUM of HK$7.3 billion (2.2% of the total).
- **Fixed income and currency (FIC) ETFs:** Track the performance of high-quality bonds in the Mainland and Asia-Pacific markets as well as money markets of HKD and USD. As of end-2019H1, there were 11 FIC ETFs with an AUM of HK$37.3 billion (11.5% of the total).
- **Commodity ETFs:** Track gold prices and crude oil prices in the global market. As of end-2019H1, there were 6 commodity ETFs with an AUM of HK$1.1 billion (0.3% of the total).
- **Active ETFs:** Hong Kong regulator started to accept applications of active ETFs from January 2019. The first active ETF, tracking USD money market, was listed on 18 June 2019[56].

---

55 The Hong Kong-listed ETFs issued by Vanguard already cover more than 80% of the exposure of global equity markets as of end-March 2018. Source: "Vanguard launches Hong Kong's most comprehensive China ETF", news release on Vanguard's website, 10 May 2018.

56 Source: "HKEX welcomes its first listing of active ETF", HKEX News Release, 18 June 2019.

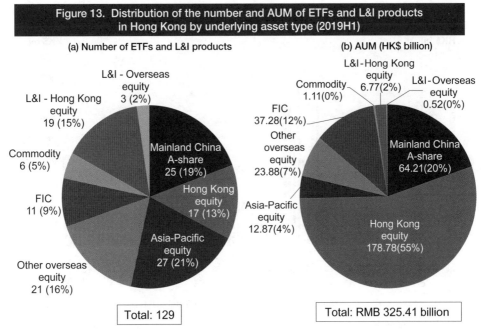

Figure 13. Distribution of the number and AUM of ETFs and L&I products in Hong Kong by underlying asset type (2019H1)

(a) Number of ETFs and L&I products

L&I - Overseas equity 3 (2%)
L&I - Hong Kong equity 19 (15%)
Commodity 6 (5%)
FIC 11 (9%)
Other overseas equity 21 (16%)
Mainland China A-share 25 (19%)
Hong Kong equity 17 (13%)
Asia-Pacific equity 27 (21%)

Total: 129

(b) AUM (HK$ billion)

L&I-Hong Kong equity 6.77(2%)
Commodity 1.11(0%)
L&I-Overseas equity 0.52(0%)
FIC 37.28(12%)
Other overseas equity 23.88(7%)
Asia-Pacific equity 12.87(4%)
Mainland China A-share 64.21(20%)
Hong Kong equity 178.78(55%)

Total: RMB 325.41 billion

Note: The listings of multiple currency counters of an ETF are counted as one product.

Source: "ETF perspective", June 2019 issue, published on HKEX's website.

The total AUM of ETFs in Hong Kong was about HK$325.4 billion as of end-2019H1 (see Figure 14). The ADT was about HK$5.3 billion during 2019H1 (about 5.4% of total securities market turnover).

Figure 14. ADT and AUM of ETFs and L&I products in Hong Kong (Feb 2013 – Jun 2019)

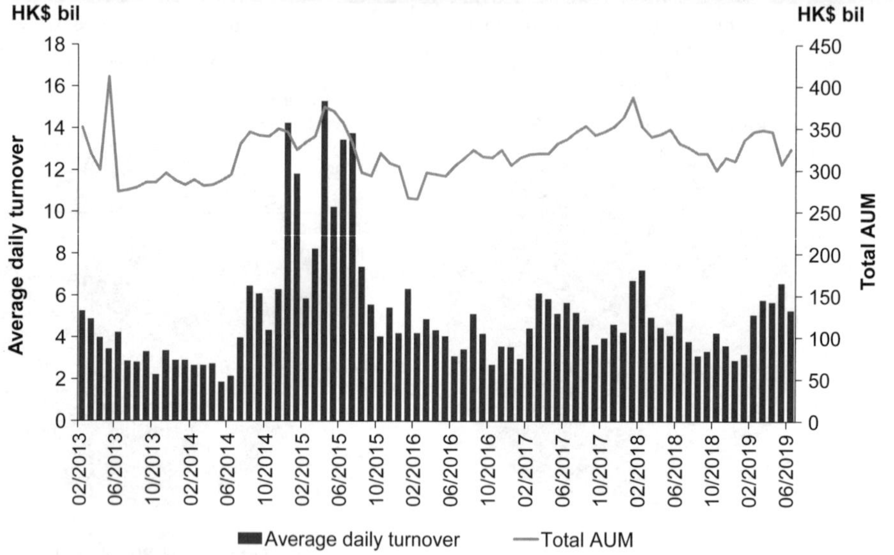

Note: L&I products were launched in Jun 2016.

Source: "ETF perspective", February 2013 to June 2019 issues, published on HKEX's website.

Institutional investors accounted for over 81% of the ETF turnover in Hong Kong during 2016 and the first half of 2017 and the average transaction size of ETFs doubled from HK$224,000 during 2007 to HK$459,000 during the first half of 2017[57]. A potential driver for institutional investor participation would be the growing number of global asset managers in Hong Kong — the number of licensed asset management companies in Hong Kong rose by 14% from 1,300 as of end-2016 to 1,477 as of end-2017 and further to 1,643 as of end-2018[58]. Besides, they are interested in fixed-income ETFs, which are treated as alternatives for fixed-income exposure — a survey in 2018[59] showed that 83% of institutional investors in Hong Kong already invested in fixed-income ETFs, which was higher than 78% for equity ETFs and 25% for commodity ETFs.

The Hong Kong ETF market is characterised by diverse asset-class exposures and a

---

57 Source: "SFC ETF Paper 2018".

58 Source: SFC, "Asset and Wealth Management Activities Survey", 2017 and 2018 issues, published on the SFC website. (Referred to as the "SFC Asset Management Survey 2018" for the 2018 issue.)

59 Source: "Ready to grow: ETF opportunities in Greater China", published on Brown Brothers Harriman's website, April 2018.

broad institutional investor base. The market's edges on the issuance and trading of ETFs further support the prosperity of the market. These are discussed in Section 3.

2.3 Complementary nature of the Mainland and Hong Kong ETF markets

The product composition, investor base and market access of the ETF market in the Mainland and Hong Kong are very different. Yet, both markets demonstrate the growing appetite of investors for ETF investments. (The features of ETFs and LOFs in the Mainland as well as ETFs in Hong Kong are summarised in the Appendix.)

In the Mainland, the growing interests in ETFs and the high proportion of ETFs on domestic assets may hint at the demand for asset diversification by the Mainland fund investors. However, the Mainland fund investors can only invest in global assets through funds under the QDII scheme and the mutual recognition of funds agreement with Hong Kong, as well as a few cross-listed Japanese ETFs (see Section 2.1). On the contrary, Hong Kong ETFs cover a wide range of Asia-Pacific and overseas equities as well as other global asset classes including fixed income, currency and commodity. These ETF products could be considered as alternatives for the Mainland investors.

In respect of investor base, retail investors are dominated in the Mainland market, who accounted for about 86% of turnover value in A-share market during 2018[60]. The relatively low degree of institutional investor participation may limit the growth of trading activities in the ETF primary market (creation/redemption of ETF units), which is important to reduce the premium or discount to an ETF's NAV, in other words, to achieve a fair pricing of an ETF. The global institutional investor base in Hong Kong could be complementary to the investor base in the Mainland to support the growth of the Mainland ETF market.

Hong Kong, on the other hand, is already an ETF marketplace with a wide range of asset classes covering Mainland and global assets. However, the range of Mainland asset classes is still rather limited. Take A-share ETFs as an example. About a half of Hong Kong-listed A-share ETFs track the popular A-share indices in Mainland — out of 25 A-share ETFs as of end-June 2019, there were 8 ETFs tracking the CSI 300 Index and 4 ETFs tracking the FTSE China A50 Index[61]. In comparison, Mainland A-share ETFs not only track the headline indices, but also indices of different sectors, ownership and smart betas (e.g. growth, value and dividend, etc). More diversified product composition in

---

60 Source: Mee, K. and G. Ralston, "Compelling characteristics of the A-shares market", published on *Insight* webpage of Schroders' website, 27 March 2019.
61 Source: "List of securities" webpage for ETPs on HKEX's website.

the Hong Kong ETF market, e.g. through access to Mainland ETFs, could contribute to attracting new investors worldwide.

In respect of trading activities, ETF turnover in the Hong Kong market is rather concentrated on a relatively small number of products — the five ETFs[62] with the highest turnover value accounted for 90% of total ETF turnover value during 2019H1. Mainland investors may have different appetite of ETF investments across different asset classes. Opening access of Hong Kong ETF market to the Mainland investors could be a potential driver of growth for the secondary market trading activities of Hong Kong ETFs.

Given the complementary nature of products and investors in the Mainland and Hong Kong, a potential increase of Mainland-Hong Kong market connectivity in the ETF market could meet investors' different investment appetite and improve the liquidity of the ETF primary and secondary markets on both sides. This could also further strengthen Hong Kong's edges as the Asia-Pacific hub for ETFs as discussed in Section 3.

3   Hong Kong's edges as the Asia-Pacific hub for issuing and trading ETFs

3.1   Issuers' considerations to issue ETFs in Hong Kong

Given the growing demand for global passive investments, the number of issuers (or ETF managers) in the Hong Kong ETF market has grown steadily. The number of ETF managers rose from 7 in 2008 to 21 in 2012 and to 27 in 2018[63], despite the withdrawal of some ETF managers from the market due to keen competition and changing market sentiment. The attractiveness of the Hong Kong ETF market to issuers lies in the factors discussed below.

---

62 These comprise five ETFs tracking FTSE China A50 Index, CSI 300 Index, HSI and Hang Seng China Enterprises Index (HSCEI) respectively. Source: HKEX.

63 The figures on 2008 and 2012 (up to May) are sourced from the presentation slides at HKEX's seminar on RQFII ETFs on 19 June 2012 (see "HKEX hosts seminar on RQFII ETFs", HKEX News Release, 19 June 2012); the figure in 2018 (as of end-2018) is sourced from the "List of securities" webpage for ETPs on HKEX's website.

First, the presence of a global investor base with increasing demand for ETFs. In 2018, 62% of the funding sources of asset and wealth management in Hong Kong came from global investors[64], including global asset managers with Asia-Pacific operations headquartered in Hong Kong. These global investors have been investing in Mainland securities through Hong Kong, which are facilitated by Stock Connect, Bond Connect, and QFII and RQFII schemes. The optimism over the inclusion of A shares into global indices would further accelerate the inflows. Besides, the demand of local institutions for ETF investments is expected to rise. For pensions in Hong Kong, the AUM of Mandatory Provident Fund (MPF) schemes reached HK$813 billion as of end-2018 with an average yearly inflow of about HK$40 billion[65]. Mandatory Provident Fund Authority (MPFA) accepted 131 ETFs as index-tracking collective investment schemes (ITCIS) into the investment scope of MPF schemes in Hong Kong[66].

Second, the rising number of market makers support the market liquidity. In Hong Kong, an ETF is required to have at least one market maker. The number of ETF market makers, which can be proprietary trading firms or investment banks, increased from 14 in 2008 to 24 in 2012 and to 33 in 2018[67]. Market makers make profits on arbitrage activities between the primary and secondary markets and their activities provide liquidity to secondary trading of the ETFs. Market makers are obliged to input market making orders upon an occurrence of a "wide spread" for at least 3 minutes. The market making order may range from 4 to 32 spreads depending on the ETF[68]. Market makers have to submit two-sided market making orders with 3 to 30 spreads (depending on the ETF) within 90 seconds for 3 minutes. In fact, certain ETFs are usually being quoted to the tightest spread in accordance to the spread table. Market makers may provide even tighter quotes if the tick sizes in the spread table can be reduced for ETFs, which have been suggested by market participants. The market making obligations would help smoothen out the demand for redemption during market correction. For example, during the Mainland stock market correction in 2015, although there were major redemptions, these redemption requests on ETF managers were met in an orderly manner without activating any redemption tools such

---

64 Source: SFC Asset Management Survey 2018.
65 Source: "Mandatory Provident Fund Schemes Statistical Digest", MPFA's website, December 2018.
66 Source: MPFA's website (as of 13 May 2019).
67 The figures in 2008 and 2012 (up to May) are sourced from the presentation slides at HKEX's seminar on RQFII ETFs on 19 June 2012 (see "HKEX hosts seminar on RQFII ETFs", HKEX News release, 19 June 2012); the figure in 2018 (as of end-2018) is sourced from the "List of securities" webpage for ETPs on HKEX's website.
68 ETPs are classified into 5 groups (A-E) which determine the market making obligations. Prior to the listing of an ETP, HKEX will consult the individual ETP issuer and market makers and set the ETP's grouping. See "Market making arrangements of exchange traded products", information sheet on the HKEX website, viewed on 26 June 2019.

as redemption suspension or gate[69].

Third, the "Trade Asia in Asia" trading philosophy highlights Hong Kong's geographical advantage for ETFs on Asian underlying assets. In addition to time zone preference, asset managers in Hong Kong are experienced in investing in Asia-Pacific assets and over 65% of assets managed in Hong Kong were invested in Asia Pacific during 2014 to 2018[70]. ETF liquidity tends to go with the availability of trading in the underlying assets due to hedging needs. The bid-ask spread, i.e. the trading cost, usually decreases with increasing liquidity. The bid-ask spread can be wider for US ETFs on underlying assets in international markets, which may be closed when the US market is still trading[71]. When the underlying market is opened, ETF managers and market makers can hedge their risks in the underlying markets and are more willing to support the liquidity of the ETF. Otherwise, they need to rely on estimating the fair value when the underlying market is closed. Therefore, the liquidity of ETFs in Hong Kong on Asian underlying assets tends to be better than those in American and European time zones.

Fourth, the market-friendly regulatory environment supports the issuance of a diverse range of ETFs. The Hong Kong market regulator, the Securities and Futures Commission (SFC), seeks to strike a balance between market development and investor protection[72]. To support product innovation, the SFC approved the launch of L&I products in February 2016 amid the rising demand in Asian markets. L&I products are designed for sophisticated trading-oriented investors with short-term investment horizon. To avoid excessive risk taking, the SFC capped the leverage factor at two times, alongside with requiring warnings as upfront disclosure in offering documents and different naming from ETFs[73]. Since July 2017, the SFC has accelerated the authorisation of investment funds with a target to limit the processing time of an application to within 2 months for "Simple Applications"

69 See "SFC ETF Paper 2018".

70 Source: SFC Asset Management Survey 2018.

71 Source: Hammond, O. H., M. Lieder, "Debunking myths about ETF liquidity", published on JP Morgan Asset Management's website, May 2015.

72 The authorisation of ETFs in Hong Kong as "collective investment scheme" is governed by the Code on Unit Trusts and Mutual Funds ("UT Code") of the SFC while authorised ETFs can be listed once they meet the requirements set out in Chapter 20 of the Listing Rules of the Stock Exchange of Hong Kong (SEHK).

73 See *Circular on Leveraged and Inverse Products*, issued by the SFC, 5 February 2016; *Supplemental Circular on Leveraged and Inverse Products*, issued by the SFC, 14 March 2019; "How do L&I Products differ from ETFs?", the Chin Family website, 8 May 2019.

and within 6 months for "Complex Applications"[74]. Effective from January 2019, the SFC extended the coverage of authorisation to include active ETFs[75]. It also issued new provisions for UCITS funds (including ETFs domiciled in Europe) in December 2018 to provide more clarity on the cross-listing process, which has been streamlined since 2007[76]. These efforts of the regulator to support market growth are accompanied by the imposition of certain safeguards to protect ETF investors' interests. For example, the SFC requires synthetic ETFs to have a marker "X" to be placed at the beginning of the short names and an asterisk (*) with an annotation of warning in English and Chinese, and imposes collateral requirements of at least 120% of gross counterparty risk exposure for equity collateral of synthetic ETFs[77].

3.2   Factors supporting the liquidity of ETFs in Hong Kong

First, the primary and secondary markets of ETFs in Hong Kong are very liquid. Liquidity attracts more liquidity. The ADT of ETFs rose from HK$1.8 billion (2.5% of the total securities market turnover) in 2008 to more than doubled in 2018 at HK$4.0 billion (3.7% of the total securities market turnover), or HK$4.4 billion (4.2% of the total securities market turnover) if L&I products are included (see Figure 15). The secondary market liquidity of the underlying assets has been sufficient to meet the needs of arbitrage activities from the ETF primary market. According to an SFC research report[78], the primary flows of HSI ETFs did not dry up the secondary market liquidity of HSI constituents towards market close (day-end turnover) during June 2012 to June 2017 — the median and 90th percentile of primary flows as percentage of HSI day-end turnover[79] was only 1% and 13% respectively. The same study found that the liquid ETF market contributed to small deviations (premium/discount) from the net asset value of the ETFs, which stayed within one percent for about 80% of ETFs listed in Hong Kong during 2013 to 2017.

---

74 See *Circular to Management Companies of SFC-authorised unit trusts and mutual funds — Launch of pilot revamped process to enhance the processing of post authorisation applications*, issued by SFC, 30 June 2017. "Complex Applications" refer to those with new and/or change of appointment of key operators, change of place of domicile of SFC-authorised funds, merger of SFC-authorised funds, termination of SFC-authorised funds involving specific issues or material issues and/or policy implications relating to the application. Otherwise, the applications will be classified as "Simple Applications".

75 See "Asia ETF Roundup (Industry) — December 2018 and January 2019", *Morningstar*, 14 February 2019.

76 The details of streamlined process of cross-listing is available in "SFC simplifies authorisation of UCITS III funds with special features", news release on the SFC's website, 30 March 2007. For details of the new provisions, see "Hong Kong — The Revised Code On Unit Trusts And Mutual Funds", *Conventus Law*, 21 February 2019.

77 See "New measures to raise investors' awareness of synthetic ETFs", SFC press release, 18 November 2010; "Enhanced investor protection for domestic synthetic ETFs", SFC press release, 29 August 2011.

78 Source: "SFC ETF Paper 2018".

79 "HSI day-end turnover" refers to the turnover (in value terms) of HSI constituents during 3:30 pm to 4:10 pm.

Figure 15. ADT of ETFs and L&I products in Hong Kong (1999 – 2018)

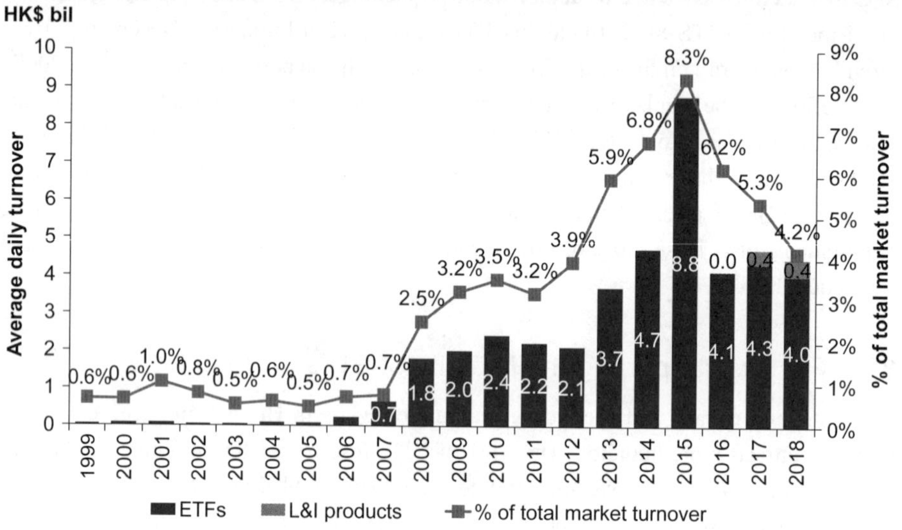

Note: L&I products were launched in Jun 2016.
Source: HKEX Factbooks.

Second, the competitive total cost of ownership (TCO) supports "Trade Asia in Asia" through the Hong Kong ETF market. According to HKEX[80], the TCO for ETF investment includes trading costs (e.g. bid-ask spread and broker commissions) and holding costs (e.g. investor-level taxation and costs associated with tracking difference[81] that include management fees, transaction costs and taxation at investment level and fund level). For trading costs, the Hong Kong Government extended the waiver of stamp duty to all ETFs in Hong Kong since February 2015. Besides, market makers in Hong Kong enjoy lower cost of trading to facilitate their market making activities as they are exempted from the exchange's trading fee (0.005%) and the regulator's transaction levy (0.0027%). As for holding costs, taxation is a key component in that investors' return is affected by three tiers of taxation — investment level, fund level and investor level. In fact, a study[82] highlighted that Hong Kong's expanding bilateral tax treaty network and domestic tax rules are

---

80 See "Exchange Traded Fund Handbook", published on HKEX's website, viewed on 8 August 2019.
81 Tracking difference is the difference in returns between an ETF and the underlying benchmark over a certain period of the time while tracking error is commonly calculated as the standard deviation of the differences in returns between an ETF and the underlying benchmark over a certain period of time.
82 Source: "ETF taxation report for investors 2019 (Hong Kong)", report commissioned by the HKEX, published on Ernst & Young's website, 2019.

favourable for cost reduction. The same study concluded that ETFs in Hong Kong should be the most tax efficient among major markets when investing in Japanese equities and Mainland corporate bonds.

Third, multiple currency counters of ETFs facilitate usage by asset managers. ETFs in Hong Kong can be traded in multiple currencies, including HKD, RMB and/or USD. This allows the Hong Kong market to list and trade ETFs on Mainland-listed securities in RMB. ETF trading in RMB is supported by the extended coverage of the HKEX-provided RMB Equity Trading Support Facility (TSF) beginning in August 2012[83]. Certain ETFs tracking US securities and global indices are not only traded in HKD and RMB, but also in USD. These currency counters expand the investment choices for exposure to different currencies. To further improve the liquidity of USD counters, suggestions were raised by market participants on either having a spread table of tick sizes harmonised with the HKD counter or allowing deviations from the "one-to-one conversion" rule (under which each unit of multiple counters of the same ETF must have the same NAV) to narrow the differences of tick size across counters of the same ETF[84].

Fourth, an ecosystem of risk management products in Hong Kong meet the needs of ETF investors. To hedge the market risk of ETFs in Hong Kong, stock futures and options of the top five ETFs by turnover value[85] are available. Index futures and options on the HSI and HSCEI as well as index futures on the CES China 120 Index and MSCI Asia ex-Japan Index are available for hedging the market risks of ETFs on these Hong Kong, Mainland and Asian equity indices. The benefits of these hedging tools in Hong Kong are recognised by market participants, who call for continued futures and options product innovation on more diversified underlyings. For gold ETFs, gold futures traded in USD and CNH are available for hedging the market risk of fluctuations in gold prices. To hedge for the currency risk of ETFs, deliverable USD/CNH futures and options as well as cash-settled futures on EUR/CNH, JPY/CNH, AUD/CNH and CNH/USD are available for investing in HKD, RMB and USD counters of ETFs.

---

83 See "Circular on extension of the product type coverage of the RMB equity trading support facility (TSF) on 6 August 2012", issued by HKEX, 30 July 2012.

84 Under the "one-to-one conversion" rule, the USD counter usually has a larger tick size as a percentage of the unit price than the HKD counter. This gives rise to spread cost concerns of brokers providing inter-counter transfer services for customers.

85 See Section 2.3 above on the top five ETFs.

4   Conclusion

Investments in ETFs have become increasingly popular. The advantages of investing in ETFs versus mutual funds include convenience, cost effectiveness, high liquidity and transparency. Among global markets, Asia Pacific is the fastest growing ETF market on rising investor demand for portfolio diversification, particularly from the Mainland investors.

Mainland ETF market has grown quickly to meet onshore investors' diversification demand, led by equity ETFs on Mainland indices and money market ETFs. In addition to ETFs, LOFs provide exposures of active returns, led by mixed strategy funds on onshore securities. Given the Mainland investors' growing desire for global asset allocation and their high acceptance of passive investments through ETFs, the limited supply of cross-border ETFs and LOFs on global underlying assets and the lack of multiple currency exposure in the Mainland market hint at the potential domestic demand for access to the Hong Kong ETF market.

The Hong Kong ETF market is a door to global investment, particularly for the Mainland investors. It offers products on diversified asset classes from global markets and a well-established institutional investor base to support market liquidity. An ecosystem of global issuers and investors have supported the development of the Hong Kong ETF market. For issuers, they enjoy the strong demand from global investors, effective market making mechanism, geographical advantage of "Trade Asia in Asia" and the market-friendly regulatory environment. For investors, they can trade ETFs in the liquid primary and secondary markets with tax advantages, multiple currency counters, and the support of an ecosystem of risk management tools.

Hong Kong is well-positioned to be Asia's ETF marketplace. Staying close to global market trends will be crucial to continue the success of the Hong Kong ETF market. Market enhancements could include further expanding the coverage of underlying asset markets and supporting product innovations to meet investors' diverse needs. The wide range of exposures covered by Hong Kong ETFs not only meet the needs of the global investors, but also the needs of the Mainland investors potentially. In comparison, the variety of exposures in Mainland assets covered by Mainland ETFs could be attractive to global investors. In addition, the potential mutual access of the two ETF markets would facilitate a more balanced mix of investors in the ETF primary and secondary markets on both sides. Further development in Mainland-Hong Kong market connectivity in respect of the ETF segment will help broaden the investor base and increase liquidity in both markets. This will be conducive to the mutual growth of the two markets.

Appendix

Comparison between ETFs and LOFs in the Mainland and ETFs in Hong Kong

Feature	ETFs in Mainland	LOFs in Mainland	ETFs in Hong Kong
Number of fund units	Variable		
Secondary trading on exchange	Yes, during trading hours		
Market makers	Usually available but not mandatory	No	Yes and mandatory
Margin trading and short selling	Allowed for some eligible funds		No limit on margin trading; short selling only for eligible ETFs
Threshold for block trading	≥2 million units or ≥RMB 2 million		No limit on size
Settlement of secondary market	Securities settlement on T; money settlement on T+1		T+2
Primary market (creation and redemption)	• Physically and cash settled • Transacted on-exchange only • For all investors • High threshold (e.g. minimum 0.5-1 million fund units)	• Cash settled • Transacted on- or off-exchange • For all investors • Low threshold (e.g. minimum 1,000 fund units)	• Physically settled • Transacted off-exchange only • Only for participating dealers (also CCASS* participants) • High threshold (e.g. minimum 1 million fund units)
Cost	• Management fee: 0.15%-0.8% of NAV	• Management fee: 0.1%-2.0% of NAV	• Management fee: 0.025%-0.99% of NAV • Stamp duty exempted • No withholding tax on dividends
Transparency	NAV updated every 15 seconds	NAV updated once a day at market close	NAV updated every 15 seconds
Strategy	Generally passive	Mostly active	Generally passive, but active and L&I products are allowed

* CCASS refers to Central Clearing and Settlement System of the HKEX securities market.

Source: SSE, SZSE, HKEX and Wind.

Remark:
This research report has made reference to views and feedback on ETF market developments sought from ETF managers in Hong Kong.

Chapter 15

Policy suggestions for the development of the Hong Kong ETF market and cross-border asset allocation

Chen DING

Chief Executive Officer
CSOP Asset Management Limited

1 The international ETF market trend

1.1 Characteristics of the international ETF market trend

1.1.1 The US market is the largest, accounting for more than 70% of the total market size; the Asian market is relatively small, but with tremendous potential

Exchange Traded Fund (ETF) is a kind of open-ended fund with variable number of fund units that are traded on the exchange. In 1993, the world's first ETF, which was based on the S&P 500 Index, was issued[1]. Over the next two decades, the global ETF market developed rapidly. At the end of 2018, the world's ETF net assets were US$4.7 trillion, of which 71% were in the United States (US), 15% in Europe, and 10% in Asia. The booming US market has become the largest ETF market in the world, accounting for more than 70% of the total market size (see Figure 1). The Asian market is relatively small, but with tremendous potential.

Figure 1. Percentage share of total net assets of ETFs by country/region (End–2018)

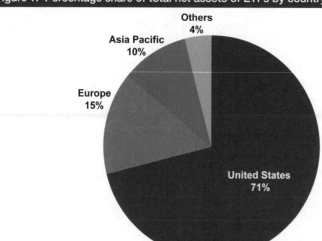

Source: 2019 ICI Fact Book, CSOP.

---

1   Source: http://online.wsj.com/ad/focusonetfs/history.html.

According to the Investment Company Institute (ICI), the net assets of US-registered investment companies in 2018 were US$21.4 trillion, of which mutual funds were US$17.7 trillion, ETFs were US$3.4 trillion, closed-end funds were US$250 billion, and unit investment trusts were US$70 billion. In terms of total net assets, ETFs were the second largest in the US market, accounting for 15.87%. (See Figure 2.)

Figure 2. Composition of total net assets of US-registered investment companies (End–2018)

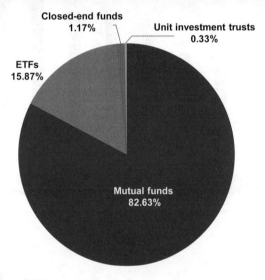

Source: 2019 ICI Fact Book, CSOP.

As of the end of 2018, there were 1,988 products in the US market. The total net assets reached US$3.37 trillion, and accounted for 71% of the world's ETF volume. Since 2009, the total net assets of US ETFs increased from US$777 billion to US$3.37 trillion, with an annualised growth of 17.71%; the number of ETFs increased from 797 to 1,988, with an annualised growth of 10.69%. (See Figures 3 and 4.)

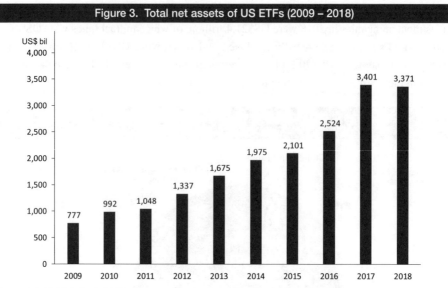

Source: 2019 ICI Fact Book, CSOP.

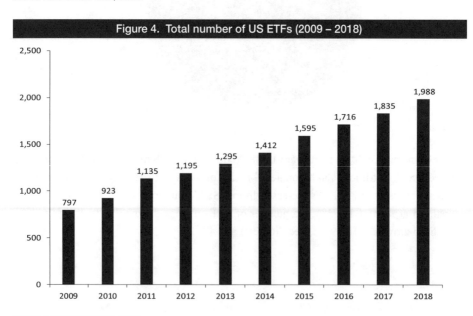

Source: 2019 ICI Fact Book, CSOP.

1.1.2 Asian Market: Japan and South Korea take the lead; Mainland China, Taiwan and Hong Kong have a smaller size; other Asian countries have late starts

Asian fund markets have dissimilar developments, and the whole market is still in an early stage. According to ICI, in 2018, in terms of total net assets of open-ended funds, Japan and Mainland China accounted for the largest shares of 40.28% and 39.48% respectively, followed by South Korea (10.34%). In terms of number of open-ended funds, South Korea and Japan accounted for the largest shares of 40.18% and 37.26% respectively, followed by Mainland China (15.02%). Japan and South Korea take the lead in market development, while other Asian regions started later with a relatively small volume.

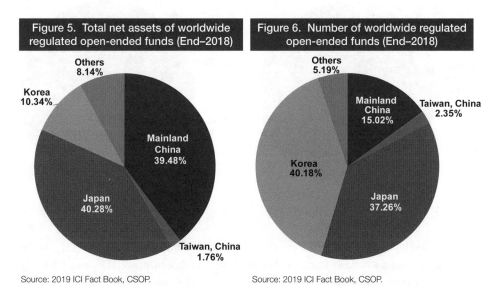

| Figure 5. Total net assets of worldwide regulated open-ended funds (End–2018) | Figure 6. Number of worldwide regulated open-ended funds (End–2018) |

Source: 2019 ICI Fact Book, CSOP.   Source: 2019 ICI Fact Book, CSOP.

1.1.3 US bond ETFs are widely accepted by investors, while major institutional investors in Mainland China cannot participate in the bond ETF market

With bonds as the underlying assets, bond ETFs purchase a basket of bonds to track the bond index or actively manage bond portfolios. In 2000, the first bond ETF was listed, and the underlying asset was Canada's five-year government bond. In 2002, four bond ETFs were issued in the US — three ETFs to track respectively Lehman 1-3 years', 3-7 years', over 20 years' treasury bond indices, and an ETF to track Goldman Sachs investment-grade corporate bond index. Before 2007, there were a total of six bond ETFs in the world, with

a total size of about US$20 billion. Since 2008, the global development of bond ETFs
speeded up. According to Bloomberg, at the end of 2018, there were 1,176 bond ETFs
globally, with a total size of US$969.4 billion. The total net assets of bond ETFs issued
in the US took up 71% of the global total, dominating other regions. In terms of number,
there were 379 bond ETFs issued in the US, accounting for 32%, which made the US in a
leading position.

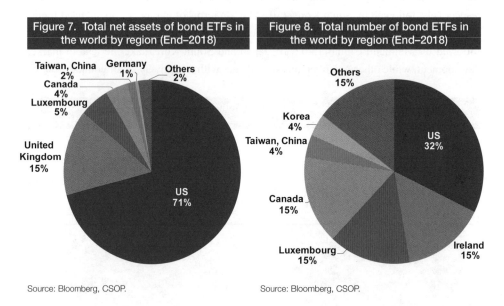

Figure 7. Total net assets of bond ETFs in the world by region (End–2018)

Figure 8. Total number of bond ETFs in the world by region (End–2018)

Source: Bloomberg, CSOP.

Source: Bloomberg, CSOP.

The development of bond ETFs in the US market has undergone three stages. During
the market incubation stage from 2000 to 2006, the number of bond ETFs was relatively
small and the development was relatively slow. At the end of 2006, there were only
six bond ETFs in the US market, with a total size of about US$20 billion. During the
product expansion stage from 2007 to 2008, investors became familiar with bond ETFs
and gradually accepted them. More and more companies began to issue bond ETFs. The
overall market scale increased substantially. The regulatory authorities issued detailed
regulations to promote the development of bond ETFs. Since 2009, the market entered the
stage of accelerated development. After the financial crisis, the liquidity of the bond market
dried up. However, the bond ETFs revitalised market liquidity. Since 2009, the trading
volume of bond ETFs has experienced explosive growth, which in turn has brought in new
funds, leading to the rapid development of bond ETFs.

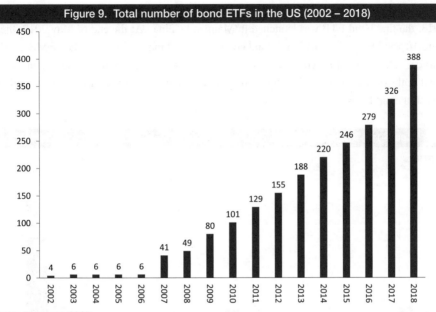

Figure 9. Total number of bond ETFs in the US (2002 – 2018)

Source: Bloomberg, CSOP.

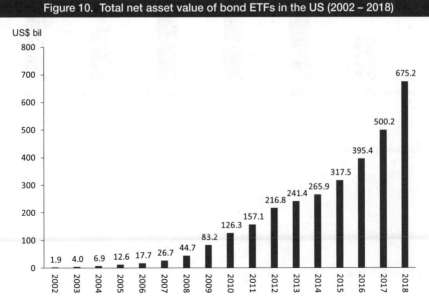

Figure 10. Total net asset value of bond ETFs in the US (2002 – 2018)

Source: Bloomberg, CSOP.

Compared with the US, bond ETFs in the Mainland market are in their infancy. In 2013, the first bond ETF was launched in Mainland China. At the end of May 2019, there were 10 bond ETFs launched by 7 fund companies, and the total size reached RMB 11,575 million. Due to the limited choice of bond ETFs and regulatory restrictions, there is no great enthusiasm for major institutional investors such as banks and insurance funds to invest in bond ETFs.

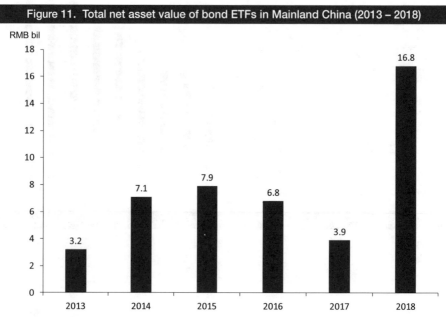

Figure 11. Total net asset value of bond ETFs in Mainland China (2013 – 2018)

Source: Wind, CSOP.

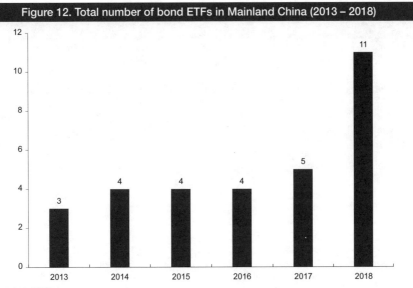

Figure 12. Total number of bond ETFs in Mainland China (2013 – 2018)

Source: Wind, CSOP.

Nevertheless, the Mainland government has been working hard to promote the development of bond ETFs. In May 2019, the China Securities Regulatory Commission (CSRC) and the People's Bank of China jointly issued the *Notice on Piloting the Innovation of Open-Ended Bond Index Securities Investment Funds*, and proposed to launch bond index publicly-offered funds that can be traded on exchanges or in the interbank market using cross-market bonds as the underlying assets. After the announcement, the CSRC is set to promote the registration and review of pilot bond index funds.

In June 2019, the General Office of the Communist Party of China and the General Office of the State Council issued the *Notice on Doing Well the Work on the Issuance of Local Government Special-Purpose Bonds and Related Project Financing*, which mentioned the enlargement of the investor base of local government bonds, the launch of exchange-traded open-ended index funds on local government bonds, and attracting investments from overseas investors through schemes such as Bond Connect. The successive introduction of related policies shows the support from Mainland China to the development of bond ETFs. Bond ETFs accounted for only 4.4% of the total size of ETFs in the Mainland market, compared with 14% in the US[2]. The Mainland bond ETF market, therefore, has a huge potential in the future.

---

2   Source: Wind.

Table 1. Summary of policies on bond ETFs in Mainland China			
Date	Policy document	Issuance agency	Key content
2013-02-08	Guide for Shanghai Stock Exchange Bond ETFs[3]	Shanghai Stock Exchange	Guidance on single-market bond ETFs that use outstanding bonds to conduct subscription and redemption
2016-04-28	Notice on Issuance of the "Shenzhen Stock Exchange Detailed Implementation Rules on the Trading, Subscription and Redemption of Securities Investment Funds (2016 Revision)"[4]	Shenzhen Stock Exchange	Continue to waive the transaction handling fee and Participation Business Unit (PBU) fee for bond ETFs and listed money market funds
2017-11-03	Notice on the implementation of the relevant code range for the cash subscription and redemption of bond ETFs[5]	Shanghai Stock Exchange	Transaction handling fee of bond ETFs is exempted for the moment
2019-01-11	Notice on the amendment of the "Shenzhen Stock Exchange Detailed Implementation Rules on the Trading, Subscription and Redemption of Securities Investment Funds"[6]	Shenzhen Stock Exchange	Investors can sell the bond ETF units that they bought through auction trading on the same day. Investors can sell through block trading the bond ETF units they bought through block trading on the same day, or sell them through auction trading the next day
2019-05	Notice on Piloting the Innovation of Open-Ended Bond Index Securities Investment Funds[7]	CSRC, People's Bank of China	It is proposed to launch publicly-offered funds on bond indices that can be traded on exchanges or in the interbank market using cross-market bonds as the underlying assets. After the announcement, the CSRC is set to promote the registration and review of pilot bond index funds
2019-06-10	Notice on Doing Well the Work on the Issuance of Local Government Special-Purpose Bonds and Related Project Financing[8]	The General Office of the Communist Party of China, the General Office of the State Council	To enlarge the investor base of local government bonds, to launch exchange-traded open-ended index funds on local government bonds, and to attract investments from overseas investors through schemes such as "Bond Connect"

---

3   Source: http://www.sse.com.cn/lawandrules/guide/zqznlc/c/c_20150912_3986154.shtml.
4   Source: http://www.szse.cn/aboutus/trends/news/t20160525_518721.html.
5   Source: http://www.sse.com.cn/lawandrules/guide/jyznlc/dealcode/c/c_20171103_4413981.shtml.
6   Source: http://www.szse.cn/lawrules/rule/tradetype/fund/t20190111_565175.html.
7   Source: http://www.csrc.gov.cn/pub/newsite/zjhxwfb/xwdd/201905/t20190520_356093.html.
8   Source: http://politics.people.com.cn/n1/2019/0610/c1001-31128317.html

1.1.4   The development of leveraged and inverse products is more mature in the US

Globally, the US is the most developed market for leveraged and inverse (L&I) products. In Asia, Japan, South Korea, and Taiwan markets have developed L&I products earlier than other markets and with a relatively large scale. In 2017, Hong Kong launched the first batch of L&I products to track the Hang Seng Index and the Hang Seng China Enterprises Index.

It should be noted that L&I products include ETFs and Exchange Traded Notes (ETNs). Their comparison is shown in Table 2.

Table 2. Comparison between ETN and ETF		
Characteristic	ETN	ETF
Issuer	Securities company or investment bank	Fund company
Nature	Obligatory right	Fund
Investor holdings	Obligatory rights issued by the issuer	Fund units
Subscription and redemption	Cash	Cash or physical equities
The use of underlying assets at issuance	Can choose whether to buy in the underlying assets	Must buy in the underlying assets
Credit risk	Yes	No
Maturity	Yes	No
Management fee	Yes	Yes

Source: CSOP.

The US is the global leader in L&I products with the largest scale in terms of total assets. As of 11 April 2019, the total number of L&I products (including ETFs and ETNs) listed in the US was 286, with a total size of US$51.87 billion. Among them, L&I products with a leverage ratio of two times (2x) reached 124, L&I products with a leverage ratio of three times (3x) reached 103, and L&I products with a leverage ratio of four times (4x) reached 10. (See Table 3.)

Table 3. Statistics on L&I products in the US (11 Apr 2019)			
	ETF	ETN	Total
Total asset value of L&I products (US$ bil)	36.9	15.0	51.9
Total number of L&I products	194	92	286
Proportion of L&I products (by value)	71%	29%	100%
Total asset value of L&I products (2x) (US$ bil)	11.6	12.6	24.2

(continued)

Table 3. Statistics on L&I products in the US (11 Apr 2019)			
	ETF	ETN	Total
The total number of L&I products (2x)	77	47	124
Proportion of L&I products (2x) (by value)	48%	52%	100%
Total asset value of L&I products (3x) (US$ bil)	20.6	2.0	22.6
Total number of L&I products (3x)	84	19	103
Proportion of L&I products (3x) (by value)	91%	9%	100%
Total asset value of L&I products (4x) (US$ bil)	0	0.035	0.035
Total number of L&I products (4x)	0	10	10
Proportion of L&I products (4x) (by value)	0%	100%	100%

Note: 0.5x and 1.5x products are not included due to their small scales.
Source: Bloomberg.

Japan and South Korea are the forerunners of L&I products in Asia. As of 11 April 2019, Japan had 39 L&I products. Compared to counterparts in the US and South Korea, Japanese regulators are relatively conservative, allowing a maximum leverage factor of only two times (2x).

In Asia, South Korea started later than Japan in the field of L&I products, but has become a well-deserved leader, not only because of its large number of products and the completeness of assets classes, but also because the products' asset value and average daily trading volume are significantly higher than those in other markets in the region. It is believed that the success of Korean L&I products has benefited from the recognition and preference of Korean investors for highly leveraged and highly volatile products. In South Korea, like Japan, L&I products mainly achieve target returns using the corresponding futures, and are managed similarly.

Taiwan market started to develop L&I products later than Japan and South Korean markets, and the market scale is smaller. As of 11 April 2019, there were 43 L&I products in Taiwan market, with a scale close to the Korean market. The products are mainly issued by Yuanda, Fubon, and Cathay. Taiwan's L&I products are all ETFs. Starting from 2019, regulators in Taiwan allow qualified brokers to issue ETNs.

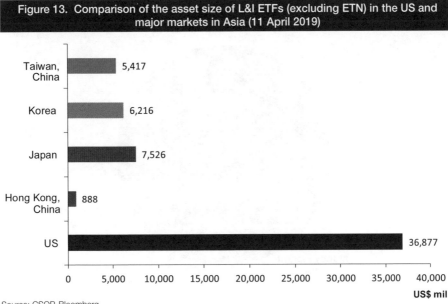

Figure 13. Comparison of the asset size of L&I ETFs (excluding ETN) in the US and major markets in Asia (11 April 2019)

Source: CSOP, Bloomberg.

1.1.5 The underlying assets of ETFs in the US are abundant, including equities, bonds, commodities, foreign exchange and other types of asset; both US assets and global assets are included

Open-ended funds include mutual funds, ETFs, and institutional funds. The composition of underlying assets of ETFs is similar to that of open-ended funds. According to ICI, in 2018, there were 118,978 globally regulated open-ended funds, of which 34% invested in stocks, 18% invested in bonds, 2% invested in money markets, and 46% invested in mixed products (balanced funds, real estate funds, capital-guaranteed funds, etc.) and others. In the US, the underlying assets of ETFs are constantly being enriched, including equities, bonds, commodities, foreign exchange, and other types of asset. Both US assets and global assets are included.

Figure 14. Distribution of regulated open-ended funds in the world by underlying asset class (End-2018)

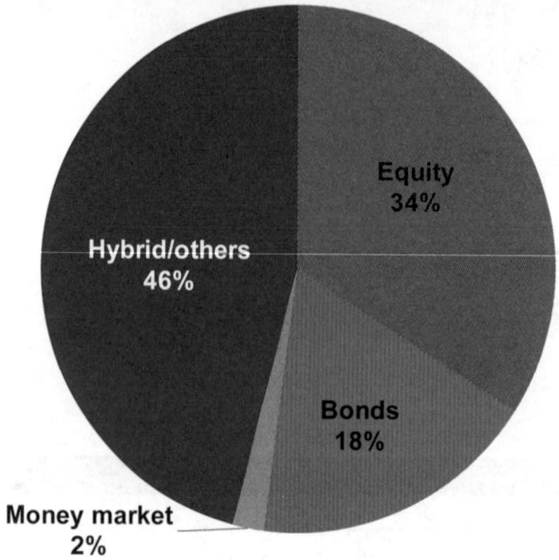

Source: 2019 ICI Fact Book, CSOP.

1.2   Reasons for the rapid development of international ETF markets in recent years

1.2.1   US investors, including institutional investors and retail investors, are more mature and more receptive of ETFs

The US financial market is more developed and complete; investors are more mature, and are more receptive of different types of investment product. In addition to institutional investors, there are also retail investors such as investors in 401k[9]. According to ICI, at the end of 2018, 7.9 million (about 6%) of American households held ETFs; 13% of households holding mutual funds also held ETFs; and 93% of households holding ETFs held stock mutual funds, stocks, and variable annuities; 66% of households holding ETFs held bond mutual funds, bonds, or fixed annuities; 48% of households holding ETFs held real estate for investment.

---

9   A 401k plan is a retirement saving plan sponsored by an employer starting from the 1980s In the United States.

Table 4. Percentage of ETF-owning households holding other investments by type (End-2018)	
Other investment concurrently held by ETF-owning households	Percentage of households
Equity mutual funds, individual stocks, or variable annuities (total)	93%
Bond mutual funds, individual bonds, or fixed annuities (total)	66%
Mutual funds (total)	90%
• Equity	87%
• Bond	53%
• Hybrid	44%
• Money market	57%
Individual stocks	74%
Individual bonds	26%
Fixed or variable annuities	31%
Real estate for investment	48%

Source: 2019 ICI Fact Book, CSOP.

1.2.2   More choices on US ETF issuers lead to more competitive prices

The asset management company issues ETFs, and participating dealers subscribe ETF units from the issuer by way of block trading. In the subscription process, the participating dealer exchanges for the ETF units with a basket of the underlying assets or cash in a volume corresponding to the ETF units. Then, the participating dealer can resell the ETF units to investors. Besides, there are market makers in the secondary market (which can be the participating dealers themselves) to ensure market liquidity.

In Hong Kong, due to the limited number of service providers for choice, there is less competition. The issuance costs including custody, legal and audit fees in Hong Kong are therefore higher. In the US, there are more choices of service providers, so the fees are more competitive.

Figure 5. Composition of ETF issuance cost (End-2018)		
Cost item	US	HK
Custody	32%	60%
Audit	6%	20%
Regulatory	1%	7%
Financial reporting/filing service	4%	7%
Distributor	7%	6%

(continued)

Figure 5. Composition of ETF issuance cost (End-2018)		
Cost item	US	HK
Director	14%	—
Compliance	13%	—
Treasurer	13%	—
Legal	7%	—
Insurance	3%	—

Source: CSOP.

1.2.3 ETF delisting is very common and widely accepted in a mature market

With more and more innovative ETFs introduced, ETFs with low liquidity and poor development face the risk of delisting. During 2013 to 2017, the number of delisted ETFs in the US continued to increase year by year. The number of delisted ETFs reached 114 in 2017 and dropped slightly in 2018 to 84.

In Asia, especially in Hong Kong, ETF issuers hesitate to delist their ETFs due to various reasons. However, with the gradual enhancement of the delisting scheme, the market began to accept ETF delisting. Recently, ETF delistings were observed in Hong Kong, signifying a move forwards the practice in mature markets.

Figure 15. Total number of ETFs listed and delisted in the US (2009 – 2018)

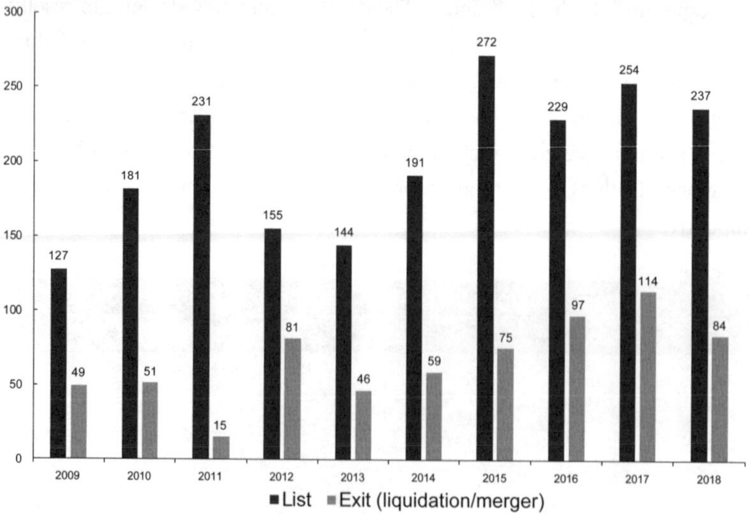

Source: 2019 ICI Fact Book, CSOP.

Table 6. Delisting process for ETFs in the US		
Task	Responsible party	Indicative timeline
Board resolutions and approval for plan of liquidation	Investment manager	L-40
Review budget and accrue needed expense in the fund	Fund administrator	L-30
Determine financial reporting and audit requirements	Financial reporting	L-15
Preliminary taxable income calculation for distributable earnings and tax considerations	Tax	L-14
Disseminate press release for liquidation information	Investment manager	L-10
Notification to New York Stock Exchange (NYSE) of the plan of delisting	Investment manager	L-10
Filing of sticker (announcement) and mailing of sticker	Transfer agency	L-10
Notification of event to the clearing house (DTCC)	Transfer agency	L-7
Review/amendment of custody, fund administration agreements, distribution agreements	Investment manager	L-5
Confirmation of the last day of trading and liquidation date	Investment manager	L-4
Finalise distribution of taxable income and report to NYSE	Transfer agency	L-1
Confirmation of liquidation rate to DTCC	Transfer agency	L-1
File Securities and Exchange Commission (SEC) Form 25 by NYSE for listing removal	NYSE	L+1
Ireland Revenue Service (IRS) diversification testing	Tax	L+10
Final Federal Form 1120-RIC submission	Tax	L+90

Note: "L" stands for Liquidation Date.
Source: NYSE, CSOP.

Table 7. Delisting process for ETFs in Hong Kong		
Task	Responsible party	Estimated timeline
Internal approval for the plan of proposed cessation of trading, proposed termination, proposed de-authorisation and proposed delisting	Investment manager	T
Review budget and accrue needed expense in the fund	Project management and operations	T+3
Determine financial reporting and audit requirements	Operations and auditor	T+3
Submit the proposed termination, de-authorisation, delisting application to the Securities and Futures Commission (SFC), the Hong Kong Exchanges and Clearing Limited (HKEX), and the Trustee for review and comment	SFC/HKEX/Trustee	T+7 ~ T+14
Board resolution for termination	Investment manager	T+25
Trustee consent letter issued	Trustee	T+30
HKEX confirmation/SFC approval	HKEX / SFC	T+40

(continued)

Table 7. Delisting process for ETFs in Hong Kong		
Task	Responsible party	Estimated timeline
Release of announcement and notice to investors regarding the proposed cessation of trading, termination, voluntary de-authorisation, and delisting and waiver from strict compliance with certain provisions of the SFC's related Code	Investment manager	T+50 (Prior notice at least 30 days before trading cessation)
Issue a circular to the Stock Exchange of Hong Kong (SEHK) participants and the Central Clearing and Settlement System (CCASS) participants respectively as soon as practicable after the publication of the Announcement to inform them of the proposed cessation of trading, termination, voluntary de-authorisation and delisting of the Sub-Fund (the "Terminating Sub-Fund")	HKEX	T+53
Last day for dealings in the units of the Terminating Sub-Fund on the SEHK and last day for creation and redemption of units of the Terminating Sub-Fund (the "Last Trading Day")	Investment manager	T+80
Dealings in the units of the Terminating Sub-Fund on the SEHK cease and no further creation and redemption of units of the Terminating Sub-Fund (the "Trading Cessation Date"), i.e. same date on which the Manager will start to realise all the investments of the Terminating Sub-Fund and the Terminating Sub-Fund will cease to be able to track the underlying index	Investment manager	T+81
The date as at which an investor needs to be recorded by the Hong Kong Securities Clearing Company (HKSCC) as the beneficial owner of the Units which are registered in the name of HKSCC Nominees Limited and held in CCASS to be entitled to the Final Distribution and further distribution, if any (the "Distribution Record Date")	HKSCC	T+85
Final Distribution announcement	Investment manager	T+100
Final Distribution, after the Manager having consulted with the Trustee and the Terminating Sub-Fund's auditor, will be paid to the Relevant Investors (the "Final Distribution Date")	Investment manager	T+105
Termination of the Terminating Sub-Fund (the "Termination Date"), when the Manager and Trustee form an opinion that the Terminating Sub-Fund ceases to have any contingent or actual assets or liabilities	Investment manager /Trustee	T+140 (Subject to the requirement of prior notice clause in the Trust Deed)
Approval of de-authorisation and delisting of the Terminating Sub-Fund; An announcement informing investors about the Termination Date, dates for the de-authorisation and delisting of the Terminating Sub-Fund	SFC/HKEX	On or shortly after the Termination Date, which is the date the SFC and SEHK respectively approve the de-authorisation and delisting
Termination Audit	Auditor	After the termination

Source: HKEX, CSOP.

1.2.4   The US capital market encourages innovation, favours the debut of new products, and discourages plagiarism

The US market encourages ETF innovation, and the underlying assets of ETFs are diversified. The underlying assets of the ETFs in the US can be divided into six main categories — commodities, hybrids, fixed income, stocks, alternative investments, and currencies. The underlying assets of commodity ETFs include agriculture, energy, precious metals, and industrial materials. The specific assets include coal, cotton, gold, iron and steel, soybeans, coffee, etc. Besides, the underlying assets of alternative investment ETFs include foreign exchange, futures, volatility, leveraged products, etc.

In addition, thematic ETFs are favoured by the market. Related hotspots include electric vehicles, artificial intelligence, blockchain and online retail. For example, a US ETF issuer has launched thematic ETFs based on technology, people and infrastructure. Under the theme of technology, there are robotics and artificial intelligence ETFs (investing in companies that conduct research and development of robots and use artificial intelligence), cloud computing ETFs (investing in companies providing cloud service software), social media ETFs (investing in social media companies), lithium battery technology ETFs (investing in the entire lithium cycle, from mining, refining to battery production), etc. Under the theme of the people, there are Millennium ETFs (investing in companies that benefit from the consumption brought about by the growing US millennial population) and long-lived ETFs (investing in companies that provide medical care, medicine, and living facilities for the elderly).

The US favours the debut of new products. Typically, the first launched fund tends to dominate the market, while the followers are left with fewer opportunities.

1.2.5   US ETFs are based on global asset allocation

US ETFs are based on global asset allocation, among which equity ETFs cover important indices of countries around the world. According to Bloomberg, US-listed ETFs covered all markets in the world at the end of June 2019. In terms of net asset value, the North American market accounted for 75.25%, the international market (all countries and regions included, except the US market) accounted for 15.46%, the global market (all countries and regions included, including the US market) accounted for 5.57%, the Asia-Pacific market accounted for 1.96% and the European market accounted for 1.40%.

Figure 16. Distribution of US-listed equity ETFs in net asset value (End-2019)

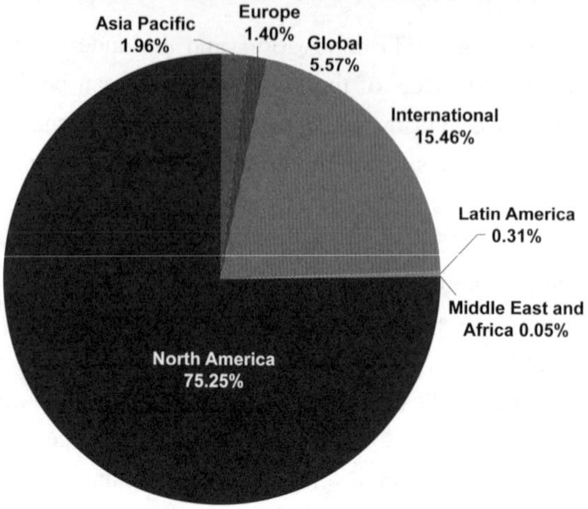

Asia Pacific 1.96%
Europe 1.40%
Global 5.57%
International 15.46%
Latin America 0.31%
Middle East and Africa 0.05%
North America 75.25%

Source: Bloomberg, CSOP.

According to Bloomberg, the underlying assets of US-listed equity ETFs have covered many countries and regions in the world, including the US, Japan, Mainland China, Brazil, India, Canada, South Korea, Taiwan, Germany, United Kingdom (UK), Hong Kong, Russia, Australia, Switzerland, France, Spain, Mexico, Saudi Arabia, Singapore, etc. In comparison, Asian ETFs emphasise more on domestic assets.

Ranking	Country/region that underlying asset belongs to	Total net asset value (US$ mil)
1	US	2,994,045.62
2	Japan	21,872.10
3	Mainland China	17,908.11
4	Brazil	9,927.35
5	India	8,706.78
6	Canada	6,735.61
7	South Korea	4,436.06
8	Taiwan, China	2,992.92
9	Germany	2,678.72
10	UK	2,423.93
11	Hong Kong, China	2,303.26
12	Russia	2,094.68

Table 8. Top 20 US-listed equity ETFs in total net asset value (End-Jun 2019)

(continued)

Table 8. Top 20 US-listed equity ETFs in total net asset value (End-Jun 2019)		
Ranking	Country/region that underlying asset belongs to	Total net asset value (US$ mil)
13	Australia	1,571.88
14	Switzerland	1,389.60
15	France	1,125.95
16	Spain	1,081.16
17	Mexico	870.15
18	Saudi Arabia	854.54
19	Singapore	613.20
20	Indonesia	543.64

Source: Bloomberg, CSOP.

1.3   Insights from global ETF market experience to Hong Kong

1.3.1   Promote investor education and encourage the use of ETFs for asset allocation

Mutual funds, including ETFs, are more widely accepted by US investors. According to ICI, in 2018, a total of 57.2 million families and 101.6 million individuals in the US held US-registered open-ended funds. The total net asset value of the pension market was US$27.1 trillion, of which US$8.2 trillion held by American Individual Retirement Accounts (IRA) and Defined Contribution Plan (DC Plan) in mutual funds, accounting for 30.26%. The Hong Kong Securities and Futures Commission (HKSFC) and ETF issuers should promote comprehensive investor education, encourage investors to accept ETFs and increase the proportion of investors utilising ETFs for diverse asset allocation.

1.3.2   Encourage more service providers to compete so as to reduce ETF costs

As analysed in Section 1.2.2, Hong Kong has high issuance costs due to the limited number of service providers available and the lack of competition. Given that low cost is one of the advantages of ETFs, if the fees of service providers remain high in Hong Kong, the resultant high costs will affect the strategic goal of Hong Kong to be an ETF centre.

Therefore, Hong Kong should encourage more ETF-related service providers such as custodians, law firms and auditing firms to compete in the market in order to reduce the operating costs of ETFs, making Hong Kong ETFs more competitive vis-à-vis mature markets.

1.3.3 Encourage market innovation and invigorate market liquidity

The Hong Kong market currently has innovative L&I products. By investing in leveraged investment tools such as stock index futures and swap contracts, L&I products achieve a certain positive or negative multiples of the daily performance of the tracked index. Currently, gold is the only underlying asset of commodity ETFs in Hong Kong. We suggest that commodity ETFs on a greater variety of commodities, including precious metals, agriculture and energy, should be introduced to the market.

1.3.4 Internationalise products and gradually include assets in other markets

Compared to the US, the underlying assets of ETFs in Hong Kong is rather homogeneous — 85% in stocks, 15% in bonds, currencies and commodities. Half of the underlying assets are concentrated in Mainland China and Hong Kong.

Table 9. Number and percentage of ETFs listed in Hong Kong, by country/region of the underlying assets (End-Jun 2019)		
Country/Region	Number of ETFs	Percentage
China (including Hong Kong and Taiwan)	80	50.63%
Global	15	9.49%
Europe	15	9.49%
Asia	11	6.96%
US	9	5.70%
Asia Pacific (excluding Japan)	6	3.80%
South Korea	5	3.16%
Japan	5	3.16%
India	4	2.53%
Malaysia	1	0.63%
Vietnam	1	0.63%
Singapore	1	0.63%
Philippines	1	0.63%
Russia	1	0.63%
Brazil	1	0.63%
Thailand	1	0.63%
Indonesia	1	0.63%

Source: Bloomberg, CSOP.

For Hong Kong to develop into an ETF centre in Asia, and even further into an ETF centre in the world, diversification of the underlying assets and the gradual inclusion of assets in other markets will be necessary.

2 Trend and new impetus in the Hong Kong ETF market

2.1   Overview of the Hong Kong ETF market

Over the past decade, ETFs, which can invest in a wide range of assets and strategies, have achieved high popularity in the market. As an international financial centre in Asia, Hong Kong has its ETF market tripled in total net asset value. On the backdrop of capital account opening in Mainland China, ETFs tracking Hong Kong equities and Mainland China's A-share market, RMB Qualified Foreign Institutional Investors (RQFII), Stock Connect (Shanghai Connect and Shenzhen Connect) have become important driving forces for capital flows. As of 8 July 2019, there were 187 ETFs listed in Hong Kong, with the total net asset value amounted to HK$457.511 million[10]. ETFs issued by many different institutions have enriched the product categories and boosted the development of the Hong Kong ETF market. In terms of market value, Hong Kong is the second largest ETF market in Asia; in terms of turnover ratio, Hong Kong ranks fourth in Asia[11].

2.2   Current structure of the Hong Kong ETF market

2.2.1   Asset composition of ETFs in Hong Kong

The underlying assets of the Hong Kong ETF market are relatively homogeneous: mostly equities. According to HKEX, as of 11 July 2019, in terms of number of ETFs, 84.49% invested in equities, 10.71% in fixed income and currencies (including the money market), 2.67% in commodities, 2.67% in commodity futures and 0.53% in multi-assets;

---

10 Source: HKEX website.
11 Source: 2019 ICI Fact Book.

in terms of total net asset value, 89.76% invested in equities, 9.63% in fixed income and currencies (including the money market), 0.56% in commodities, 0.04% in commodity futures and 0.01% in multi-assets. (See Figures 17 and 18.)

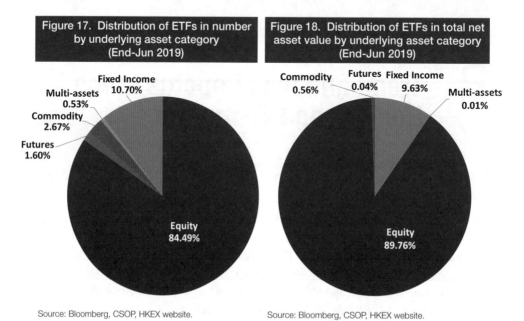

Figure 17. Distribution of ETFs in number by underlying asset category (End-Jun 2019)

Figure 18. Distribution of ETFs in total net asset value by underlying asset category (End-Jun 2019)

Source: Bloomberg, CSOP, HKEX website.

Source: Bloomberg, CSOP, HKEX website.

In respect of asset categorisation by physical assets, synthetic assets and futures, physical assets take the dominant position in the Hong Kong ETF market. According to HKEX, as of 11 July 2019, in terms of number of ETFs, physical assets took up 94.65%, while synthetic assets and futures took up 3.74% and 1.60% respectively; in terms of total net asset value, physical assets took up 99.75%, while synthetic assets took up 0.21% and futures took up 0.04%.

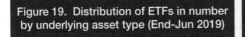

Figure 19. Distribution of ETFs in number by underlying asset type (End-Jun 2019)

Figure 20. Distribution of ETFs in total net asset value by underlying asset type (End-June 2019)

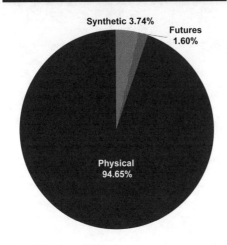

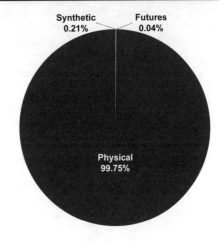

Source: Bloomberg, CSOP.

Source: Bloomberg, CSOP.

2.2.2   Current ETFs in the Hong Kong market

There are a great variety of ETFs listed in the Hong Kong market, including A-share ETFs, H-share ETFs, L&I products, smart beta ETFs, industry ETFs and thematic ETFs, etc.

According to HKEX, at the end of June 2019, ETFs covered 91 indices, including 11 products tracking CSI 300 Index, 8 products tracking FTSE China A50 Index, 5 products tracking Bloomberg Barclays China Treasury and Policy Bank Bond Index, 5 products tracking NASDAQ 100 Index and 4 products tracking MSCI China A Inclusion Index. Furthermore, there are also ETFs tracking the Hang Seng Index, the MSCI China Index, the Euro STOXX 50 Index, the Xtrackers MSCI Indonesia Index, the S&P Dow Jones Indices, etc.

Table 10. Top 20 ETFs by transaction amount (Jan-Jun 2019)					
Stock code	ETF name	ETF manager	Transaction currency	Net asset value (HK$ mil)	Turnover (HK$ '000)
2800	Tracker Fund of Hong Kong	State Street Global Advisors Asia Limited	HKD	87,786	1,442,255,350
3188	ChinaAMC CSI 300 Index ETF	China Asset Management (Hong Kong) Limited	HKD	16,600	1,050,695,868
2822	CSOP FTSE A50 ETF	CSOP Asset Management Limited	HKD	17,269	980,422,993
2823	iShares FTSE A50 China Index ETF	Blackrock Asset Management North Asia Limited	HKD	24,375	854,177,827
2828	Hang Seng China Enterprises Index ETF	Hang Seng Investment Management Limited	HKD	28,989	365,881,777
7500	CSOP Hang Seng Index Daily (-2x) Inverse Product	CSOP Asset Management Limited	HKD	1,249	244,134,749
7300	CSOP Hang Seng Index Daily (-1x) Inverse Product	CSOP Asset Management Limited	HKD	3,525	231,846,544
7200	CSOP Hang Seng Index Daily (2x) Leveraged Product	CSOP Asset Management Limited	HKD	244	62,069,909
82822	CSOP FTSE A50 ETF	CSOP Asset Management Limited	RMB	17,269	20,565,059
83188	iShares FTSE A50 China Index ETF	China Asset Management (Hong Kong) Limited	RMB	16,600	16,118,673
3053	CSOP Hong Kong Dollar Money Market ETF	CSOP Asset Management Limited	HKD	2,713	13,330,407
7288	CSOP Hang Seng China Enterprises Index Daily (2x) Leveraged Product	CSOP Asset Management Limited	HKD	178	12,826,638
82833	Hang Seng Investment Index Funds Series II - Hang Seng Index ETF	Hang Seng Investment Management Limited	RMB	53,151	11,821,619
83010	iShares Core MSCI AC Asia ex Japan Index ETF	Blackrock Asset Management North Asia Limited	RMB	462	11,304,521

(continued)

Table 10. Top 20 ETFs by transaction amount (Jan-Jun 2019)					
Stock code	ETF name	ETF manager	Transaction currency	Net asset value (HK$ mil)	Turnover (HK$ '000)
9010	iShares Core MSCI AC Asia ex Japan Index ETF	Blackrock Asset Management North Asia Limited	USD	3,610	11,061,239
3147	CSOP SZSE ChiNext ETF	CSOP Asset Management Limited	HKD	525	10,916,298
2801	iShares Core MSCI China Index ETF	Blackrock Asset Management North Asia Limited	HKD	2,807	8,802,568
3167	ICBC CSOP S&P New China Sectors ETF	CSOP Asset Management Limited	HKD	914	8,398,372

Source: Bloomberg, CSOP.

2.3   Existing gaps in the Hong Kong market

2.3.1   Higher requirements for issuance compared with the US market

When compared with the US market, the issuance requirements are relatively strict in Hong Kong. The minimum size for ETF issuance starts at US$2 million in the US while it starts at tens of million US dollars in Asia. After the financial crisis in 2008, the US enacted several laws restricting the proprietary trading of financial institutions, aiming to tackle the problem of financial institutions that are "too big to fail", preventing systemic risks effectively, protecting vulnerable groups in the financial market and protecting financial consumers from fraudulent activities. As a result, the European and US investment banks gradually lost the ability to provide seeding funds for ETFs, which has become an obstacle for ETF issuers in Hong Kong in finding seed money.

2.3.2   Different stages of development in the ETF market compared with the US market

After more than 20 years of development, the US ETF market is in a mature stage. In the process of ETF issuance, subscription and redemption, as well as trading, each department has a clear division of labour in serving the market. These services can be outsourced to different professional institutions. Compared with the US, ETFs in Hong Kong are still in the start-up stage and the roles are rather ambiguous. ETF issuers need to master multiple skills to deal with various situations in the process of issuance, subscription and redemption as well as trading. In the future, the Hong Kong ETF market will follow and develop with the insights from the US market.

2.3.3 No longer competitive in transaction fees when compared with the Mainland market

The Shenzhen Stock Exchange currently exempts bond ETFs and currency ETFs from transaction handling fees. The Shanghai Stock Exchange also currently exempts bond ETFs and currency ETFs from Participation Business Unit (PBU) fees. However, the HKEX still charges currency ETFs fees on the same basis as for common stocks. This increases the transaction costs for ordinary investors and may hinder the development of currency ETFs in Hong Kong.

The trading volume of bond ETFs and currency ETFs in Hong Kong is limited — the total volume in 2018 was only HK$1.65 billion and HK$800 million respectively; and the total volume in 2019 up to May was only HK$460 million and HK$820 million.

Table 11. Summary of trading volumes of bond ETFs and currency ETFs		
	Bond ETF (HK$)	Currency ETF (HK$)
Annual turnover in 2018	1,652,393,536	803,993,300
Turnover from January to May 2019	459,365,134	817,574,130

Source: Bloomberg, CSOP.

In the secondary market, in addition to brokerage commission fees, investors also need to pay transaction levy, trading fee, trading tariff and settlement fees. See Table 12 for details.

Table 12. ETF transaction fees in Hong Kong			
Type of fee	Charging institution	Charging basis	Charging standard
Transaction levy	SFC	Trading value	0.0027%
Trading fee	HKEX	Trading value	0.005%
Trading tariff	HKEX	Per transaction	HK$0.50

Source: HKEX.

From this, it can be calculated that the fees investors pay to trade bond ETFs and currency ETFs (excluding brokerage commission) are modest. In 2018, the total transaction fees of bond ETFs and currency ETFs were HK$320,000 and HK$156,000 respectively. In 2019 up to May, the transaction fees were only about HK$89,000 and HK$159,000 respectively. See Table 13 for details.

Table 13. Total fees regarding ETF transactions in Hong Kong (2018 & Jan – May 2019)					
Type of fee	Charging Institution	Total transaction fees of bond ETFs (HK$)		Total transaction fees of money market ETFs (HK$)	
		2018	Jan-May 2019	2018	Jan-May 2019
Transaction levy	SFC	89,229	24,806	43,416	44,149
Trading fee	HKEX	165,239	45,937	80,399	81,757
Settlement fee	HKSCC	66,096	18,375	32,160	32,703
Total		320,564	89,117	155,975	158,609

Note: Excluding trading tariff.

Source: Bloomberg, CSOP.

3   How to further develop the Hong Kong ETF market

3.1   Focus on investors' demand for ETFs

3.1.1   Increase the multiples of L&I products

Over the past decade, L&I products have become increasingly popular among investors in mature markets. From a global perspective, the US market is the most developed one for L&I products. In Asia, Japan, South Korea and Taiwan markets have also developed L&I products earlier than other markets and with a relatively large scale. (See Section 1.1.4 above.)

There are relatively large number of choices for L&I products listed in the US market, e.g. there are 103 L&I products with a leverage ratio of three times and 10 L&I products with a leverage ratio of four times.

The development of L&I products in Hong Kong market started a bit later than Japan, South Korea and Taiwan markets, but the growth has been rapid, signifying an immense potential. Currently, L&I products in Hong Kong mainly track the Hang Seng Index and the Hang Seng China Enterprises Index. The underlying assets are stock index futures and cash management tools (including high-quality money market funds in Hong Kong dollar as well as high-liquidity and high-credit-rating government bonds). ETF portfolios are reallocated every day at the end of trading hours, and the cash position is systematically managed. Therefore, the operation is stable with no unanticipated incidents.

Currently, only leveraged products with a leverage ratio of two times (2x) and inverse products with a leverage ratio up to two times (-2x) are allowed to list in Hong Kong. The underlying assets of listed L&I products are exclusively the Hang Seng Index and the Hang Seng China Enterprises Index. According to market feedbacks, to meet the demand of the market, Hong Kong can increase the allowed leverage multiples and enlarge the scope of eligible underlying assets, including A shares.

3.1.2 More diversified underlying assets

As analysed in Section 2.2.2, Hong Kong should vigorously develop the bond ETF market. This includes the issuance of RMB Treasury Bond ETFs with various bond tenors to help build the offshore RMB yield curve and lay the foundation for the further development of bond ETFs in Hong Kong.

In respect of equity ETFs, Hong Kong should encourage more diversified underlying assets so that investors could allocate assets to more markets.

3.2 Innovative design and development of ETFs

3.2.1 Examples of innovative ETFs existing in the market

Currently, there are L&I products to track the market performance with multiples of -1x, -2x and 2x times. According to Bloomberg, during January to June 2019, of the total turnover value of L&I products, CSOP products accounted for 98%, the products of Samsung, Mirae, E Fund and ChinaAMC together accounted for close to 2%; as of end-June 2019, of the total assets under management of L&I products, CSOP accounted for 82%, Samsung accounted for 8%, ChinaAMC accounted for 5%, Mirae accounted for 3%, E Fund accounted for 2%.

Figure 21. Distribution of turnover value of L&I products by issuer (Jan-Jun 2019)

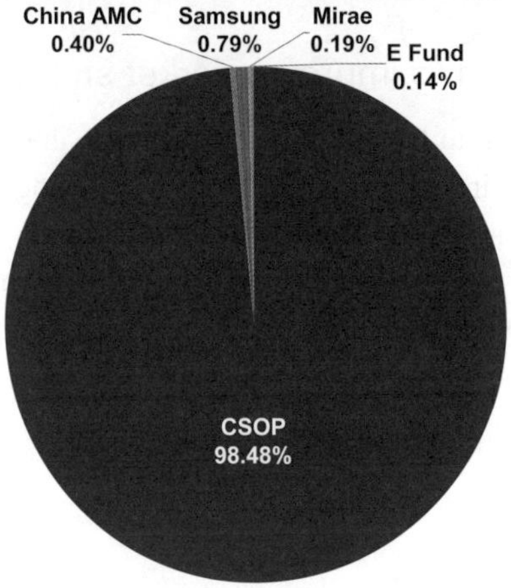

China AMC 0.40%
Samsung 0.79%
Mirae 0.19%
E Fund 0.14%
CSOP 98.48%

Source: Bloomberg, CSOP.

Figure 22. Distribution of net asset value of L&I products by issuer (End-Jun 2019)

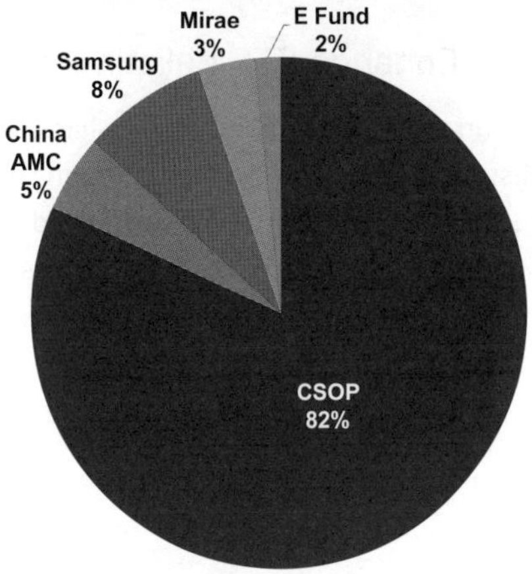

Mirae 3%
E Fund 2%
Samsung 8%
China AMC 5%
CSOP 82%

Source: Bloomberg, CSOP.

We believe that cross-border underlying assets could be introduced to connect with domestic and international ETFs. Cross-border ETFs could make reference to the experience of the China-Japan ETF Connectivity Scheme.

In October 2018, the CSRC and the Japan Financial Services Agency Commission signed a Memorandum of Understanding (MOU) to strengthen the cooperation of the two securities markets. On 22 April 2019, China and Japan signed an ETF Connectivity Agreement during the China-Japan Capital Market Forum. The two parties agreed to cooperate to establish a mutual exchange mechanism and establish funds that invest in ETFs in the counterpart's market. On 25 June 2019, the Shanghai Stock Exchange held the opening ceremony of the China-Japan ETF Connectivity Scheme, and four China-Japan ETF Connect products successfully listed on the Shanghai Stock Exchange. On the same day, Japan had four ETF Connect products listed on the Tokyo Stock Exchange.

Under the China-Japan ETF Connectivity Scheme, fund management companies from China and Japan set up cross-border funds through Qualified Domestic Institutional Investors (QDII) and Qualified Foreign Institutional Investors (QFII), and invest all or most of the assets in typical ETFs in each other's market. Unlike Shanghai Connect and Shenzhen Connect, which allow investors from either side of the Mainland and Hong Kong to directly invest in stocks listed on the other side, the China-Japan ETF Connect allows investors to purchase local fund products to achieve cross-border investment. The China-Japan ETF Connect links the financial market between Mainland China and Japan via feeder funds. Hong Kong could learn from this model to connect with the ETF market in Mainland China.

3.3 Enhance ETF delisting mechanism to improve market structure

Currently, if the underlying A shares of an ETF to be delisted are suspended, the ETF delisting process will be more complicated and time-consuming. During the elongated delisting process, ETF issuers still need to pay high index service fees, custody fees, etc. This brings unnecessary costs to ETF issuers. It is recommended that Hong Kong should simplify the delisting mechanism when the A shares underlying the A-share ETF are suspended in order to ease the burden on ETF issuers.

4 Summary of policy suggestions

(1) Exempt currency funds from CCASS and other miscellaneous fees

As analysed in Section 2.3.3, compared with Mainland China, Hong Kong is less competitive in trading and settlement fees and still has transaction levy and other fees. Although the amount may appear trivial, the reduction or exemption of fees will enhance market attractiveness. Therefore, it is suggested that the HKEX and related institutions could make reference to the practice of the Shanghai Stock Exchange and the Shenzhen Stock Exchange to exempt trading and settlement fees on bond ETFs and currency ETFs to benefit retail investors.

(2) Expand MPF investment categories

Officially implemented on 1 December 2000, the Mandatory Provident Fund Schemes (MPF) is a policy to require all employees in Hong Kong to set up investment funds for retirement purposes. MPF currently does not allow investors to invest in A-share ETFs, which is equivalent to excluding all Hong Kong employees from the A-share market. It is suggested to expand the MPF investment target categories, such as allowing investment in A-share ETFs, to benefit from China's economic development.

(3)   Expand ETF underlying asset categories

Currently, most of the ETF underlying assets are concentrated exclusively on equities. It is suggested that the categories could also include commodities, futures, fixed income products, bonds and foreign exchange so that investors in Hong Kong can achieve global asset allocation.

(4)   Increase the multiples of L&I products

Currently, L&I products listed in Hong Kong have a leverage ratio of up to two times only, and the underlying assets are exclusively the Hang Seng Index and the Hang Seng China Enterprises Index. According to market feedbacks, to meet the demand of the market, Hong Kong could increase the leverage multiples of L&I products and include A shares as the underlying assets.

Note: CSOP Xiaoya XU, Chenning LIU also contributed to the article.

Afterword

Promoting mutual market access and innovation in new-economy financing by riding on the change in global financial landscape

Against a backdrop of global financial turbulence, the Mutual Market Access (MMA) programme, comprising currently the Shanghai Connect, the Shenzhen Connect and the Bond Connect, has solidified Hong Kong's status as a unique financial hub connecting the East and the West and, in this process, has explored a new path for the opening-up of China's capital market.

After the inclusion of A shares into MSCI's global indices and of Chinese bonds into major international bond indices like Bloomberg Barclay's, there is increasing participation from Mainland and overseas market participants in trading through the Connect schemes under the MMA programme for satisfying their growing demand of investing in Chinese or global assets. This, together with the continuous wealth accumulation by Mainland residents and their increasing global asset allocation, has gradually promoted the MMA programme to become a new platform of significance for cross-border investment and global asset allocation. Meanwhile, in the face of economic structural changes as witnessed across the globe, Hong Kong seeks to look into innovations in financing the new economy, and has built up valuable experiences for reference in the Asian time zone. Hong Kong is gradually establishing a new ecosystem around new-economy enterprises and made them new targets of investment for both Mainland and overseas investors trading under MMA.

In response to the market spotlight on MMA and new-economy financing, we extended special invitations to experts and research teams who are experienced in these areas to compose this book, *Mutual Market Connectivity and Hong Kong Market Innovations for New-Economy Financing*, with in-depth discussions on the key functional drivers of innovations in Hong Kong and the Mainland, namely the "MMA", "new-economy financing" and "global asset allocation". Elaborating the unique concepts of the MMA model and summarising what HKEX has learned in its reforms in new-economy financing, this book shall come in handy for market participants to make better use of the MMA model and benefit from the latest reforms of Hong Kong's listing regime.

As the first bilingual book in the market that explains in detail the MMA model between Hong Kong and Mainland China and innovations in new-economy financing, this book is unique and innovative in the following ways:

Firstly, contents are contributed by people who are "participants" and "designers" of the MMA model, giving systematic and in-depth reviews and summaries of the model based on their first-hand experience. These could hopefully help market participants better understand how this connectivity model benefits the opening-up of China's capital market.

As an innovative institutional arrangement to facilitate the opening-up of China's capital market, the MMA programme promotes two-way opening-up of the Mainland capital market with minimum institutional costs, fostering an effective connection between the financial systems of Mainland China and the rest of the world without adding to the risk of capital outflows. The smooth operation and expansion of the programme over the years, from the Shanghai Connect to the Shenzhen Connect and then to the Bond Connect, has played a vital role in promoting the inclusion of Chinese assets into the global indices.

Against this backdrop, users in both the Mainland market and overseas markets imminently need to gain a systematic understanding of the overall framework and operational logics of the MMA model, thereby helping in the search of more convenient investment channels for trillions of dollars of international and Chinese capital, as well as in looking for better arrangements to channel financial resources to new-economy enterprises. Chapter 1 of Part 1 of the book was contributed by Charles LI, the Chief Executive of HKEX and the designer-cum-promoter of the MMA model. He systematically summarised the operation of the model and shared his rationale and thoughts behind Hong Kong's listing regime reforms and new-economy financing that he has been promoting all these years, reflecting his macro vision as an internationally renowned financier. Other contributors in Part 1 are LU Ting, the Chief China Economist of Nomura International (Hong Kong) Limited and HONG Hao, Head of Research at BOCOM International Holdings Company Limited. As veteran experts in overseas markets, they discussed important topics of market focus, including Hong Kong's unique positioning in innovative financing for new-economy enterprises, the impact of the MMA model on the Mainland and overseas markets, and its special implications for the reform and opening-up of China's capital market. Their thorough analysis based on solid market data provides a multi-dimensional lens through which market participants can see how crucial MMA is in the opening-up of the Mainland capital market.

Secondly, the book gives a systematic introduction of the HKEX's listing regime and the listing process, especially the reforms that is introduced in April 2018 for the listing of new-economy companies, to give a comprehensive reference for the market to understand the new listing regime and practices at HKEX.

The rise of new-economy enterprises has become a leading force that brings about structural transformation of global capital markets. HKEX's listing regime reforms in April 2018 have greatly facilitated the financing of new-economy enterprises, driving the Hong Kong stock market to return to the top of the world rankings in terms of the amount of funds raised through initial public offerings (IPOs). As more and more new-economy enterprises are planning a listing in Hong Kong, the need for a clear and comprehensive understanding of HKEX's new listing regime becomes imminent.

In order to provide comprehensive and easily accessible reference materials for Mainland and overseas users, Part 2 of the book gives an all-round introduction to the general requirements for a Main Board listing, features of the listing review process, IPO pricing and issuance on the Main Board, and other related points of interest of the market. Listing procedures and practices related to HKEX's listing regime reforms in 2018 and biotech companies' listing applications are also highlighted. There are two other chapters in this part that examine the legal implications and impacts of weighted voting rights (WVR) from different perspectives, shedding light on the unique role of WVR structures in fostering the growth of new-economy enterprises.

In order to help the market apprehend Hong Kong's innovative initiatives in financing the new economy, we also invited the research teams at Citigroup Global Markets Asia Limited and Deloitte China to share empirical cases of overseas bond issuances and mergers and acquisitions (M&As) by Chinese companies, demonstrating how multi-tiered financing channels in Hong Kong support the development of new-economy enterprises and overseas M&As.

Thirdly, Part 3 of the book illustrates, in multiple dimensions, the significance of the MMA model in providing a platform for Mainland and overseas capital to invest globally, broadening market participants' horizons in understanding the practical implications of the MMA model in the global allocation of funds.

The MMA model, in an innovative way, promotes the flows of international capital into Mainland China and provides a low-cost platform for Mainland investors to invest in global assets. Alongside, innovations in new-economy financing provide new target assets for cross-border investments. How new ways can be explored for cross-border asset allocation and for related product innovations under "MMA" and "new-economy financing" has been a hot topic of discussion in the Mainland and overseas markets.

In Part 3, WEI Zhen, Head of China Research at MSCI Inc., and DING Chen, Chief Executive Officer of CSOP Asset Management Limited, shared their professional views on cross-border asset allocation by way of investing in A shares and exchange traded funds

(ETFs) under MMA. Indices related to the "Belt and Road" initiative and biotech sectors are introduced, in the hope of enabling investors to make good use of available tools and platforms. Possible enhancements to Stock Connect by allowing block trading and introducing ETF Connect are also discussed.

The book revolves around the three inter-connected themes of "MMA", "innovation in new-economy financing" and "global asset allocation". The contents are organised and empowered mainly by the research team at the Chief China Economist's Office of HKEX, in cooperation with related business departments at HKEX and invited persons-in-charge at well-known Mainland and overseas financial institutions specialised in the respective business areas, without whom this professional and authoritative piece of work covering these three themes would not have come into existence.

I have to thank in particular Charles LI, the Chief Executive of HKEX, for his generous support, especially that he has kindly written the Preface and Chapter 1 of the book to share his profound knowledge and insights on the MMA model and new-economy financing which he has spared no effort to promote. My thanks also go to HKEX colleagues in regulatory compliance, legal services, corporate communications, translation and other business teams for their generous support and advice, without which this book would not have been published. And thanks too to the efficiency and professionalism of our publisher, the Commercial Press (HK) Limited, we have been able to release the book onto the market at the right time. To all these parties, we sincerely express our gratitude.

Today, Hong Kong and Mainland China are inextricably intertwined and interdependent upon each other. The launch of the MMA model in Hong Kong offers China a unique, open and risk-controlled platform to connect with the global market, a win-win for both the prosperity of Hong Kong and the opening-up and development of our country. Under this model, Hong Kong is not only the Mainland's gateway to the world, it can also further integrate with the Mainland financial system, thereby providing new impetus for bringing capital flows into the Mainland for its sustainable development, and helping to turn a new page in the opening-up of China's capital market.

This book may have imperfections given that MMA is an innovative business model involving multiple facets of the market and is still undergoing continuous evolution. Any comments you may have are most welcome.

Professor BA Shusong
Chief China Economist, Hong Kong Exchanges and Clearing Limited
Chief Economist, China Banking Association
October 2019